Materials Chemistry

ADVANCES IN CHEMISTRY SERIES **245**

Materials Chemistry

An Emerging Discipline

Leonard V. Interrante, EDITOR
Rensselaer Polytechnic Institute

Lawrence A. Caspar, EDITOR
University of Wisconsin—Madison

Arthur B. Ellis, EDITOR
University of Wisconsin—Madison

Developed from a symposium sponsored
by the Division of Industrial and Engineering Chemistry, Inc.,
at the 204th National Meeting
of the American Chemical Society,
Washington, D.C.,
August 23–28, 1992

American Chemical Society, Washington, DC 1995

Library of Congress Cataloging-in-Publication Data

Materials chemistry: an emerging discipline / Leonard V.
Interrante, editor, Lawrence A. Casper, editor, Arthur B.
Ellis, editor.

p.   cm.—(Advances in chemistry series, ISSN 0065–2393; 245)

"Developed from a symposium sponsored by the Division of
Industrial and Engineering Chemistry, Inc., at the 204th National
Meeting of the American Chemical Society, Washington, D.C., August
23–28, 1992."

Includes bibliographical references and indexes.

ISBN 0–8412–2809–4

1. Chemical engineering—Congresses.

I. Interrante, Leonard V., 1939–   . II. Casper, L. A. III. Ellis,
Arthur B., 1951–   . IV. American Chemical Society. Meeting (204th:
1992: Washington, D.C.) V. Series.

QD1.A355   no. 245
[TP5]
540 s—dc20   94–23580
[660]   CIP

FOREWORD

The ADVANCES IN CHEMISTRY SERIES was founded in 1949 by the American Chemical Society as an outlet for symposia and collections of data in special areas of topical interest that could not be accommodated in the Society's journals. It provides a medium for symposia that would otherwise be fragmented because their papers would be distributed among several journals or not published at all.

Papers are reviewed critically according to ACS editorial standards and receive the careful attention and processing characteristic of ACS publications. Volumes in the ADVANCES IN CHEMISTRY SERIES maintain the integrity of the symposia on which they are based; however, verbatim reproductions of previously published papers are not accepted. Papers may include reports of research as well as reviews, because symposia may embrace both types of presentation.

ABOUT THE EDITORS

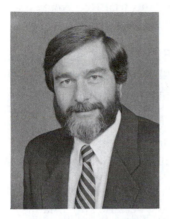

LEONARD V. INTERRANTE is Professor of Inorganic and Materials Chemistry at Rensselaer Polytechnic Institute. He received his Ph.D. degree in inorganic chemistry with J. C. Bailar, Jr., at the University of Illinois in 1964. He was a National Science Foundation Postdoctoral Fellow at University College, London, and an assistant professor in the Chemistry Department at the University of California at Berkeley from 1964 to 1968. Before coming to Rensselaer Polytechnic Institute in 1985, he spent 17 years as a staff scientist at the General Electric Research and Development Center in Schenectady. He has served as the chair of two Gordon Research Conferences (Inorganic and Chemistry of Electronic Materials) and as Program Chair, Secretary–Treasurer, and Chairman of the Inorganic Division of the ACS. He is currently Editor-in-Chief of the ACS journal, *Chemistry of Materials*. His research areas include molecular precursor routes to ceramic materials, inorganic polymer chemistry, chemical vapor deposition using organometallic precursors, and high-temperature structural composites. He has published more than 140 papers, holds seven patents, and has edited three other books in these various areas of materials-related chemistry.

LAWRENCE A. CASPER received his B.S. degree in chemistry from Juniata College and his Ph.D. degree in Chemistry from Lehigh University, both in Pennsylvania. He also earned an M.S. degree in Environmental Science from the University of Alaska. From 1977 until 1982 he conducted research on advanced materials for energy systems at the Idaho National Engineering Laboratory of the U.S. Department of Energy, specializing in surface and thin-film chemistry problems.

From 1982 until 1990, Casper was an Engineering Fellow on the technical staff of the Honeywell Solid State

Electronics Center in Minneapolis, where he developed a materials and device characterization laboratory and worked on development of advanced integrated circuit devices.

In 1990, Casper moved to the University of Wisconsin—Madison, where he is Assistant Dean of Engineering for Industrial R&D and Associate Director of the University—Industry Research program.

 ARTHUR B. ELLIS is Meloche–Bascom Professor of Chemistry at the University of Wisconsin—Madison. From 1990 to 1993, he chaired the UW—Madison's interdisciplinary graduate Materials Science Program. Ellis received his Ph.D. degree from Massachusetts Institute of Technology in inorganic chemistry and his B.S. degree in chemistry from Caltech. His research interests are in materials chemistry, specifically in the electro-optical properties of solids, and he has co-authored more than 100 research papers and holds eight patents. Ellis has been the recipient of Exxon, Sloan, and Guggenheim fellowships. He has helped develop a variety of instructional materials for integrating solids into the chemistry curriculum, including a superconductor levitation kit and an optical transform kit. Since 1990, Ellis has chaired an ad hoc committee that has produced a book to help chemistry teachers integrate materials chemistry into introductory chemistry courses. This volume, published by ACS Books, is entitled *Teaching General Chemistry: A Materials Science Companion.*

DEDICATION

This volume is dedicated to the memory of Kenneth G. Hancock, an author in this volume and the former Director of the Chemistry Division of the National Science Foundation.

As Director of the Division of Chemistry at the National Science Foundation, Hancock pioneered many new initiatives, especially those that encouraged interdisciplinary and international collaborations. He had an expansive view of chemistry as a science and urged chemists to pursue imaginative research on the discipline's traditional boundaries that would expand chemistry's frontiers. In materials chemistry, he saw a vital and growing area to which chemists could make important contributions. Under his leadership, the Division's support for research in materials chemistry grew to over $20 million in 1993.

Hancock died in Budapest, Hungary, in September 1993 while on official travel. He had served as Division Director since 1990, and since 1987 had provided direction either as Acting Division Director or Deputy Division Director. He guided the development of joint programs with the Division of Chemical and Thermal Systems in Electrochemical Synthesis and in Environmentally Benign Synthesis and Processing. Hancock recognized very early the opportunities for U.S. scientists in Eastern Europe after the demise of the Soviet Union, and the Chemistry Division responded by sponsoring visits by groups of U.S. chemists to Eastern Europe, by granting supplements to promote collaborations, and by bringing an intern from the Soviet Academy of Sciences to the Division. At the time of his death, Hancock was attending a workshop on Environmental Chemistry under joint United States–Hungarian–French sponsorship, a workshop that had been his idea.

Hancock advocated collaboration among government agencies and between government and the private sector. For example, he negotiated pioneering agreements for NSF with the Environmental Protection Agency and the National Institute of Standards and Technology, and helped design a unique pilot program with a private entity, the Electric Power Research Institute. The environmental program he helped fashion included the Council on Chemical Research as a partner.

Hancock received his B.A. from Harvard and his Ph.D. from the University of Wisconsin in 1968. After a National Institutes of Health Postdoctoral Fellowship at Yale, he worked as assistant and associate professor in the chemistry department of the University of California—Davis from 1968 to 1979, where he taught graduate and undergraduate chemistry and did research in organic and organometallic photochemistry. His work opened a new field of study in organoboron photochemistry.

Hancock joined the NSF as a Visiting Scientist in 1977 and held a variety of positions within the Chemistry Division, including Program Director for Chemical Dynamics, for Organic and Macromolecular Chemistry, and for Chemical Instrumentation. He served for two years as Senior Manager of Cooperative Science Programs with Southern Europe in the Division of International Programs, during which time he negotiated, established, and administered a new joint science and technology research agreement with Spain. Several years later, he again worked with International Programs as Interim Office Head, NSF—Europe. To broaden his knowledge of the legislative process Hancock served for one year as a legislative assistant to Senator John C. Danforth, while on leave from NSF as a LEGIS Fellow. He served on numerous NSF-wide committees and task forces, including the Director's Long Range Planning Task Force for Disciplinary Research and Facilities. In 1992, he was awarded the Director's Award for Excellence in Management.

His broad public service included membership on the National Research Council's Committee on Chemical Industry, Chemical and Engineering News' Editorial Advisory Board, the American Chemical Society's Committee on Science, and numerous other posts.

Marge Cavanaugh
National Science Foundation
Washington, DC

CONTENTS

Preface xv

THE SOCIETAL AND EDUCATIONAL CONTEXT

1. **Government, Academic, and Industrial Issues** 3

Critical Technologies and U.S. Competitiveness: The Materials
Connection 4
    Robert M. White
Materials Science and Engineering for the 1990s: A National
Academies Study 12
    Praveen Chaudhari
The National Science Foundation's Program in Materials Science:
New Frontiers, New Initiatives, New Programs, and New Prospects 18
    Kenneth G. Hancock
Development and Commercialization of Advanced-Performance
Materials 28
    Mary L. Good
Partnerships with Universities 37
    Mark Wrighton

2. **Educational Issues** 43

Materials Education for and by Chemists 44
    Rustum Roy
Industrial Perspective on Materials Chemistry Education 55
    Glen H. Pearson
Funding Opportunities for Materials Science Education 60
    C. T. Sears and S. H. Hixson
Chemistry of Materials Courses at Rensselaer Polytechnic Institute 62
    Gary E. Wnek and Peter J. Ficalora
New Curricular Materials for Introducing Polymer Topics
in Introductory Chemistry Courses 66
    John P. Droske
General Chemistry as a Curriculum Pressure Point: Development
of *Teaching General Chemistry: A Materials Science Companion* 71
    Arthur B. Ellis
How Scientists and Engineers Can Enhance Science Education
in Grades K–12 76
    Kenneth H. Eckelmeyer

3. New Directions in the Design of Lithographic Resist Materials: A Case Study 85
   Elsa Reichmanis and Larry F. Thompson

4. High-Conductivity, Solid Polymeric Electrolytes 107
   Michael S. Mendolia and Gregory C. Farrington

5. Preceramic Polymers: Past, Present, and Future 131
   Dietmar Seyferth

6. Molecular Magnets: An Emerging Area of Materials Chemistry 161
   Joel S. Miller and Arthur J. Epstein

7. Optimization of Microscopic and Macroscopic Second-Order Optical Nonlinearities 189
   Seth R. Marder

8. Materials Chemistry of Organic Monolayer and Multilayer Thin Films 211
   Christine M. Bell, Huey C. Yang, and Thomas E. Mallouk

9. Orientation-Dependent NMR Spectroscopy as a Structural Tool for Layered Materials 231
   Mark E. Thompson, David A. Burwell, Charlotte F. Lee, Lori K. Myers, and Kathleen G. Valentine

10. Nanoscale, Two-Dimensional Organic–Inorganic Materials 259
    E. Giannelis

11. Nanoporous Layered Materials 283
    Thomas J. Pinnavaia

12. Catalytic Materials 301
    Bruce C. Gates

13. Molecular Sieves for Air Separation 321
    John N. Armor

14. Nanomaterials: Endosemiconductors and Exosemiconductors 335
    Geoffrey A. Ozin

15. Molecule-Based Syntheses of Extended Inorganic Solids 373
    Michael L. Steigerwald

16. Organometallic Chemical Vapor Deposition of Compound Semiconductors: A Chemical Perspective 397
    Klavs F. Jensen

17. Interfaces, Interfacial Reactions, and Superlattice Reactants 425
     Thomas Novet, David C. Johnson, and Loreli Fister

18. Oxide Superconductors 471
     Arthur W. Sleight

19. Characterization of Complex Materials by Scanning Tunneling
     Microscopy: A Look at Superconductors with High Critical
     Temperatures 479
     Zhe Zhang and Charles M. Lieber

20. Biomimetic Mineralization 509
     Patricia A. Bianconi

21. Inorganic Biomaterials 523
     Larry L. Hench

INDEXES

Author Index 551

Affiliation Index 551

Subject Index 552

PREFACE

MATERIALS CHEMISTRY is receiving increasing recognition worldwide as a key area of chemical research and technology. If we define materials chemistry as chemistry related to the preparation, processing, and analysis of materials[1], it is apparent that "materials chemistry" has always been an integral part of chemistry and that a substantial fraction (approximately one-third by one estimate[2]) of chemists are in fact "materials chemists".

On the other hand, this label is not always applied to the wide range of activities that constitute the component subjects of materials chemistry. Instead, terms such as "polymer chemistry", "solid-state chemistry", and "surface chemistry", are more usually employed by chemists when referring to their background, interest, or research and development (R&D) activity relating to materials. In this context, one might question the need for a relabeling of these various activities under the common heading of materials chemistry. The thesis of this volume, and of a growing number of scientists, engineers, and educators, is that there is indeed a real benefit to be gained by viewing these activities in the broader context of materials chemistry or the "chemistry of materials". In part, this benefit relates to the well-developed concepts of "strength in numbers" and "critical mass". The effectiveness of a group that includes one-third of the entire chemistry profession in promoting changes in policy, the distribution of funding, the education of chemists, etc., can hardly be compared with even that commanded by the largest subgroup of materials chemists.

Beyond this pragmatic line of reasoning, a materials chemistry subdiscipline combines the various components of the subject in a way that makes sense from both operational and educational viewpoints. Many concepts relating to structure, bonding, and properties are common to materials composed of organic molecules, inorganic networks, or polymer chains, and a more integrated perspective could aid both in fundamental understanding and practical applications of new materials.

Although many internal and external obstacles remain to be overcome, considerable evidence supports the idea that the chemistry profession worldwide is gradually accepting the view that there is a distinct chemistry of materials. In the past few years, the American Chemical Society, the Royal Society of Great Britain, and VCH Publishers have established journals in this area that feature titles such as *Chemistry of Materials, Journal of Materials Chemistry,* and *Advanced Materials.* All of these journals appear to be thriving; the number of papers being published, issues per year, and subscribers are increasing. In general, the number of publications in the materials-related areas of chemistry has grown at a significantly higher rate than the total in all areas of chemistry during the past 20 years or so, and this growth suggests a substantial increase in materials chemistry research over this period.[3]

This revolution in thinking about materials chemistry is beginning to have an impact on the education of chemists. In the United States, several universities have instituted courses at the graduate and undergraduate levels in materials chemistry, and a large number of faculty members in chemistry departments all over the country are engaged in research programs in this area, typically involving interdisciplinary efforts with faculty from other departments. These efforts are in addition to the long-standing programs in many universities in the various component areas of materials chemistry, such as polymer, solid-state, and surface chemistry. In some cases these efforts are being stimulated by funding agencies such as the National Science Foundation, which have established special research initiatives and have funded efforts to evaluate and change the way in which chemists view materials science, and materials scientists view chemistry.

The symposium from which this volume derives was sponsored by the Industrial and Engineering Division of the ACS and was designated, by the ACS Joint Board–Council Committee on Science, as one of the first in a series of pedagogical symposia relating to the 22 National Critical Technologies identified by the White House Office of Science and Technology Policy in 1991, based on previous critical technology studies by the Departments of Defense and Commerce. The symposium was also cosponsored by the Federation of Materials Societies and endorsed by the Materials Research Society. In addition to the ACS Committee on Science, it was supported financially by the Chemistry Division of the National Science Foundation and nine industrial organizations (*see* Acknowledgments). Its five sessions were attended by a large and diverse audience, including scientists, engineers, educators, and members of the press.

The symposium's location in the nation's capital, Washington, DC, was viewed as a unique opportunity to address the role of materials chemistry in national science and technology policy in general; consequently, the first session of this symposium was devoted to the discussion of broad issues relevant to the needs, opportunities, and problems confronting materials chemistry R&D today. This plenary session featured representatives from the U.S. Government, the National Academy of Sciences, industry, and academia. The second session was directed at educational issues and covered a range of topics relating to education of—and communication between—chemists, materials scientists, and the general public regarding materials chemistry. The last three sessions featured internationally recognized leaders in materials chemistry R&D from universities, government laboratories, and industries throughout the United States. These individuals were asked to highlight a few of the many specific topics that characterize current materials chemistry R&D and to indicate the problems, prospective solutions, and opportunities for new technology in these key areas.

The impetus for this volume came from the symposium, and many of the presentations from that symposium are represented here as chapters or parts of chapters. To broaden the scope of this volume, other individuals were invited to contribute chapters relating to their own topic of interest. Our chief objective is to provide, for chemists, materials scientists/technologists, and the science and engineering community in general, an overview of this emerging "new" subdiscipline of chemistry.

Acknowledgments

It remains only to thank the many individuals and organizations who have made the ACS symposium and volume possible. The ACS Committee on Science and the NSF Chemistry Division in particular provided both financial and moral support, without which we could not have managed a symposium, and a volume, of this type. This support was supplemented by individual grants from nine companies, representing a wide range of products and services in materials chemistry. The companies and the other organizations who provided financial support to the symposium are Allied Signal, Inc.; Air Products and Chemicals; AKZO America, Inc.; AT&T Bell Laboratories; Corning Inc.; General Electric Corporate Research and Development; Hoechst Celanese Corp.; Milliken Research Corp.; and Schumaker.

Next, the authors of the chapters that make up this volume deserve particular thanks for their excellent contributions. We also gratefully acknowledge the help of the individuals who served as reviewers for the papers in this volume. Finally, we thank the ACS Books staff who so ably managed the many details involved in the preparation and publication of a volume of this size.

LEONARD V. INTERRANTE
Department of Chemistry
Rensselaer Polytechnic Institute
Troy, NY 12180–3590

LAWRENCE A. CASPER
University of Wisconsin—Madison
Engineering Hall
1415 Johnson Drive
Madison, WI 53706–1691

ARTHUR B. ELLIS
Department of Chemistry
University of Wisconsin—Madison
1101 University Avenue
Madison, WI 53706

February 4, 1994

[1]Interrante, L. V. "Materials Chemistry—A New Subdiscipline?" *MRS Bull.,* January 1992.

[2]"National Materials Policy", Proceedings of a Joint Meeting of the National Academy of Sciences and National Academy of Engineering; National Academy of Sciences Press: Washington, DC, 1975, p 125.

[3]Based on a comparison of *Chemical Abstracts* by section and publications type for the years 1970 and 1992.

THE SOCIETAL AND EDUCATIONAL CONTEXT

1

Government, Academic, and Industrial Issues

The importance of materials science to the U.S. Government, academia, and industry is discussed in five subchapters by five different authors. First, materials science is shown to be a critical technology, and its importance to U.S. competitiveness is explained. The U.S. Government's roles in supporting research and development and in technology transfer are described. The materials science and engineering study of the National Research Council is reviewed. Implementation of the Nation Science Foundation's program in materials science, the Advanced Materials and Processing Program (AMPP), is discussed. The unique issues in the development and the challenges of bringing a new, advanced-performance material into the marketplace as a profitable article of commerce are detailed. Areas of opportunity for advanced-performance materials are outlined. The Department of Defense's support of research is reviewed. Finally, the need for Government and industry partnerships with universities is stressed.

0065–2393/95/0245–0003$16.75/0

Critical Technologies and U.S. Competitiveness: The Materials Connection

Robert M. White

Department of Electrical and Computer Engineering, Carnegie Mellon University, Pittsburgh, PA 15213–3890

MATERIALS SCIENCE IS A CRITICAL TECHNOLOGY for America. In 1987 and again in 1990, the U.S. Department of Commerce included advanced materials such as ceramics, polymers, advanced composites, and superconductors in a short list (1) of very important emerging technologies. The world market based on these advanced materials was estimated conservatively at $600 million by the year 2000.

The U.S. Department of Defense (DOD) has its own list (2) of critical technologies. Semiconductor materials rank high on this list, followed by composite materials and superconductors. In the private sector, the Council on Competitiveness released a report called *Gaining New Ground* (3), which listed technological priorities for America's future. Again, materials science and its associated process technologies were among their five major headings. Furthermore, a report (4) of the National Critical Technologies Panel highlighted an array of materials synthesis and processing technologies and singled out three advanced classes of materials: ceramics, composites, and high-performance metals. Japan (5) and Europe (6) have generated similar lists. The general consensus seems to be that materials science and its associated technologies constitute a growth industry.

Except for a few minor differences, these lists are really all the same. The relatively recent proliferation of these critical technologies reflects a growing realization that they play a critical role in our strategic and economic well-being. Technology is increasingly recognized as a resource that we have to learn to manage.

A Critical Technology

The general agreement on which technologies belong on the list suggests similar agreement on the attributes of a critical technology. One such attribute is the support of other technologies. Materials are the building blocks for electronics, aircraft, automobiles, etc. Supporting materials include semiconductors, optical materials, ceramics, polymer composities, and many others. But in some sense, this widespread use can also be a problem. The benefit from investment in materials, although highly leveraged for society, is diluted for any particular investing company. This situation often tends to discourage investment. On the other hand, it is the primary reason for U.S. Government support of an area such as this.

A critical technology also tends to be vertical, that is, its real application requires a series of advances at all stages of development. For example, high-temperature superconductors constitute an interesting class of materials. To use these materials, process technologies for making wires will need to be developed. The process technologies will have to be followed by manufacturing technology. And certainly the applications themselves will stimulate progress. The use of new classes of materials in a motor, for example, will require new concepts in power generation.

Effective utilization of a critical technology requires a much closer collaboration throughout the whole chain in which the material is used. Critical technologies evolve and develop over a long period, and so they can support a broad industrial base. This relationship creates a certain inertia with respect to revolutionary change, but it certainly indicates that it is very important to stimulate and maintain evolutionary change.

The importance of materials science to U.S. competitiveness can hardly be overstated. Key materials science areas underlie virtually every facet of modern life. Semiconductors underpin our electronics industry. Optical fibers are essential for communications. Superconducting materials will probably affect many areas; ceramics, composites, and thin films are having a big impact now in transportation, construction, manufacturing, and even in sports—tennis rackets are an example.

More than just intuition tells us how important materials are. In 1989 the National Research Council (NRC) prepared a report, *Materials Science and Engineering for the 1990s* (7) that examined in detail the impact of materials science on our national competitiveness. The NRC study surveyed eight major industries that together employed 7 million people in 1987 and had sales of more than $1.4 trillion. Additional millions of jobs in ancillary industries depend on the materials industry. Despite the very different needs for particular materials, the NRC survey also showed a remarkable consistency in generic and technological needs. Every industry surveyed expressed a clear need to produce new and traditional materials more economically and with a higher reproducibility and quality than is currently possible.

In particular, the industry survey revealed a serious weakness in U.S. research efforts involving the synthesis and processing of materials. Moreover, industry has the major responsibility for maintaining the competitiveness of its products and product operations. Collaboration with research efforts in universities and U.S. Government laboratories tends to enhance the effectiveness of those research and development (R&D) programs for the involved company.

These efforts will result in renewed emphasis on the effective long-range R&D capabilities of the industry. The materials science industry must take the lead in developing a strong competitive position for this country. However, the U.S. Government also has a clear role in supporting that effort.

Government's Role

Beyond the knowledge, technology management, and manufacturing skills, competitive commercial technologies depend heavily on a mix of complex issues and economic trade and regulatory policies. The Government can do many things to affect our global competitiveness in these areas. Examples of such activities are education, procurement, antitrust regulations, intellectual property rights, product liability, tariff barriers, nontariff barriers such as international standards, regulatory uncertainty, and the general financial climate.

The Department of Commerce has active programs in all of these policy areas. Among other areas, the Federal Government is playing an important role in the funding of research and the transfer of federal technology.

As an example, consider a test section of a prototype crush tube for the frame of a Ford Escort. The composite material consists of glass fibers in a thermoset resin. Complete frame members made of this component weigh less than one-third as much as the steel that would be used in the original component. The lighter weight of the composite material produces a corresponding savings in fuel. The Automotive Components Consortium (a joint venture of Ford, General Motors, and Chrysler) is working in a cooperative program with the National Institute of Standards and Technology (NIST, the Department of Commerce's R&D organization) to develop an improved process technology, structural reaction injection molding, to produce such parts. This part is not a particularly high-tech example of the materials revolution. It contains no carbon or ceramic fibers and no exotic polymer blends. But it is a practical technology resulting in a real product.

The Federal Government has made very significant progress in developing creative and cooperative relationships among the different departments and agencies. Federal agencies have realigned and enhanced their R&D programs. They have coordinated their activities with other agencies and share common resources. Nowhere has this been more successful than in the area of materials science. The forum for much of this coordination during the Bush administration was the Federal Coordinating Council for Science, Engineering, and Technology (FCCSET).

The FCCSET Committee on Industry and Technology, which I chaired, undertook to inventory what the Federal Government was doing in a variety of critical technology areas. Materials science is obviously one of these areas. In fiscal year 1991–1992 the Federal Government funded $1.66 billion in materials R&D. That figure has remained fairly constant since approximately 1976. Materials now account for about 2.3% of the federal R&D expenditures.

Our inventory also showed that funding levels for materials vary widely across differing materials classes. R&D on advanced metals received 13% (the largest fraction in 1992), composites were 11%, electronic materials were 10%, and biomaterials were also 10%. The FCCSET process proved enormously successful in focusing the federal program on materials research.

Presidential Initiative

The NRC report expressed concern about U.S. weakness in synthesis and processing. These areas were given high priority as the FCCSET inventory became the foundation for a presidential initiative that was referred to as the Advanced Materials and Processing Program (AMPP). This program was prepared by the FCCSET Committee and published in a January 1993 report (8). It represents the first effort to coordinate the approach to materials research across all the federal agencies. The goal of AMPP is to improve the manufacture and performance of materials and thus to enhance the nation's quality of life, security, industrial productivity, and economic growth. AMPP adopted four strategic objectives:

- to establish a leadership position in advanced materials and processing
- to bridge the gap between science and applications in materials
- to support the mission agencies
- to encourage university and private R&D activities, particularly in the area of applications

The AMPP identified four areas that required critical efforts. Not surprisingly, synthesis and processing was the first area. AMPP also decided to capitalize on U.S. leadership in computational techniques by emphasizing theory, modeling, and simulation (the second area). The third area is materials characterization, and the fourth is education.

A unified program that was submitted to the U.S. Congress requested $1.8 billion in the 1993 budget. That amount represented a 10% increase over 1992. The final Congressional appropriations only added up to a 2% increase. This budget reflected both direct research and the construction and operation of user facilities such as the cold neutron facility at NIST. The increases would have gone to specific AMPP program enhancements chosen by an interagency committee for their technical content and the extent to which they support the chosen goals. The enhancements generally follow the critical program areas of synthesis and processing, simulation, characterization, and education. The program had an enormous impact on coordination of efforts within the Federal Government. Fifty-five memoranda of understanding or agreements had been established between different agencies in the materials area at the time the budget was submitted. This number represented 55 collaborations that did not exist a few years previously. Obviously, if part of the goal of such a program is to increase industrial productivity and economic growth, there must be strong private sector involvement. There was a great deal of involvement at the technical level; 105 cooperative research and development agreements (CRADAs) between the private sector and federal agencies existed in the materials area.

Many academic and industrial scientists serve as advisors to various federal agency programs in materials science. For example, the evaluation board for the Materials Science and Engineering Laboratory (MSEL) at NIST contains representatives from Bell Laboratories, DuPont, General Electric, Allied Signal, and numerous universities. The private sector would like to improve the interactions at the strategic policy level, but some bureaucratic handicaps must be overcome.

Under the Clinton administration, the FCCSET Committee has become the National Science and Technology Committee (NSTC). The AMPP no longer exists as a stand-alone initiative, but its components have been folded into specific programs such as the Partnership for a New Generation of Vehicles (the "clean car" initiative).

Advanced Technology Program

Considering the crush tube example, you might ask why we should not improve the material itself, as well as the process technology. Thermoset plastics may not be the best practical choice for automobiles, because they are difficult to recycle. That, in fact, was the subject of a new research project announced in the spring of 1992 by the Department of Commerce's Advanced Technology Program (ATP). ATP is co-funding research at Ford Motors and General Electric to develop the basic technologies and materials data needed to use "cyclictherm" plastic composts in automobiles. Like most ATP projects, this one is somewhat risky. It requires advanced work in chemistry and materials, but the rewards can be great. The ATP is one of the Department of Commerce's key efforts in support of industrial R&D. Materials research has fared quite well in this program; in the first four rounds, materials received 13% of the funding.

Because these programs were not targeted at any particular industry or technology, advanced materials research was a natural for ATP. Like materials science, the characteristics of the critical technologies closely match the philosophy of ATP. The goal is to assist U.S. industry to carry out R&D technologies that are enabling and also have high values.

This Government program is unique on several accounts. Input originates from industry itself, from business leaders, venture capitalists, R&D directors, and also economists. The ATP program is also very highly leveraged, requiring cost sharing from industry or matching funds for grants that go to consortia. After four general competitions, ATP has committed $247 million to 89 projects; industry has committed an additional $258 million. The ATP awards completely match the lists of critical technologies. That type of comparison is the best way to determine what is a critical technology.

Under the Clinton administration, a portion of ATP funding will be "focused" in certain areas. One of these, manufacturing composite structures, relates directly to materials. Does ATP work? It is too early after just a few years to make a valid judgement about the success or failure of a program that, by its nature, is supposed to support long-term R&D. Still, some early indications are very encouraging. The program has spurred both increased investment and better leveraged investment in important technologies. Small companies with ATP awards report that they have an improved ability to raise private capital and to develop strategic partnerships with larger firms. ATP has also been quite successful in encouraging the formation of research collaborations and strategic alliances. During the first four competitions, joint ventures accounted for 26% of the awards with 134 participants, including 42 small businesses. Small businesses themselves accounted for 48% of the funding (43 companies).

Government and Industrial Cooperation

The challenge we still face is how to fund what I would call "megatechnologies". Just as we have megascience projects, we now have megatechnology projects. In some sense they are a subset of a critical technology, but they require something on the order of $100 million to fund. Flat panel display technology could be an example. The Clinton administration is using flat panel displays as a test for an entirely new way to encourage industry to engage in the manufacture of such high-risk technologies. The

Government, through ARPA, will provide R&D funds to match what a company invests in the manufacturing. This is a new concept of "matching".

I mentioned that the automotive composites consortium is working with NIST, under one of the CRADAs. CRADA, once an obscure acronym, has become very popular in Washington, DC. If federal research initiatives are to be fully successful, the results of this research must be transferred to industry, where they can form the basis for new and enhanced products and processes. Traditional mechanisms exist (such as contracts, patent licensing, and the like), but I don't think any of them are quite as powerful as the CRADA.

NIST has perhaps more experience than many Government agencies in working with private industry. Close to 90 years of collaboration in research programs with industries has taught NIST technologists one lesson in particular: There is no substitute for hands-on cooperative R&D programs if your goal is to transfer technology.

More and more companies are discovering the advantages of access to federal laboratories through CRADAs, including a wide range of technical facilities and world-class experts. CRADAs provide a way to leverage a company's resources with the research power of more than 700 federal laboratories. In 1988 there were probably only about 100 of these cooperative research and development agreements; today there are several thousand.

The increasing use of the CRADA as a tool for technology transfer was one of the key motivations behind the National Technology Initiative (NTI) at the end of the Bush administration. These were regional meetings held across the country to identify specific areas for collaboration between business, Government, and the university. These NTIs prompted dramatically increased contact and cooperation between private industry and Government laboratories. The NTI has also promoted increased collaboration between Government agencies, to make more effective use of our resources.

Much can be done by both Government and industry to enhance America's competitive position and to advance materials research, even in this time of fiscal restraint. But we should certainly make sure that we are using all of the existing resources to the best advantage, eliminating duplication of effort, and avoiding lost or wasted opportunities. Through mechanisms such as CRADAs and resource consortia, Government and industry can join forces to maximize our return on investment.

References

1. "Emerging Technologies: A Survey of Technical and Economic Opportunities", U.S. Department of Commerce, Technology Administration, 1990.

2. "Critical Technologies Plan", U.S. Department of Defense, 1990.
3. *Gaining New Ground: Technology Priorities for America's Future;* Council on Competitiveness: Washington, DC, 1991.
4. Report of the National Critical Technologies Panel, Washington, DC, 1991.
5. "Trends and Future Tasks in Industrial Technology", Japanese Ministry of International Trade and Industry: Tokyo, Japan, 1988.
6. "First Report on the State of Science and Technology in Europe", Commission of the European Communities, 1988.
7. Chaudhari, P.; Flemings, M. *Materials Science and Engineering for the 1990s: Maintaining Competitiveness in the Age of Materials;* National Academy Press: Washington, DC, 1989.
8. "Advanced Materials and Processing: The Federal Program in Materials Science and Technology", A Report by the FCCSET Committee on Industry and Technology to Supplement the President's Fiscal Year 1993 Budget.

Materials Science and Engineering for the 1990s: A National Academies Study

Praveen Chaudhari

IBM Research Division, T. J. Watson Research Center, P.O. Box 218, Route 134, Yorktown Heights, NY 10598

THE MATERIALS SCIENCE AND ENGINEERING STUDY of the National Research Council was initiated by a letter from U.S. Representative Don Fuqua (who was at the time Chairman of the House Science and Technology Committee) to the presidents of two academies of science and engineering. Fuqua asked about the definition of materials science and about its priorities. After deliberations and several meetings, the academies approached Merton Flemings and me in 1988 to chair a study that would involve scientists from the universities, industry, and Government laboratories.

Description of the Study

The study, concluded in December 1988, was formally unveiled in the fall of 1989 with a report entitled *Materials Science and Engineering for the 1990s (1)*. Funding was provided by a number of agencies. About 400 people contributed to this study, and about 100 of these were formally appointed by the two academies.

We soon decided that we would not follow the pattern of earlier studies, such as the physics and chemistry surveys. Instead we would focus on issues associated with materials science and engineering and its relationship to industry. We organized ourselves into five issues-related panels. The goal of the first panel was basically to examine opportunities and needs. They sought the opinions of our colleagues at universities, Government laboratories, and industrial laboratories about the intellectually exciting areas in this field. Our colleagues in industry, particularly the senior technical people, were asked what they thought was needed and what would give them a competitive advantage in their business. We asked the second panel to explore ways by which all of this knowledge could be utilized. The third panel considered what our colleagues abroad were doing, particularly our trading partners such as Germany and Japan.

The fourth panel examined research resources within the United States, and the fifth looked at education.

The study results were integrated into the report that was published. The project surveyed eight industries that collectively employ more than 7 million people and account for more than $1.4 trillion in business annually. The influence of this survey permeates our study.

We also found that in the field of materials science education, both education and work force needed considerable reorientation. For historical reasons (listed in the study) synthesis processing had been neglected at university and Government laboratories. We also looked at the debate about the role of a principal investigator versus large facilities, such as the synchrotron radiation laboratories. We concluded that materials science needs all of these different modes of research, and no particular approach is more decisively effective than another.

At the time of the study we tried to get the best possible information about the financial support available from the Federal Government. Over the past decade, funding had declined in terms of constant dollars. In contrast, materials science and engineering, information science, and biotechnology had been targeted as the three main growth areas by most of our trading partners. Government had played a proactive role in materials science and engineering in these countries. Rather than covering a broad range, as we tend to do here, these countries focused on specific areas that complemented their economies. We recommended a substantial increase of federal funding in selected areas, with a $160 million increase targeted for 1993 as the first step in a multiyear program.

Vitality of Materials Science

Materials science and engineering is a vital field of endeavor for industry and defense and also with regard to its sheer intellectual content. This field is no longer a disparate collection of disciplines. Its unity and coherence can be seen in a tetrahedron; this image captures the essence of the field. The four aspects to any materials-related activity are as follows:

1. performance—provides a yardstick for measuring whether a material is good or bad, and what is needed to make it better. This yardstick can be technical, financial, or both.
2. properties—in physics they would be called phenomena.
3. structure—where the atoms are and what kinds of atoms are present.
4. the making (synthesis and processing) of the material

All of these factors are interconnected parts of the tetrahedron.

We found no exception to this description in any materials class or materials problem. The perspective of the tetrahedron shows that the

United States is relatively weak in the area of synthesis and processing. Other themes common to all four corners are instrumentation and computers. Instruments are essential for making advances in science. The United States had a unique advantage in being at the forefront in computer use. We felt that this advantage should be exploited in materials science.

What measures do we have for our assertion that this field is vital? We tried to generate a diverse data set from varied areas.

The first measure concerns the structural strength of materials. The prevailing strength-to-density ratio has increased over time and has been accelerating throughout the past two decades. The quality of products used in permanent magnets has improved a great deal over the past decade. This measure of energy efficiency is important in designing motors or generators. Even in a more prosaic area, such as cutting tool materials, superior materials are making it possible to cut faster. In more complex areas such as aircraft engines, there are continuous advances in operating temperatures, which determine thermodynamic efficiency.

Materials science has often led to major changes in the way we live and the way we carry out our day-to-day transactions. A recent example is the impact of glass fibers for optical communication. The most dramatic improvement occurred in the mid-to-late 1970s. During this period transmission through glass fibers increased by several orders of magnitude. Now, 96–99% transmission is possible through a thick glass windowpane. These glass fibers, when they reach our homes, will change the way we live and entertain ourselves. The products of a number of industries are nothing more than a very sophisticated assemblage of materials. The car and aircraft industries are notable examples. Advances in materials such as composites could play important roles in replacing materials such as structural steel. This development, of course, would make the car more efficient and conserve energy.

Materials are central not only in the transportation industry but even in electronics. The rate of progress in hardware is determined by materials and their processing. Nothing in the laws of nature that says that we cannot build a device that is about 700 atom diameters wide. In fact, we have built devices smaller than these and they all operate well. But this achievement was a laboratory demonstration. To manufacture these devices, we need a steady advance across a broad front of materials processing, new tools and techniques, and materials properties.

These observations apply not just to traditional industries, but even to those that are still evolving and in which the United States has a commanding lead. For example, the biomaterials industry is in its infancy compared to the transportation and electronics industries. I expect it to become a major industry in the future.

This is all well and true for the past and present, but are exciting developments happening in materials science now that could change the future? The scanning tunneling microscope (STM), a great tool for making very precise measurements of surfaces, is an example. It is used extensively for surface topography. Recently scientists have been able to pick atoms up, deposit them where they want to, and write images such as the 35-atom image of IBM. Why is this interesting? It demonstrates not only that we can deliberately change a surface at the atomic level, but also that we are reaching a level of complexity that allows use of the STM or the atomic force microscope (AFM) to store information. Researchers at the Almaden laboratory are enthusiastic about this possibility and have demonstrated high-density storage.

The STM is a demonstration of spatial control and resolution. Comparable advances in temporal resolution have been accomplished by using, for example, the pulse compression technique. As a pulse of light propagates through an optical fiber, it responds (if the pulse intensity is large enough) in a nonlinear way. The output pulse is chirped, similar to the chirping of a bird. The red component of the pulse appears first and the high-frequency blue end later. This pulse is then passed through a grating, which forces the red component to traverse a longer path. Given that the speed of light is constant, the frequency components pile on top of each other, resulting in pulse compression.

This technique will allow compression of a 100-femtosecond pulse down to 12 femtoseconds or even to 8 femtoseconds. (A femtosecond is a millionth of a billionth of a second or 1×10^{-15} s.) Pulse compression can be used to study chemical reactions, particularly intermediate states, at very high speeds. Alternatively, these optical pulses can be converted to electrical pulses to study electrical phenomena. This aspect, of course, is of great interest to people in the electronics industry because of their concern with the operation of high-speed electronic devices. It also is of great interest to people who are trying to understand the motion of biological objects such as bacteria.

In addition to these two examples of very short temporal pulses and very short spatial resolution techniques, other notable advances have been made. For example, many new materials have been discovered over the past decade, for example, the quasi-crystals with their unique and totally unexpected fivefold symmetry, the fullerenes, and the new high-temperature superconductors. Progress in the use of high-temperature superconductors for wires is entirely dependent on how well we can understand grain boundaries and the role of other defects that control the critical current density. We still do not understand the pairing mechanism responsible for this class of superconductors, and we have no idea if today's 125 K value is the limit or simply a limit until we exceed it. Most recently, interest in the fullerenes has been growing, now that their structure has been unraveled, and their properties are demonstrating a variety of interesting phenomena. These few examples hint at the sense of vitality and of excitement that we tried to capture in the study.

After the Study

A very important aspect of any major study is the follow-up activity. It affects three communities, each of which speak a "different language". The message is the same, although the words and the emphasis differ for each group. The three communities are as follows:

1. The technical community; it is important to communicate to them the scholarly aspect of your findings because their support and agreement are important.
2. The policy makers; in the United States, this means the agency heads and Congress.
3. The public—particularly the business community.

The second group may not respond unless the third group takes an interest.

In dealing with the technical community, we spoke at national meetings and also organized regional meetings. We divided the United States into four regions and asked the local industry, university, and Government groups to tell us what made sense for materials science in that particular region. The report from these regional meetings, published by the Materials Research Society, was used by the Federal Coordinating Council for Science, Engineering, and Technology (FCCSET) as part of their analytical process.

We also saw a number of policy makers, starting with Allan Bromley. Allan was convinced of the importance of this field before we finished our talk, and he subsequently provided the leadership that led to Bob White overseeing the FCCSET initiative. We also went to see Eric Bloch. He responded positively, and NSF took the initiative and proposed an increase in their 1992 budget. We also met with Admiral James D. Watkins, the Secretary of the U.S. Department of Energy (DOE) and with a host of congressmen and senators.

For the public, we made a video documentary of materials science education in which the National Academy of Sciences collaborated with WQED. This program was broadcast on public television as part of the Infinite Voyage series. The video on materials science education, called "Miracles by Design", is useful for showing to students who are not sure what this field is all about. It covers the spectrum: physics, chemistry, metals, ceramics, and polymers. An article in *Business Week* called "The New Alchemy" covered this field in some detail for the business community.

All of this activity culminated with the presidential initiative that Robert White described in the previous section. The follow-up activities are not over, and much still remains to be done. In particular, money has to be appropriated by the legislative bodies and made available to the funding agencies. In these uncertain times we cannot take for granted that a presidential initiative will automatically result in new funding. We need our community to continue to convince our congressmen of the national importance of this field.

References

1. Chaudhari, P.; Flemings, M. *Materials Science and Engineering for the 1990s;* National Academy Press: Washington, DC, 1989.

The National Science Foundation's Program in Materials Science: New Frontiers, New Initiatives, New Programs, and New Prospects

Kenneth G. Hancock[†]

Division of Chemistry, National Science Foundation,
Washington, DC 20550

IN WASHINGTON, REPORTS ARE PLENTIFUL. Most of them are on important topics and reflect careful analysis by well-informed, thoughtful people. But there are too many: with few exceptions, these reports conclude with a request for new programs and more money from agencies, departments, and legislators that hear the same litany over and over again; only the supplicants change. Rarely, a report commands major attention and holds it over some period of time. Praveen Chaudhari and Merton Flemings produced a report that has received the attention it deserves. *Materials Science and Engineering for the 1990s* (*1*), published by the National Research Council in 1989, has been the foundation for planning on a coordinated and sustained level not often seen in Washington, outside of major national efforts such as the space program or the superconducting supercollider.

The importance of materials science in tomorrow's world cannot be overstated. Technologies from microelectronics and nanostructures to spacecraft and biomedical prostheses depend absolutely on amazing materials created through the ingenuity of scientists and engineers. Kevlar composites, high-temperature superconductors, and buckeyball-based structures did not exist a short time ago. But we are already on the edge of an even more astonishing materials fixture: Just in the past few years, we have developed techniques to assemble materials molecule by molecule and atom by atom; we literally have the ability to move single atoms and place them where needed. Imagine the possibilities that power brings to the design, synthesis, and processing of newer, "smarter" materials for applications we have not yet considered. Truly, the opportunities are not only mind-boggling, they are also only mind-limited.

The socioeconomic impact of materials in the United States is no less staggering. The eight industries of aerospace, automobiles, biomaterials,

[†]Deceased

chemicals, electronics, energy, metals, and telecommunications—all critically dependent on materials—together generate $1.4 trillion in sales (1987 figures) and employ 7 million people. No wonder that *Materials Science and Engineering for the 1990s* attracted such attention!

Robert White, chair of the Federal Coordinating Council for Science, Engineering, and Technology (FCCSET) Committee on Industry and Technology, has already discussed the materials research and development activities of 10 federal departments and agencies that were coordinated into a coherent, cross-agency 1993 Presidential Initiative known as the Advanced Materials and Processing Program (AMPP). The AMPP Initiative as proposed for fiscal year 1993 was roughly a $2 billion enterprise, including both the existing base and the proposed 1993 increases. This section will focus on implementation of the AMPP at the National Science Foundation (NSF).

Early Precursors to AMPP at NSF

Although the NSF has had a Division of Materials Research (DMR) and has supported a network of materials research laboratories (MRLs) since 1972, additional recognition of the emerging opportunities on the molecular scale in materials science prompted a 1984 workshop at the NSF. Sponsored by NSF's Divisions of Chemistry and Materials Research, the conference explored the scientific opportunities and programmatic needs in the area of materials chemistry (MC), roughly defined as the region of overlap between the macroscopic frontier of chemistry (the molecular science) and the microscopic frontier of real materials (a macroscopic science). Out of that workshop report grew a small but catalytic program to support collaborative projects involving both chemists and materials scientists or engineers, trying to bridge the gap between the colligative and molecular worlds. The role of chemists in materials science is now as it was then: To synthesize next-generation materials atom by atom, the chemist must understand all dimensions of molecular interactions and their impact on macroscopic properties in order to know which atoms to put where. Moreover, synthesis is the heart and soul of chemistry. Physicists, mathematicians, and engineers analyze, characterize, and process materials; chemists synthesize them.

Materials chemistry proposals were jointly reviewed and split-funded, and in 1987 and 1988, 33 cooperative research projects were initiated. In 1989, the partnership was expanded to include the Division of Chemical and Thermal Systems (NSF's home for chemical engineering), and the program was renamed Materials Chemistry and Chemical Processing (MCCP). In 1989, 1990, and 1991, each of the three participating NSF divisions invested about three-quarters of a million dollars in additional projects, most of which were later mainstreamed into existing program rubrics, and a high fraction of which have been successfully renewed.

The total 5-year investment in the materials chemistry programs (MC and MCCP) was more than $18 million. That number might be considered modest, but MC–MCCP accomplished two things: (1) it initiated about 60 collaborative research projects in the chemistry research community, where individualism was the overwhelming norm; and (2) it established within NSF a paradigm for interdivisional cooperation in the review and funding of interdisciplinary research.

NSF's 1992 Materials Synthesis and Processing Initiative

During 1990–1991, while the FCCSET Committee on Industry and Technology was carrying out the extensive analysis, coordination, and planning necessary to implement a Presidential Initiative, the NSF was carrying out its own component of this analysis. An inventory of support for materials science and engineering at NSF (1991 actual expenditures) includes the following:

- $216 million for materials research and development (R&D) research project support, principally in DMR, the Engineering Directorate (ENG), and the Chemistry Division (CHE)
- $31 million (additional) for national user facilities (nanofabrication, synchrotrons, magnet labs, and supercomputers)
- materials research laboratories (nine) and groups ($47 million)
- science and technology centers (7 out of 25 have materials as their focus)
- engineering research centers (6 out of 18 have materials as their focus)
- industry–university cooperative research centers (15 out of 26 have materials as their focus)

Recognizing the criticality of materials science and engineering, the NSF moved to get a head start on the materials programs being planned through FCCSET by establishing its own Materials Synthesis and Processing (MS&P) Initiative for fiscal year 1992. An increment of $25 million was requested in the 1992 budget as the first phase of a 5-year effort to strengthen research in materials synthesis and processing. The 1992 MS&P Initiative had two aims: (1) molecular-level approaches to the design and synthesis of new materials based on fundamental principles and a developing base of molecular structure–property–performance relationships; and (2) new and improved processing methods, including reactor design, kinetics, and applications to manufacturing, looking to produce materials with improvements in efficiency, properties, and quality.

The MS&P Initiative was launched as planned in 1992, although the 1992 congressional appropriation was less than requested. Features of the program (2) included the following:

- They focused on synthesis and processing (including relevant theory and characterization).
- They included five eligible materials classes, two favored (electronic–photonic and biomolecular) and three others (structural, magnetic, and superconducting materials).
- They included single- and multidisciplinary projects.
- They accepted proposals from single investigators and groups.
- Nine NSF divisions cooperated in review and funding.

Biomolecular materials were defined as those substances, natural or synthetic, with novel materials properties that use or mimic biological phenomena. A sample menu of ideas was generated to provide some sense of the envisioned scope of the MS&P program:

Electronic and Photonic Materials

- new materials with unique properties (semiconductors, superconductors, insulators, and composites)
- methods for deposition and growth (films, layered structures, bulk crystals, and fibers)
- low-temperature synthesis and preparation
- combining materials growth and processing techniques
- laser, electron, ion, and plasma-assisted processing
- real-time, in situ diagnostics

Biomolecular Materials

- genetic modification of natural synthetic pathways
- biomolecular self-organization and phase behavior
- novel catalyst, sensor, and transducer materials
- materials aspects of in vivo biopolymer processing
- synthetic structures mimicking natural composites
- biodegradable or biorecyclable materials

Structural Materials

- new metallic alloys, polymers, ceramics, and composites
- origin and evolution of phases, defects, and microstructures
- solid-state behavior controlling multiphase materials properties (phase transformations and grain boundaries)
- processing methods (particle consolidation, sol–gel conversions, rapid solidification, and powder synthesis)

- direct conversion of precursors to finished forms (reaction bonding and injection molding, microwave sintering, and net-shape manufacturing)

Magnetic Materials

- design and synthesis of new magnetic materials
- artificially structured multilayer magnetic materials
- enhanced properties in hard and soft magnets, thin films, and magneto-optics
- surface and two-dimensional magnetic behavior
- new processing methods for magnetic materials

Superconducting Materials

- superconduction in bulk materials, thin films, and reduced-geometry structures
- low-temperature, in situ processing and fabrication methods
- improved structure–property relationships and theory
- single-crystal growth of superconducting materials
- properties of surfaces and interfaces: connections, contacts, and passivation
- crystal structure, microstructure, and morphology

All multi-investigator proposals were due by November 1, 1992, because it was anticipated that the large majority of them would have to be co-reviewed and co-funded by two or more disciplinary NSF divisions. Single-investigator proposals were accommodated within regular programmatic boundaries and guidelines. A matrix-managed review procedure was established. Proposals were to be addressed to the NSF division appropriate to the principal technical thrust of the proposal (its "center of gravity"), where a divisional coordinator carried out preliminary screening for suitability, negotiated with other divisions where required for joint review, and then managed the review itself. To a large extent, each NSF division used its usual review procedures, although several divisions reviewed all MS&P proposals with specially assembled review panels instead of using ad hoc mail review.

Approximately 700 proposals were received in response to the MS&P announcement in fiscal year 1992; some divisions had no deadlines for individual investigator proposals, so this inventory was not complete until the end of the 1992 fiscal year. The breakdown between collaborative proposals from groups and those from single investigators was approximately 2:1. About 50% of proposals had a center of gravity in the DMR (principally solid-state chemistry, polymers, and electronic materials) and were managed by DMR. Another 33% were managed by five engineering divisions, 12% by chemistry, and 5% by two biosciences divisions. As expected, most of the proposals fell into the categories of electronic and optical or photonic materials; fewer proposals than expected were received with a biomolecular materials focus.

Data for proposals and awards in which the Division of Chemistry was involved are as follows:

- Reviewed 122 out of 700 proposals; managed 82; 52 were single investigators; 70 were groups; 92 required interdisciplinary review; 30 were reviewed within CHE.
- Focus was on electronics (48%) and photonics (25%); biomaterials was 10%; magnetic was 8%; structural was 7% of the proposals.
- Funded 26 awards (21%), $3.1 million; 13 were single investigator; 13 were groups; 16 out of 26 were co-funded with four different divisions.

Data for proposals and awards in which the Division of Materials Research was involved are as follows:

- Reviewed 351 out of 700 proposals.
- Funded 57 awards and co-funded 28.
- Total investment was $6.4 million in 85 grants (16% success rate).
- Award distribution was as follows: 40%, electronics; 23%, optical–photonics; 18%, structural; 3%, biomolecular; 9%, magnetic; and 7%, superconducting.

What has been learned from MS&P about multidisciplinary program management? Program management must be kept simpler by taking a "varietal wine" approach to the labeling and review of proposals. It is quite cumbersome to matrix-manage a large number of proposals. In the future, NSF will have to assign proposals to a given program on the basis of the scientific "center of gravity", have that program solicit assistance as needed, but make review and award decisions more locally.

Looking Ahead to the AMPP

The AMPP is a coordinated interagency effort to exploit opportunities in materials research and development to meet significant national goals and to extend U.S. leadership in materials-dependent critical technologies. The goal (*3*) is "to improve the manufacture and performance of materials to enhance the Nation's quality of life, security, industrial productivity, and economic growth". To achieve this goal, a set of strategic objectives was established:

1. maintain U.S. leadership in advanced materials and processing
2. bridge the gap between innovation and application of technologies

3. support agency mission objectives to meet national needs
4. encourage university and private sector R&D related to AMPP

Implementing priorities were also established:

1. support strategic objectives through R&D effort
2. plan federal programs to incorporate needs of strategic, industrial, and social sectors
3. promote applications through university–industry–Government co-operation in generic, competitive technology development
4. support the human resource base to meet future needs
5. maintain healthy infrastructure (e.g., facilities)
6. focus R&D on materials and processes that are most important to achieving AMP strategic objectives

Although the AMPP is an R&D program, its purpose goes well beyond curiosity-driven research. Success is going to be measured not only by new discoveries, but also by successful application of new knowledge and technology. Thus, a significant ambition within AMPP is to strengthen productive interaction between the Government, industry, and academic sectors. All participating federal agencies share the same AMPP goals and objectives consistent with their missions. NSF's mission is the generation and dissemination of fundamental knowledge and the training and development of scientists and engineers.

The AMPP has three conceptual tiers: (1) an inventory of current materials R&D; (2) targeted program enhancements; and (3) conceptual opportunities for technical breakthroughs. The inventories are by agency, by materials class (the terms are familiar to chemists), and by program component. AMPP has four program components: (1) synthesis and processing; (2) theory, modeling, and simulation; (3) materials characterization; and (4) education and human resources. National user facilities are included in the inventories, but they are not a program component in the sense of being subject to the same priority-setting practices.

The priority-assigned program components increase roughly as their applicability to the national needs identified in the AMPP program goal. Within synthesis and processing, process integration takes a higher priority than basic research; similarly, application-specific theory, modeling, or simulation takes precedence over more fundamental research. Bigger budgetary increments were proposed for synthesis and processing than for materials characterization. These priorities are for the overall interagency program. In synthesis and processing, for example, process integration may be emphasized at the National Institute of Standards and Technology (NIST), and basic research would be emphasized at NSF. Or, even within NSF, process integration and applied research may be centered in ENG, and basic synthesis is centered in CHE or DMR.

The prioritization of research objectives and classes is is not necessarily what a "curiosity-driven" researcher likes to hear. However, these are Government-wide priorities. They apply to all participants in the AMPP, but not all agencies have exactly comparable missions. AMPP research carried out with NSF support will probably have a more fundamental flavor, on average, than R&D sponsored by a mission agency. Objectives for 1993 within the NSF component are

- synthesis of advanced materials
- fundamental physics and chemistry of materials
- links between synthesis and processing and materials structure, properties, and performance
- development of novel processing and manufacturing methods
- creation of linkages with industry for knowledge and technology transfer
- emphasis on academic research for education and training

The AMPP represents a major response to the needs and opportunities spelled out in the materials science and engineering (MS&E) report (*1*). Major opportunities exist for scientific breakthroughs in materials science, and many of them, perhaps most, will need chemists for the key finding or concept. Everyone who thinks about chemistry and materials can generate her or his own list. Some ideas and possibilities that seem particularly challenging and ripe for plucking by chemists are in the areas of polymers, biomolecular materials, and electronics–photonics.

New "natural" polymers based on synthesis from renewable resources, improved recyclability based on retrosynthesis to reusable precursors, and molecular "suicide switches" to initiate biodegradation "on demand" are the exciting areas in polymer science. In the area of biomolecular materials, new materials for implants with improved durability and biocompatibility, light-harvesting materials based on biomimicry of photosynthetic systems, and biosensors for analysis and artificial enzymes for bioremediation will present the breakthrough opportunities. Finally, in the field of electronics and photonics, the new challenges are molecular switches, transistors, and other electronic components; molecular photoaddressable memory devices; and ferroelectrics and ferromagnets based on nonmetals.

Although the AMPP represents a set of real opportunities—both intellectual and financial—for chemists, two important constraints must be recognized. The first constraint is also financial: The federal budget will not be everything the scientific community might hope for. Congressional spending caps and competition among many different funding demands will restrict budget growth. In some situations, agency or program budgets may not exceed those of 1992. At the NSF, at least, the AMPP will move ahead in 1993 at some level, because chemists and engineers are seizing on the fundamental intellectual challenges and basic questions posed by materials problems, whether or not funds are "set aside". In all three divisions that are the principal supporters of the chemistry aspects of materials, materials chemistry and chemical engineering have already been identified as major intellectual frontiers in long-range planning exercises. Hence, the AMPP represents an intellectual thrust as well as a fiscal one.

The other constraint is that AMPP is a goal-oriented research program. Even at the NSF, it is not quite "business as usual". Policy issues at the national level are pushing the NSF to take a broader view of its mission in education and research, relating those traditional strengths to national needs, especially in the area of economic competitiveness. NSF will increasingly look for opportunities to contribute to the nation's priorities through its unique programs. NSF, for example, is particularly well-suited to support fundamental research at academic institutions because that activity couples the research and education missions; that is, NSF is contributing to the nation's human resource infrastructure through research support.

However, NSF is also moving to contribute to more effective partnership in research between Government, industry, and academia. This partnership is important to speed knowledge transfer from the basic research laboratory to application and commercial development, and maps well onto the strategic priorities of the AMPP. For example, some quantitative measures of performance have been proposed to exist in monitoring the effectiveness of NSF programs and activities, such as the number of interdisciplinary research projects, the number of industry–university collaborations, the number of centers and groups, and the number of Memoranda of Understanding or Cooperative Research Agreements with other Government agencies.

The number of interdisciplinary projects supported is a useful indicator because it is in keeping with the AMPP goal to bridge the gap between different disciplines. An increase in the number of industry–university collaborations might speed knowledge transfer between those research sectors. The very existence and purpose of a fair number of centers hinge on industry–university partnerships. Extending the partnership concept from industry–university to include Government research laboratories is important to get maximum return on investment from these national treasures of scientific talent; that step, too, is already underway. Such criteria are not substitutes for the old standbys of important results and education of tomorrow's students, but they may be viewed as value-added measures for some situations.

Within the Division of Chemistry, several initiatives to improve intersectoral cooperation have already been established. New in 1992 were (1) a cooperative program with the Electric Power Research Institute on electrochemical synthesis: joint review and joint funding; and (2) a cooperative program with the Council for Chemical Research (CCR) on environmentally benign chemical synthesis and processing. In this CCR–NSF activity, university-based research projects are required to have industrial intellectual partnership in order to speed knowledge transfer and to ensure applicability of the research to real-world problems. Other new and experimental ventures are likely to follow. Many of today's important fields of chemistry grew out of basic research carried out in the years after World War II to answer important practical questions. Those applications of chemistry to the real world made chemistry the central science that it is today. The AMPP will be an important force for renewing existing links between basic research and application and for building the new ones for chemistry's tomorrow.

Acknowledgments

This chapter represents only the views of the author and is not intended to represent official views or policy of the NSF. Data cited are drawn from NSF sources and publications.

References

1. Chaudhari, P.; Flemings, M. *Materials Science and Engineering for the 1990s;* National Academy Press: Washington, DC, 1989.
2. *Materials Synthesis and Processing: Research at the Interfaces of Materials Research, Engineering, Chemistry, and Biology;* NSF 91–75, National Science Foundation: Washington, DC, 1991.
3. *Advanced Materials and Processing: The Federal Program in Materials Science and Technology;* FCCSET Committee Report to supplement the President's fiscal year 1993 budget; Washington, DC, 1992.

Development and Commercialization of Advanced-Performance Materials

Mary L. Good

Allied Signal Inc., Des Plaines, IL 60017

Bringing a new, advanced-performance material into the marketplace as a profitable article of commerce presents many problems. However, I'm reminded of a cartoon I have saved over the years that has two scientists in white lab coats looking at their test tubes, and one of them says, "Remember, our job is to push science to the state of the art and make a buck in the process." So, my remarks will be directed to the problems associated with "making a buck" out of all this forefront research in materials science.

I define advanced-performance materials, development, and commercialization for the purposes of this chapter as follows:

- *Advanced performance materials* are materials (metals, ceramics, polymers, etc.) whose functional and structural properties impart improved performance to specific products, that is, an enabling technology.

- *Development* is the "proof of principle" of material design and product applications.

- *Commercialization* is scale-up, process development, and marketing either as a material or as an integral part of a product.

Thus, advanced-performance materials are not necessarily a product in their own right, but rather an enabling technology that allows the design of new products with new utility or the improvement in performance of an existing product. As we progress from the research needed to discover a new material, to the development of a prototype product, to the commercialization of the product, the associated cost goes up exponentially with each step. The research is relatively inexpensive, development will cost on average 10 times the research costs, and a commercial launch can easily cost 10 times the development expenses. So, the issue is that a $200,000 piece of research may require a $2 million development program culminating in $20 million for a commercialization launch. Thus, if we are to capitalize on our research activities, we need to understand the system totality for commercialization. So, although we talk about the research base, unless we have an economic model for the development and commercialization phases, the competitiveness issues are moot.

To understand the system, let's look at the generic characteristics of advanced-performance materials and their utility. These characteristics can be summarized as follows:

1. Specific properties such as heat resistance, strength, and inertness are "designed in".

2. Utility depends on product applications and proof of superior performance.

3. Initial applications are usually niches and require a limited amount of material.

4. Attractive commercialization schemes require a material of intrinsic value that will justify a high margin or applications for which the value can be captured from end-product margins.

5. The dilemma then is high development costs and high-risk returns.

Examples of the utility and need for new, advanced-performance materials are numerous. For example, in turbine engines today, the need is to be able to increase operating temperatures by 100–150 °C. The laws of thermodynamics allow significant fuel efficiency to be gained as the temperature increases. However, the material, particularly for the turbine blades, must be able to handle these increased temperatures. This material need illustrates the connection between product application and material performance.

Development of Advanced-Performance Materials

One of the unique issues in the development of advanced-performance materials is that they are very product-specific, and their development requires expensive prototype iteration and performance testing. The product development is people- and design-intensive and usually results in a niche market for the material; that is, the specific product slate for which the material has been designed and tested. Many of the applications are in high-tech industrial products like aerospace components, so the total volume of material used will be small. Thus, attractive commercialization schemes require that the material have intrinsic value that will justify a high margin, or there must be a product application for which the value can be captured in the end product.

Designing materials for a proprietary product as a vertical integration process, in many cases, provides a superior product for which the material development costs are derived from final product margins. In this area, the Japanese manufacturers seem to have an edge. They frequently target material development for specific products and use a "market pull-through" philosophy to get the right properties to enhance their product line. This approach is in contrast to many materials efforts that are "technology-push", that is, new performance materials are looking for a product home. In this case, the material development costs are high with high risk returns. Most chief executive officers do not favor these odds in their prioritization of development resources.

These *life-cycle dynamics* for advanced-performance materials were described in 1987 by Eckstut (*1*) (Figure 1). His model still has relevance to the commercialization dilemma of advanced-performance materials. In Figure 1, development costs are going up from the molecular invention to material utilization steps. Early on, the new molecular system may have a specific application in a small and fragmented market in this technology-push model in which a new material with unique properties is "looking for a home". If this small market materializes and processing costs are successfully driven down, new applications can be found that contribute to other products' functionality or product life when direct substitution of the material can be made in existing part or product designs. An example is a new light-weight composite that is substituted directly for a metal in a weight-sensitive part.

If the material has sufficient intrinsic property enhancement, newly designed parts can allow for the optimal utility of the material. Again, if sufficient market develops, resources can be deployed to optimize the

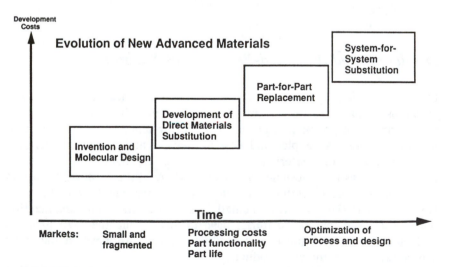

Figure 1. The life-cycle dynamics of advanced-performance materials. (Reproduced with permission from reference 1.)

material processing and the final product design. Now the possibility exists for system-for-system substitution using the new material as the foundation for a superior performance substitute system. An example of a new material that was successfully commercialized through these stages is optical fiber. Currently, communication systems are being designed on the basis of optical fiber technology, which has little relationship to copper wire systems. Optical fibers have become a major materials business in which there is intrinsic value in the material itself and the volume demand has grown to the extent that processing costs, including capital, can be recovered from material sales.

The number of such examples, however, is not high. In many other examples of advanced-performance materials, such as DuPont's Kevlar and Allied Signal's SPECTRA, the volume applications associated with system-for-system substitution has not yet occurred at a level necessary to pay back the development and commercialization costs already expended. High-performance ceramics is another area in which the early promise has yet to materialize. The consequences of Eckstut's life-cycle dynamics have been overcapacity and severe rationalization in high-performance carbon fiber businesses, some specialty alloy activities, and high-performance polymer composites. Thus, with critical technologies that involve advanced-performance materials, we need to better understand how to exploit their value in a commercially viable way.

Areas of Opportunity for Advanced-Performance Materials

To begin to understand the need and the areas of opportunity for advanced-performance materials, we need to analyze where they fit. First, several unique "drivers" exist for advanced-performance materials for which enhanced properties translate into next-generation products:

- aerospace industry (structural materials): military, space, and civilian aircraft
- electronics–communications (electronic materials): miniaturization, higher performance, and new technologies (electronics–electro-optics–optical)
- ground transportation:/ automotive, ships, and shipping containment

Clearly, aerospace has been a major driver for advanced structural materials. It will continue to be a driver, although perhaps at a slower pace, particularly for weight reduction and higher temperature reliability. The slowdown of military aircraft development presents a unique problem because many of these programs, such as the advanced tactical fighter (ATF), were "test beds" for advanced-performance materials for which

U.S. Government funds were used to defray research and development costs. The aerospace plane is another opportunity for companies to develop new materials with Government support. However, military support for advanced-performance materials development is rapidly decreasing, both in appropriate test beds and in manufacturing support such as the U.S. Defense Department's Mantech program.

Space technology development has also provided advanced-performance materials test beds in both communications and structural areas. The value of these programs has been passed on in many cases from space and the military to civilian aircraft. Many of the advanced-performance materials in the new generation of airline transports, such as structural composites, were first developed for spacecraft or advanced military aircraft.

In electronics and communications, the drivers are the need for further miniaturization, higher performance, and new optical technologies that provide entirely new products. For example, in aircraft, control systems have progressed from mechanical hydraulic components to fly-by-wire electronic systems to the new concept of fly-by-light optical systems. This progression has depended on the development of the appropriate materials to design the performance systems.

In ground transportation, the push for advanced-performance materials has been less dramatic because cost is the discriminator in the selection of components and systems. However, incremental improvement in automotive materials has been substantial in many areas, particularly in structural steels and plastics.

The list of high-performance materials currently in some stage of development or commercialization is quite extensive. Some of the most promising in terms of potential product enhancement include high-temperature and high-strength ceramics, rapidly solidified metals (metallic glasses), high-temperature and high-strength fibers, electronic polymers, optical polymers, inorganic electronic materials, high-strength polymer composites, metal matrix and ceramic matrix composites, and high-temperature alloys. All of these areas have specific products of high promise in terms of product enhancement. However, many of them are offered by small, start-up companies who have yet to truly capture the markets they seek, and others have been developed by "deep pocket" chemical companies who are reassessing the long-term opportunities. The disincentives for continued development are high:

- Users are not generally the developers.
- High development costs and extensive product application work are required for market entry.
- Initial markets are small and specific.

- Long development times are required for certification and market penetration.
- Capital costs for production can be large.

When the users are not the developers, a major mismatch can result between the material design and the end use. Thus, the developer will have extensive product application development to do, which is both expensive and time-consuming. Initial market potentials can be small, and manufacturing capital costs can be high. Thus, the development of new performance materials as the foundation of a materials business does not look very attractive to materials suppliers. Yet, these enabling technologies are very important to the future development of many basic industries. Some models for the successful development of advanced-performance materials are the following:

- U.S. Department of Defense (DOD) development and demonstration of performance products
- NASA programs
- in-house development of proprietary materials for vertical integration of new product development
- joint partnerships between materials developers and users

The DOD and NASA programs generally involve an aerospace firm that works with material suppliers to develop a material for a specific performance goal. In addition, the Government provides some research and development (R&D) funds so the risk to the materials developer is minimized. In a sense, this is a case of vertical integration in which DOD or NASA is the end user and manages the complete product development.

Cases of the development of proprietary materials for in-house use are found in certain industries. For example, Pratt—Whitney initially developed high-temperature alloys for their own jet engines. Since that time, they have licensed some of their materials technology to other manufacturers, but the market advantage to them early on was significant.

In-house materials development is most prevalent in the microelectronics industry, even down to polymer substrates for circuit lithography. Here, companies such as AT&T and IBM have established some of the most impressive polymer science laboratories in the world to design and develop polymer systems for their own microelectronic products. They recover their development costs from the margins on final products.

However, in today's world of rapid cycle time and cost-efficient manufacturing, perhaps the best examples are in joint development partnerships between chemically or materials-based companies and the end users. One such example is the joint composite development partnership between the Dow Chemical Company and Sikorsky, a subsidiary of United Technologies. Here is a clear case of materials know-how being used in a technology-pull environment guided by the performance needs of the end users. Clear targets are provided to the materials designers, and product application and testing can be iterated rapidly with the user.

A look at the current commercial scene in advanced-performance materials is not very encouraging. Although a great deal of R&D resources have been used for materials development, and materials research is very strong in universities and Government laboratories, the commercialization of the results has been disappointing. For example, structural ceramics has received essentially no play in the United States; most components are from Japan. Carbon fibers are at the point of excess capacity. Many players getting out of this area because of strong Japanese competition. Advanced composites are also at the point of excess capacity. Players are dropping out, and fabrication is still too expensive for mass markets. Advanced metal products is receiving mostly small efforts by users and start-up companies. Electronic materials are mostly developed by in-house R&D by component manufacturers.

Some notable exceptions to these generalities have occurred, especially in electronic materials. Several chemical companies, including Du-Pont and Hoechst–Celanese, have major successful businesses in electronic materials.

The lessons to be learned are to determine the requirements for success. These include the following:

- Early integration of material modification, product application, and process optimization. This integration reduces cycle time and up-front risk. Today's fast-moving markets cannot accommodate a 20-year development cycle and still ensure commercial success. Concurrent engineering with discovery and manufacturing is required to be a leader.

- Understanding of markets and the need for a supplier–user partnership. Where is the value added? If the market needs are understood, the value of the performance material can be estimated, and market-pull will ensure that the technology is cost-effective and commercially viable. This result will almost always require a working partnership between the material developer and the end user.

- Improved processing technology. In many cases, properties are conveyed to advanced-performance materials through specific processing routes, for example, single-crystal alloys and molecularly oriented polymer films. Sometimes, the ultimate cost structure of the materials depends on processing parameters. When the demand is relatively small, it is difficult to improve cost position through economies of scale. Another issue relates to the manufacturing processes to make the final product. For example, the use of polymer composites has been hampered by the lack of cost-effective lay-up equipment and rapid prototyping capability. These processing questions beg for new approaches to materials production and their subsequent processing into final products. The solutions will include new, flexible production systems in which plant overheads are carried by several materials' products and new concurrent product design processes that are optimized for the material of choice. Process research in both areas should be priorities for chemical and mechanical engineering departments.

I conclude with a few suggestions. First, continued Government support for advanced-performance materials test beds, processing, and manufacturing should be a high-priority item for the Department of Defense and for Government laboratories. The potential for civilian spillover should also be a high priority. To be successful, these programs must be joint partnerships with industry so that concurrent R&D can rapidly create new product opportunities. Materials processing and product development must be of equal (or greater) importance to basic materials research.

Second, partnerships between materials developers and end users must be facilitated and encouraged. In some cases, this step may mean an R&D consortium between university researchers, a chemical or materials company, and a product manufacturer. The National Institute of Standards and Technology's Advanced Technology Program (NIST ATP) is a good model that should be expanded. Several projects that have been funded are examples of this integration of research–development–commercialization by universities, a materials supplier, and an end-user product manufacturer.

Third, engineering education should be reconsidered in both product design and materials processing. Most design engineers have limited materials expertise. Thus, their ability to optimize product designs to use an advanced-performance material (normally a substitute for metal) is not high. In addition, real "hands-on" experience in property enhancement through processing is minimal for most students. Thus, new materials-design courses should relate design parameters to materials properties and explore the materials property enhancements that can be obtained by controlled processing.

These issues should be regarded as an important part of research policy and Government support of critical technologies involving advanced-performance materials. In addition to basic research support, other mechanisms should foster the engineering and development activities that will make the materials candidates for commercialization. The next generation of high-technology products will, undoubtedly, require the enabling technology associated with advanced-performance materials. Our competitive position will, in many cases, depend on our ability to put together all elements for a successful commercial product—research, development, and commercialization—in a coherent, integrated fashion. This ability will require creative scientists, innovative engineering, and "street smart" commercial launches. Generally, no one organization has all of the necessary components, so partnerships and "concurrent engineering" will be the "name of the game".

References

1. Eckstut, Michael, *Outlook 11*, Booz-Allen and Hamilton: New York, 1987.

Bibliography

Other information and overviews of parts of this topic can be found in the following sources.

1. *New Technology Week*, June 1, 1992.
2. *Chem. Eng. News*, August 3, 1992, p 16.
3. *The New York Times*, September 13, 1989, article by Eric Weiner on Dow—Sikorsky venture.
4. *New Materials International*, November 1989 (Vol. 4, No. 38).

Partnerships with Universities

Mark Wrighton

Department of Chemistry, Massachusetts Institute of Technology, Cambridge, MA 02139

RESEARCH UNIVERSITIES ARE IN THE MIDST OF MAJOR CHANGE. Historically, the research universities have been supported by the Government with two theories in mind: (1) national security is important, and science and technology are critical to a strong defense; and (2) human health is important. The interest in human health persists, an interest in national security persists, but the adversary has given up. The Soviet Union no longer exists. The question now is, What is the rationale for the support of universities—support in the post-Cold War era? The Department of Defense, which has nurtured an important set of activities, has a role in electronics and devices, structural materials, and high-performance or advanced-performance materials.

Support by the Department of Defense

Academic institutions have been included, and in many instances, there have been commercial consequences, although that has not been the mission of the Department of Defense. The Department of Defense mission is defense and national security, not the development of compact disk players. But in fact, for example, in electronics and devices, fundamental materials research was sponsored by the Department of Defense. Various organizations and activities in parallel in industry (at Lincoln Laboratory, IBM, and General Electric) led to the development of the semiconductor laser in the early 1960s.

The Department of Defense encouraged long-term efforts that led to this practical development. In 1962, no one had the vision that for a few dollars, every compact disk player would have a semiconductor laser as an important component.

What have we learned about the modus operandi of the Department of Defense that should prove useful as we move through a transition to a new era for support of science and engineering in the universities? First, the Department of Defense, despite its mission-oriented character, has had a very long-term view. It has consistently supported research and development (R&D) in support of its mission, and it has included in that mission the support of education. Many highly educated individuals have stepped out of the university scene and made important contributions in the industrial sector. A number of people as well stepped out of one university and right into another and continued in the footsteps of their research mentors.

There is a recognized need for continuity in support of research by the Department of Defense. National security is something that we have embraced as a nation. Moreover, the Department of Defense has provided a substantial level of support consistent with the nature of the project under investigation. This fact stands in striking contrast to programs supported by other agencies, in which all are leveraged, one against the other, no one of which is providing adequate support to accomplish very much of anything. An extreme view, but the Department of Defense has been better in terms of long-term consistency and providing more than subcritical support for its projects.

Finally, the Department of Defense has in fact has taken the posture that planning, integration, and systems orientation are important. The Department of Defense probably will maintain "tech base". That means that we will have a strong science and technology base, which should bode well with the universities. However, a tremendous down-sizing has already occurred in active forces, and more is planned. Moreover, the larger projects that had been awarded to industry contractors are dissolving, and less R&D funding is available to those industries. This decrease in funding in turn means that there will be more competitors at the resource table to perform the basic science and technology efforts for the Department of Defense. More "players" for the same dollar means a more competitive world.

Why should we maintain strength in science and technology? Science and technology are critical for a variety of reasons. Science and technology are critical to solving international conflicts. Our ability to determine whether airplanes are moving around southern Iraq hinges on advanced science and technology and communications capability. Furthermore, solving global-scale problems will rely more and more on science and technology. Pollution is a global-scale problem, but telecommunications is another, and energy systems is yet another. Achieving and maintaining economic vitality and economic viability and enhancing the quality of life are important reasons for maintaining strength in science and technology.

The entire federal policy is predicated on the notion that education and research are tightly coupled and intertwined, and that language is in the pamphlet that describes how universities are funded by the Federal Government. Here are the products and services of the research universities. Our most important set of products and services is the people we educate and our commitment to human resource development.

Contributions of Chemistry

Chemistry really should be at the heart of this revolution in materials. Chemistry is the discipline that has been associated with the study of matter; that is the science of chemistry. Moreover, chemistry is also the discipline associated with the purposeful manipulation of matter at the atomic and molecular level. But, in terms of materials chemistry, the time is right because of the ability to do analysis at an unprecedented level of resolution. However, our academic system has not yet responded for our students because our laboratory and lecture subjects have not yet included the dramatic advances in analytical capability. This is an important charge to the academic community.

One unique contribution for chemistry is analysis. Analysis is really making it possible to make giant strides. It is fueling the revolution. Chemistry has one important thing in its educational belt that the other disciplines do not have, and that is synthesis. Synthesis and processing relative weaknesses in the area of materials in the United States. This is an area of unquestioned strength in chemistry, but synthesis has not been brought to bear, at least as we see it as molecular scientists, to the potential possible, especially by the most gifted and talented people in the area.

Fundamentally, chemists have not been educated or have not learned on their own where chemistry plays a role in materials systems. Educational laboratories, including those at the most expensive universities, are lacking state-of-the-art equipment that would introduce young people or people who are going to enter college later in life to state-of-the-art technology that would intrigue them with the opportunities about complicated systems like materials.

Finally, too many leading chemists have dismissed materials chemistry as too applied. In a chemistry department, generally the largest subgroup is the organic chemists. The leading people in the department, the people most highly regarded, and most highly rewarded are organic chemists who have in mind a practical synthesis. They want to develop a method, or they actually have a target in mind, something that looks good, smells good, and tastes good, like flavorings, fragrances, molecules with a purpose, or a method to obtain them. That's practical. Synthesis is the heart of chemistry. But these individuals, the most capable people, need to be turned on to the opportunities in these complicated, messy materials systems where progress is just being made because of our ability to establish structure and to relate that to function.

Need for Partnerships

The traditional academic units that might be called on to contribute here, chemistry first, but also chemical engineering, physics, electrical and mechanical engineering, and materials science, are all standard academic departments at institutions that have strength in science and engineering. All of these in their own way have an educational prospective to bring to bear on this materials chemistry issue. One can quite legitimately ask if there are better ways to organize ourselves to address these problems. Materials research laboratories that are a quarter of a century old have served us well in some ways, but new organizational structures are needed to assist in executing research programs.

For example, there is an increasingly important role for parallel input to both communities of the social sciences and the physical and engineering sciences. That is particularly true in the problems I mentioned earlier, global environment, telecommunications and information, and energy systems. Dean Lester Thoreau of the Sloan School of Management points out quickly that two organizational systems have survived for about a thousand years: the Catholic Church and the university. Our longevity and strong traditions are our strengths, and our curse is that we change slowly. Despite opportunities for change, some important strengths that are associated with our current structure should be maintained. The most important is that this large number of academic units that can be brought to bear on these complicated problems includes people with unique educational perspectives, each of which would be necessary to solve some of these problems. The current structure is not so bad, provided that we are able to learn each other's language and to see the other contributions that others might make.

We have been very weak in fostering teamwork in universities. In interdisciplinary or multidisciplinary collaborative projects, individual contribution can be recognized, appreciated, prized, and nurtured. But we have done a very poor job at developing systems for teaching teamwork. This is something that we can learn from industry and from other organizations that work in a mode that promotes teamwork. I said that our greatest product is the people we educate. The majority of our graduates will go into the private sector, and this is where they are going to need to know teamwork.

There is a need to develop renewed partnerships, and particularly so when the technology for doing certain kinds of science and engineering is so expensive. If for no other reason, we must develop working relationships that work. At one time it was possible for every university to have gas chromatographs, IR spectrometers, and NMR spectrometers. But it is not possible for every university in the nation to have a full complement of complicated and expensive instrumentation for the study of complicated materials.

I have come to appreciate a role for social sciences, and I also believe there is an important role for the business community. The kinds of solutions that we are trying to find in science and engineering are not going to be implemented in a vacuum. For example, Mario Molina at MIT is associated with studies that led to an understanding that chlorofluorocarbons might be a problem. But doing something about that problem is much more complicated in some ways than doing the fundamental science that led to that observation.

Finally, we need to form partnerships while maintaining our traditional values, meaning that we must remain committed to this notion that individual scholarship is something to be nurtured in the university setting. Ideas and knowledge are creations, not unique to universities, but that are happening everywhere. However, in universities, there still needs to be the opportunity for people to pursue what would apparently be the useless, because discovery and innovation stem from the unexpected. The best examples of this truism are in materials chemistry: high-temperature superconductors from ceramics, weird; a new form of carbon, I thought everything had been discovered, but now we have C_{60}. These are just two of many examples of where unfettered investigation needs to be nurtured and continued. There is a great basis for optimism as we move through a period of enormous change.

2

Educational Issues

Educational issues related to raising the chemistry community's consciousness of materials chemistry are discussed in seven sub-chapters by seven sets of authors. A historical overview provides a context for an industrial perspective on materials chemistry and for the initiatives of a government funding agency, the National Science Foundation (NSF), that is supporting efforts to integrate materials chemistry into the undergraduate curriculum. Specific NSF-funded projects that are materials chemistry intensive are described: an undergraduate course in materials chemistry for engineers, instructional materials for mainstreaming polymers into introductory chemistry courses, and instructional materials for illustrating fundamental chemical principles with a variety of high-technology solids. The importance of pre-college outreach activities is also discussed.

EDUCATIONAL ISSUES associated with materials chemistry are discussed in this chapter. Several key questions provided the framework for the chapter's structure: How has materials chemistry education been defined to date? How can materials chemistry be brought into the mainstream of our educational system? How can interdisciplinary thinking and teamwork, so critical to progress in materials chemistry, be nurtured by our educational system? What are industry's needs in materials chemistry education? What role can a principal funding agency for science education, the National Science Foundation, play in materials chemistry education? What can be done at the pre-college level?

We have addressed only some of the issues associated with materials chemistry education. Additional information and views are presented in the September 1992 issue of the *MRS Bulletin*, which is devoted to materials education.

The views expressed by the authors in their personal contributions to this chapter are their own and may or may not be shared by the other contributing authors.

0065–2393/95/0245–0043$17.00/0
© 1995 American Chemical Society

Materials Education for and by Chemists

Rustum Roy

Pennsylvania State University, Materials Research Laboratory, University Park, PA 16802

Introduction: A Historical Perspective

The empirical fact is that the science of chemistry began as the science of materials—inorganic at first, then organic. Extractive and process metallurgy (recall alchemy, too), the preparation of the vast tonnages of industrial chemicals from ammonia and soda ash to fertilizers—this was what chemists did for 150 years. At the beginning of this century, they turned to the organic world. Perkin's dyestuffs, aspirin, Salvarsan, and rayon led to the giant organic chemical industries of today: polymers, pharmaceuticals—all materials. And on that economic base chemistry had the luxury to start studying the fine structure of matter and its interactions at the subatomic level. The progression of the content of what was taught by "chemistry departments" in the United States (very different from Europe and Asia), which is not by any means the same as what constitutes chemistry, can be summarized in Table I. At the turn of the century the study of chemical reactions and their fundamentals embodied in the phase rule was firmly in the purview of chemistry departments. Applications were directly linked to the basic science that the applications "pulled." Josiah Willard Gibbs, arguably the country's greatest chemist, recipient of Yale's first Ph.D. (in engineering), taught Latin for 3 years before embarking on understanding steam engines. Thermodynamics owes more to the steam engine than vice versa. Van't Hoff and his study of phases was concerned with the extraction of pure salts from mixed solutions for industry.

By the 1930s, the fragmentation of American academic life was in full swing. As far as the inorganic world was concerned, the chemistry of the earth, of metals and ceramics, all became the province of new specializations that eventually called themselves departments, even "disciplines". In chemistry departments, inorganic (materials) chemistry became equated with the chemistry of coordination compounds. Although the profound impact of DuPont's discovery of nylon and the opening of the polymer world was felt throughout the world, by 1948 (when I finished my Ph.D.) U.S. chemistry departments had opted out of the world of materials science and engineering.

Table I. Teaching and Research in Materials Chemistry in the Twentieth Century

1900	1930	1950–1990
Chemists • Extraction of metals and ores, heavy chemicals • Physical and inorganic chemistry of the earth • Analysis of all matter • Phase rule (thermodynamics)	Metallurgists • Chemistry, physics, and properties of metals and alloys Ceramicists • Preparation and uses of nonmetallic compounds Geochemists • Analysis of solids • Fundamental physical chemistry and synthesis of earth-forming solids Mineralogists • Crystallography • Structural analysis of solids Chemists • Coordination complexes • Thermochemistry	Physicists • Theory of properties of solids • Utilization of new materials Materials Scientists • Structure–composition–property relations • Synthesizing new materials • Characterization of materials • Utilization of one or more classes Mineralogist–Geochemist • Synthesis and stability under extreme pressure and temperature conditions • XRD and characterization of-structure Chemists • Crystal field theory • Surface and specialty characterization

Although the emphasis on organic chemistry and synthesis was very suitable for the pharmaceutical industry, and although hundreds of graduates and Ph.D.s went to work for the polymer materials industries, astonishingly, for 15–20 years, essentially all faculty in chemistry departments with notable exceptions like C. S. Marvel and Paul Flory ignored the field or, worse, regarded it as second-rate chemistry. It was no wonder then that when the semiconductor revolution started, physicists and electrical engineers took over the chemists' role of synthesis, structure, and property analysis of this defining new class of materials.

By 1960, when the Materials Research Laboratories of the Pennsylvania State University were formed, the role of chemistry departments was minor. Indeed their funding agency, the Advanced Research Projects Agency (ARPA), had to fund a special program to do materials synthesis and processing research in its new interdisciplinary laboratory, which was not in the chemistry department. Since the early days of the development of the materials research field, the relevance of chemistry education to the materials field has improved with respect to polymers. In metallurgy, semiconductor science, and ceramics, U.S. chemistry departments do not contribute significantly. Participation by many chemistry departments has been on the increase as money has become tighter and "materials" has appeared to become a desirable label. However, there has been a disappointing and largely unsuccessful effort to force-fit materials into the mold of pre-existing specializations in chemistry.

Two examples stand out: research on ceramic precursors and research on biomaterials via so-called "biomimetic" approaches. I had started the systematic organometallic precursor work for making ceramic powders in a series of a few dozen papers starting in 1948 (1, 2). By the mid-1950s I had shown that inorganic sols did essentially as well and were orders of magnitude cheaper. The very sophisticated ceramic precursor work, now a decade old, has yet to demonstrate a special niche for itself in any real examples with unique properties.

The biomimetics theme is more current. Here again, E. W. White et al.'s (3–5) early success in making the only genuinely biomimetic commercialized material used both in human prostheses and in electroceramic technology is more than 20 years old. Work on a wide front summarized by Heuer et al. (6) is singularly innocent of any success in making a real material with special properties or potential use in science or technology. Indeed not one other real biomimetic material has ever been prepared. Furthermore, in a remarkable concession, the dozen authors of this paper (6) conceded that the use of the term "biomimetics" was an exaggeration, but instead of choosing an accurate term, they suggested that a special dispensation be granted them to let the word mean something other than what it does!! I suggested an accurate term: "materials derived from biognosis", that is, using learning or knowledge from nature. Here also the only real example of technologically significant materials, based on biognosis, are the transducers and actuators by Newnham et al. (7).

I cite this history including these two recent examples of attempts to "insert" chemistry into the thicket of materials research already densely populated by physicists, ceramists, metallurgists, and chemical and civil engineers, because they form the background for my recommendations in the following sections. It is, of course, directly relevant to the initiative in education of the American Chemical Society. First let me list good reasons for such an initiative:

- To make the proper contribution of chemical sciences to the materials research community, noting the fact that chemists outside chemistry departments have already been doing this for decades.

- To inform and educate chemistry faculty and students in materials science.

- To compensate for the fact that, for the past 50 years, U.S. chemistry departments (uniquely in the world) have ignored the chemical science of solids.

There are also poor reasons for the ACS to launch such an initiative without proper assessment of the real situation:

- There is money available under the label "materials".

- The materials research community is unaware of particular insights or research now being done in chemistry departments and should be exposed to them. (This erroneous view ignores the facts that every industry utilizes all disciplines in attacking research problems, and chemists are already fully involved in industry.)

The problems confronting the ACS initiative in 1992 were very, very substantial.

1. Very few chemistry faculty (perhaps 2–5%) are interested in, and even fewer are appropriately trained in, the chemistry basic to materials research (discussed later).

2. Who will retrain the professoriat and how? (Indeed, who will admit that they need retraining?)

3. There is an oversupply of very adequately trained personnel in materials research, although the situation may not be as severe in chemistry.

A Balanced Content for Degrees in the Science of Solids

In 1959, Pennsylvania State University launched the first Ph.D. degree program in the United States that was administered by an interdisciplinary committee of the graduate school through participation by senior faculty from all departments. Chemistry was represented by J. G. Aston and J. J. Fritz and me (counted as a "geochemist"). This very successful program, the largest such in the nation, continues to the present (renamed the "materials" degree in 1991–1992). It defined the content of the field and broke it up into appropriate courses (*see* Table II).

Chemists clearly contribute to both synthesis and analysis of materials, less so in the physical and mechanical properties of materials and their applications in systems.

In the three loops of Figure 1, modified from my paper written for the physics community (*8*), the role of chemistry is dominant in the "preparation" and "characterization" functions.

The role of materials chemistry education will therefore have to have as its core the knowledge base adequate for

1. preparation

    - the theoretical basis in crystal chemistry and phase equilibria
    - the experimental capacity to synthesize new materials (including those that never existed before, which has expanded enormously in the past 20 years). The chief variables used are pressure (to ~1 megabar) and temperature (to 4000 K). These are combined in many families of apparatus.

2. characterization

    - the appropriate characterization of solids demands a rational scheme based on the level of detail needed (*see* ref. 8).
    - theory and practice for the use of tools for analysis of composition.
    - theory and practice for the use of tools for analysis of structure.

Basic Subject Matter Chemists Must Acquire To Work in Materials Chemistry

Materials Synthesis or Preparation. It would be regarded as foolhardy for any physicist to attempt to work in materials physics without a grounding in quantum mechanics and the standard 1-year course in solid-state physics. The chemistry community does not fully appreciate that the equivalent courses to be grounded adequately in materials chemis-

Table II. Contents of Solid-State Science Ph.D. Program

Subject Matter	Usual Courses To Cover This Subject Matter
1. Preparation: synthesis, growth, special forms	• Crystal chemistry • Phase equilibria • Kinetics of (solid-state) reaction mechanisms
2. Characterization: position and nature of atoms and ions (chemistry and structure)	• Optical microscopy • X-ray diffraction theory and practice • Elemental analysis: XRD, emission spectroscopy • Defect structure determination
3. Properties: their relation to composition and structure	• Quantum mechanics, statistical mechanics • Introduction to solid-state physics (Kittel) • Relation of properties to structure and composition

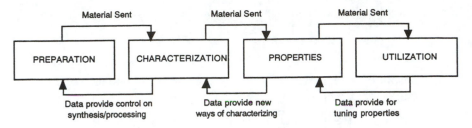

Figure 1. The logical idealized structure of materials research.

try are crystal chemistry (not only crystallography) and phase equilibria. Crystal chemistry deals with the relationship of (crystal) structure to composition and (thermodynamic) environment. It can be distinguished from crystallography, as shown in Figure 2. Crystallography, the study of the analysis of crystal structure, is a necessary prerequisite for crystal chemistry but is far removed from it. Linus Pauling laid the groundwork for crystal chemistry and, together with V. M. Goldschmidt, clearly defined the general field. The power of crystal chemical concepts can be illustrated by the fact that Goldschmidt's 1926 book, *Geochemische Verteilungsgesetze der Elemente*, lists on page 144 all the structural analogues of Si and Ge and the III–V and II–VI compounds, which were laboriously discovered one by one by the semiconductor community. Likewise, Pauling's structural chemistry of the silicates provides insights still current in today's synthesis research. It is inconceivable that a materials chemist not be grounded in this material.

The second major component of the education of a modern materials chemist in materials synthesis is, of course, in phase equilibria, possibly through the treatment of quaternary systems and P–T–X (pressure–temperature–composition) equilibria. This subject was essentially eliminated from chemistry departments and is taught in the materials and geological science departments. Every student (and professor) aspiring to be a materials chemist will have to master this subject.

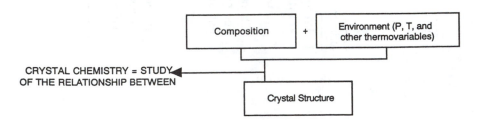

Figure 2. A representation of the field of crystal chemistry.

The information contained in a phase diagram is as follows:

1. number of compounds formed and stoichiometry: whether one or more phases form from a given composition

2. absolute magnitude of one-component solid-liquid, solid-vapor curves: bond strength and bond nature

3. data on crystalline solubility (CS): defect concentration and its temperature dependence; CS with vacuum

4. solidus and liquidus: heat of formation ΔH_f, "strength" of "compound", liquid structure; "clustering" in crystalline solutions

5. pressure–temperature curves for solid–vapor reactions in binary systems: nature of defects—anisodesmicity of bonds and energy relations

6. pressure–temperature curves for solid–solid reactions: sign of the change of volume (ΔV); magnitude of the change of enthalpy at the transition (ΔH) or ΔV; nature of phase transition

Unfortunately, many of those who can read the thermodynamic aspects of the phase diagram do not realize the enormous amount of quantitative crystal chemical information that is contained (or is it concealed?) therein. In two reviews (*9, 10*), I detailed my view on the intellectual processes necessary for successful materials synthesis.

Materials Characterization. Regarding education in the characterization or analysis of materials—a central topic of materials chemistry—there is a similar hierarchy of importance of subjects that chemistry students (and faculty) will need to have learned. Reference 7 treats this topic systematically, and Roy and Newnham (*11*) presented a comprehensive (albeit somewhat outdated) presentation of the architecture of materials characterization. Thus Rutherford backscattering and extended X-ray absorption fine structure (EXAFS) are excellent characterization research tools, but in the sequence of tools used every day on every sample, they are insignificant. Thus for structural characterization, X-ray powder diffraction reigns supreme, yet the full power of the modern automated search routines that can be universally applied are taught only to a minuscule fraction of even the materials science student body.

Pedagogical Aids for Teaching Materials Chemistry

The Pennsylvania State University is the base for the Materials Education Council (MEC), the national and international center for the creation, collection, and distribution (at not-for-profit prices) of teaching aids for college-level students.

Two classes of materials will be invaluable to any group contemplating teaching materials chemistry:

1. print

   - The *Journal of Materials Education (JME)* is a must for the departmental library. It is in fact a continuously changing textbook of teaching modules on new and emerging materials topics. Free reprint rights make the articles easily available.

   - Clusters of modules (from *JME*) on selected topics such as polymer experiments or cement chemistry are available as short textbooks.

   - Special autotutorial modular texts on crystallography and on phase diagrams are available. (Nearly 10,000 units are sold every year.)

   - The materials science and engineering of wood is thoroughly covered in four texts.

2. video

   - The National Materials Science Film series includes an excellent treatment of ternary phase diagrams and of mechanical and electrical properties.

   - TV and text courses on materials synthesis and related topics are available. The use of such offers by far the most cost-effective way to give faculty the wherewithal to teach new topics.

Recommendations for Improvements in Materials Chemistry Education

Most chemistry faculty will not undertake a serious study of materials science. Hence I propose the following:

- Those few chemistry departments interested in materials chemistry should start hiring faculty trained in materials sciences.

- Chemistry departments should start offering a concentration area in materials chemistry, and the requirements for this degree include the substitution of 15 credits of phase equilibria, crystallography, crystal chemistry, and materials characterization to be taken in the materials or geoscience departments. Where local chemistry faculty can teach these topics, students would, of course, take them under chemistry labels.

- Beyond these beginning basics, the local chemistry or materials faculties' own specialization will determine both the research and specialized courses that may be available to the student.

At this time, the interdisciplinarity of the materials field is being emphasized as much as the topic itself. Materials chemistry will bring to the academic chemistry community an excellent opportunity to practice what they often preach (or agree with) regarding the importance of interdisciplinarity, as they incorporate more courses from the materials science and physics departments as part of their requirements.

Acknowledgments

My research on materials synthesis has recently been supported by the Air Force Office of Scientific Research, Office of Naval Research, and the National Science Foundation, Division of Materials Research.

References

1. Roy, R. "Aids in Hydrothermal Experimentation: II. Methods of Making Mixtures for Both 'Dry' and 'Wet' Phase Equilibrium Studies." *J. Am. Ceram. Soc.* **1956,** *49,* 145–146.
2. Roy, R. "Ceramics by the Solution-Sol-Gel Route." *Science (Washington, D.C.)* **1987,** *238,* 1664–1669.
3. White, E. W.; Weber, J. N.; Roy, D. M.; Owen, E. L.; Chiroff, R. T.; White, R. A. "Replamineform Porous Biomaterials for Hard Tissue Implants." *J. Biomed. Mater. Res. Symp.* **1975,** *6,* 23–27.
4. White, E. W.; Weber, J. N.; White, E. W. "Replamineform: A New Process for Preparing Ceramics, Metal, and Polymer Prosthetic Materials." *Science (Washington, D.C.)* **1972,** *176,* 922.
5. Roy, D. M.; Linnehan, S. "Hydroxyapatite Formed from Coral Skeleton Carbonate by Hydrothermal Exchange." *Nature (London)* **1974,** *247,* 220.
6. Heuer, A. H.; Fink, D. J.; Laraia, V. J.; Arias, J. L.; Calvert, P. D.; Kendall, K.; Messing, G. L.; Blackwell, J.; Reike, P. C.; Thompson, D. H.; Wheeler, A. P.; Veis, A.; Caplan, A. I. "Innovative Materials Processing Strategies: A Biomimetic Approach." *Science (Washington, D.C.)* **1992,** *255,* 1098.
7. Newnham, R. E.; Ruschau, G. "Smart Electroceramics." *J. Am. Ceram. Soc.* **1991,** *74,* 463.
8. Roy, R. "Parameters Necessary for Adequate Characterization of Solid State Materials." *Phys. Today* **1965,** *18,* 71–73.
9. Roy, R. "Synthesizing New Materials to Specification." *Solid State Ionics* **1989,** *32/33,* 3–22.

10. Roy, R. "New Materials: Fountainhead for New Technologies and New Science." Intl. Science Lecture Series; U.S. National Academy of Sciences and the Office of Naval Research, National Research Council: Washington, DC, 1994.

11. Newnham, R. E.; Roy, R. In *Treatise on Solid State Chemistry: The Chemical Structure of Solids;* Hannay, Bruce, Ed.; "Structural Characterization of Solids"; Plenum: New York, 1974, Vol. 3, pp 437–529N.

Industrial Perspective on Materials Chemistry Education

Glen H. Pearson

Manufacturing Research and Engineering Organization, Eastman Kodak Company, Rochester, NY 14652–3208

A STUDENT STARTING AN INDUSTRIAL POSITION in a materials-based manufacturing company will find a vastly different environment from the one that existed 10–15 years ago. The pressures on most companies to compete in today's global marketplace demand that they operate differently. In a high-technology business such as Kodak's, innovative products are essential. Today we must bring those innovations to market as quickly as possible. Once the concepts are proven and a manufacturing route is determined, the manufacturing process must be competitive to produce products with the required features at the lowest possible cost. Product technologies change rapidly compared to the career lifetimes of the employees, so the work force has to remain flexible to adapt to new technologies.

The views expressed here are my own, and they are based upon 17 years of industrial experience, 12 of which were in research management, so they should prove representative, as they are driven by a common set of pressures present in the current industrial climate. The three basic questions that I address are

1. What does industry seek in new hires?
2. What do we find we need to provide after hiring?
3. Where would a materials chemistry background help?

Needs of Industry

Innovation has historically been a strength of U.S. industry. In recent years, the United States appeared to be losing its position to foreign competition, particularly the Japanese. Table I (*1*) illustrates, for 1991, the number of patents obtained by several firms. Japanese companies hold the top few positions, but several U.S. companies have shown continued progress.

Table I. Number of U.S. Patents in 1991

Company	Number
Toshiba (Japan)	1156
Hitachi (Japan)	1139
Mitsubishi Electric (Japan)	959
General Electric (U.S.)	923
Eastman Kodak (U.S.)	887
General Motors (U.S.)	863
Canon (Japan)	828
Philips (Netherlands)	768
Fuji Photo Film (Japan)	742
Motorola (U.S.)	631
DuPont (U.S.)	631
Subtotal, U.S. companies	3935
Subtotal, Japanese companies	4824
Subtotal, Netherlands	768
Total	9527

Although a good record of innovation is a key ingredient for success, the progression from innovation to successful manufacturing is perhaps even more important. A product such as photographic color film, where multiple layers of material are coated onto a plastic support, where literally hundreds of materials are involved, where the photographic system shows a high sensitivity to impurities, and where the product must be manufactured at high rates in dark conditions, presents unique challenges to a materials chemist and engineer. An effective development cycle from innovation to manufacturing will require careful attention to all aspects of manufacturing, starting at the design phase.

Reducing the development cycle time will depend upon talented, well-trained people. Employees must possess a broad knowledge of the entire manufacturing process, whether their job is product design or manufacturing process research and development (R&D). They must focus on understanding, not empiricism. Workers must understand all aspects of the product's behavior, and they must integrate their knowledge with that of others on the team and work well within the team. Individual excellence is a key ingredient for success, but teamwork is essential. Today's emphasis on quality requires that employees understand statistical principles and tools and use them. Problems that show up in parts per million or parts per billion require special experimental plans to detect and to correct. Today's environment really requires workers to integrate their materials knowledge throughout all aspects of the process to achieve the desired levels of system performance in ways that were not required several years ago.

As a simple example of this integration, consider a new product design engineer who might be given the problem of designing a plastic part for a new camera or film system. Starting with the development of customer requirements and specifications, the individual must design the part, choose the appropriate materials, select a manufacturing process, and then demonstrate that the new part and manufacturing process work, leading to a final verification that the part functions as the customer requires. To reduce the product development cycle, both analysis and testing become crucial ingredients in the concept stage rather than after the part is manufactured, as has been the case in the past. The key notion is get it right the first time and not spend time and money fixing it later. To perform a proper analysis, the engineers must understand all aspects of the system from start to finish.

Job Skills Required

Getting back to the issue of people and materials chemistry training, consider the job skills required for functional excellence in the type of environment described. Four key dimensions come into play:

1. technical
2. business
3. quality—statistics and planning
4. people—teamwork and communication

Of these, the academic environment seems to concentrate on the first two, technical and business. To be sure, for students in technical curricula, exposure to business concepts and training often is minimal. The other two areas, quality tools and people skills, are seldom formally addressed.

D. Kezsbom (2) conducted a survey of some 285 managers and project specialists who were asked to list the sources of conflict in their current projects and to assign a prioritized rating to each. Table II illustrates one set of results from the survey. Technical options is found very low on the list, twelfth of thirteen. According to the prioritized ratings, technical problems are not an issue, but goal-setting, people problems, communications, and politics play a much larger role in project conflict. The findings of this work differ greatly from the typical notion of academic training.

Table II. Causes for Project Conflict

Cause	Score
Goals–priority definition	1877
Personality	1201
Communication	1182
Politics	655
Administrative procedures	633
Resource allocation	571
Scheduling	527
Leadership	453
Ambiguous roles–structure	185
Costs	174
Reward structure	151
Technical options	132
Unresolved prior conflicts	96

Need for Materials Chemistry Background

One way to examine the problem that academia faces in preparing one of its products, students for industry, is to compare the preparation in terms of the environment found in industry today. In Table III, I attempt to compare the academic preparation students receive to what they will find in industry.

The contrast in environment versus preparation is interesting. Education has a lot to do with the creation of habits, and students do receive considerable technical input, but they are not generally given the chance to develop people skills and to practice them in a nonthreatening environment. The academic focus on correct answers to problems is at odds with industrial practice, because most industrial problems do not have correct answers, meaning that many approaches are doomed to failure. The pain of failure has been cited as one impediment to learning in an organization (3). Individuals will become defensive and stop listening or communicating with each other if they feel their projects may fail.

As suggestions for the future, the following are opportunities for materials chemistry training. My suggestions certainly are not specific to materials chemistry but should be germane to other technical disciplines that train students for industrial careers.

1. Start by providing a solid foundation, one that will suffice for a career of learning.
2. Develop the habits required for continuous learning.
3. Teach from the system perspective from the outset.

Table III. Academic Preparation versus Industry Environment

Academic Preparation	*Industry Environment*
Individual performance is tested	Team performance is essential
Problems have correct answers	Problems have solutions
Projects have short time lines	Projects can take years
Constant performance feedback	Feedback is much less frequent
Courses are independent	Problems require integration
Failure is painful	Failure happens
Statistics are not routine	Statistics are essential

4. Encourage flexibility in thinking about alternative solutions to problems. Perhaps it might be possible to develop a set of open-ended problems, much like the case-study approach used so effectively by business schools.
5. Provide experience and training to develop group work skills. This is happening in grades K–12 where it is known as cooperative learning.
6. Finally, teach and require the use of statistics.

A focus on materials chemistry will lead to more system-oriented thinking and integration. It should be possible to build curricula in a way that encourages the development of the people skills needed in the industrial environment.

References

1. Buderi, R.; Casey, J.; Gross, N.; Lowry Miller, K. *Business Week* **1992**, *August 3*, 68–69.
2. Kezsbom, D. Project Management Institute seminar, Dallas, TX, 1991.
3. Argyris, C. *Harvard Business Review* **1992**, *69*, No. 3, 99.

Funding Opportunities for Materials Science Education

C. T. Sears and S. H. Hixson

National Science Foundation, Washington, DC 20550

SUPPORT FOR THE DEVELOPMENT AND IMPLEMENTATION of new courses and laboratories in materials science is available through National Science Foundation programs in both the Division of Undergraduate Education and the Division of Materials Research. The Division of Undergraduate Education has separate programs targeting laboratory, curriculum, and faculty.

The Instrumentation and Laboratory Improvement (ILI) Program aids in the purchase of laboratory equipment for use in undergraduate laboratories at all levels. Annual funding has been $23 million for the past 5 years and is anticipated to remain at this level for the near future. Typically, 2300 proposals are received, resulting in approximately 600 awards per year. ILI has two components: The major one accepts proposals for equipment only; the other, known as Leadership in Laboratory Development, seeks to support the development of exemplary national models for laboratory curricula by providing funds for personnel and supplies as well as for equipment. Five percent of the ILI budget is devoted to Leadership projects, and preliminary proposals are required. A 50% institutional match for equipment costs is necessary for all ILI proposals. The maximum allowable request from NSF is $100,000. In the 1992 competition, 60 proposals to initiate or improve materials science laboratories were received; 15 were from departments of chemistry, the remainder from engineering units.

The Undergraduate Course and Curriculum (UCC) Program focuses on the development of introductory courses for both science and nonscience majors. Eligible activities include the production of textbooks, lecture modules, software, and other media materials and the pursuit of alternate teaching strategies. The program is funded at $18 million for fiscal year 1992. Typical awards are for 2 years with annual budgets in the range of $75,000–$150,000. Two materials science projects supported by UCC are (1) "Development of a Materials-Oriented General Chemistry Course," under the direction of A. B. Ellis; and (2) "Development of Instructional Materials in Polymer Chemistry for General and Organic

Chemistry Courses," under the direction of J. P. Droske. These projects are described elsewhere in this chapter.

Short courses and workshops for college faculty primarily engaged in undergraduate instruction are supported under the Undergraduate Faculty Enhancement (UFE) Program. The purpose of the program is to ensure the vitality of the teaching faculty by assisting them in learning new ideas and techniques in their fields and using the knowledge gained to improve undergraduate instruction. Several short courses devoted to the study of polymers have been sponsored by the program during the past several years. Proposals for the development of more broadly based materials science workshops suitable for chemistry faculty are encouraged.

The Division of Materials Research launched the Undergraduate Materials Education Initiative during the spring of 1992 with a closing date of May 15. The goal of the Initiative is the development of advanced undergraduate courses, including laboratories, in materials synthesis and processing. The courses should focus on fundamental principles, modeling and simulation, characterization, and property evaluation. Subject to the availability of funds, the Foundation expects to make 7–11 3-year awards with annual budgets of $150,000.

Chemistry of Materials Courses at Rensselaer Polytechnic Institute

Gary E. Wnek[1] and Peter J. Ficalora[2]

[1]Department of Chemistry, Rensselaer Polytechnic Institute,
Troy, NY 12180–3590
[2]Department of Materials Engineering, Rensselaer Polytechnic
Institute, Troy, NY 12180–3590

FRESHMAN CHEMISTRY is arguably an important course, one that needs to be viewed as a contribution beyond a "service" level. It affords the opportunity to make the case, to many students of varied disciplines, of why chemistry is the central science and is responsible for virtually all of the high-tech developments they encounter or read about. The course should be a vehicle to attract more students to chemistry. More importantly, it should instill greater respect for and appreciation of chemistry by students who will not necessarily specialize in it. In our view, this function is particularly important for engineering students, as they will frequently use the basic ideas in freshman chemistry in their professional lives, yet they often wonder where the connection is while they are exposed to these ideas in the classroom.

To make the connection between chemistry and engineering more immediate, we have developed, with support from Rensselaer's administration, the General Electric Foundation, and NSF, a two-semester freshman course that emphasizes solid-state chemistry and materials science and that is now taken by all of our engineering freshmen. The course is co-taught by faculty from the chemistry and materials engineering departments, and is in the truest sense a cooperative venture. Here we briefly summarize our motivations for moving in this direction and outline the course as it is now constituted.

The Shift toward Materials Chemistry

Chemistry and materials science are inextricably linked. In fact, in the National Research Council's "Opportunities in Chemistry," familiarly known as the Pimentel Report (1), can be found the following definitions from Webster: A material is "the substance or substances out of which a thing is constructed," and chemistry is "the science that deals with the composition, properties, and changes of properties of substances." However, a common perception is that chemistry is concerned primarily with microscopic phenomena. Nowhere is this attitude more evident than in a freshman chemistry textbook. Occasional chapters on the solid state can be found, but these typically dwell on atomic packing. Virtually no mention is made of material properties such as modulus, electrical conductivity, and transparency, and how these are dictated by atomic and molecular composition and structure. If solids are discussed, imperfections such as point and line defects are almost never mentioned, and yet these dictate, for example, mechanical properties.

This approach is unfortunate, as it has been pointed out quite clearly that tremendous opportunities exist for chemists to participate in the exciting area of materials science. For example, in the Pimentel Report, materials chemistry played a prominent role, and the interdisciplinary nature of materials chemistry was stressed. It was stated that "chemists are increasingly joining and expanding the specialist communities concerned with glasses, ceramics, polymers, alloys, and refractory materials." Furthermore, the report predicted that "coming years will see entirely new structural materials, liquids with orientational regularity, self-organizing solids, organic and ionic conductors, acentric and refractory materials." A companion report, "Frontiers in Chemical Engineering" (2), expressed similar opportunities in the materials area for chemical engineers.

The fact that more and more chemical research activity is indeed being directed toward the materials area was responsible for the American Chemical Society's launching of a new journal, *Chemistry of Materials*, in 1989. Mary Good, former ACS President, edited a book published by the American Chemical Society that directly addresses the point, *Biotechnology and Materials Science: Chemistry for the Future* (3). To meet these challenges, changes in chemical education may be needed "to keep chemists in the center of revolutions in materials and biological sciences" (4). Changes are also necessary to thwart the decline of interest in science by prospective students and, more broadly, by the public at large. What is being done to respond to this particularly serious need has been nicely summarized in a *Chemical and Engineering News* article (5).

Perhaps the most compelling argument for at least considering a shift toward a more macroscopic focus in freshman chemistry comes from engineering faculty. At a number of institutions, engineering faculty have complained that introductory chemistry is not terribly relevant to the needs of their undergraduates. What they mean (in part) is that their students are not learning enough about the solid state and, hence, about materials. For example, many electrical engineering faculty believe that students should be exposed to semiconductors and how simple devices can be constructed from them. Another way of saying it is that freshman chemistry is not applied enough for engineering needs. But is it not, after all, the application of chemistry that fascinates chemists?

The reluctance of chemistry faculty to respond to these needs has led to the initiation of alternative chemistry courses taught by engineering faculty. For example, for the past approximately 15 years, freshmen at the Massachusetts Institute of Technology (MIT) may elect to take their one-semester requirement in chemistry in the form of Introduction to Solid-State Chemistry. This course is taught exclusively by the Materials Science and Engineering Department. Roughly one-half of the freshman class takes this course instead of the Chemistry Department's offering. Northeastern University now teaches a second-semester freshman course emphasizing materials chemistry for honors students. At the University of Pennsylvania and the University of Arizona, freshmen engineering majors can take their second-semester chemistry course in solid-state chemistry, again taught by materials science and engineering faculty. Most likely many more such programs are in progress or in the planning stages.

Should this shift toward materials chemistry be taken seriously? Yes, because materials topics bring vitality to freshman chemistry, which should be fascinating but many times has fallen short of this goal.

An Approach to the Microscopic–Macroscopic Merger

We strongly believe in a hierarchical approach that begins with the structure of the atom and continues to molecules and then collections of atoms or molecules into various superstructures (i.e., condensed phases). This idea is not new, as such an approach can be found in the first few chapters of Pauling's classic introductory text (6). The fundamentals of thermodynamics, kinetics, and the solid state are discussed next to prepare students for the remainder of the course. Students learn, for example, in detail why metals are electrical conductors and malleable, whereas most ceramics are insulators and are brittle. They also learn why structural metals are not as strong as one might predict on the basis of an ideal (perfect) crystal lattice, and how it is possible for solid-state chemical reactions to alter the strength of an alloy. We then move onto ideal and real (i.e., defect-containing) solids, phase diagrams, kinetic processes in solids (e.g., diffusion and sintering), and a discussion of the materials classes, and conclude the course with a few case studies that underscore most of the course material. Two examples include the fabrication of semiconductor devices and the synthesis of diamonds. The response of previous students has in general been supportive of this approach (7).

No single textbook is adequate, but we have had success using a first-year chemistry text along with a materials science text, although more recently we are placing heavy emphasis on lecture notes that we have written. Concerning operation, students now meet for two lectures per week, two 1-hour recitations, and a 3-hour lab every other week. Many traditional experiments can be readily employed, but the course content also calls for new experiments. Several experiments that have been developed are relevant to the lecture material and can be performed with the large throughput of students in a freshman course. Examples include the growth of a metal–semiconductor junction and its electrical properties, absorption spectroscopy of transition metal ions in glasses, and construction of a phase diagram. Each is intended to provide a conceptual link between a macroscopic property and its atomic–molecular level origin.

Long-range goals include the development of a textbook along with a laboratory manual, modification of lecture notes, demonstrations, and experiments for use in high school science courses. The latter goal is particularly important to us in view of the declining interest of students in science. Students frequently find high school chemistry (and other science courses) to be too abstract; in short, they are demanding more immediate relevance to the real world. Perhaps a dose of a microscopic–macroscopic merger can meet this demand.

References

1. Pimentel, G. C. *Opportunities in Chemistry;* National Research Council: Washington, DC, 1985.
2. Amundson, N. *Frontiers in Chemical Engineering;* National Research Council: Washington, DC, 1988.
3. *Biotechnology and Materials Science: Chemistry for the Future;* Good, M. L., Ed.; American Chemical Society: Washington, DC, 1987.
4. Hileman, B. *Chem. Eng. News* **1987,** *October 26,* 34.
5. Krieger, J. *Chem. Eng. News* **1990, June** *11,* 27.
6. Pauling, L. *General Chemistry,* 2nd ed.; Freeman: San Francisco, CA, 1953.
7. Wnek, G. E.; Ficalora, P. J. *Chemtech* **1991,** *November,* 664.

New Curricular Materials for Introducing Polymer Topics in Introductory Chemistry Courses

John P. Droske

POLYED National Information Center, University of Wisconsin—Stevens Point, Department of Chemistry, Stevens Point, WI 54481

INFAMOUS QUOTES that are purported to be advice that was given to H. Staudinger in 1925 told him to "Leave the concept of large molecules well alone," and "There can be no such thing as a macromolecule" (1). Today we know that much of our world is made of macromolecules and that they are an integral part of today's science. Whoever gave the advice to Staudinger probably is glad that the quotes are anonymous. It is unlikely that anyone would feel slighted by not having it attributed to them. Even though the existence of macromolecules is well-established today, macromolecules have been largely ignored in the chemistry curriculum. Fortunately, this situation is changing, but the first part of this quote has been all too true of the treatment that polymers received, until recently, in the chemistry curriculum.

Another oft-quoted line is the advice given to Dustin Hoffman in the movie, "The Graduate". He had just graduated from college, and a friend of his parents gave him this advice: "I want to say one word to you, just one word: plastics" (2). This remains good advice today, yet, like Dustin Hoffman, most of our undergraduates still do not hear about polymers until after they graduate.

Action by ACS Divisions

Recognizing that insufficient attention was being given to polymer topics in the chemistry curriculum, the ACS Division of Polymer Chemistry (POLY) formed the Polymer Education Committee in 1972. Shortly thereafter they were joined in this effort by the ACS Division of Organic Coatings and Plastics (now known as the Division of Polymeric Materials: Science and Engineering or PMSE), and the committee was called JPEC, the Joint Polymer Education Committee. Over the years, the Polymer Education Committee was very active and instituted a variety of programs targeted toward including polymers in the chemistry curriculum. In 1989, recognizing that there continued to be a need for pro-active efforts in polymer education, the two divisions (POLY and PMSE) renewed their commitment to polymer education. To reflect this renewal in both financial as well as human resources, the committee was renamed POLYED.

Today, POLYED has more than 50 active members serving four Directorates: Pre-College Faculty, College and University Students, College and University Faculty, and Government and Industrial Professionals. Each of these Directorates offers polymer education programs for these groups. Within the Pre-College Directorate, POLYED sponsors an annual Excellence in Polymer Education Award which is given to an outstanding high school teacher. These awards typically are presented at national ACS meetings, which often feature symposia for teachers that are organized by POLYED.

The College and University Students Directorate offers several awards programs, including an award for outstanding performance in undergraduate organic chemistry courses, an award for excellence in undergraduate research, summer research scholarships, and awards to graduate students such as the Sherwin–Williams Award and the Unilever Award.

The College and University Faculty Directorate has been offering hands-on polymer chemistry demonstrations and experiments workshops for faculty for nearly a decade. In addition to this active program, the Directorate also oversees the Curriculum Development Award as well as the Textbook Author Program. Both of these programs are targeted toward increasing the availability of curricular materials in the polymer area.

The Government and Industrial Professionals Directorate is expanding its activities and recently published a Short Course Catalog listing institutions that offer short courses in the polymer area. This Directorate also is working closely with the new ACS Division of Chemical Technicians.

Because of the wide variety of POLYED programs, in 1989 the POLYED National Information Center for Polymer Education was established at the University of Wisconsin—Stevens Point. The purpose of the Center was to serve as a clearinghouse for POLYED programs and, in particular, to provide a single site that interested individuals could contact to obtain information about POLYED programs. Since its inception, about 800 faculty have contacted the Center for information regarding the inclusion of polymers in their courses. Letters to the Center from college and university faculty show that schools are developing new courses in the polymer area and that there is considerable interest in including polymer topics in existing courses. However, these communications also frequently cite limitations in realizing these goals, such as administrative constraints on the number of new courses that can be offered, an already very heavy

course load for students, insufficient room in existing courses for introducing new material, as well as faculty concerns about their limited familiarity
with polymers and the time involved in "getting up to speed" in this area.

Scholar Program To Develop Lectures

These concerns led to a grant proposal to the National Science Foundation from the POLYED Center entitled, "Incorporating Polymeric Materials Topics into the Undergraduate Chemistry Core Curriculum." The
purpose of this grant was to address the need for new curricular materials
in the polymer area that were ready to use and that provided the necessary
background information for faculty. With primarily NSF support as well
as some matching funds from POLYED, the Center named a team of five
NSF–POLYED Scholars. The Scholars were college and university faculty, from assistant to full professors, with a range of experience in the polymer area. Some of the Scholars had considerable prior experience with
polymers, and others were relatively new to the field. All of the Scholars
had experience in the development of curricular materials for college
chemistry courses.

Implementation of the program began in the spring of 1992 with each
of the Scholars reviewing the University of Massachusetts video course,
"Introduction to Polymer Chemistry." In May, the five NSF–POLYED
Scholars (Guy Mattson from University of Central Florida, Ann Nalley
from Cameron University (OK), Karen Quaal from Siena College (NY),
Chang-Ning Wu from the University of Massachusetts—Dartmouth, and
Joe Young from Chicago State University) met in Stevens Point, WI with
their host–mentors: Lon Mathias from the University of Southern Missis-
sippi, Gary Wnek from Rensselaer Polytechnic Institute, and me. During
this meeting, the Scholars and host–mentors set goals for their 4-week
summer residencies to be held at the three host institutions.

The Scholars, working with their mentors, did an outstanding job
during their residencies, and many laboratory experiments and new "lecture snapshots" were developed. Lecture snapshots are short discussions
of timely or fundamental polymer topics. They are designed to minimize
the time necessary for faculty to familiarize themselves with the topic and
to prepare their presentation on it. In general, the lecture snapshots require only a few minutes of lecture time and setup and interface with topics that already are covered in general chemistry courses. The snapshot
format is as follows:

- title
- keywords
- context

- brief description
- packet contents
  — transparency masters
  — demonstration(s)
  — sample questions, problems, and solutions
  — related experiments
- approximate lecture time, with and without demonstrations
- background
- transparency masters
- demonstration(s), with details
- related topics and suggested extensions
- references

The following lecture snapshots were prepared:

- molecular weights of polymers
- solutions of macromolecules
- physical properties of matter
- small molecules vs. large molecules
- thermal properties of polymers
- rubber-like elasticity and entropy
- effects of temperature on vinyl polymerization
- calculation of degree of polymerization
- silicon, silicates, silicones
- adhesives
- common polymeric materials
- transdermal drug delivery
- chitosan, a natural polymer
- composites, manufactured and natural

In addition to the lecture snapshots, a variety of experiments suitable for general chemistry lab classes were developed:

- microscale bulk polymerization of styrene
- dependence of film quality on molecular weight of polystyrene
- recycling by selective dissolution
- determination of molecular weight by thin-layer chromatography (with recycling of plates)
- molecular weight determination of polyethylene glycol by titration
- film casting
- molecular weight determination of a urethane prepolymer by titration
- preparation of a synthetic metal
- a kinetics experiment on the effect of temperature on the curing of an epoxy resin
- water-proofing filter paper with a silicone polymer

- PCMODEL, molecular structures laboratory exercise
- separation of metal ions using a natural biopolymer, chitosan
- preparation of nylon 11, fiber drawing, and tensile properties
- thermal properties of polymers

The experiments were tested by student assistants at the University of Wisconsin—Stevens Point with good success. Most of the experiments do not require any special equipment beyond that found in a typical general chemistry lab.

The next phase of this effort will involve three steps:

1. review of the snapshots and experiments by an expert advisory panel, comprising 14 leading polymer scientists and educators
2. editing of the manuscripts
3. field testing

The new curricular materials probably will be disseminated by publication in the *Journal of Chemical Education* and by distribution from the PO-LYED Center at the University of Wisconsin—Stevens Point.

We hope that these new curricular materials will not only facilitate the introduction of polymer topics into introductory chemistry courses but, in conjunction with other general chemistry curricular efforts, will also serve as a catalyst for revitalization of our introductory chemistry curriculum.

Acknowledgments

The fine efforts of many individuals, especially the NSF–POLYED Scholars and their host–mentors, have contributed to making this endeavor a success. They are greatly appreciated. Also, the financial support of the National Science Foundation (Grant Nos. 91–50497 and 92–54351) and POLYED is gratefully acknowledged.

References

1. Olby, R. *J. Chem. Ed.* **1970**, *47*, 168.
2. "The Graduate", screenplay by Calder Willingham and Buck Henry. Based on the novel by Charles Webb. Copyright 1967 by Mike Nichols–Laurence Turman Productions for UA/Embassy.

General Chemistry as a Curriculum Pressure Point: Development of *Teaching General Chemistry: A Materials Science Companion*

Arthur B. Ellis

Department of Chemistry, University of Wisconsin, Madison, WI 53706

IN 1989, A PANEL CONVENED BY THE NATIONAL SCIENCE FOUNDATION examined introductory college chemistry courses and concluded that "the historic bias of chemistry curricula toward small-molecule chemistry, generally in the gaseous and liquid states, is out of touch with current opportunities for chemists in research, education, and technology" (*1*). Moreover, the report noted that "the attractiveness of chemistry and physics for undergraduate majors could be enhanced by greater emphasis on materials-related topics which would help students better relate their studies to the real world."

A Materials Chemistry Resource Book for Teachers

Shortly thereafter, the NSF solicited proposals for projects that would bring about comprehensive curriculum reform in introductory chemistry courses. During this period, a group of about a dozen chemistry researchers and teachers, many of whom were working in various areas of modern solid-state research, had organized themselves into what was subsequently called the Ad Hoc Committee for Solid-State Instructional Materials (SSIM). With funding from the American Chemical Society's Society Committee on Education (SOCED), the group met for the first time at an ACS National Meeting in the spring of 1990. As a result of the meeting, the Ad Hoc Committee decided to submit a proposal in response to the NSF solicitation. The thrust of the proposal would be to help revitalize introductory chemistry courses by providing examples from materials chemistry that would illustrate basic chemical concepts. Pooling the committee's collective expertise led the group to conclude that, in fact, essentially all of the concepts typically covered in introductory chemistry courses could be illustrated with solids, be they polymers, ceramics, semiconductors, superconductors, or biocompatible materials, to list a few examples.

In many respects, the NSF solicitation for projects affecting introductory chemistry courses provides an ideal strategy for effecting broad curricular change: pre-college chemistry courses have traditionally reflected the content of the freshmen chemistry course, and upper-level collegiate courses use this course as a foundation. Thus, the introductory collegiate chemistry course can sensibly be regarded as a curriculum "pressure point" because it influences the entire curricular content.

Moreover, the introductory chemistry course, along with introductory calculus, serves as a foundation course to technical careers. Unpleasant experiences in these courses or poor performance in them often results in the loss of these students from the future technical labor-force pipeline. Enhancing the appeal of these courses is a viable strategy for kindling interest in technical careers across the spectrum of student "customers". For the many students who will not pursue technical careers, a materials-oriented chemistry course can provide a sense of relevance by connecting chemistry to advanced materials and devices that we increasingly encounter in everyday life.

Both the NSF and the Camille and Henry Dreyfus Foundation provided funding for the project proposed by the Ad Hoc Committee. The committee recognized that several problems had to be overcome if materials chemistry were to be mainstreamed into the chemistry curriculum. First was the recognition that most college teachers of general chemistry have been trained as molecular chemists. Much of the language of solid-state chemistry is unfamiliar to this teaching community, coming as it does from other disciplines, including physics, engineering, and materials science. The jargon associated with solids thus needs to be translated. Second, solids typically involve extended three-dimensional structures that can be hard to visualize for students and teachers. These structures need to be made comprehensible. Finally, even if these problems can be solved, a critical question is how to convince teachers to try the new materials!

A multifaceted strategy was developed to overcome these obstacles. Rather than writing a textbook on materials chemistry suitable for introductory chemistry courses, the committee determined that a resource volume was needed. Entitled *Teaching General Chemistry: A Materials Science Companion*, this volume was written for teachers and presents solid-state examples to complement the molecular examples typically given in introductory chemistry courses. The Companion covers topics paralleling those in traditional chemistry texts, facilitating incorporation of the materials in "traditional" chemistry courses. At the same time, a matrix connecting materials and devices with core concepts is provided to serve as a guide for use of the materials in unconventional course treatments. For example, light-emitting diodes could be used to illustrate spectroscopy, substitutional stoichiometry, bonding, and periodic properties.

The objective of the Companion, published by the ACS in 1993, is to empower teachers by providing them with exciting examples from materials chemistry that students can see, hear, and touch, firmly grounding the introductory course in concrete examples. Moreover, these examples can illustrate the key role chemistry plays in developing high-tech materials and advanced devices. The philosophy of the committee is to make it possible to obtain a balance in the introductory course between molecular and solid-state examples. By using both to illustrate concepts, teachers can also demonstrate the universality of scientific thinking.

Other Instructional Materials

The committee recognized that supporting instructional materials for the Companion would be needed. The Institute for Chemical Education (ICE), a NSF-funded entity whose mission is the enhancement of chemical education, is serving as the distribution arm for supporting instructional materials. The first product is an Optical Transform Kit that shows how diffraction is used to determine relative atomic positions. The kit essentially scales up the X-ray diffraction experiment by use of a small laser as a visible light source and 35-mm slides bearing photographically reduced, laser-written patterns that mimic atomic packing arrangements (2). (For ordering information, contact ICE, Department of Chemistry, University of Wisconsin, Madison, Madison, WI 53706.)

A second product is the ICE Solid-State Model Kit, developed by L. A. Mayer and G. C. Lisensky, which makes it possible to build extended three-dimensional structures: Using a base with holes, templates for some 60 different structures, rods, and four sizes of spheres in radius ratios, common crystal structures can be assembled in a matter of minutes (3). Furthermore, many structures can be assembled from different perspectives by teams of students: For example, the cubic NaCl unit cell can be assembled with its orientation on the face of the cube or on the body diagonal. Natural cleavage planes can be found with the kit: Lifting one sphere will separate atomic planes from one another. (Contact ICE for ordering information.)

Suppliers for other products are collected in the Companion. An example of a "smart" material that is rapidly gaining popularity among teachers is "memory metal," a NiTi alloy. This solid has several remarkable features derived from a martensitic phase change. Thin wires of NiTi are easily bent. After bending, the wire can be gently heated to restore the initial linear shape. Heating to higher temperature in a candle flame permits the wire to be "re-trained" to remember a new shape. Rods of the alloy, also commercially available, have strikingly different mechanical and acoustical properties in the two phases: the high-temperature phase is rigid and "rings" when dropped; the low-temperature phase is more flexible and "thuds" when dropped. (Samples of memory metal are available from ICE, which can be contacted for ordering information; or from Shape Memory Applications, Inc., 1034 W. Maude Avenue, Suite 603, Sunnyvale, CA 94086.)

As noted, light-emitting diodes can be used to illustrate a variety of basic chemical concepts. Substitutional solid solutions like $GaAs_xP_{1-x}$ (0 < x < 1) effectively extend the periodic table by providing a tunable band gap, which translates to tunability in the color of emitted light (4).

Re-entrant foam provides a counter-intuitive demonstration of processing (5). Polyurethane can be isotropically compressed in a mold and heated to about 170 °C. The microstructure of the resulting solid yields a material that bulges in cross section when stretched! More information on polymers will be available from John Droske's complementary NSF-funded project (described in the preceding section).

Even given this new and user-friendly package of instructional materials represented by the Companion, the critical question is still how to get teachers to try it—how to have them "buy in". Several strategies are proving to be effective methods for acquainting the academic community with this project. First, the broad constitution of the committee itself has provided substantial visibility for the project; as new interested participants were identified, the committee's membership grew to its present size of about two dozen, comprising individuals with a broad range of research and teaching expertise. Second, field testing has provided a sense of ownership of the project for participating institutions. The structural model kit, for example, was tested at some dozen colleges and universities. User comments not only provided valuable feedback, leading to improved kit design, but made faculty at the institutions aware of and willing to try the kit.

A particularly effective way to identify potential users of these materials has been through presentations at professional meetings and informal networking. Individuals have volunteered to field-test the materials at a variety of institutions across the country that will expose some 5% of all students taking introductory chemistry courses to the material this year. Moreover, colleagues in other disciplines—physics, engineering, geology, and biology—have also expressed interest in and begun to use materials from the project that are of particular interest to them.

In short, momentum for the project has been established. Publication of the Companion by the ACS represents a key step in making this material an integral part of the curriculum. The ACS will encourage use of the instructional packet not only by teachers, the primary targeted audience, but also by textbook authors and publishers. Incorporation of materials chemistry throughout introductory chemistry texts will be a certain sign that the discipline has entered the mainstream. As one of our committee members noted at the outset of this project, if 10% of all of the examples used in introductory chemistry courses are solids, this project will have accomplished a lot. It remains to be seen whether the project will achieve this goal, but we believe that the infrastructure is now in place to make it possible.

Acknowledgments

Brian Johnson and Margret Geselbracht are thanked for critically reading this manuscript. On behalf of the Ad Hoc Committee for Solid-State Instructional Materials, it is a pleasure to acknowledge the National Science Foundation (Grant USE–9150484), the Camille and Henry Dreyfus Foundation, the American Chemical Society, the Dow Chemical Company Foundation (Solid-State Model Kit), the University of Wisconsin—Madison Outreach Program (Solid-State Model Kit), and the Institute for Chemical Education for their generous support of this project.

References

1. "Report on the National Science Foundation Undergraduate Curriculum Development Workshop on Materials" (October 11–13, 1989); National Science Foundation: Washington, DC, April, 1990.
2. Lisensky, G. C.; Kelly, T. F.; Neu, D. R.; Ellis, A. B. *J. Chem. Ed.* **1991**, *68*, 91–96, and references therein.
3. Mayer, L. A.; Lisensky, G. C. The ICE Solid-State Model Kit (based on Mayer, L. A., U.S. Patent 4 014 110, 1977); ICE, Department of Chemistry, University of Wisconsin, Madison, WI 53706.
4. Lisensky, G. C.; Penn, R.; Geselbracht, M. J.; Ellis, A. B. *J. Chem. Ed.* **1992**, *69*, 151–156.
5. Lakes, R. *Science (Washington, D.C.)* **1987**, *235*, 1038–1040.

How Scientists and Engineers Can Enhance Science Education in Grades K–12

Kenneth H. Eckelmeyer

Sandia National Laboratories, Albuquerque, NM 87175

MATH AND SCIENCE EDUCATION IS A CRITICAL ISSUE. Scientists and engineers are learning that the technical community can make a real difference.

We at Sandia have always had educational outreach programs. Until several years ago, these programs were relatively small. Then Admiral Watkins issued a Department of Energy (DOE) directive: "We must expand our involvement in science education to inspire the youth of America to either enter, or feel more comfortable in, the fields of math, science, and engineering. With our labs and facilities, we are uniquely well positioned to provide major assistance in strengthening science and engineering motivation and education, making it come alive for the main body of students who too often fear these disciplines or who cannot relate to them. I intend to lead this effort personally." (1)

School Partnership Program

This proclamation encouraged the national laboratories and DOE facilities to become more heavily involved in pre-college education. On the basis of my experience working with youth groups, I was asked by management to initiate our expanded effort of science and math education enrichment. I first gathered together a small group of our technical staff who had experiences similar to mine and who were experienced in working with kids through activities such as youth work, coaching, or scouting. We asked ourselves the following questions: What are the problems, and how can we affect them? We decided that our goal should be to inspire increased student interest in math and science. This decision was based on our belief that two keys to learning are positive impressions and excitement—if you get kids interested and excited, then learning will happen.

A strong point of our School Partnership Program is that it does not consist solely of exciting programs that we imposed on the schools. Instead, we teamed with teachers and emphasized that we wanted to fit in with their curriculum. We asked them, "What science topics will you be covering, and what types of activities might we do to get kids excited about them?" We learned through our discussions and experiences that there is a big difference between the needs in elementary and high schools. Elementary school teachers frequently had very little background in science, and as a result, often avoided science instruction. However, they recognized their shortcomings in this area, and were eager for any kind of help we could give. High school teachers, on the other hand, were well versed in science content, but had more need for activities that demonstrated real-world applications of particular topics. Middle school teachers were highly varied; some had elementary school and others high school credentials. Key to our success was that, in all cases, we designed our activities around the curriculum and the expressed needs of the teacher.

Albuquerque Public Schools picked three pilot schools for implementation. They had a good ethnic mix and were below the 50th percentile in socioeconomic level. My assignment was in a middle school with the highest turnover rate in the city. I contacted the middle school teacher for physical science classes and said, "I'm from the government and I'm here to help you." Initially she was not enthusiastic. After explaining to her that I wanted to know about her curriculum and intended to develop some exciting activities to go along with her topics, she reluctantly agreed that I could try an activity on Newton's law—force and acceleration.

As I thought about developing a demonstration, I remembered that Sandia had been involved in the design of nuclear waste shipping containers and had some videotapes of dramatic crash tests, such as tractor trailers running into concrete bridge abutments. Knowing that middle school kids are fascinated by destruction, but would be bored by a 45-minute videotape, I knew I had to incorporate some additional activities that engaged the students in fun and exciting ways. We accomplished this goal by wrapping the kids up in bubble wrap, outfitting them with helmets, and then putting them on skateboards and letting them run into the wall. This got their attention!

Discussions afterwards focused on how the levels of pain would vary, how the force would change, and how the suddenness of the change in speed would vary with the number of layers of bubble wrap. Out of this directed discussion we developed and discussed the formula $F = ma$ (force equals mass times acceleration), and then used it to do a simplistic calculation of what the force would be on a skier running into a tree. We then viewed the most dramatic 10 minutes of the video of the crash tests of nuclear waste shipping casks. Afterward, I explained that each test cost hundreds of thousands of dollars, but that using Newton's law we were able to calculate the forces and damage that would occur in each crash. After doing a few tests to make sure our calculations were correct, we were then able to do many subsequent "crashes" on the computer. These are the elements of a good in-class activity. I first got them excited and actively involved, helped them develop the key principle in their own minds, applied it to a situation that they could relate to, and finally, showed them a concrete real-world application. I was invited back!

Out of experiences like this, we developed continuing relationships with the teachers and students at our pilot schools. We conducted in-class activities to complement major curriculum topics every 3 weeks or so. In addition, we assisted in other ways, such as helping elementary school teachers understand science content and working with science fairs.

We evaluated our effectiveness through a questionnaire designed to assess attitudes toward science. We administered this questionnaire to the students prior to any of our activities and again following one semester of activities. The "before" results were consistent with previous studies, indicating that substantial declines in attitude occur between grades 3 and 7. The "after" results showed substantial improvements in attitudes toward science following our semester's activities with them. In addition, more than 95% of the students at every elementary and middle school grade level said they would like us to continue and expand our program next year.

When we considered how to expand this program, we decided that we could have an impact on a much larger number of schools by concentrating on teacher support, rather than direct interaction with students. This approach also broadened the range of potential Sandia participants, as most of our staff are more adept at communicating with adults than interacting motivationally with students. Prior to the first year of this Science Advisors Program, we wrote letters to all elementary and middle schools in Albuquerque and offered to assign a technical employee to work with their teachers 1 day a week doing whatever they thought would be helpful. About 80% responded positively; in the second year the others also joined.

We prepared our "science advisors" with training on how to interact effectively with teachers and students, and we developed an Education Resource Center where they could borrow both activity plans and equipment to help their teachers do interesting hands-on science activities in their classes. This resource center was stocked both with commercial educational materials, as well as selected surplus equipment. Computers turned out to be particularly popular items—we now loan out on an annual basis about 200 computers that are outmoded for our technical purposes, but that the schools are delighted to have. At the request of teachers, our science advisors have ended up doing many of the same things as the School Partnership participants. About 50% of them do at least some activities directly with students. Others provide help in understanding science content, coordinate access to the resource center, assist with science fairs, and provide support for teachers in a variety of other ways.

In addition to the School Partnership and Science Advisor programs, we also sponsor efforts in which women's and minority outreach groups provide mentoring and science enrichment activities with students who have been historically underrepresented in the technical community. High school students receive tutoring from our staff, and high school students can be mentored through part-time or summer employment experiences. One of the reasons so many of our employees are involved is that we offer them this variety of options.

We are also involved in attempting to propagate programs in which technical professionals enhance grades K–12 science education in areas beyond Albuquerque. We have assigned science advisors to 80 public and Bureau of Indian Affairs schools in rural portions of New Mexico. In addition, we are helping other communities set up their own science advisor programs. Finally, we are working through technical professional societies to involve technical professionals in their local communities nationwide.

How Scientists Can Help Teachers

One of Sandia's primary contributions to the national effort involves helping technical professionals understand how to be effective. On the basis of our experiences, we have developed training materials that we are making available to technical professionals through the coordinated grades K–12 education programs of a large number of professional societies. Some of the key points made in these training materials are outlined in the following list.

- Adopt a productive attitude. Don't go into schools with an attitude that tells the teachers you think they're doing a bad job. The fact is that most of our schools and teachers are doing a terrific job with limited support and resources and in spite of a lot of societal baggage. Don't alienate the teachers—try to help them make a good thing even better.

- Do activities that complement and enhance the existing curriculum. Don't waste time with things that aren't curriculum-related. Teachers have goals and competency requirements imposed on them. If you help them meet these they will be appreciative. If not, you are failing to help them where they most need it. Remember, they are your customers.

- Make teachers' lives simpler, not more complex. Teachers are pulled in many different directions and are very busy. If you become a time and work saver for them, they will love you. If you become a time sink and another person competing for their attention, they are likely to resent you.

- Help integrate excitement and fun into science education. We should not mislead people to believe that science is nothing but excitement and fun, but neither should we start off with a hard-line diatribe about science requiring great personal sacrifice and self-discipline. Instead, begin by doing a fun activity that illustrates the scientific principle, such as the skateboard experiment. Then make the transition into topics activities that require greater concentration. Once interest is aroused, concentration becomes more achievable.

- Promote hands-on, discovery-based activities. Kids learn by doing! And hands-on activities that let kids discover things for themselves are the best of all. Kids forget most of the things their teachers tell them, but when adults lead kids in experiences where they wrestle with an interesting personal observation and then figure it out "by themselves", kids remember those things forever.

- Help integrate principles and applications. Our traditional educational paradigm says, "First you learn the principles and theory, then you learn what it's good for and how to apply it." That is boring! Very few of us are motivated to learn in this sequence. Applications are not only more interesting than theory, they also provide a concrete context through which we are better able to understand abstractions, particularly when we are first learning about a topic. I wish that when I was first learning differential calculus someone had helped me understand that the first derivative was velocity and the second derivative was acceleration. It not only would have given physical meaning to what seemed to me to be nothing but an abstraction, but it would have helped me understand why it was important to learn it. We need to help change the teaching paradigm to give much greater emphasis to applications.

- In working with students, do activities that are age-appropriate. This is second nature to most teachers, but we technical professionals violate this principle frequently. Scientists and engineers who work with the schools need to understand the basics of Piaget and the progression of the way kids learn. Most importantly, they need to know that essentially all elementary school students are concrete thinkers— they understand what they detect with their physical senses, but not abstract representations of reality, such as graphs and algebraic equations. Only in middle and high school do students begin to develop abstract thinking abilities. Even there, it is best to start with the concrete and move toward the abstract to the extent that the students remain engaged.

- Emphasize activities in which everyone can succeed. Don't do things that you think will be challenging and rewarding only to the top 20%.

The others will be discouraged and decide that science is just too hard for them.

- Plan in advance to eliminate sources of confusion and demotivation. Don't try to impress students or teachers with how much you know; that attitude results in intimidation and discouragement. Deal with topics to which students can relate. (Fifth graders are not engaged by integrated circuits, no matter how important and exciting we think they are.) Consciously scrub your language of technical jargon and acronyms. Remember that familiarity with technical content isn't all that is needed. To teach in a clear and compelling way takes careful planning.

- Take advantage of existing educational resources. Lots of outstanding hands-on activity plans are available. Take some time to learn about and become familiar with some of these. (I'll send you a listing and descriptions of some of the ones that we've found most useful if you request it.) Avoid the syndrome of reinventing all your activities from scratch; they probably won't be nearly as good as if you capitalized on the ideas and experiences of others. We hope that at some point in the future, a database will be available for searching existing materials by topic and age level.

- Emphasize scientific method and logic, not just science content. Science is not knowing all of the right answers. Science involves wondering what would happen if ..., making guesses, doing experiments, figuring out what the results mean, and changing your mind when the experiment proves you wrong. These are the type of logical process skills that all citizens need to develop and employ. Helping promote such technical literacy is perhaps the most important challenge the scientific and engineering communities face. The alternative to a society that uses logical thinking to arrive at conclusions is one in which policies are set and resources are committed on the basis of who has the slickest advertising or the most emotionally heart-rending appeal.

- Be an encourager. Praise people when they get something right. Take advantage of opportunities to turn even incorrect answers and hypotheses into something positive. When people are encouraged to share and discuss their reasoning, they frequently uncover their incorrect assumption or logic flaw themselves—then you can praise them. And if they don't, use an experiment to help them discover where they are wrong. Remember, science is doing experiments and figuring things out, not knowing all the right answers.

- Build positive relationships and be fun to be with. Your best technical efforts probably won't be very effective unless you also build positive relationships. Learn the teachers' and students' names. Eat lunch with them. Get to know their hobbies and interests. Give them lots of positive feedback. Laugh with them. Convey with your time, attention, and attitudes that you like them and value them. Pretty soon they will start to like you, too. Then you will have won the important "right-to-be-heard."

- Maintain control when working with students but without stifling fun. It's best to have an agreement with the teacher that he or she will always be present and will take responsibility for any disciplinary problems. Discuss with the students your expectations, and expect them to behave responsibly. If criticism is needed, criticize the behavior rather than the person, and never embarrass kids in front of their peers.

- Finally, make a long-term commitment and follow through on it. Don't do a one-shot presentation with a school and then disappear; you'll never get beyond the steepest part of the learning curve or build the trusting relationships out of which the most important interactions occur. The beginning activities are usually the most difficult. Subsequent interactions become more relaxed and fruitful as relationships develop and your understanding of their real needs increases. Remember that improving the educational process is a marathon, not a sprint.

References

1. Admiral James D. Watkins, U.S. Secretary of Energy, September 15, 1989.

RECEIVED for review May 7, 1993. ACCEPTED revised manuscript July 27, 1993.

SELECTED RESEARCH TOPICS

New Directions in the Design of Lithographic Resist Materials: A Case Study

Elsa Reichmanis and Larry F. Thompson

AT&T Bell Laboratories, Murray Hill, NJ 07974

In the past decade, major advances in fabricating very large scale integrated (VLSI) electronic devices have placed increasing demands on microlithography, the technology used to generate today's integrated circuits. In 1976, state-of-the-art devices contained several thousand transistors with minimum features of 5–6 μm. Today, they have several million transistors and minimum features of less than 0.7 μm. Within the next 10–15 years, a new form of lithography will be required that routinely produces features of less than 0.25 μm. Short-wavelength (deep-UV) photolithography and scanning electron-beam, X-ray, and scanning ion-beam lithography are the possible alternatives to conventional photolithography. However, each needs new resists and processes. When deep-UV photolithography is implemented, it will represent the first widespread use in manufacture of a lithographic technology that requires an entirely new resist technology. We describe the processes involved in the development of a resist system for this lithographic technique.

A MODERN INTEGRATED CIRCUIT is a complex three-dimensional structure of alternating, patterned layers of conductors, dielectrics, and semiconductor films (*1*). This structure is fabricated on an ultrahigh-purity wafer substrate of a semiconducting material such as silicon. To a substantial degree, the performance of the device is governed by the size of the individual circuit elements. As a general rule, the smaller the ele-

0065–2393/95/0245–0085$12.25/0

ments, the higher the device performance. These high-performance devices are faster and they require less energy per operation (*2*). The device
structure is produced through a series of steps that precisely pattern each
layer. These patterns are formed by lithographic processes (Figure 1) that
consist of two steps:

1. delineation of the patterns into a radiation-sensitive thin polymer
   film (i.e., a resist)
2. transfer of those patterns into the substrate using an appropriate
   etching technique (*1, 3, 4*).

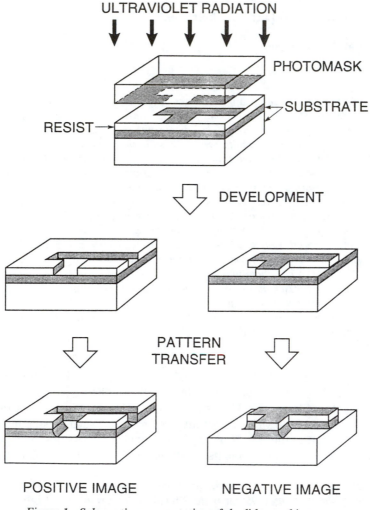

Figure 1. Schematic representation of the lithographic process.

Technology Trends

Remarkable progress has been made in the fabrication of microelectronic devices, and especially in the lithographic technology used to generate the high-resolution circuit elements that are characteristic of today's integrated circuits. In 1976, state-of-the-art devices contained 4000–8000 transistor elements and had circuit elements with dimensions of 5–6 μm. These devices were patterned by photolithography using either contact printing or, the then relatively new, one-to-one projection printing. Today, devices with several million transistor cells are commercially available and fabricated with minimum features of 0.5 μm or smaller. This trend is illustrated in Figure 2 (*2*). Surprisingly, photolithography is still the technology used to fabricate microelectronic chips. Step-and-repeat 5× or 10× reduction cameras or highly sophisticated, one-to-one projection printers are the dominant printing tools.

Photolithography is similar to printing a photograph (not surprising to Greek buffs) from a negative (*3*). First a mask is produced (equivalent to the negative) that contains the integrated circuit patterns. Mask pro-

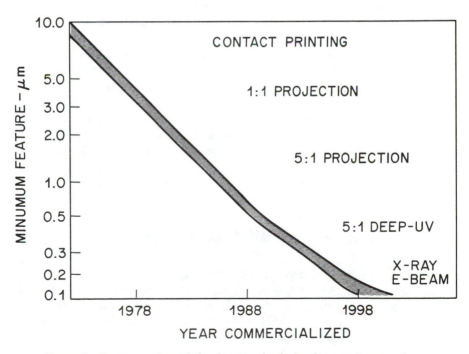

Figure 2. Representation of the decrease in device feature size as a function of the year the device was first commercially available. (Reproduced with permission from reference 12. Copyright 1990 AT&T.)

duction is usually accomplished by translating a computer drawing of the circuit onto a glass plate by using a sophisticated electron-beam writing system. The mask is placed in an optical imaging tool that projects the pattern with UV light onto a semiconductor substrate that has been coated with a photosensitive polymer film known as a resist. The solubility of the polymer is either increased or decreased by photoinduced reactions, and the more soluble material is removed in a subsequent developing step to provide a three-dimensional relief image of the circuit pattern.

There is, perhaps, no better example than lithography to illustrate the uncertainty associated with predicting technological direction and change. In 1976, it was generally believed (albeit, not by everyone) that photolithography could never produce features smaller than about 1.5–2.0 μm with high chip yields in a production environment. The current belief is that conventional photolithography (that is, g and i line, where g line = 436 nm and i line = 365 nm) will be able to print features as small as 0.35 μm and will remain the dominant technology well into this decade (5). This achievement has been accomplished through incremental improvements in tool design and performance. However, the ultimate resolution of a printing technique is governed, at the extreme, by the wavelength of the light (or radiation) used to form the image. Shorter wavelengths yield higher resolution. The same physical principles also govern the resolution limits in conventional optical microscopy (3).

Throughout this era (beginning in the mid-1970s) of enhanced tool performance, the same basic, positive photoresist has been in pervasive use. This resist consists of a photoactive compound that belongs to the diazonaphthoquinone chemical family and a novolac (phenol–formaldehyde polymer) resin (Scheme I) and will remain the resist of choice for several more years. The cost of introducing a new technology, which includes the cost associated with the development and implementation of new hardware and resist materials, is a strong driving force pushing photolithography to its absolute resolution limit and extending its commercial viability.

Alternatives to Conventional Photolithography

The technological alternatives to conventional photolithography are essentially the same as they were a decade ago; short-wavelength photolithography, scanning or projection electron-beam, or 1:1 proximity X-ray lithography (3). This chapter will not discuss these lithographic technologies in detail. However, each are briefly summarized.

Deep-UV Photolithography. In the past decade, the major advances in short-wavelength (i.e., deep-UV) photolithography were achieved

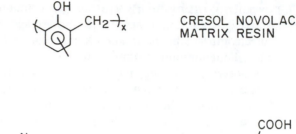

CRESOL NOVOLAC
MATRIX RESIN

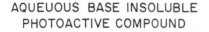

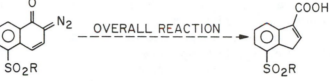

AQUEUOUS BASE INSOLUBLE
PHOTOACTIVE COMPOUND

AQUEOUS BASE SOLUBLE
PHOTOPRODUCT

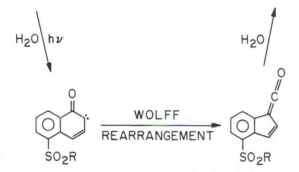

Scheme I. Chemistry and processes associated with conventional positive photoresists.

with improved quartz lenses and high-output light sources for reduction projection cameras. Several step-and-repeat reduction systems that use high-brightness laser sources were designed and built. Because it is not practical to correct for chromatic aberrations in quartz lenses (necessary to provide high transmittance of 248-nm light), the systems that use refractive optics require a light source that has a very narrow bandwidth (i.e., less than 0.03 Å). Laser sources can provide such narrow bandwidths with enough intensity to accommodate resists that have sensitivities of 50–100 mJ/cm^2 (*6, 7*). Aggressive development is also being done on one-to-one projection systems using reflective optics and conventional, wide-bandwidth, mercury arc sources in the 200- to 260-nm region (*8*). Because the intensity of these sources is considerably less than that of the laser sources, more sensitive resists (i.e., less than 15 mJ/cm^2) will be required for high throughput.

X-ray Lithography. Early work in X-ray lithography followed the general strategy of conventional, one-to-one optical projection systems and used electron-beam bombardment X-ray sources. However, thermal considerations limit the power from such sources. Also, to minimize the penumbral shadowing and other physical effects that limit resolution, the distance between the source and mask–wafer combination must be relatively large, which makes the incident flux on the wafer small. These limitations require the use of resists with sensitivities of less than 5 mJ/cm^2 (3). Both the wavelength used and the extremely high sensitivity requirement limit the chemistries available for resist design. Several negative resists that are available demonstrate submicrometer resolution, but are limited to features of at least 0.5 μm or larger.

Recent work in X-ray lithography has been directed toward step-and-repeat systems that use high-intensity synchrotron (9) or laser-based radiation sources (10). Both types of sources are capable of producing X-rays in the wavelength region of interest. When compared to other lithographic hardware options that have submicrometer capabilities, the tools that use these X-ray sources are economical for manufacturing devices. In addition, the brightness of these sources is high enough to permit the use of resists with sensitivities of 20–50 mJ/cm^2.

Scanning Electron-Beam Lithography. For well over two decades, electron-beam lithography has been investigated at many industrial and university laboratories. Thus, this technology is the most mature of all the alternatives to conventional photolithography. This form of lithography uses a focused beam of electrons that is scanned under computer control across the resist-coated substrate. Early systems operated at a modulation rate of 10–40 MHz and used a Gaussian, round beam that was 0.5–2.0 μm in diameter. These early machines used a tungsten emitter as a source of electrons and required about 1 h to "write" a pattern onto a 4-inch diameter wafer with a resist that had a sensitivity of 1–3 μC/cm^2. The most advanced of the new systems use high-brightness electron sources, shaped beams, and modulation frequencies greater than 400 MHz and can produce three to eight 4-inch wafers per hour. These systems require resists with sensitivities of 1–5 μC/cm^2 at 20 kV.

Electron-beam lithography offers high resolution (less than 0.1 μm) and extraordinary registration accuracy, which is important when writing very large scale integrated (VLSI) patterns directly onto the resist (11). The major disadvantages of electron-beam lithography are the high overall cost it introduces into the lithographic step, low throughput, and size of the manufacturing equipment.

No matter which technology becomes dominant after photolithography has reached its limit, new resists and processes will be needed, and enormous investment in research and process development will be required. The introduction of new resist materials and processes will also require considerable lead time. Probably, more than 3–6 years will elapse before the new resists reach the performance level currently realized by conventional positive photoresists.

Resist Design Requirements

The focus of this chapter relates to the methodology employed in the design of polymer–organic materials and chemistry that will prove useful in radiation-sensitive resist films for advanced lithographic methods and is adapted from an article that appeared in the *AT&T Technical Journal* (*12*). Lithographic resists must be carefully designed to meet the specific requirements of each lithographic technology. Although these requirements vary according to the radiation source and device-processing sequence, the following resist properties are common to all lithographic technologies.

Sensitivity, measured in units of energy per unit area, is a direct measure of how efficiently a resist responds to a given amount of radiation. The sensitivity must be great enough to allow a satisfactory image (pattern) to be produced in the required exposure time. The new lithography tools have relatively low brightness sources, and therefore resists with very high sensitivity are required to minimize exposure time and to ensure a high wafer throughput.

Contrast is an indirect measure of the rate of chemical change (e.g., solubility change) upon exposure to radiation and is also an indication of the resolution capability of a resist. After resist exposure and development, a plot of the normalized thickness remaining as a function of log dose generates a "sensitivity curve". The contrast for either a positive or negative resist, γ_p or γ_n, is determined from the slope of the linear portion of that curve.

Resolution is the smallest lithographically useful image that can be formed and is a property that is difficult to quantify because many external variables affect the size of the minimum feature that can be resolved. The resolution is generally defined as the smallest feature that can be perfectly and cleanly resolved over a large area in a useful resist thickness.

Optical density determines how much light is absorbed per micrometer of resist thickness and is very critical for UV resists. Uniform resist imaging can generally be attained if the absorption is less than 0.4 μm^{-1}.

The *etching resistance* of a resist is a measure of how well the patterned polymer withstands the etching (pattern-transfer) step (Figure 1). Acidic and basic liquid etching processes have been replaced with "dry" etching methods that rely on high-energy, gas-phase, plasma-enhanced chemical reactions, in which the patterned resist is exposed to high levels of radiation and heat. Pattern transfer must be accomplished with well less than 10% change in the finest feature size, which represents an extraordinary demand on the polymer. Essentially, sensitivity requirements necessitate a system that responds strongly to radiation, but high plasma etching resistance demands radiation and thermal stability. These two requirements are mutually opposed and represent a dilemma to the resist designer.

Purity is of paramount importance because semiconductor device performance can be significantly affected by tiny quantities of impurity atoms or ions. Known, or even just perceived, device contaminants must be controlled to the part-per-billion level.

Manufacturability must be considered at the early stages of resist design. Because "cost per level" is of extraordinary importance, any new resist must be produced at a reasonable cost. To achieve the requirements just discussed, the molecular properties must be controlled very tightly, and the use of reagents containing potential device contaminants must be eliminated.

These properties can be achieved by careful manipulation of polymer structure, molecular properties, and synthetic methods. References 13 and 14 are reviews describing chemistries that have been adapted for lithographic applications.

The materials issues that must be considered in designing resists with the appropriate properties include the following. The polymer resins must

1. exhibit solubility in solvents that allow the coating of uniform, defect-free, thin films, or be amenable to vapor-deposition to achieve the same result
2. be sufficiently thermally stable to withstand the temperatures and conditions used with standard device processes
3. exhibit no flow during pattern transfer of the resist image into the device substrate
4. possess a reactive functionality that will facilitate pattern differentiation after irradiation
5. for UV exposure, have absorption characteristics that will permit uniform imaging through the thickness of a resist film

In general, thermally stable (>150 °C) materials with high glass-transition temperatures (T_g 90 °C) and low absorption at the wavelength of interest are desired. If other additives are to be employed to effect the desired reaction, similar criteria apply. Specifically, they must be nonvolatile, be stable up to at least 175 °C, possess a reactive functionality that will allow a change in solubility after irradiation, and have low UV absorbance.

As already noted, resists function by altering the solubility of the polymer through radiation-induced chemical reactions. The reactions can either increase the solubility (i.e., positive tone) or decrease the solubility (i.e., negative tone) of the irradiated region. The terms *positive resist* and *negative resist* reflect this change in solubility, as illustrated in Figure 1. Both tones of resist have been developed for all the lithographic options. However, it is generally accepted that positive resists exhibit the best overall performance.

The most widely used positive resists (also known as conventional positive photoresists) operate through a dissolution inhibition mechanism. Such resists are generally two-component materials. They consist of an aqueous alkali-soluble matrix resin (typically a novolac) that is rendered insoluble in aqueous alkaline solutions by the addition of a hydrophobic, radiation-sensitive material (a diazonaphthoquinone) known as a photoactive compound (PAC). When irradiated (i.e., exposed to UV light), the PAC undergoes a Wolff rearrangement and is converted to an acid that is alkali-soluble. This change permits selective removal of the irradiated portions of the resist by an alkaline developer (*14*). Scheme I outlines the chemistry and processes associated with these resists.

Deep-UV photolithography is likely to be the first of the alternatives to conventional photolithography to be used in manufacture, and thus will be used as an example to define the methodology involved in designing a new resist material chemistry.

Resist Design for Deep-UV Lithography

On the basis of initial evaluations, we have established some specific performance criteria for an optimum deep-UV resist, and they are given in Table I (*15*). When one examines the available light sources and associated optical systems, it appears unlikely that a material with an overall quantum efficiency of 1 or less will provide the required sensitivity. A

Table I. Performance Criteria for Deep-UV Resists

Parameter	Criteria
Sensitivity[a]	<100 mJ/cm^2
Contrast	>5
Resolution	0.25 μm
Optical density	<0.4 μm^{-1}
Etching resistance	approximately equal to novolac-based positive photoresist
Shelf life	>1 year

[a]For tools that use conventional Hg sources, a sensitivity of <15 mJ/cm^2 will be required.

new class of resists that achieve differential solubility from acid-catalyzed chemical reactions was discovered by Ito et al. (16–19), who utilized the findings of Crivello and co-workers (20–22) that arylonium salts efficiently produce strong acids upon radiation. These resists are nominally classified as chemically amplified resists and are compatible with the exposure tools currently available.

Reference 23 is a review describing the many chemistries associated with chemically amplified resist materials. Chemically amplified resists generally exhibit high contrast, good process latitude, excellent thermal stability, and good dry-etching resistance. The process sequence in Figure 3 for these chemically amplified materials is similar to that for conventional positive resists, although the postexposure bake (PEB) assumes a different role. One chemically amplified material will be used to examine how each of the performance criteria is achieved through the careful design and manipulation of the detailed chemical structures that will form the final resist.

As previously discussed, sensitivity is a direct measure of how efficiently a resist responds to a given amount of radiation. To have radiation sensitivity, a chemical compound must have bonds that undergo bond cleavage or rearrangement when exposed to radiation (photons, for the photoresist example presented here). In addition, these materials must contain chemical moieties whose structure allows a specific wavelength of light to be absorbed and the photon energy to be transferred efficiently to the appropriate bonds. A material that undergoes one or fewer chemical events per photon absorbed is not likely to provide the required sensitivity, that is, <100 mJ/cm^2 for deep-UV photolithography.

Chemical Amplification

Several years ago, a new class of resist was discovered that achieves differential solubility through a catalyzed chemical reaction (16–19, 24). The term *chemical amplification* was coined to describe this mechanism. In a chemical amplification scheme, a photogenerated species (A^+ in Figure 3) catalyzes many chemical events such as the deblocking of a protective group (P in Figure 3) from a hydroxy-substituted polymer that is soluble in an aqueous base. The inherent sensitivity associated with the positive-acting, chemically amplified resists results from the regeneration of the acid (A^+) (a chain reaction), which is available for additional deblocking reactions that occur during the postexposure baking step. Typical turnover rates (i.e., the catalytic chain length) for each acid molecule in a working resist formulation are in the 800–1200 range (25). The "deprotection" and turnover rates are critically dependent on postexposure bake temperature, time, and method of bake (15).

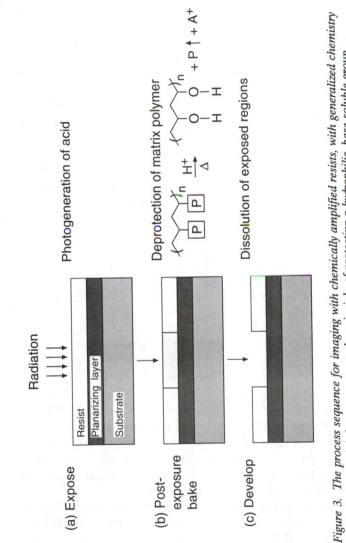

Figure 3. *The process sequence for imaging with chemically amplified resists, with generalized chemistry depicting a material that operates on the principle of protecting a hydrophilic, base-soluble group.*

Achieving Material System's Characteristics

One chemically amplified resist system currently being investigated uses (4-*tert*-butoxycarbonyloxystyrene sulfone) (PTBSS) as the matrix resin (*15, 26, 27*) and 2,6-dinitrobenzyl tosylate (*28*) as the photo acid generator (PAG). The motivation that lead to the development of the nitrobenzyl ester acid generators was based upon the desire to have a nonionic, metal-ion-free PAG (*28, 29*). The requirement for a metal-ion-free material relates to the issue of resist purity (vide supra), and the preference for a nonionic material stems from a desire to minimize the probability of phase incompatibility of the resist components.

Nitrobenzyl esters absorb light efficiently and then undergo a series of rearrangements and bond-cleavage reactions to generate an acid, tosic acid in this case (III in Scheme II). Tosic acid is now available as a catalyst for the deprotection reaction (vide supra). The relevant subgroups of PTBSS are highlighted and labeled IV–VII in Scheme II to illustrate how this system responds to acid and radiation. PTBSS was designed to be used with PAG materials such as the tosylate ester described. The *tert*-butoxycarbonyl moiety (IV) is hydrophobic and prevents dissolution of the polymer in an aqueous base.

The inclusion of sulfur dioxide (VI) into the backbone of the polymer affords a high T_g (165 °C for PTBSS vs. 120 °C for its non-SO_2-containing counterpart) that provides greater flexibility for performing the PEB at elevated temperatures for improved sensitivity. Also, the introduction of sulfur dioxide into similar polymers has effected improved sensitivity to deep-UV and electron-beam radiation as a result of radiation-induced scission of the carbon–sulfur bonds (*30, 31*). These features are important when arylsulfonic acid materials are contemplated for use as the catalysts for deprotection. These acids are generally weaker than the inorganic acids available from the onium salts and thus inherently lead to resists with lower sensitivity (*32*). In the presence of acid, PTBSS is thermally deprotected, losing carbon dioxide and isobutylene to form poly(hydroxy-styrene sulfone). The hydroxyl (OH) group (VII) is the hydrophilic moiety that imparts base solubility in this system. The combined use of PTBSS with a nitrobenzyl ester PAG allows the definition of high-resolution circuit elements with sensitivities <50 mJ/cm^2.

Achieving a contrast value of 5 or greater is also a difficult requirement. The contrast of a resist is a measure of how the resist responds to incident radiation and is related to the resolution that can be achieved. Higher contrast materials generally exhibit higher resolution. To obtain high contrast, materials must exhibit an exceptionally nonlinear dissolution response as a function of radiation dose. For the deep-UV resist in Scheme II this goal is accomplished by converting the hydrophobic *tert*-butoxycarbonyl group (IV) to the hydrophilic hydroxyl group (VII). This

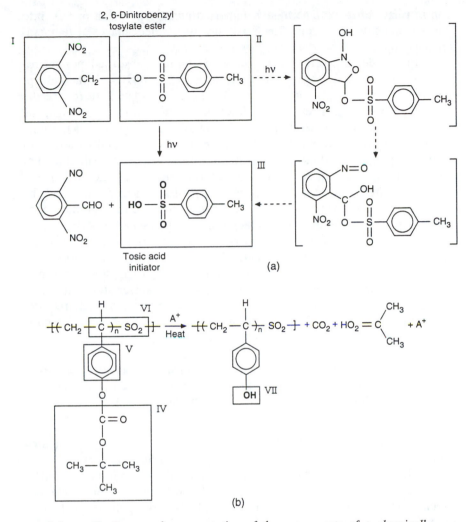

Scheme II. *Structural representation of the components of a chemically amplified resist based on PTBSS–nitrobenzyl ester chemistry.*

conversion, coupled with the fact that aqueous base solubility is achieved only when nearly 100% of the protecting groups have been removed (*16*), results in a resist that has very high contrast.

Many physical and chemical properties of the matrix polymer affect resolution. In addition to having high contrast, the polymer must neither swell during development nor deform (i.e., flow) at the elevated temperatures encountered during later processes such as postdevelopment bakes. Achieving these properties entails careful design of the basic polymer structure and of the solvents used as the developer (an aqueous base solution in this example). The high T_g imparted by incorporation of SO_2 into the polymer affects thermal flow resistance of defined images at temperatures as high as 140 °C (27).

Optical density determines how much light is absorbed per micrometer of resist thickness and is critical for deep-UV resists. If the absorption is too low, few photons are available in the resist film to induce the desired photochemical reactions. However, if absorption is too high, the photons are not deposited uniformly throughout the thickness of the film, and thus they degrade the shape (i.e., the profile) of the final pattern. An optical density value of 0.4 μm^{-1} is the maximum value that will allow uniform resist imaging in resists of 0.8–1.0-μm thickness. The specific organic structures contained in the resist control its optical density, and the detailed structure must be carefully designed to provide the optimum optical density and quantum efficiencies. For the deep-UV resist being discussed, the optical density (0.3 μm) is determined primarily by the structure of the PAG; the matrix polymer, PTBSS, has minimal absorption.

The etching resistance of a resist is a measure of how well the patterned polymer withstands the etching or pattern-transfer step (Figure 1). In the example presented, the dichotomy between the materials requirements necessary to achieve high sensitivity and those required for good etching resistance is dealt with in two ways:

1. The photochemistry required for high sensitivity occurs in the PAG (Scheme IIa), which is a small percentage of the total resist mass.
2. The etching resistance is achieved with the photochemically stable, PTBSS matrix polymer.

Radiation and thermal stability is provided by careful design of the matrix polymer, and the aromatic organic structure (V in Scheme IIb) has excellent radiation and thermal stability. In addition, to minimize thermal flow, SO_2 (VI in Scheme II) was incorporated into the polymer backbone, which increases the glass-transition temperature of the polymer by hindering polymer chain flexibility.

Process Optimization

Once the materials chemistry is defined, the process for using the resist must be optimized. Figure 4 outlines each of the stages or steps involved in the lithographic process (3). Each step may not be required in every resist process (dashed box). The steps are described as follows.

1. *Cleaning.* A key requirement for any device substrate is that it be cleaned before deposition of the resist. To avoid adversely affecting

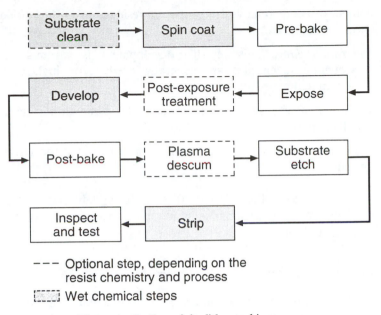

--- Optional step, depending on the
    resist chemistry and process

[▒▒▒▒] Wet chemical steps

Figure 4. Outline of the lithographic process.

the lithographic process, a substrate must be atomically clean, that is, free (to the limits of detection) of any contamination.

2. *Spin-coating.* After cleaning, the resist is spin-coated onto the substrate to obtain a uniform, adherent, defect-free polymeric film over the entire substrate.

3. *Pre-baking.* The resist is dried to remove any residual solvent.

4. *Exposure.* The photochemical reaction required to change the solubility of the resist takes place during this step.

5. *Postexposure baking.* For the deep-UV resist described here, the postexposure treatment is a baking step. This bake induces the deblocking reaction that changes the polarity and, thus, the dissolution characteristics of the polymer.

6. *Development.* After the latent image has been formed in the polymeric resist film, the image must be developed to produce the final, three-dimensional relief image. The developer selected must produce images that deviate less than 5% from the desired size.

7. *Postbaking.* After development, the structure often undergoes a bake step to dry and harden the resist image.

8. *Plasma de-scumming.* Before substrate etching, a plasma de-scum step

may be used to remove trace organic residues in the developed regions of the resist.

9. *Substrate etching.* This step, like development, is extremely critical. Dry-etching techniques are generally needed for high-resolution lithographic processing.

10. *Stripping.* The resist may be stripped by using either liquid- or oxidizing-plasma processes.

11. *Inspection and testing.* The substrate is then inspected for defects and tested.

These steps are repeated as many as 20 times during fabrication of a complex VLSI device. (For example, a static random-access memory (SRAM) requires 14 iterations.)

Chemically Amplified Resist Process Issues

As mentioned earlier, the inherent sensitivity associated with most chemically amplified resists emanates from the catalytic action of the photogenerated acid during the postexposure baking step. The extent of deprotection (or cross-linking) and turnover rate are critically dependent on postexposure bake temperature, time, and the method of bake itself. A typical chemically amplified resist process involves conversion of the PAG molecule to a strong acid upon absorption of a photon. The rate of this reaction is fast, the extent of reaction being governed by the quantum efficiency of the particular acid generator and exposure energy.

The subsequent reactions of the photogenerated acid proceed at a characteristic rate, which is a function of the acid concentration, the temperature, and the diffusion rate of the acid in the polymer matrix (*33–35*). The diffusion rate in turn, depends on the acid structure, the temperature, and the polarity of the polymer matrix. At room temperature, the rate of reaction is typically low, and generally the film must be heated to well above room temperature to increase the rate to acceptable levels. As discussed, the acid (H^+) is regenerated and continues to be available for subsequent reaction, hence, the amplification nature of these resists.

For this system to work satisfactorily, the radiation-generated acid concentration should be uniform through the thickness of the irradiated portion of the resist film. However, in most chemically amplified systems, undesired side reactions, that is, reactions with contaminants such as amines, water, ions, or reactive sites on the polymer, prematurely destroy the acid, (*27, 36, 37*). The most notable reactions occur at the resist–air interface.

Part-per-billion (ppb) levels of amines ubiquitous to device-processing facilities will efficiently neutralize the photogenerated acid at the surface of acid-catalyzed chemically amplified resist films (*36, 37*). For instance, an HMDS (hexamethyldisilazane) concentration of only 15 ppb reduces resist sensitivity by at least a factor of 2, severely degrades image quality, and generates substantial amounts of surface residue for a PTBSS-acid generator formulation (*27*). A decrease in HMDS concentration to <5 ppb leads to enhanced sensitivity, and images with straight wall profiles are obtained. However, even with such low contaminant concentrations, the time the resist resides in the atmosphere between exposure and PEB (called the postexposure delay or PED) can be critical; the quality of resist images is demonstrably affected by the PED time.

Other examples of air-borne basic contaminants include *N*-methylpyrrolidone (NMP, a solvent typically used to strip resist) and *N,N*-dimethylformamide (DMF, solvent commonly found in "weld-on" chemicals for poly(vinyl chloride) (PVC) water piping). Even materials such as fresh latex paints contain sufficient quantities of amines to have pronounced effects on resist performance. The disparity in time delays available for different PAG materials (ionic or nonionic) is probably a function of the inherent reactivity of the photogenerated acids with contaminants in the ambient environment.

Because organic acids are generally weaker, they allow more of a time delay between exposure and PEB than acids typically generated from onium salts. Furthermore, polymer structures can have a significant effect on diffusion of amines into the resist (*38*). Clearly, an understanding of fundamental structure–property relationships of various PAG molecules and host polymers is needed to effect the design of a radiation-sensitive, chemically amplified resist system that is insensitive to environmental issues.

Because ppb levels of basic vapors in the resist-processing environment can affect process performance profoundly, solutions to this problem are required (*27, 36*). Some alternatives include

- removal of the surface residue during development with developer additives that are capable of dissolving the resist surface uniformly
- addition of additives to the resist formulation that can migrate to the resist surface during processing to quench the basic contaminants in the air (*39*)
- performing resist processing in a tightly controlled environment that is devoid of any basic contaminants (*36*)
- isolation of the resist from the immediate processing environment by means of a thin cover coat (or overcoat) on the top surface of the resist (*27, 37*)

Some combination of these measures would, in fact, be expected to pro-vide wider process latitude and protection from atmospheric contaminants than any of the alternatives taken on its own. As depicted in Figure 5, high-resolution imaging is readily achieved with chemically amplified resist processes given an appropriate process environment.

Future Trends

As devices with features below 0.2 μm become a reality, process and con-tamination control become increasingly important issues. Dry processes have replaced many conventional, wet-chemical processes. Pattern etching is a good example. Plasma-assisted etching provides excellent control of critical dimensions.

A natural extension is to convert all the fabrication steps into vacu-um processes and integrate them into a continuous production line. Re-

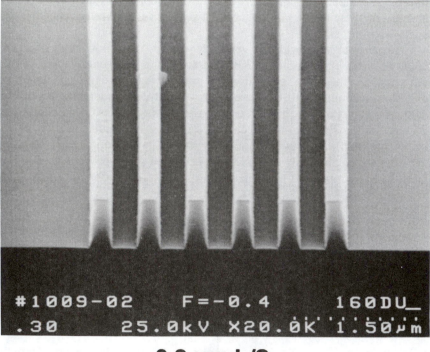

0.3 μm L/S

Figure 5. SEM micrograph depicting 0.3-μm image obtained in a deep-UV resist based on PTBSS–nitrobenzyl ester chemistry.

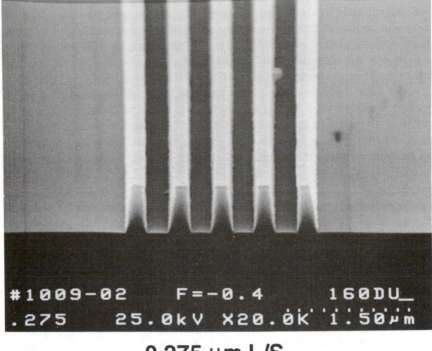

0.275 µm L/S

Figure 5. Continued. *SEM micrograph depicting 0.275-µm image.*

sist processes represent the most difficult group to convert because new materials and methods must be invented to replace existing ones. Several schemes that involve pattern formation in plasmas are currently under investigation. In addition, direct material growth, removal, and modification are all possible during exposure to patterned radiation. Totally dry lithography represents a considerable challenge, and much research remains to be done.

Summary

The unabated progress in design and integration of VLSI devices continues to demand increasingly smaller and more precise device features. Currently, almost all commercial devices are made by photolithography using UV radiation in the wavelength range of 365–436 nm. But within the next 5–8 years, new lithographic strategies will be required to meet resolution needs that will likely extend below 0.25 µm.

Technologies under development include electron-beam, X-ray, and short-wavelength lithographies. Electron-beam lithography is already an important technology in the manufacture of chromium masks and custom VLSI devices. Each of these alternative technologies will require new polymeric resist materials and processes. This chapter discussed one approach to resist design applicable to deep-UV lithography.

The future of microlithography contains many challenges in the areas of resist research and associated processing. The methods we described for the design and processing of the deep-UV resist are being applied to the other lithographic technologies. Within 10 years, many new materials will be commonplace in manufacture.

References

1. Sze, S. M. *VLSI Technology;* McGraw-Hill: New York, NY, 1983.
2. Powell, M. W. *Solid State Technol.* **1989,** *332*(3), 66.
3. Thompson, L. F.; Willson, C. G.; Bowden, M. J. *Introduction to Microlithography;* ACS Symposium Series 219; American Chemical Society: Washington, DC, 1983; pp 2–85.
4. Wolf, S.; Tauber, R. N. *Silicon Processing for the VLSI Era;* Lattice Press: Sunset Beach, CA, 1986.
5. McCoy, J. H.; Lee, W.; Varnell, G. L. *Solid State Technol.* **1989,** *32*(3), 87.
6. Pol, V.; Bennewitz, J. H.; Escher, G. L.; Feldman, M.; Firtion, V. A.; Jewell, T. E.; Wilcomb, B. E.; Clemens, J. T. *Proc. SPIE* **1986,** *633,* 6.
7. Pol, V. *Solid State Technol.* **1987,** *30*(1), 71.
8. Buckley, J. D.; Karatzas, C. *Proc. SPIE* **1989,** *1088,* 424.
9. Murphy, J. B. *Proc. SPIE* **1990,** *1263,* 116.
10. Peters, D. W. *Proc. SPIE* **1988,** *923,* 28.
11. Alles, D. S.; Biddick, C. J.; Bruning, J. H.; Clemens, J. T.; Collier, R. J.; Gere, E. A.; Harriott, L. R.; Lione, F.; Liu, R.; Mulrooney, T. J.; Nielsen, R. J. Paras, N.; Richman, R. M.; Rose, C. M.; Rosenfeld, D. P.; Smith, D. E. A.; Thompson, M. G. R. *J. Vac. Sci. Technol. B* **1987,** *5*(1), 47.
12. Reichmanis, E.; Thompson, L. F. *AT&T Technical Journal* **1990,** *Nov/Dec,* 32–45.
13. Reichmanis, E.; Thompson, L. F. *Chem. Rev.,* **1989,** *89*(6), 1273.
14. Willson, C. G. In *Introduction to Microlithography;* ACS Symposium Series 219, Thompson, L. F.; Willson, C. G.; Bowden, M. J., Eds.; American Chemical Society: Washington, DC, 1983; pp 88–159.
15. Houlihan, F. M.; Reichmanis, E.; Thompson, L. F.; Tarascon, R. G. In *Polymers in Microlithography;* Reichmanis, E.; MacDonald, S. A.; Iwayanagi, T., Eds.; ACS Symposium Series 412; American Chemical Society: Washington, DC, 1989; pp 39–58.
16. Ito, H.; Willson, C. G. In *Polymers in Electronics;* Davidson, J., Ed.; ACS Symposium Series 242: Washington, DC, 1984; pp 11–23.
17. Frechet, J. M. J.; Eichler, E.; Ito, H.; Willson, C. G. *Polymer,* **1980,** *24,* 995.
18. Ito, H.; Willson, C. G.; Frechet, J. M. J.; Farrall, M. J.; Eichler, E. *Macromolecules,* **1983,** *16,* 1510.

19. Ito, H.; Willson, C. G. *Polym. Eng. Sci.* **1983**, *23*, 1012.
20. Crivello, J. V.; Lee, J. L.; Coulon, D. A. *Makromol. Chem. Makromol. Symp.* **1988**, *1314*, 145.
21. Crivello, J. V.; Lam, J. H. W. *Macromolecules*, **1977**, *10*, 1307.
22. Crivello, J. V.; Lam, J. H. W. *J. Polym. Sci. Polym. Chem. Ed.* **1979**, *17*, 977.
23. Reichmanis, E.; Houlihan, F. M.; Nalamasu, O.; Neenan, T. X. *Chem. Mater.* **1991**, *3*, 394.
24. Crivello, J. V. In *Polymers in Electronics;* Davidson, J., Ed.; ACS Symposium Series 242; American Chemical Society: Washington, DC, 1984; pp 3–10.
25. McKean, D. R.; Schaedeli, U.; MacDonald, S. A. In *Polymers in Microelectronics;* Reichmanis, E.; MacDonald, S. A.; Iwayanagi, T., Eds.; ACS Symposium Series 412; American Chemical Society: Washington, DC, 1989; pp 27–38.
26. Tarascon, R. G.; Reichmanis, E.; Houlihan, F. M.; Shugard, A. *Polym. Eng. Sci.* **1989**, *29*(13), 850–855.
27. Nalamasu, O.; Reichmanis, E.; Cheng, M.; Pol, V.; Kometani, J. M.; Houlihan, F. M.; Neenan, T. X.; Bohrer, M. P.; Mixon, D. A.; Thompson, L. F. *Proc. SPIE,* **1991**, *1446*, 13–25.
28. Houlihan, F. M. Shugard, A.; Gooden, R.; Reichmanis, E.; *Macromolecules,* **1988**, *21*, 2001.
29. Neenan, T. X.; Houlihan, F. M.; Reichmanis, E.; Kometani, J. M.; Bachman, B. J.; Thompson, L. F. *Macromolecules,* **1990**, *23*, 145.
30. Bowden, M. J.; Chandross, E. A. *J. Electrochem. Soc.* **1975**, *122*, 1370.
31. Novembre, A. E.; Tai, W. W.; Kometani, J. M.; Hanson, J. E.; Nalamasu, O.; Taylor, G. N.; Reichmanis, E.; Thompson, L. F. *Chem. Mater.* **1992**, *4*, 278.
32. Houlihan, F. M.; Reichmanis, E.; Thompson, L. F.; Tarascon, R. G. In *Polymers in Microlithography;* Reichmanis, E.; MacDonald, S. A.; Iwayanagi, T., Eds.; ACS Symposium Series 412; American Chemical Society: Washington, DC, 1989; pp 39–56.
33. Schlegel, L.; Ueno, T.; Hagashi, N. Iwayanagi, T. *J. Vac. Sci. Technol. B* **1991**, *9*(2), 278–289.
34. Schlegel, L.; Ueno, T.; Hayashi, N.; Iwayanagi, T. *Jpn. J. Appl. Phys.* **1991**, *30*(11B), 3132.
35. Nakamura, J.; Ban, H.; Deguchi, K.; Tanaka, A. *Jpn. J. Appl. Phys.* **1991**, *30*(10), 2619.
36. MacDonald, S. A.; Clecak, N. J.; Wendt, H. R.; Willson, C. G.; Snyder, C. D.; Knors, C. J.; Deyoe, N. B.; Maltabes, J. G.; Morrow, J. R.; McQuire, H. E.; Holmes, S. J. *Proc. SPIE,* **1991**, *1446*, 2.
37. Nalamasu, O.; Reichmanis, E.; Hanson, J. E. Kanga, R. S.; Heimbrook, L. A.; Emerson, A. B.; Baiocchi, F. A.; Vaidya, S. *Polym. Eng. Sci.* **1992**, *32*(21), 1566.
38. Hinsberg, W. D.; MacDonald, S. A.; Clecak, N. J.; Snyder, C. D.; *Proc. SPIE* **1992**, *1672*, 24.
39. Chatterjee, S.; Jian, S.; Lu, P. H.; Khanna, D. N.; Potvin, R. E. McCaulley, J. A.; Rufolko, J. *Polym. Eng. Sci.* **1992**, 32.

RECEIVED for review November 9, 1992. ACCEPTED revised manuscript June 16, 1993.

High-Conductivity, Solid Polymeric Electrolytes

Michael S. Mendolia[1] and Gregory C. Farrington

Department of Materials Science and Engineering, University of Pennsylvania, Philadelphia, PA 19104

This review of ionically conductive polymers describes the various polymeric media developed to improve upon the characteristics of first-generation polymer electrolytes formed with poly(ethylene oxide) (PEO). Criteria for an ideal solid polymeric electrolyte are discussed, and several systems, including networks and comb-branch polymers, are surveyed. The focus of this work is on plasticized systems, electrolytes in which the highest ionic conductivities (~10^{-3} S/cm at room temperature) have been attained.

POLYETHERS CONTAINING DISSOLVED SALTS can be used as solid electrolytes (*1*). A considerable scientific effort (*2–5*) has been dedicated to exploring and understanding the characteristics of these electrolyte systems. Many investigations have focused on developing polymeric electrolytes with high ionic conductivities (10^{-3} S/cm or higher) at ambient temperature. The most obvious applications of ionically conductive polymers are in all-solid-state batteries, especially high energy density, rechargeable lithium batteries (*6*).

Several advantages result from replacing the aqueous electrolytes of conventional batteries with solid polymeric electrolytes. First, polyethers have good chemical stability and are generally unreactive with lithium, which is a very attractive battery anode because of its high reducing power

---

[1]Current address: Colgate–Palmolive Company, Piscataway, NJ 08855

0065–2393/95/0245–0107$13.00/0

and low equivalent weight. Second, solid polymeric electrolytes can be used as thin layers; thus, the battery volume and weight devoted to the electrolyte are decreased. In addition, their mechanical properties are very attractive: Polymer electrolytes are deformable, flexible, and easily fabricated by conventional manufacturing processes. The typical polymer-based battery is a layered thin-film structure in which a thin polymer electrolyte ($\sim$20–50 mm thick) is sandwiched between a lithium anode and an intercalation cathode (e.g., V_6O_{13}) (Figure 1). The entire cell can be produced as a continuous tape and can be rolled or folded into its finished shape.

Initial investigations of ionically conductive polymers were principally focused on poly(ethylene oxide) (PEO). The strong interaction between high-molecular-weight PEO and inorganic salts was first reported more than 20 years ago (7, 8), and it is now known (3) that PEO can solvate a wide variety of salts, even at very high salt concentrations. The solvation of salts occurs through the association of the metallic cations with the oxygen atoms in the polyether backbone (Figure 2). The solvated ions can be

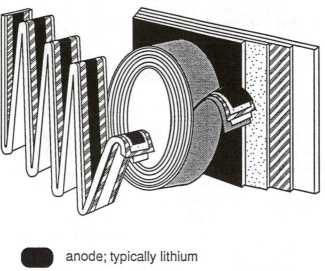

 anode; typically lithium

solid polymeric electrolyte

intercalation cathode

current collector

Figure 1. Schematic of polymer-based batteries. The thin-film composites are very flexible and can be folded or rolled into various geometries.

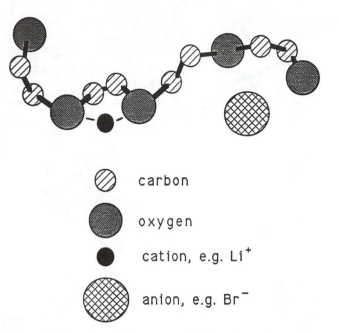

carbon

oxygen

cation, e.g. Li^+

anion, e.g. Br^-

Figure 2. Solvation of a salt by a PEO chain. The primary interaction is between the ether oxygens and the cation.

quite mobile in the polymeric solvent, and thus give rise to significant bulk ionic conductivities.

Pure PEO is a semicrystalline polymer, possessing both an amorphous and a crystalline phase at room temperature. Significant ionic transport occurs only within the amorphous phase (*3, 9*). This feature explains the dramatic decrease in ionic conductivity seen in many PEO-based systems (*10*) for temperatures below the melting point of pure crystalline PEO (T_m ~66 °C) (Figure 3); the crystalline PEO regions are nonconductive and serve to hinder bulk ionic transport. Clearly, the inherent crystallinity of PEO is not very attractive for applications in solid electrolytes.

For several completely amorphous polyether network systems (*11–13*), the temperature dependence of the conductivity can be closely modeled by the Williams–Landel–Ferry (WLF) equation, and free volume theory has been extremely useful in describing the behavior of these salt–polyether systems. Many researchers (*2, 3*) prefer to model their data with the semi-empirical Vogel–Tamman–Fulcher (VTF) equation, which, although mathematically equivalent to the WLF equation, is derived differently. An excellent discussion of the various theoretical models used to describe polymer electrolytes is given in an overview by Ratner (*14*).

In any case, the mobility of the ions in polymer electrolytes is linked

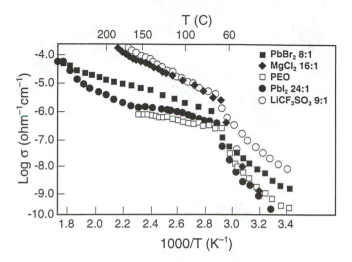

Figure 3. Conductivities of various PEO-based electrolytes containing divalent cations. The composition of each is indicated as a molar ratio of polymer repeat units to salt. The knee in the curves at around 60 °C is attributed to the melting of crystalline PEO. (Reproduced with permission from reference 10. Copyright 1989.)

to the local segmental mobility of the polymer chains. Significant ionic conductivity in these systems will occur only well above the glass-transition temperature (T_g) of the amorphous phase; therefore, one of the requirements for the polymeric solvent is a low glass-transition temperature (e.g., the T_g for PEO is around −67 °C). At room temperature, then, while the polymeric electrolyte is macroscopically solid, there is liquid-like mobility within the amorphous phase on a molecular scale, and this condition allows rapid ion transport. This union of solid-like and liquid-like properties is what makes polymer electrolytes so fascinating technologically and scientifically.

One problematic feature of polymer electrolytes arises from the opposing effects of increasing the salt concentration: Higher salt contents generally imply higher numbers of charge carriers, which should result in increased ionic conductivity. However, through their interaction with the polyether chains, inorganic salts increase the T_g of the system as their concentration is increased. Such effects can be formidable; James et al. (*15*) found that divalent metal chlorides could increase the T_g of poly(propylene oxide) by more than 100 °C (Figure 4). This increase in the T_g of the system tends to lower the mobility of the ions and thus cause decreased ionic conductivity. The balance of these two effects leads to a maximum in the conductivity at a specific salt concentration, which varies from system to system.

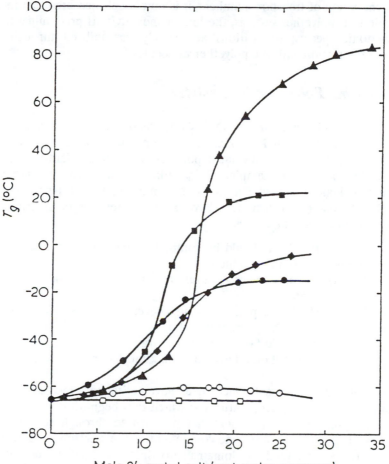

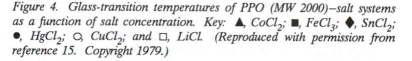

Figure 4. Glass-transition temperatures of PPO (MW 2000)–salt systems as a function of salt concentration. Key: ▲, $CoCl_2$; ■, $FeCl_3$; ◆, $SnCl_2$; ●, $HgCl_2$; ○, $CuCl_2$; *and* □, *LiCl. (Reproduced with permission from reference 15. Copyright 1979.)*

The dielectric constant of high-molecular-weight polyethers is quite low (probably around 5–10; compare 78.5 for water at room temperature), and extensive ion-pairing is expected in such media. The existence of significant ion-pairing has been deduced for several salt–polyether systems through many techniques, including Raman spectroscopy (*16–18*), infrared spectroscopy (*19*), and analysis of conductivity data (*20, 21*). In our laboratory, UV–visible spectroscopic studies of transition metal halides in low-molecular-weight poly(ethylene glycol) (*22, 23*) indicated that complex ionic species are prevalent. Although some controversy surrounds the determination of transport numbers, it is now widely accepted that in most simple salt–polyether systems, the ionic conductivity is predominantly due to anionic species; the cations are largely immobilized through their strong interaction with the polyether molecules.

Optimizing Polymeric Electrolytes

The ionic conductivity of a PEO–LiClO$_4$ electrolyte at room temperature is only on the order of 10^{-6} S/cm (2). The previous discussion shows that unmodified PEO is not the ideal polymer host for ambient-temperature solid electrolytes; for example, its crystalline nature alone is a great hindrance to ionic transport. In attempts to improve on the performance of PEO–salt systems, researchers have sought alternate polymeric media based on the following criteria:

1. The polymer host should be completely amorphous because crystalline phases are nonconductive.
2. A polymer with a lower T_g should result in greater ionic mobilities at room temperature.
3. The system should promote a high level of salt dissociation so that the number of charge carriers is large.
4. The system should facilitate cationic transport.
5. The electrolyte should have good bulk mechanical properties.

In general, PEO–salt electrolytes have satisfactory mechanical properties, except at high temperatures at which they begin to flow. Good mechanical properties in polymers can be ensured through several mechanisms: by keeping at least one component of the electrolyte below its T_g, by incorporating crystalline domains into the system, by the use of inert fillers, or by cross-linking the polymer. Clearly, methods of improving mechanical properties sometimes do so at the expense of ionic conductivity.

As a final note in this discussion of desirable characteristics of polymer electrolytes, we must stress the importance of stability toward the electrode materials; as Scrosati (5) pointed out, even the most conductive electrolyte is of no practical use if not compatible with useful electrodes.

We have grouped the various polymeric electrolyte systems that have been investigated over the years into seven categories: (1) various simple homopolymers, (2) networks, (3) copolymers and comb-branch polymers, (4) polyelectrolytes, (5) polymer blends, (6) polymers with inert fillers, and (7) plasticized polymers. In each category, some improvement over simple PEO–salt systems has been attained, and the major accomplishments of each category are briefly reviewed here. An enormous body of scientific literature on polymeric electrolytes exists, and this discussion is not meant to be exhaustive. We pinpoint only a few particularly notable or interesting investigations in each category and hope this account will give a flavor of the research done in this field.

Our discussion of plasticized systems will be much more thorough, partly because these systems have been most successful in achieving significant conductivities at room temperature and partly because considerable effort at the University of Pennsylvania has been devoted to investigating materials of this type. For reference, several of the chemical formulas of the most important polymers are given in Table I.

Categories of Polymeric Electrolytes

Simple Homopolymers. The ability of PEO to dissolve inorganic salts is not unique among polymers. For example, certain nitrate salts are quite soluble in poly(vinyl acetate) (T_g ~30 °C), poly(vinyl alcohol) (T_g ~99 °C), poly(methyl methacrylate) (T_g ~105 °C), and poly(methyl acrylate) (T_g ~5 °C) (*24*). However, these polymer hosts, which have T_gs near or above room temperature, would not be expected to produce particularly conductive electrolytes.

A number of polyethers of low T_g have been investigated as polymeric solvents. Curiously, poly(methylene oxide) (repeat unit CH_2O) and poly(trimethylene oxide) ($CH_2CH_2CH_2O$) do not dissolve ionic salts (*1*), despite their close similarity with PEO (CH_2CH_2O). Salts do dissolve in low-molecular-weight poly(tetramethylene glycol) ($CH_2CH_2CH_2CH_2O$), but the resulting electrolytes are very poor conductors (*23, 25*). Our spectroscopic studies of $CoBr_2$ dissolved in poly(tetramethylene glycol) (*23*) indicated that neutral species predominate in this solvent; although salts dissolve, they do not produce a high concentration of charge carriers. These results and many others clearly indicate that the backbone repeat unit found in PEO, $-C-C-O-$, is an exceptionally favorable geometry for the solvation of salts.

Another related polyether, poly(propylene oxide) or PPO ($-CH_2CH(CH_3)O-$), retains the $C-C-O$ backbone and does indeed complex salts. This polymer has the advantage that it is completely amorphous at all temperatures; unfortunately, as a result of the steric hindrance of the methyl groups, it is less effective in dissolving salts than PEO (*4*), and, in general, PPO-based systems have lower conductivities than amorphous PEO–network systems (*26*). In spite of these drawbacks, PPO–salt systems are frequently investigated in fundamental studies (*18, 26–29*), primarily because working with a single-phase system is simpler.

Networks. Several advantages are achieved in using cross-linked polyethers, or polyether "networks", as media for ionic conduction (*13,*

Table I. Chemical Formulas of Selected Polymers

Polymer	Repeat Unit
Poly(ethylene oxide) (PEO)	$-CH_2CH_2O-$

Poly(propylene oxide) (PPO)	$-CHCH_2O-$
	    |
	   CH_3

Poly(tetramethylene oxide) (PTMO)	$-CH_2CH_2CH_2CH_2O-$
Poly[bis(methoxyethoxyethoxide)- phosphazene] (MEEP)	$O(CH_2CH_2O)_2CH_3$ | $-N=P-$ | $O(CH_2CH_2O)_2CH_3$
Poly(methyl methacrylate) (PMMA)	CH_3 | $-CCH_2$ | $C=O$ O CH_3
Poly(vinylene carbonate)	

Polyacrylonitrile (PAN)	$-CH_2CH-$
	    |
	   $C{\equiv}N$

Poly(vinylidene fluoride) (PVdF)	$-CH_2CF_2$
Poly(dimethylsiloxane)	CH_3 | $-SiO-$ | CH_3

30–33). First, the systems have good mechanical properties and are not prone to creep. In addition, sufficient cross-linking reduces the crystallinity of the polymer and thus removes one impediment to high conductivity. In PEO networks, if the molecular weight between cross-links exceeds 1000, crystalline domains (with melting points typically ~50-60 °C) are still found at room temperature, although the addition of salt can completely suppress the crystallinity (*34*).

Too much cross-linking has deleterious effects; as the cross-link density increases, so does the T_g. The ionic conductivity decreases as the cross-link density increases and as the rigidity of the cross-linking group increases (*35, 36*); thus the ideal network should have dilute, flexible cross-links. PEO-based network electrolytes can generally attain conductivities of the order 10^{-5} S/cm at room temperature (*31, 32, 35*), although a PEO-based network incorporating $LiClO_4$ was reported (*37*) to reach a room-temperature conductivity of 1.3×10^{-4} S/cm. As noted earlier, the conductivities of PPO-based networks at room temperature are much lower (*26*). PEO-based networks are certainly superior to simple linear PEO electrolytes, but their conductivities are still too low for effective ambient-temperature use.

Copolymers and Comb-Branch Polymers. The repeat unit of PEO seems to be optimal for salt solvation, but the crystallinity inherent to linear PEO is highly undesirable. One method of reducing crystallinity in PEO-based systems is to synthesize polymers in which the lengths of the oxyethylene sequences are relatively short, such as by incorporating oxyethylene sequences into the backbone of linear copolymers or as side-groups in comb-branch polymers (*38*).

The most notable linear copolymer of this type is oxymethylene-linked poly(oxyethylene), or, as it is more commonly known, amorphous poly(ethylene oxide) (aPEO) (*38–40*). The general formula of its repeat unit is $\{-(OCH_2CH_2)_x-OCH_2-\}_y$, where x is typically ~8; the $-OCH_2-$ segments in the backbone serve to disrupt the stereoregularity of the chains and depress the melting temperature of the crystalline phase below room temperature, while not greatly affecting the T_g of the polymer (the T_g of aPEO is ~−65 °C). As such, aPEO and its solutions of inorganic salts are completely amorphous at room temperature. Electrolytes based on aPEO generally have ambient-temperature conductivities around 10^{-4} S/cm (*39, 40*), an improvement of about 2 orders of magnitude over linear PEO-based electrolytes.

Considerable research has been conducted to develop polymers with very flexible backbones and pendant ethylene oxide branches. If the branches are kept short enough, their crystallization is suppressed. In this way, the low T_g of the parent polymer can be combined with the salt-solvating ability of PEO in a completely amorphous system. Several of these comb-branch systems have been investigated (41), but those based on polysiloxanes and polyphosphazenes are most well-documented.

Simple polysiloxanes generally have extremely low glass-transition temperatures (poly(dimethylsiloxane) has a T_g of ~−123 °C), but unfortunately, they do not dissolve salts (4) because the donor properties of the oxygen atoms in the polymer backbone are suppressed and because spacing between oxygen atoms is unfavorable for salt complexation. However, if oligomeric ethylene oxide side chains are anchored onto a polysiloxane backbone (42–44), salt solvation occurs. The highest conductivities of these systems involving lithium salts are around 10^{-4} S/cm at room temperature.

Comb-branch polymers based on polyphosphazenes have also been studied. The advantages of polyphosphazenes include low T_gs, possibilities for easy chemical modification, and good chemical stability. The most successful polymer of this type is poly[bis(methoxyethoxyethoxide)phosphazene] (MEEP) (45–48). MEEP-based electrolytes generally reach conductivities around 10^{-5} S/cm near room temperature.

Polyelectrolytes. As noted earlier, in many polymer electrolytes, the ionic conductivity is established mainly through anionic transport. In an attempt to increase the fraction of charge carried by cations, some researchers have synthesized polymers in which anions are covalently attached to the polymer backbone. These polymers are called "polyelectrolytes". Because the anions are effectively immobilized, all ionic conductivity is due to cationic transport. Some of the most successful polyelectrolyte systems have involved polyphosphazene backbones with both ethylene oxide and anionic sulfonate groups attached covalently as side chains (49, 50). These systems exhibit room-temperature conductivities around 10^{-6} S/cm. Other polyelectrolyte systems (51–54) are often worse, rarely reaching conductivities above a dismal 10^{-7} S/cm at ambient temperature. The low conductivities of polyelectrolytes can be attributed to immobilization of the cations by extensive tight ion-pairing with the bound anions.

Polymer Blends. In some investigations, an inorganic salt has been dissolved in a blend of PEO and another polymer, where the second polymer generally has a high T_g. This mixing is done mainly to improve mechanical properties or to reduce crystallinity, but in some cases, the two polymers interact such that the ionic conductivity is significantly improved.

Tsuchida et al. (55) dissolved $LiClO_4$ in a blend of PEO and poly(methacrylic acid) (PMAA). The system was much more rigid physically, but the presence of the PMAA greatly hindered ionic conduction (resulting in conductivities around 10^{-9} S/cm). Another study (56) discussed the blending of a poly(methyl methacrylate-co-methacrylic acid) copolymer

and then began to reach a limiting value around 10^{-4} S/cm at room temperature. For ratios higher than 2, the conductivity was not especially dependent on the nature of the plasticizer. However, at lower plasticizer–salt fractions, the effect of the plasticizer on the conductivity enhancement at constant plasticizer–salt was the following: DMF > EC > PC. This result is interesting because, although DMF has the lowest viscosity of the three, it does not have the highest dielectric constant (Table II). This observation hints that the mobility of the ions rather than their concentration determines conductivity in these systems. Because the conductivity did not correlate well with the PAN content, it was concluded that the PAN host was not active in the ionic transport mechanism, but was instead simply a source of structural stability. The plasticizers both dissociate the salt and provide a low viscosity medium through which the ions can move.

Plasticized poly(vinylidene fluoride) (PVdF) systems yield similar results (*81, 82*). In systems with 30 mol% $LiClO_4$, plasticizers increased the conductivity in the following order: DMF > γ-butyrolactone > EC > PC > poly(ethylene glycol), MW 400 > poly(propylene glycol), MW 1000 (Figure 5). Again, the viscosity rather than the dielectric constant of the plasticizer seems to be the controlling parameter (Table II); although the plasticizers aid in salt dissociation, their primary role in enhancing the conductivity seems to be the increasing of ionic mobility. Low-molecular-weight polyethers with low dielectric constants and high viscosities are the least effective in increasing the conductivity. Finally, although the addition of DMF produces the most highly conductive systems, DMF is reactive with lithium electrodes and thus is not a suitable plasticizer for battery applications.

Yang (*83*) studied the effects of the PEG, poly(ethylene glycol) dimethyl ether (PEGDE), and EC plasticizers on the $ZnBr_2$–PEO system. For systems of stoichiometry $ZnBr_2(0.5PEO + 0.5A)_{16}$ where PEO is a poly(ethylene oxide) repeat unit and A is a plasticizer molecule, the T_g decreased in the following order with the variation of the plasticizer A: no plasticizer (A = PEO) (-16 °C) > PEG (-24 °C) > PEGDE (-47 °C) > EC (-62 °C). The conductivity followed the opposite trend: EC > PEGDE > PEG > no plasticizer. The EC-plasticized system reached a conductivity of 10^{-4} S/cm at room temperature, about 5 orders of magnitude higher than unplasticized $ZnBr_2(PEO)_{16}$. This dramatic effect on the conductivity is not surprising: In this extreme case, the molar content of the plasticizer is quite substantial (50%). The large decrease in the T_g of the system due to the EC (almost 50 °C) reflects a large gain in free volume and segmental mobility, facilitating ion transport. Yang concluded that the plasticizers' primary effect is to increase ionic mobility. Her hypothesis is supported by the observation that PEGDE enhances the conductivity more than PEG; PEGDE, less viscous and possessing a lower T_g

Table II. Properties of Selected Plasticizers at Room Temperature

Plasticizer	Chemical Formula	Dielectric Constant	Viscosity θ (cP)
γ-Butyrolactone		39.1	1.75
Dichloromethane	CH_2Cl_2	9.1	0.45
N,N-Dimethylformamide (DMF)	$HC=ON(CH_3)_2$	36.7	0.80
Ethylene carbonate (EC)		89.0^a	1.90^a
Poly(ethylene glycol), MW 400 (PEG)	$HO-(CH_2CH_2O)_n-H$	$-^b$	95.92
Poly(ethylene glycol) dimethyl ether, MW 400 (PEGDE)	$CH_3O(CH_2CH_2O)_nCH_3$	$-^b$	40.09
Poly(propylene glycol), MW 1000 (PPG)	$HO-(\underset{\underset{CH_3}{\mid}}{C}CH_2O)_n-H$	$-^b$	171.13
Propylene carbonate (PC)		64.4	2.51
Tetrahydrofuran (THF)		7.4	0.46
Water	H_2O	78.5	1.0

^{a}Measured at 40 °C (above the melting point of the substance).
^{b}The dielectric constants of oligomeric polyethers are thought to be in the range of 5–10.

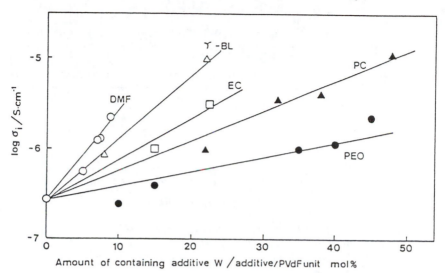

Figure 5. Effect of plasticizers on the ionic conductivity of poly(vinylidene fluoride)–LiClO$_4$ systems. The LiClO$_4$ content was 30 mol% in all cases. γ-butyrolactone is γ-BL. (Reproduced with permission from reference 82. Copyright 1983.)

than PEG, would certainly impart more free volume to the system. On the other hand, the PEGDE lacks the hydroxy end groups and the excellent salt-solvating ability of PEG, and thus is less likely to improve the salt dissociation.

From this discussion, one might conclude that the dielectric constant of the plasticizer is relatively unimportant and that the most effective plasticizers are those that decrease the microscopic viscosity of the media the most. However, Yang's results are a bit ambiguous because the most effective plasticizer, EC, had the lowest viscosity and the highest dielectric constant. To investigate this issue further, we studied CoBr$_2$–PPG systems plasticized with PC and dichloromethane (*84*). We chose PPG, MW 2000, as the polymeric solvent because it is completely amorphous and thus simplifies the determination of the T_g. CoBr$_2$ was interesting mainly because its solvation can be probed rather effectively using UV–visible spectroscopy (*23*). Finally, the plasticizers PC and dichloromethane are interesting to pair because PC possesses both a higher viscosity and a higher dielectric constant than dichloromethane (Table II).

A 0.3 molal concentration of CoBr$_2$ raised the T_g of PPG from −72 to −66 °C. In the plasticized systems, dichloromethane reduced the T_g more effectively than PC (Figure 6), but the conductivities were enhanced more by PC than by dichloromethane (Figure 7) at 30 °C. For this case, quite clearly, the plasticizer does more than simply decrease the system's micro-

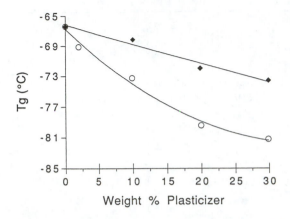

Figure 6. Glass-transition temperatures of PPG (MW 2000)–CoBr₂– plasticizer systems as a function of plasticizer content. The concentration of $CoBr_2$ was 0.3 molal in each case. The plasticizers were dichloromethane (○) and propylene carbonate (◆).

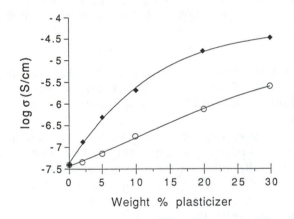

Figure 7. Conductivities of PPG (MW 2000)–CoBr₂–plasticizer systems as a function of plasticizer content. The concentration of $CoBr_2$ was 0.3 molal in each case. The plasticizers were dichloromethane (○) and propylene carbonate (◆).

scopic viscosity. The enhanced conductivities with PC must be attributed to the larger dielectric constant of PC and its consequent ability to create more charge carriers.

This speculation regarding the role of the PC molecules is consistent with recent computer simulations (*85*) of PEO–LiBr–PC systems. In these studies, the presence of PC (in this case, five PC molecules per 25 PEO repeat units) enhances salt dissociation. However, the simulations do not indicate strong interactions between cations and PC molecules; the Li^+ cations still associate predominantly with polyether oxygens. The Br^- anions become less associated with the cations and attain higher mobilities in PC-rich regions. Such computer simulations should not be blindly trusted, but these studies are certainly intriguing and thought-provoking.

Fundamental studies by Cameron et al. (*25*) also support our conclusion that one of the important functions of a plasticizer in polymer electrolytes is to increase the number of charge carriers. For both poly(tetramethylene glycol) and an ethylene oxide–propylene oxide copolymer as macromolecular solvents for $LiClO_4$, the plasticizer PC increased the conductivity much more than tetrahydrofuran (THF) despite THF's more-pronounced reduction of the T_g of the system; it was concluded that the higher dielectric constant of PC promoted ion dissociation. The highest conductivity enhancements were seen at high ratios of plasticizer to salt.

These investigations indicate that plasticizers in polymer electrolytes enhance ionic conductivity both by increasing ion mobility and by increasing salt dissociation, although in different systems, either of these functions may be the more significant. For example, in certain cases, the plasticizer can lower the T_g substantially. In conventional polymer–plasticizer systems, such as polystyrene–benzene (*86, 87*), the decrease in T_g can be greater than 100 °C; the extent of the plasticization by benzene is, in part, due to a high T_g value of the pure polymer (the T_g of polystyrene is ~80 °C). On the other hand, in polymer electrolytes in which the value of the T_g is low at the start, the reduction of the T_g by additives can be much less pronounced (*see* Figure 6, for example).

This reasoning may explain the importance of the plasticizer viscosity in some of the electrolytes previously discussed. In considering the plasticization of PAN-based electrolytes (*79, 80*), it may be important to recall that the T_g of pure PAN is 85 °C, a very high value. Similarly, pure poly(vinylidene fluoride) has a T_g of −39 °C and, with 30 mol% $LiClO_4$, the T_g would certainly be elevated significantly, though its value is not reported in the plasticization studies of $PVdF–LiClO_4$ (*81, 82*). Finally, Yang's studies (*83*) show that $ZnBr_2(PEO)_{16}$ has a T_g roughly 50 °C higher than pure PEO. In these electrolytes, which all have moderately high unplasticized glass-transition temperatures, it is reasonable that the dominant route through which the additives enhance conductivity is by increasing ionic mobility. On the other hand, in systems of relatively low T_g, and especially at low salt content, the ability of a plasticizer to induce salt dissociation becomes more important. This condition would be especially true of polymers known to exhibit significant ion-pairing, such as PPO and poly(tetramethylene oxide).

The additive PC, with its ability both to increase salt dissociation and to improve ionic mobility, has been used successfully in several investigations of polymer electrolytes. Bohnke et al. (*88*) reported a conductivity of 2.3 $\times$ 10^{-3} S/cm at 25 °C for a system incorporating 20 wt% PMMA into a LiClO$_4$–PC solution. The presence of the high-molecular-weight PMMA imparted a very high macroscopic viscosity ($\sim$335 Pa s) to the system, while not diminishing the conductivity much. It was concluded that the PMMA acts primarily as a stiffener, and that fast ion transport occurs through a continuous conduction path of PC molecules.

Xia and Smid (34, 36) achieved conductivities of 2 $\times$ 10^{-3} S/cm at 25 °C with cross-linked poly(ethylene glycol) networks containing LiClO$_4$ and 30 wt% PC. The network structures ensured good mechanical properties. Similar results were reported by Huq et al. (89) for a radiation-cross-linked ethylene oxide–acrylate network containing LiClO$_4$ and PC. In a later study, Huq at al. (90) showed that mixtures of PC and EC led to better overall properties. In a system comprising a radiation-cross-linked ethylene oxide–acrylate network with 1 M LiAsF$_6$ and with a 1:1 weight ratio of EC–PC, an ambient conductivity of 8.4 $\times$ 10^{-4} S/cm was attained. This electrolyte had good mechanical properties and was fully amorphous from −90 to 150 °C with a T_g $\sim$−94 °C. It was completely transparent and rubbery. Systems plasticized with only PC had higher conductivities, but the EC–PC mixture imparted better mechanical and thermal stability and better lithium cycling characteristics. Another report (91) also described excellent results using mixed plasticizers; the solid polymeric electrolyte LiClO$_4$(PEO)$_{10}$(PC)$_5$(EC)$_5$ exhibited a conductivity of 2 $\times$ 10^{-3} S/cm at room temperature, and had good electrochemical stability.

Alamgir et al. (92) discussed a polymer electrolyte that couples a polyacrylonitrile (PAN) host with mixed plasticizers. A LiClO$_4$–PAN–EC–PC electrolyte (8, 21, 38, and 33 mol%, respectively) possessed a conductivity of 1.7 $\times$ 10^{-3} S/cm at 20 °C as well as good stability with electrodes and good discharge performance in an electrochemical cell. In this completely amorphous system, the PAN contributes minimally to the ionic transport and acts mainly as a support matrix. Compared with the PAN-based plasticized systems of Watanabe et al. (80), the electrolytes described by Alamgir et al. (92) incorporate much lower molar contents of PAN.

The extremely high contents of "plasticizers" used in many polymer electrolytes raises the question of where the boundary between plasticized polymers and solvent-swollen gels lies. In several of the systems described, the polymer serves mainly to provide a means of immobilizing the low-molecular-weight solvent on a macroscopic scale. Thus, we conclude this discussion with two investigations discussing systems unabashedly labeled "gel electrolytes". The first of these (93) involves a network formed by cross-linking 2-acrylamido-2-methyl-1-propanesulfonic acid (AMPS). This network is swollen with the solvent dimethyl sulfoxide (DMSO) containing a lithium salt. The resulting systems have acceptable ambient-temperature conductivities ($\sim$10^{-3} S/cm) but limited electrochemical stability; passivation layers developed at interfaces between the electrolytes and lithium and vanadium oxide electrodes.

Another lithium gel electrolyte (*94*) was prepared from poly(*N,N*-dimethylacrylamide) (PDMA), lithium salts, and the solvent *N,N*-dimethylacetamide (DMAc). To retain mechanical stability at high solvent contents, the electrolyte was cross-linked. In this system, the plasticization due to DMAc is appreciable; the addition of 15% DMAc lowers the T_g roughly 100 °C. These systems achieved ambient-temperature conductivities higher than 10^{-3} S/cm, but the electrochemical stabilities were not reported.

In summary, the addition of low-molecular-weight plasticizers to polymer electrolytes can result in highly conductive systems. The ideal plasticizer should possess the following qualities: (1) miscibility with the polymer-salt system, (2) low viscosity, (3) high dielectric constant, (4) good electrode stability, and (5) low volatility. Clearly, high plasticizer contents will have a dramatic effect on the physical, chemical, and electrochemical properties of the system. However, not all of these effects are desirable; high levels of plasticizers generally soften polymer systems and require the use of cross-linking to regain mechanical stability. Of more concern, low-molecular-weight species tend to be more reactive with battery electrodes and can encourage unwanted reactions at electrode interfaces.

Conclusions

Since the first studies of PEO–LiClO$_4$ electrolytes, considerable advances have been made in the development of polymer electrolytes for use at ambient temperatures in lithium batteries. The ionic conductivities of some of the more notable systems are compared in Figure 8. Indeed, the best systems begin to approach the conductivities of conventional nonaqueous liquid electrolytes; an 80 mol% EC, 20 mol% PC, 1.0 M LiClO$_4$ liquid electrolyte has a conductivity of 8.6×10^{-3} S/cm at room temperature (*95*), only about 4 times higher than that attained by some polymer electrolytes. The most conductive polymer electrolytes are those that incorporate low-molecular-weight plasticizers having low viscosities and high dielectric constants. The properties of many systems have been further improved through the use of plasticizer mixtures (such as EC–PC).

Regrettably, plasticizers are not without their own drawbacks: They tend to be volatile, and at high contents they can diminish the mechanical and electrochemical stability of polymer electrolytes. Some of the plasticized systems perform well enough to be used in primary lithium batteries and secondary batteries with limited cycle life. However, the dream of achieving sufficient cycle life (1000 cycles at deep discharge) so that a plas-

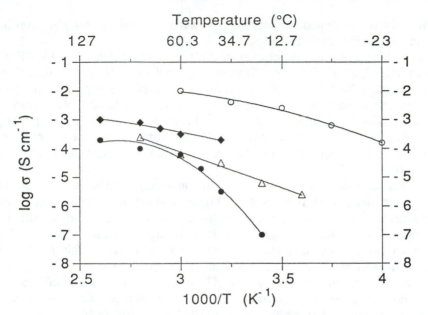

Figure 8. Conductivities of various polymer electrolytes as a function of temperature. Key: ●, *homopolymer:* PEO–LiClO$_4$ *(data from reference 2);* △, *network: cross-linked* PEO–LiClO$_4$ *(data from reference 32);* ◆, *blend: polycrown–poly(vinylene carbonate)–LiCF$_3$SO$_3$ (data from reference 64); and* ○, *plasticized system:* PEO–LiClO$_4$–*propylene carbonate–ethylene carbonate (data from reference 91).*

ticized polymer electrolyte battery can power, for example, an electric car will surely require further improvements in electrolyte stability, particularly at the lithium anode–electrolyte interface.

Acknowledgments

This work was supported by the Defense Advanced Research Projects Agency through a subcontract monitored by the Office of Naval Research. M. S. Mendolia also thanks the Office of Naval Research for their generous graduate fellowship. Additional support from the National Science Foundation, Materials Research Laboratory program, under Grant DMR–8819885, is gratefully acknowledged.

References

1. Armand, M. B.; Chabagno, J. M.; Duclot, M. J. In *Fast Ion Transport in Solids;* Vashista, P.; Mundy, J. N.; Shenoy, G. K., Eds.; Elsevier: New York, 1979; p 131.

2. Armand, M. B. In *Polymer Electrolyte Reviews 1;* MacCallum, J. R.; Vincent, C. A., Eds.; Elsevier: New York, 1987; p 1.
3. Vincent, C. *Prog. Solid St. Chem.* **1987**, *17*, 145.
4. Cowie, J. M. G.; Cree, S. H. *Annu. Rev. Phys. Chem.* **1989**, *40*, 85.
5. Scrosati, B. *Mater. Sci. Eng.* **1992**, *I12*, 369.
6. Gauthier, M.; Belanger, A.; Kapfer, B.; Vassort, G.; Armand, M. In *Polymer Electrolyte Reviews 2;* MacCallum, J. R.; Vincent, C. A., Eds.; Elsevier: New York, 1989; p 285.
7. Lundberg, R. D.; Bailey, F. E.; Callard, R. W. *J. Polym. Sci. Part A1* **1966**, *4*, 1563.
8. Fenton, D. E.; Parker, J. M.; Wright, P. V. *Polymer* **1973**, *14*, 589.
9. Berthier, C.; Gorecki, W.; Minier, M.; Armand, M.; Chabagno, J.; Rigaud, P. *Solid State Ionics* **1983**, *11*, 91.
10. Farrington, G. C.; Linford, R. G. In *Polymer Electrolyte Reviews 2;* MacCallum, J. R.; Vincent, C. A., Eds.; Elsevier: New York, 1989; p 255.
11. Killis, A.; Le Nest, J.; Cheradame, H.; Gandini, A. *Makromol. Chem.* **1982**, *183*, 2835.
12. Watanabe, M.; Ogata, N. *Br. Polym. J.* **1988**, *20*, 181.
13. Le Nest, J.; Gandini, A.; Cheradame, H. *Br. Polym. J.* **1988**, *20*, 253.
14. Ratner, M. A. In *Polymer Electrolyte Reviews 1;* MacCallum, J. R.; Vincent, C. A., Eds.; Elsevier: New York, 1987; p 173.
15. James, D. B.; Wetton, R. E.; Brown, D. S. *Polymer* **1979**, *20*, 187.
16. Schantz, S.; Sandahl, J.; Borjesson, L.; Torell, L. M. *Solid State Ionics* **1988**, *28–30*, 1047.
17. Schantz, S. *J. Chem. Phys.* **1991**, *94*, 6296.
18. Schantz, S.; Torell, L. M.; Stevens, J. R. *J. Chem. Phys.* **1991**, *94*, 6862.
19. Wendsjo, A.; Lindgren, J.; Paluszkiewicz, C. *Electrochimica Acta* **1992**, *37*, 1689.
20. MacCallum, J. R.; Tomlin, A. S.; Vincent, C. A. *Eur. Polym. J.* **1986**, *22*, 787.
21. Bruce, P. G.; Gray, F. M.; Shi, J.; Vincent, C. A. *Philos. Mag. A* **1991**, *64*, 1091.
22. Cai, H.; Farrington, G. C. *J. Electrochem. Soc.* **1992**, *139*, 744.
23. Mendolia, M. S.; Farrington, G. C. *Electrochimica Acta* **1992**, *37*, 1695.
24. Wissbrun, K. F.; Hannon, M. J. *J. Polym. Sci. Polym. Phys. Ed.* **1975**, *13*, 223.
25. Cameron, G. G.; Ingram, M. D.; Sarmouk, K. *Eur. Polym. J.* **1990**, *26*, 1097.
26. Watanabe, M.; Ogata, N. In *Polymer Electrolyte Reviews 1;* MacCallum, J. R.; Vincent, C. A., Eds.; Elsevier: New York, 1987; p 39.
27. Wintersgill, M. C.; Fontanella, J. J.; Greenbaum, S. G.; Adamic, K. J. *Br. Polym. J.* **1988**, *20*, 195.
28. McLin, M. G.; Angell, C. A. *J. Phys. Chem.* **1991**, *95*, 9464.
29. Frech, R.; Manning, J. P. *Electrochimica Acta* **1992**, *37*, 1499.
30. Cheradame, H.; Le Nest, J. F. In *Polymer Electrolyte Reviews 1;* MacCallum, J. R.; Vincent, C. A., Eds.; Elsevier: New York, 1987; p 103.
31. Giles, J. R. M.; Greenhall, M. P. *Polym. Comm.* **1986**, *27*, 360.
32. Watanabe, M.; Nagano, S.; Sanui, K.; Ogata, N. *Polymer J.* **1986**, *18*, 809.
33. LeNest, J. F.; Callens, S.; Gandini, A.; Armand, M. *Electrochimica Acta* **1992**, *37*, 1585.
34. Xia, D. W.; Smid, J. *Polym. Preprints* **1991**, *32*, 168.

35. Le Nest, J.; Gandini, A. In *Second International Symposium on Polymer Electrolytes;* Scrosati, B., Eds.; Elsevier: New York, 1990; p 129.
36. Xia, D. W.; Smid, J. In *Recent Advances in Fast Ion Conducting Materials and Devices;* Chowdari, B. V. R., Liu, Q., Chen, L., Eds.; World Scientific: Singapore, 1990; p 249.
37. Baochen, W.; Li, F.; Xinsheng, P. *Solid State Ionics* **1991,** *48,* 203.
38. Booth, C.; Nicholas, C. V.; Wilson, D. J. In *Polymer Electrolyte Reviews 2;* MacCallum, J. R., Vincent, C. A., Eds.; Elsevier: New York, 1989; p 229.
39. Gray, F. M. *Solid State Ionics* **1990,** *40/41,* 637.
40. Lemmon, J. P.; Lerner, M. M. *Macromolecules* **1992,** *25,* 2907.
41. Cowie, J. M. G. In *Polymer Electrolyte Reviews 1;* MacCallum, J. R., Vincent, C. A., Eds.; Elsevier: New York, 1987; p 69.
42. Hall, P. G.; Davies, G. R.; McIntyre, J. E.; Ward, I. M.; Bannister, D. J.; Le Brocq, K. M. F. *Polym. Comm.* **1986,** *27,* 98.
43. Fish, D.; Khan, I. M.; Smid, J. *Makromol. Chem., Rapid Commun.* **1986,** *7,* 115.
44. Tsutsumi, H.; Yamamoto, M.; Morita, M.; Matsuda, Y.; Nakamura, T.; Asai, H. *Electrochimica Acta* **1992,** *37,* 1183.
45. Blonsky, P. M.; Shriver, D. F.; Austin, P.; Allcock, H. R. *J. Am. Chem. Soc.* **1984,** *106,* 6854.
46. Blonsky, P. M.; Shriver, D. F.; Austin, P.; Allcock, H. R. *Solid State Ionics* **1986,** *18/19,* 258.
47. Greenbaum, S. G.; Adamic, K. J.; Pak, Y. S.; Wintersgill, M. C.; Fontanella, J. J. *Solid State Ionics* **1988,** *28-30,* 1042.
48. Palmer, D. N. In *Rechargeable Lithium Batteries;* Subbarao, S., Koch, V. R., Owens, B. B., Smyrl, W. H., Eds.; The Electrochemical Society: Pennington, NJ, 1990; p 245.
49. Ganapathiappan, S.; Chen, K.; Shriver, D. F. *Macromolecules* **1988,** *21,* 2299.
50. Zhou, G.; Khan, I.; Smid, J. *Polym. Comm.* **1989,** *30,* 52.
51. Tsuchida, E.; Kobayashi, N.; Ohno, H. *Macromolecules* **1988,** *21,* 96.
52. Ohno, H.; Kobayashi, N.; Takeoka, S.; Ishizaka, H.; Tsuchida, E. *Solid State Ionics* **1990,** *40/41,* 655.
53. Doan, K. E.; Ratner, M. A.; Shriver, D. F. *Chem. Mater.* **1991,** *3,* 418.
54. Liu, H.; Yeh, T. F.; Skotheim, T. A.; Okamoto, Y. *J. Polym. Sci.: Part A: Polym. Chem.* **1992,** *30,* 879.
55. Tsuchida, E.; Ohno, H.; Tsunemi, K.; Kobayashi, N. *Solid State Ionics* **1983,** *11,* 227.
56. Xu, X.; Wang, Q. *Chin. Sci. Bull.* **1991,** *36,* 1171.
57. Liu, Y.; Cai, Z.; Fang, C.; Zhang, Y.; Gao, H. In *Recent Advances in Fast Ion Conducting Materials and Devices;* Chowdari, B. V. R., Liu, Q., Chen, L., Eds.; World Scientific: Singapore, 1990; p 279.
58. Gray, F. M.; MacCallum, J. R.; Vincent, C. A. *Solid State Ionics* **1986,** *18/19,* 252.
59. Gray, F. M. In *Polymer Electrolyte Reviews 1;* MacCallum, J. R., Vincent, C. A., Eds.; Elsevier: New York, 1987; p 139.
60. Wixwat, W.; Stevens, J. R.; Andersson, A. M.; Granqvist, C. G. In *Second International Symposium on Polymer Electrolytes;* Scrosati, B., Eds.; Elsevier: New York, 1990; p 461.

61. Wieczorek, W.; Such, K.; Przyluski, J.; Florianczyk, Z. *Synth. Met.* **1991**, *45*, 373.
62. Florianczyk, Z.; Such, K.; Wieczorek, W.; Wasiucionek, M. *Polymer* **1991**, *32*, 3422.
63. Such, K.; Florianczyk, Z.; Wieczorek, W.; Przyluski, J. In *Second International Symposium on Polymer Electrolytes;* Scrosati, B., Eds.; Elsevier: New York, 1990; p 119.
64. Kaplan, M. L.; Reitman, E. A.; Cava, R. J. *Polymer* **1989**, *30*, 504.
65. Weston, J. E.; Steele, B. C. *Solid State Ionics* **1982**, *7*, 75.
66. Przyluski, J.; Such, K.; Wycislik, H.; Wieczorek, W.; Florianczyk, Z. *Synth. Met.* **1990**, *35*, 241.
67. Capuano, F.; Croce, F.; Scrosati, B. *J. Electrochem. Soc.* **1991**, *138*, 1918.
68. Capuano, F.; Croce, F.; Scrosati, B. *J. Power Sources* **1992**, *37*, 369.
69. Hardy, L. C.; Shriver, D. F. *J. Am. Chem. Soc.* **1985**, *107*, 3823.
70. Spindler, R.; Shriver, D. F. *Macromolecules* **1986**, *19*, 347.
71. Ito, Y.; Kanehor, K.; Miyauchi, K.; Kudo, T. *J. Mat. Sci.* **1987**, *22*, 1845.
72. Takeoka, S.; Horiuchi, K.; Yamagata, S.; Tsuchida, E. *Macromolecules* **1991**, *24*, 2003.
73. Iwatsuki, S.; Kubo, M.; Ohtake, M. *Chem. Lett.* **1992**, 519.
74. Wahg, C.; Yu, D.; Liu, Q.; Yang, L.; Wang, Z. In *Recent Advances in Fast Ion Conducting Materials and Devices;* Chowdari, B. V. R., Liu, Q., Chen, L., Eds.; World Scientific: Singapore, 1990; p 237.
75. Kelly, I. E.; Owen, J. R.; Steele, B. C. H. *J. Power Sources* **1985**, *14*, 13.
76. Sandner, B.; Steurich, T.; Wiesner, K.; Bischoff, H. *Polym. Bull.* **1992**, *28*, 355.
77. Nagasubramanian, G.; Di Stefano, S. In *Rechargeable Lithium Batteries;* Subbarao, S., Koch, V. R., Owens, B. B., Smyrl, W. H., Eds.; The Electrochemical Society: Pennington, NJ, 1990; p 262.
78. Kaplan, M. L.; Reitman, E. A.; Cava, R. J.; Holt, L. K.; Chandross, E. A. *Solid State Ionics* **1987**, *25*, 37.
79. Watanabe, M.; Kanba, M.; Matsuda, H.; Tsunemi, K.; Mizoguchi, K.; Tsuchida, E.; Shinohara, I. *Makromol. Chem., Rapid Commun.* **1981**, *2*, 741.
80. Watanabe, M.; Kanba, M.; Nagaoka, K.; Shinohara, I. *J. Appl. Polym. Sci.* **1982**, *27*, 4191.
81. Tsuchida, E.; Ohno, H.; Tsunemi, K. *Electrochimica Acta* **1983**, *28*, 591.
82. Tsunemi, K.; Ohno, H.; Tsuchida, E. *Electrochimica Acta* **1983**, *28*, 833.
83. Yang, H. "Poly(ethylene oxide)-Based Zn(II) Polymer Electrolytes", Ph.D. thesis, University of Pennsylvania, Philadelphia, 1991.
84. Mendolia, M. S.; Farrington, G. C. Unpublished data.
85. Xie, L.; Farrington, G. C. Unpublished data.
86. Ferry, J. D. *Viscoelastic Properties of Polymers;* Wiley: New York, 1980.
87. Jenckel, V. E.; Heusch, R. *Kolloid Zeitschrift* **1953**, *130*, 89.
88. Bohnke, O.; Rousselot, C.; Gillet, P. A.; Truche, C. *J. Electrochem. Soc.* **1992**, *139*, 1862.
89. Huq, R.; Koksbang, R.; Tonder, P. E.; Farrington, G. C. In *Recent Advances in Fast Ion Conducting Materials and Devices;* Chowdari, B. V. R., Liu, Q., Chen, L., Eds.; World Scientific: Singapore, 1990; p 63.

90. Huq, R.; Koksbang, R.; Tonder, P. E.; Farrington, G. C. *Electrochimica Acta* **1992,** *37,* 1681.
91. Yu, D.; Qiu, W.; Liu, Q.; Yang, L.; Qiu, B.; Liang, W. *Synth. Met.* **1992,** *47,* 1.
92. Alamgir, M.; Moulton, R. D.; Abraham, K. M. In *Primary and Secondary Lithium Batteries;* Abraham, K. M., Salomon, M., Eds.; The Electrochemical Society: Pennington, NJ, 1991; p 131.
93. Prasad, P. S. S.; Owens, B. B.; Smyrl, W. H.; Selvaggi, A.; Scrosati, B. In *Primary and Secondary Lithium Batteries;* Abraham, K. M., Salomon, M., Eds.; The Electrochemical Society: Pennington, NJ, 1991; p 170.
94. Dobrowski, S. A.; Davies, G. R.; McIntyre, J. E.; Ward, I. M. *Polymer* **1991,** *32,* 2887.
95. Tobishima, K. S.; Yamaji, A. *Electrochimica Acta* **1984,** *29,* 267.

RECEIVED for review November 9, 1992. ACCEPTED revised manuscript April 5, 1993.

Preceramic Polymers: Past, Present, and Future

Dietmar Seyferth

Department of Chemistry, Massachusetts Institute of Technology, Cambridge, MA 02139

A review of nonoxide preceramic polymer chemistry is given. Preceramic polymers are needed because of high-technology applications that arose in our aerospace age. A discussion of the early work in Germany and Japan leads to an enumeration of the requirements that must be met if a polymer is to be a useful precursor in pyrolytic ceramic synthesis. Seven current research directions in preceramic polymer chemistry are discussed, and the uncertain future of preceramic polymer technology is pointed out.

Why Preceramic Polymers?

Although Aylett (*1*) and Chantrell and Popper (*2*) suggested in 1964 that inorganic polymers might serve as precursors for ceramics, active research on inorganic and organometallic polymer pyrolysis as a route to useful nonoxide ceramics began only in the early 1970s. The initiation of inorganic–organometallic preceramic polymer chemistry as a new subarea of polymer science was prompted by the demands of modern high technology, in particular, technology as practiced by the defense–aerospace industry. There was a pressing need for new structural materials that could serve as replacements for metals and metallic alloys in many structural applications in advanced aircraft, spacecraft, and weapons systems. These materials need to be as light as or lighter than the metal that they replace; able to survive in harsh environments; have exceptionally high thermal stability (above 1200 °C), high strength, high fracture toughness, and high thermal shock resistance; and be resistant to high-temperature air oxidation, atomic oxygen, and chemical corrosion.

0065–2393/95/0245–0131$14.50/0

The outstanding properties of carbon–carbon composites, their superior strength and thermal stability, had already been recognized (3, 4), but, of course, their facile oxidation at higher temperatures in air was a serious flaw. Replacing the carbon fibers with appropriate ceramic fibers and the carbon matrix with a ceramic matrix was expected to give a composite with greatly superior properties, one that would be stable to high-temperature oxidation and that would meet the requirements of the aerospace age (5). This goal required the availability of a polymer that could be spun into fibers whose subsequent pyrolysis gives the needed ceramic fibers. Because of their superior high-temperature properties, silicon carbide (6, 7) and silicon nitride (8, 9) were of immediate interest in this application, and as a result, the initial focus was on organosilicon and inorganic silicon polymers. For such polymers, which could serve as useful precursors, through their pyrolysis, to ceramics, the term "preceramic polymers" was applied (10). In this application, the ceramic matrix also could be a product of polymer pyrolysis, and the use of ceramic fibers as reinforcing materials in metal, glass, and organic polymer matrix composites also is possible.

The use of preceramic polymers in the fabrication of fiber-reinforced composites aroused the greatest interest, but other applications were recognized: precursors for high-temperature oxidation- and corrosion-resistant ceramic coatings for structural parts fabricated from carbon–carbon composites or from reactive metals or metallic alloys, coatings for carbon and ceramic fibers designed to give the proper fiber–matrix interfacial interaction, and low-loss binders for use in conventional fabrication of parts from ceramic powders. As low-loss binders, the preceramic polymer would replace the organic polymers heretofore used, and the result would be a stronger part that has far fewer defects because the binder would not have to be pyrolyzed out completely, but rather for the most part would be converted to ceramic.

These are just the most obvious applications of preceramic polymers. The fact that they may be processed by conventional polymer-processing techniques presents opportunities for other applications as well: for their use in ceramic–ceramic and ceramic–metal joining and in the preparation of ceramic foams and membranes, to mention just a few.

Early Developments

The results of two early research and development efforts, one in Germany, the other in Japan, serve to define the new field of preceramic polymer chemistry. In the laboratories of Bayer AG, Verbeek (11) used the known (12) reaction of methyltrichlorosilane with methylamine to produce $CH_3Si(NHCH_3)_3$. This product had been shown to undergo condensation with loss of methylamine when heated in the presence of H_2SO_4, giving a

bicyclic silazane, structure **1** *(13)*. Verbeek found that flow thermolysis of
$CH_3Si(NHCH_3)_3$ at around 520 °C through a Raschig ring-filled glass tube
gave volatile compounds and a brittle, solid carbosilazane resin that was
soluble in organic solvents. The resin could be melt-spun at 220 °C to
give carbosilazane fibers. Before pyrolysis of the green fibers, it was
necessary to render them infusible, and this was accomplished by heating
them in moist air at 110 °C for 20 h. Subsequent pyrolysis in nitrogen to
1500 °C gave black, glistening, amorphous ceramic fibers in 45 wt% yield.

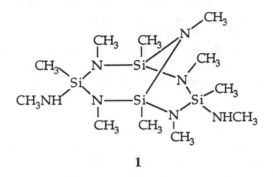

1

When they were heated to 1800 °C, crystalline material, mostly β-SiC and
a little α-SiC and β-Si_3N_4, was obtained. A detailed study of this carbosilazane preparation and pyrolysis was published by Penn et al.*(14)*. The
thermolysis of $CH_3Si(NHCH_3)_3$ appears to form a cross-linked material of
higher molecular weight, and weight-average molecular weights (derived
by gel permeation chromatography) of up to 4200 were reported by Penn
et al. Structure in the classical sense is not meaningful for a preceramic,
nonuniformly cross-linked network polymer. At best, the component
building blocks can be determined and often quantified by means of the
integrated NMR spectra of the polymers. Some local components (e.g.,
cyclic and linear portions) may be identified. Generally, idealized or composite structures are drawn when the subject of structure is discussed. For
the $CH_3Si(NHCH_3)_3$-derived carbosilazane resin, the idealized structure **2**,

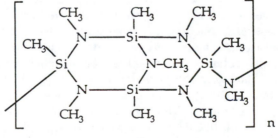

2

in which 1 is the building block, was suggested in a review (15). However, the structure of the polymer must be more complex than this in view of the well-known thermal lability of the Si–N bond.

In a variation of this procedure, Verbeek and co-workers (16) used ammonia in place of methylamine. In this case the product was a poly(methylsilsesquiazane) of type $(CH_3SiN_{1.5})_n$. [This reaction also had been reported previously (8).] A concentrated solution of the ammonolysis product could be dry-spun (with poly(ethylene oxide) as a spinning aid) to give polysilazane fibers. Their pyrolysis in a stream of nitrogen to 1200 °C resulted in silicon carbonitride ceramic fibers. Recourse to the less desirable dry-spinning process was necessary because the polysilsesquiazane did not form a stable melt. Ceramic films and shaped ceramic bodies also were fabricated by polysilsesquiazane pyrolysis. These processes never were commercialized, although one of them was reported (15) to reach the pilot-plant stage.

Verbeek and Winter (17) also developed a preceramic polymer process for silicon carbide based on the prior investigations of Fritz and co-workers (18) on the thermal decomposition of tetramethylsilane and the methylchlorosilanes. Such thermolysis, effected at around 700 °C with provision for recycling of starting material, gave a carbosilane resin that could be dry-spun from dichloromethane solution with an organic polymer as spinning aid. Pyrolysis of fibers prepared in this manner to 1500 °C in a stream of nitrogen resulted in black ceramic fibers that were amorphous by X-ray diffraction and contained, in addition to SiC, about 10 wt% of free carbon.

Pioneering work on polymeric precursors for silicon carbide was initiated in the early 1970s by Yajima and co-workers (19–25) at the Research Institute for Iron, Steel, and Other Metals of Tohoku University in Orai, Japan. The starting material in this process is the readily available dimethyldichlorosilane. Its reductive dechlorination by sodium, studied earlier by Burkhard (26) at the General Electric Company, results in formation of poly(dimethylsilane) [or poly(dimethylsilylene)], $[(CH_3)_2Si]_n$, a white, intractable powder. The Japanese workers found that when this material is heated at around 450 °C in an autoclave for 14 h (or in the presence of a few weight-percent of a poly(borodiphenylsiloxane) at 350 °C under nitrogen for 10 h), it is converted to a polycarbosilane in which the original Si–Si backbone has been changed to a Si–CH_2 backbone.

These polycarbosilanes, as isolated after removal of volatile compounds, are glassy, resinous materials that are meltable and soluble in organic solvents. They were shown by means of NMR (^{1}H and ^{29}Si) spectroscopy to contain $(CH_3)_2Si$ and $CH_3(H)Si$ groups bound together by CH_2 and CH bridging units. The major building blocks in these polycarbosilanes are 3 and 4, but the structure of these polymers also has been drawn only as an ideal one, as shown in 5 (25). They appear to contain

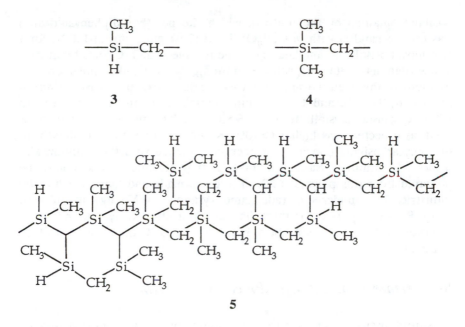

3   **4**

5

both linear and cyclic components and also are nonuniformly cross-linked. They, like the Bayer polysilazanes, are not high-molecular-weight materials ($M_n \sim 1250–1750$). (This appears to be the case for most organosilicon preceramic polymers.) The Yajima polymers could be melt-spun at 250–350 °C to give green polymer fibers.

The pyrolysis of these fibers to ceramic fibers required another processing step: They had to be rendered infusible so that the fiber form would be retained when they were pyrolyzed. Such curing of the green fibers could be effected by heating them in air between 110 and 190 °C. Pyrolysis in an inert atmosphere at 1200 °C subsequently gave ceramic fibers 10–20 μm in diameter. These, by elemental analysis in one experiment, had the nominal molar composition 1 SiC:0.78 C:0.22 SiO$_2$. These fibers are a commercial product of the Nippon Carbon Company and are sold under the trade name Nicalon. They have excellent properties at temperatures up to about 1200 °C: tensile strengths of 2.5–3.2 GPa, Young's modulus of 180–200 GPa, acid and base corrosion resistance, and resistance to oxidation. However, above 1200 °C, SiC crystallization and grain growth occur, and solid-state reactions between the SiO$_2$ formed in the cure step and elemental carbon and SiC generate CO and SiO, both gases whose evolution causes significant damage. Pores, flaws, and larger crystallites form everywhere in and on the fibers. As a result, the mechanical strength of the fibers is lost.

A variation of the Nicalon polycarbosilane process was developed by Yajima and co-workers (*27, 28*). Heating a xylene solution of the polycarbosilane obtained by the route in which the poly(borodiphenylsiloxane) was used as catalyst with $(n\text{-}C_4H_9O)_4Ti$ (10:7 weight ratio) at 130 °C for 1 h, removal of solvent, and heating of the residue at 220 °C for 30 min gave a new titanium-containing, soluble, more highly cross-linked polymer (27). Pyrolysis of this material in an inert gas stream to 1200 °C gave an amorphous Si-, Ti-, C-, and O-containing ceramic. Heating this material to 1700 °C caused crystallization of β-SiC and TiC phases. The ceramic yield, as expected, was higher (>70% vs. ~60%) than that obtained with the polycarbosilane that had not been cross-linked with the titanium alkoxide. Continuous Si–Ti–C–O-containing ceramic fibers could be prepared from this polymer (28). These are made and marketed by Ube Industries Ltd. under the trade name Tyranno. A Young's modulus of 220 GPa was reported for these fibers. Their strength was found to be superior to that of Nicalon fibers, but above 1200 °C there was almost no difference.

Requirements for a Useful Preceramic Polymer

This brief discussion of the early preceramic polymer systems makes obvious some of the requirements that are important for the preparation of a useful preceramic polymer. These requirements are dictated by the intended application of the preceramic polymer and can vary from application to application. For instance, its use in the fabrication of ceramic fibers will require different properties than when its intended use is as a binder in ceramic powder processing. Thus the preceramic polymer must be tailor-made for successful use in the intended application. This approach requires close collaboration between the synthetic chemist, the ceramist, and the end-user. Successful preceramic polymer technology is an interdisciplinary effort.

Some general principles have been recognized as the field of preceramic polymer chemistry has developed. Although at present the aerospace industry is by far the major end-user of preceramic polymer-derived products, one would hope that the technology based on this new approach will penetrate far broader markets, especially those in the civilian sector, for instance, the automobile industry. Thus cost factors are important, and critical nontechnical requirements are that the starting material be reasonably cheap and readily available and that the synthesis of the preceramic polymer be as uncomplicated as possible in terms of processing.

In the two examples cited, the starting materials both are products of the reaction of methyl chloride with Si–Cu (Rochow–Müller direct synthesis), which produces methylchlorosilanes (mainly $(CH_3)_2SiCl_2$, but also CH_3SiCl_3, CH_3SiHCl_2, and $(CH_3)_3SiCl$ as by-products), which are the foundation of the worldwide silicones industry. Thus both are readily available and not expensive. In either case, however, the chemical processing is not as simple as would be preferred. Rough calculations of costs indicate that a price of around $50/lb is a realistic estimate for volume production for some preceramic polymers of interest. This amount compares with typical prices of $10–25/lb for reasonable to high-quality Si_3N_4 and SiC powders.

The steps involved in the preceramic polymer technology, starting with the monomer and ending with the ceramic, are shown in Scheme I. In the preparation of inorganic and organometallic polymers, the condensation processes involving the monomer often give oligomeric, usually cyclic products. These then must, in one way or another, be converted to materials of higher molecular weight. In organic polymer technology, the final major step is the forming operation. In preceramic polymer chemistry, the shaped polymer (fiber, coating, matrix in a composite, and binder in powder-derived part) must be pyrolyzed to the final ceramic. If the shape is to be retained (e.g., the fiber form), then the polymer must be infusible or rendered infusible, or it will melt with loss of its shape. For melt-spun fibers, a cure step is required, as has been noted in the discussions of both the Bayer carbosilazane resin and the Nicalon polycarbosilane. However, many preceramic polymers are self-curing; that is, they become infusible during the early stages of the pyrolysis without going through a melt stage, and thus the shape is retained.

If a preceramic polymer is to be a useful one, first and foremost, it must be processable by conventional polymer-processing techniques. For most applications it must be soluble, preferably in common organic solvents. If it is a solid, it should be fusible, forming a stable melt, if ceramic fibers are the goal; that is, it should be melt-spinnable. (Melt-spinning is

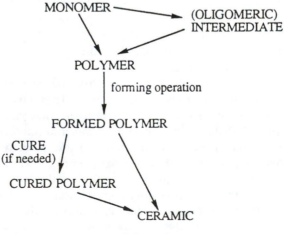

Scheme I.

preferable to dry-spinning: No solvent is required and, in general, the resulting green fibers have better properties.) A solid polymer is not always necessary or desirable. For other applications, such as chemical vapor deposition (CVD), powder synthesis, and injection molding, oligomeric fluids of various viscosities are useful or required.

The preceramic polymer should be stable at room temperature. Specifically, it should not undergo further cross-linking on storage that could lead to decreased solubility or even insolubility. An asset would be stability to atmospheric moisture and oxygen, but that is not always possible (e.g., the polysilazanes, which are more or less readily hydrolyzed on exposure to air).

Another important requirement is that pyrolysis of the inorganic or organometallic preceramic polymer give as high a yield of ceramic residue as possible. Ceramic yields (yield = weight of ceramic residue $\times$ 100/weight of pyrolysis charge) of 60–75% are acceptable, but higher ceramic yields of 80% or greater are desirable. Most polymers that have an inorganic backbone contain substituents on some or all of the backbone atoms. These often are hydrogen atoms or organic groups, but they can be of other types. During the pyrolysis of the polymer these substituents may become incorporated into the ceramic residue or they may be eliminated in the form of volatile, low-molecular-weight compounds that escape as gaseous products at the high pyrolysis temperatures. Thus H and CH_3 are preferred substituents because their loss will leave a high ceramic residue yield (if the backbone atoms are retained).

The pyrolysis of a polymer may give the theoretical yield of a ceramic as represented by the backbone atoms, but if the ceramic yield is low because of the large contributing weight to the pyrolysis charge of rather heavy substituents that will end up as volatile compounds, then useful applications of the polymer may be questionable. High ceramic residue yields are of some importance from an economic standpoint, but they are of greater importance from a technical standpoint, which can involve the type of body that is being fabricated and its function, quality, geometry, microstructural character, and strength.

A high ceramic yield means that the quantity of gases evolved during pyrolysis will be small. These gases can be very destructive, especially if they are released over a narrow temperature range; they can cause cracking or even rupture of the ceramic part. So the fewer evolved gases, the better for many applications in which bodies of low or zero porosity are the goal. Lower ceramic yields, however, can be tolerated if the porosity generated in the developing ceramic during the pyrolysis is of the right kind: open porosity that lets the pyrolysis gases escape. Also, the greater the weight loss of a preceramic polymer on pyrolysis, the greater will be the shrinkage of the residual body. However, shrinkage is unavoidable when a polymer is pyrolyzed to a ceramic because the starting material has a low (1 g/cm^3 or less) density and the material being formed has a greater (2–3 g/cm^3) density.

The requirement of high ceramic yield on pyrolysis determines the nature of the polymer that will be satisfactory. Ideally, the polymer should already be cross-linked (but not to the point of insolubility). More important, whether cross-linked or not, it should undergo extensive cross-linking during the early stages of the pyrolysis so that evolution of volatile compounds is minimized and the elements of interest are retained in the hot zone. To accomplish this extensive cross-linking, the polymer must contain reactive functionality ["latent functionality" in the words of Wynne and Rice (10)] that on heating will react rapidly to cause the needed further cross-linking. The design of a useful preceramic polymer thus requires the introduction of reactive functional groups in the monomer or, later, in the polymer so that the initial stage of the pyrolysis results in formation of a highly cross-linked, nonvolatile, probably insoluble network, hence in a high ceramic residue yield.

Another important requirement is that the polymer have the right rheological properties for the intended application. This consideration is important when melt-spinnable fibers are desired, when thick or thin film coatings are to be prepared, and when the preceramic polymer is to be used as a binder for ceramic powder formulations for injection or compression molding. In this connection, the ability to control the molecular weight of the polymer is important. A given molecular-weight range will give rheological properties that may be satisfactory for one application but not for another.

Since this early work by Verbeek and Yajima and their co-workers, the field of preceramic polymer chemistry has become very active, with participation by scientists in industrial, academic, and government laboratories throughout the world. The field has advanced considerably during the past 20 years. Research and development efforts have been directed in the main at the synthesis of polymeric precursors for silicon-containing ceramics: silicon carbide, silicon nitride, and silicon carbonitride, but precursors for boron nitride and carbide, aluminum nitride, and early transition metal carbides and nitrides also have been of interest. Numerous reviews covering this work have been published (10, 15, 25, 29–38). Rather than summarizing this work, it may be more useful to point out and discuss briefly some of the current research directions in the preceramic polymer area.

Current Research Directions in the Preceramic Polymer Area

Single-Phase Ceramics via Polymer Pyrolysis. Much effort
has been devoted to synthesis. Until recently, the preparation of pure

(i.e., single-phase) ceramics by polymer pyrolysis has not been straightforward, and the tendency has been to take what you get in the pyrolysis of a given polymer. "What you get" is not always optimum in terms of ceramic composition and may even be rather unsatisfactory. A surprising amount of effort has been devoted to the synthesis and study of preceramic polymers that clearly are impractical with respect to eventual commercialization in terms of expense, unavailability, or instability of the starting material used; expense or complexity of the chemical processing; properties of the polymer prepared; or the properties and composition of the ceramic formed on pyrolysis of the polymer.

However, more and more synthetic efforts have been aimed at preparation of preceramic polymer systems in which what you get on pyrolysis is really what is wanted or needed. Some of these efforts are directed at the preparation of polymeric precursor systems whose pyrolysis will give single-phase ceramic products. The ways of accomplishing such preparations will be described next. However, ceramic products consisting of more than one phase are also useful. We shall return to this point later on.

For pure, nonoxide silicon ceramics, the target ceramics are silicon carbide, SiC, and silicon nitride, Si_3N_4. However, in most cases, pyrolysis of a preceramic organosilicon polymer does not give a composition approaching these pure ceramic phases. In most organosilicon polymers (e.g., the Nicalon polycarbosilane), the C:Si ratio is greater than 1. However, in the preparation of SiC by polymer pyrolysis, a 1:1 C:Si ratio in the ceramic is desired, or, at the most, SiC plus only a small amount of free carbon, because carbon is a sintering aid for SiC.

With most preceramic organosilicon polymers, a substantial amount of free carbon remains in the final ceramic product after pyrolysis. At the least, a large amount of free carbon in the ceramic product, for instance, in fibers, will bring problems of high-temperature oxidation. If there also is oxygen in the ceramic, introduced in the synthesis of the polymer or in its subsequent processing, the high-temperature reactions of the SiO-containing phases with free carbon that generate gaseous CO and SiO cause serious problems, as noted earlier in the discussion of the Nicalon polycarbosilane. On the other hand, an organosilicon polymer whose pyrolysis results in formation of substantial amounts of elemental silicon in addition to SiC also is unsatisfactory. Such free silicon (mp 1414 °C) would compromise the high-temperature applications of the ceramic part.

A 1:1 C:Si ratio in the preceramic polymer does not guarantee that stoichiometric SiC will be produced in its pyrolysis, because the pyrolysis chemistry might be such that one of these elements will be lost in part in the form of gaseous products. Polymer composition and pyrolysis chemistry both are important. A notable example of a polymer in which the C:Si ratio is 1 and in which the pyrolysis in a stream of nitrogen gives essentially pure SiC in high yield is the polycarbosilane $(H_2SiCH_2)_n$ prepared by Wu and Interrante (*39*).

On the other hand, pyrolysis in a stream of argon of the poly-(methylsilane) $[(CH_3SiH)_x(CH_3Si)_y(CH_3SiH_2)_z]_n$ ($x + y + z = 1$; $x >> y$, z), developed independently by Brown-Wensley and Sinclair (*40*) at the 3M Company and by Wood (*41*) at the Massachusetts Institute of Technology (MIT), which also has a C:Si ratio of 1, evolved substantial amounts of CH_4 and left, in low yield, a ceramic residue composed of 76 wt% SiC and 24 wt% elemental silicon (*41*). However, further chemical manipulation of this system was possible, and heating this poly(methylsilane) in hexane with 0.5 mol% (based on Si) of $[(\eta^5\text{-}C_5H_5ZrH_2]_n$ or $(\eta^5\text{-}C_5H_5)_2Zr(CH_3)_2$ resulted in a cross-linked polymer whose pyrolysis gave near-stoichiometric SiC in high ceramic yield (*42*). In one example, use of the zirconium hydride catalyst gave as final nominal ceramic composition 98 wt% SiC, 1.6 wt% ZrC, and 0.4% Si. The poly(methylsilane) was prepared by the simply effected reaction of sodium with CH_3SiHCl_2; hence, this approach represents an economical route to SiC. The catalytic cross-linking chemistry is based on Harrod's Ti or Zr complex-catalyzed dehydrogenative coupling of primary silanes, $RSiH_3$, to give oligomeric polysilanes (*43*).

Silicon nitride of good purity is accessible by pyrolysis in a stream of ammonia of the complex perhydropolysilazane prepared by ammonolysis of the H_2SiCl_2–pyridine adduct in pyridine solution, a process developed by chemists of the Tonen Corporation (*44, 45*; *see also* ref. 46). The 1:1 Si:N ratio in the polysilazane should lead to a ceramic product containing an excess of elemental Si and does indeed do so when the pyrolysis is carried out in an inert gas stream. However, when the pyrolysis is carried out in a stream of ammonia, reaction of NH_3 (or of N_2 formed in its decomposition) with the elemental silicon converts this to Si_3N_4, so that a pure ceramic product is obtained.

This example illustrates the importance of the gas stream used in the pyrolysis of a preceramic polymer. Polymer pyrolysis usually is carried out in a furnace through which a gas is flowing. The gas stream serves two purposes: to protect the system from the atmosphere and to sweep away the volatile compounds produced in the pyrolysis. The gas used may be either an inert gas or a reactive gas. The pyrolysis of our polysilazane of composition $[(CH_3Si(H)NH)_a(CH_3SiN)_b]_n$ (*47–49*) serves to illustrate the use of both types of pyrolysis gas stream. Pyrolysis of this polysilazane in a stream of argon (an inert gas) to 1000 °C gives a *black* ceramic residue in 80–85% yield. This material is amorphous, so composition in terms of pure species such as SiC and Si_3N_4 is meaningless. However, the analytical results (percent C, N, and Si) can be used to calculate a hypothetical composition in terms of SiC, Si_3N_4, and free C. The results of such a calculation in one case were as follows: 67% by weight of Si_3N_4, 28% SiC, and 5% C. This result may be close to the actual composition of this material when it becomes crystalline above 1450 °C.

In marked contrast, when this polysilazane is pyrolyzed to 1000 °C in a stream of ammonia, a *white* solid residue is obtained. This residue usually contains less than 0.5% by weight of carbon and is almost pure silicon nitride. The chemistry that takes place at temperatures of 400–600 °C to remove the carbon must be complex. Nucleophilic cleavage of methyl groups from silicon by NH_3 to give CH_4 and generate $SiNH_2$ functions may be involved at the lower temperatures, but at high temperatures NH_2 and NH radicals derived from NH_3 may be the reactive species. Such pyrolysis in a stream of ammonia serves also to remove diamine ligands from $[B_{10}H_{12} \cdot diamine]_n$ polymers, giving BN as the final ceramic product (*50*) and to remove organoamino substituents from titanium in $[(CH_3)_2N]_2Ti$-$[(RNH)_2Ti]_x[N(CH_3)_2]_2$ polymers, giving TiN (*51*). Pyrolysis of a polycarbosilane that contains no SiN bonds at all gives silicon nitride as the ceramic product when effected in a stream of ammonia. For example, such NH_3 pyrolysis of the Nicalon polycarbosilane was reported (*52*) to produce Si_3N_4. Thus, what we refer to as the "carbon kickout reaction" appears to be generally applicable in the preparation of nitride ceramics by NH_3 pyrolysis of silicon (and boron and aluminum) polymers that contain organic substituents.

A single monomer is not necessarily required to obtain near-stoichiometric SiC. One may make use of the fact that elemental carbon and silicon react to give SiC, especially so if they are generated *in situ* as finely dispersed pyrolysis products. As mentioned, most organosilicon polymers when pyrolyzed in argon give SiC plus an excess of free carbon and, on the other hand, pyrolysis of our $[(CH_3SiH)_x(CH_3Si)_y(CH_3SiH_2)_z]_n$ poly(methylsilane) gives SiC plus an excess of free silicon. Thus if we take our "silicon-rich" poly(methylsilane) and a "carbon-rich" organosilicon polymer in the appropriate quantities, either in chemical or physical combination, the excesses of Si and C formed react and close to stoichiometric SiC is obtained in the pyrolysis. Predicting the elemental composition of the derived ceramic obtained on pyrolysis of a preceramic polymer or a mixture of preceramic polymers is not always possible; for the most part it is a matter of doing experiments until pyrolysis trends are recognized. This empirical approach does, however, work.

An example of a chemical combination of poly(methylsilane) and a "carbon-rich" organosilicon polymer is the hydrosilylation reaction product of $[(CH_3SiH)_x(CH_3Si)_y(CH_3SiH_2)_z]_n$ with the poly(silaacetylide) $[(CH_3(CH_2=CH)SiC\equiv C]_n$ (*53*). An azobisisobutyronitrile- (AIBN)-catalyzed reaction (SiH to $CH_2=CH$ addition) between appropriate quantities of these two polymers gave a hybrid polymer whose pyrolysis in a stream of argon resulted in formation of SiC, 99% pure by elemental analysis. An example of a physical combination whose pyrolysis gave near-stoichiometric SiC is the appropriate mixture of the poly(methylsilane) and the Nicalon polycarbosilane.

Multiphase Ceramics via Polymer Pyrolysis. As noted, a mixture of different ceramic phases may be a useful product because of its superior properties compared with those of the pure components. A striking example is that of composites of SiC and TiC that have greater fracture toughness (maximum at a 50:50 wt% mixture) than pure SiC and TiC (*54, 55*). Such a composite was prepared earlier by Yajima et al.(*27*), as already noted. At MIT, we prepared polymeric precursors for SiC–MC (M = Ti, Zr, or Hf) composites by UV-induced reaction of the $[(CH_3SiH)_x(CH_3Si)_y(CH_3SiH_2)_z]_n$ poly(methylsilane) with stoichiometric quantities of the respective $(\eta^5\text{-}C_5H_5)_2M(CH_3)_2$ compounds (*56*). The SiC–MC ceramic product obtained in their pyrolysis in a stream of argon contained significant amounts of free carbon (up to 40 wt% when M = Ti). This result could be avoided by pyrolyzing a composite of the Si- and M-containing polymer with the respective M metal powder. A low-carbon SiC–MC composite resulted on high-temperature reaction of the free carbon formed in the pyrolysis with the M metal powder.

For SiC–MC composites, the ceramic products were heated sufficiently high and long to cause crystallization of β-SiC and MC. However, crystalline ceramic products are not always desirable, and amorphous ceramic products in which grain growth cannot occur may be advantageous in some applications. Thus the amorphous, covalent silicon carbonitrides obtained by pyrolysis in an inert gas stream of diverse poly(organosilazanes) to around 1200–1300 °C appear to be a quite acceptable for many applications.

We also prepared polymeric silazanylborazines by reaction of the cyclic $[CH_3Si(H)NH]_n$ oligomers obtained by ammonolysis of CH_3SiHCl_2 with a BH_3-Lewis base adduct such as $H_3B \cdot S(CH_3)_2$ and $H_3B \cdot (CH_3)_2NH$ (*57*). Their pyrolysis in a stream of ammonia gave amorphous borosilicon nitrides in high ceramic yield. This work was prompted by reports that Si_3N_4–BN composites had, in some respects, properties superior to those of pure Si_3N_4 and BN (*58–61*). The Tonen polysilazane also was modified by incorporation of boron (*62*). Also of interest is the preparation by Interrante and co-workers (*63–65*) of various mixed ceramics (SiC–AlN, Si_3N_4–AlN, Si_3N_4–BN, AlN–BN, and TiN–BN) by the polymer pyrolysis route. The preparation of such ceramic–ceramic composites is expected to become an active research direction in preceramic polymer chemistry.

The Problem of Shrinkage in Polymer Pyrolysis. In general, shaped, three-dimensional ceramic parts (in contrast to fibers and coatings) are not fabricated directly by pyrolysis of a preceramic polymer alone because, as noted earlier, the shrinkage is enormous on going from the green polymer part to the final ceramic part. Furthermore, the ceramic part thus obtained is very porous and far from theoretical density, and hence is not very strong. Instead, the preceramic polymer is used as a low-loss binder for ceramic powders in the fabrication of parts by compression or injection molding with subsequent (usually pressureless) sintering. Such parts, despite the fact that they contain significant porosity and are not fully dense, can be quite strong.

Thus, Semen and Loop (66) at the Ethyl Corporation laboratories demonstrated strengths of >650 MPa for parts made from submicrometer-sized SiC particles bonded with an amorphous silicon carbonitride matrix generated by pyrolysis of a polysilazane of type $[(CH_3Si(H)NH)_a$-$(CH_3SiN)_b]_n$. Additionally, these parts exhibited excellent oxidation resistance and strength retention at temperatures up to 1300 °C. The overall porosity in these parts was 20–25%; however, the porosity in the matrix is closer to 50% because the SiC particles were fully dense. These results clearly demonstrate the feasibility of achieving superior properties in porous, polymer-derived ceramics.

On the other hand, in work carried out at the Dow Corning laboratories (67), nearly fully dense SiC parts were fabricated by using polysilazane or polysiloxane binders for SiC powder. In these procedures, enough excess carbon for high-temperature (>2000 °C) sintering of the part was introduced via pyrolysis of the binder. In another approach, Riedel et al. (68) prepared a Si_3N_4 composite containing 24 wt% SiC particulates by sintering in nitrogen to 1750 °C (with Al_2O_3 and Y_2O_3 as sintering aids) an amorphous silicon carbonitride powder obtained by pyrolysis of a Chisso Corporation polysilazane in an inert gas stream. A ceramic part with 97% relative density that had fracture strength and toughness similar to those of conventionally processed SiC–Si_3N_4 composites was obtained. A special advantage of this approach was an observed better reproducibility of these properties due to improved homogeneity in the microstructure of the part.

Alternatively, if this polysilazane is first converted thermally to an infusible powder, subsequent pyrolysis of the green part can give 93% dense, crack-free silicon carbonitride parts at temperatures as low as 1000 °C (69). Also, dense (>99% of theoretical) fire-grained ceramics have been fabricated by hot-pressing very fine crystalline SiC powders obtained by pyrolysis of a polycarbosilane (70). A disadvantage, which can be decisive with respect to possible commercial application, is that using the preceramic polymer to make ceramic powder will (at least at present) be very expensive.

The problem of shrinkage when a preceramic polymer alone, or even a preceramic polymer–ceramic powder composite, is pyrolyzed was addressed by Greil and Seibold (*71, 72*). In an approach that they call active filler-controlled pyrolysis, a commercial polysilsesquioxane, $[RSiO_{1.5}]_n$–Ti powder composite was pyrolyzed in a stream of argon to 1400 °C [R is mostly Ph, minor contents (for thermal cross-linking) of R = H, CH_2=CH, and CH_3]. Titanium carbide was formed by reaction of the titanium powder with carbon from the decomposition products of the polymer, so that the final ceramic was composed of TiC particles dispersed in a microcrystalline silicon oxycarbide matrix (β-SiC, SiO_2, and graphitic C). This procedure results in very significantly less shrinkage in formation of the final ceramic part in comparison with that observed when the filler-free polymer is pyrolyzed, and it may find useful application in the preparation of ceramic parts by polymer pyrolysis.

In independent studies at MIT, with different intentions, we used pyrolysis of organosilicon preceramic polymer–metal powder composites in argon to generate ceramic–ceramic composites such as SiC–TiC, SiC–ZrC, SiC–TaC, SiC–NbC, SiC–TiN, SiC–ZrN, as well as transition metal silicides such as W_5Si_3, Mo_5Si_3, WSi_2, $MoSi_2$, and Mo_3Si (Tables I–III) (*73*). The polymers used included the Nicalon polycarbosilane, the MIT $[(CH_3Si(H)NH)_a(CH_3SiN)_b]_n$ polysilazane and the MIT $[(CH_3SiH)_x$-$(CH_3Si)_y(CH_3SiH_2)_z]_n$ poly(methylsilane). Different phases were obtained when the initial pyrolysis was carried out in ammonia to 800 °C before continuing to 1500 °C in argon because all carbon was removed by this procedure. This approach could be extended to the preparation of ceramic–ceramic composites containing transition metal borides (e.g., TiB_2, HfB_2, CrB, CrB_2, MoB, Mo_2B, WB, W_2B, and LaB_6) using the $[B_{10}H_{12}(CH_3)_2NCH_2CH_2N(CH_3)_2]_n$ polymer as the boron source (*73*).

Table I.   Crystalline Phases from Metal Powder–Polysilazane Composites Pyrolyzed to 1500 °C under a Flow of Argon

Metal	M:Si	Ceramic Yield (%)	XRD Results
Ti	1.2:1	72	TiN + SiC
Zr	1.3:1	82	ZrN + SiC
V	1:1	54	V_5Si_3 + SiC
Nb	1:1	80	NbC + $NbSi_2$
Mo	1:2	73	$MoSi_2$ + Mo_5Si_3
Mo	1:1	80	Mo_5Si_3
W	1:1	87	WC + WSi_2 + W_5Si_3 + SiC
W	5:3	93	WC + W_5Si_3

Table II. Crystalline Phases from Metal Powder–Polysilazane Composites Pyrolyzed at 800 °C for 4 h under Ammonia Flow, Then at 1500 °C Under a Flow of Argon

Metal	M:Si	Ceramic Yield (%)	XRD Results
V	1:1	49	$V_2N + V_5Si_3$
Mo	1:1	73	Mo_5Si_3
W	1:2	76	$W_5Si_3 + WSi_2$
W	5:3	89	$W_5Si_3{}^a$

[a]Anal.: W, 93.32; Si, 4.42; and C and N, <5.

Table III. Crystalline Phases from Metal Powder–Nicalon PCS Composites Pyrolyzed to 1500 °C under a Flow of Argon

Metal	M:Si	Ceramic Yield (%)	XRD Results
Mo	1:1	91	α-$Mo_2C + Mo_5Si_3 + SiC$
W	1:2	85	$WC + WSi_2 + SiC^a$
Ti	1:1	80	$TiC + SiC$
Zr	1:1	87	$ZrC + SiC$
V	1:1	79	$V_8C_7 + SiC$
Nb	1:1	85	$NbC + SiC$
Ta	1:1	83	$TaC + SiC$

[a]Anal.: W, 68.73; Si, 16.77; and C, 10.80 (equivalent to ~0.38 WC, 0.59 SiC, and 0.03 WSi_2).

Laser Pyrolysis of Oligomer Aerosols. In an interesting application of preceramic polymer chemistry, Gonsalves and co-workers (74, 75) prepared nanophase ceramic particles by pyrolysis of an aerosol of a liquid mixture of oligomeric cyclosilazanes, $[CH_3Si(H)NH]_n$ (obtained by ammonolysis of CH_3SiHCl_2) in a laser plume. As shown in Figure 1 (76), the silazane precursor is injected into the plume of a high-power CO_2 laser using an ultrasonic nozzle to generate the aerosol. Important features of this process are the three-dimensional cross-linking and polycondensation reactions that result in ultra-rapid condensation of molecular species from the laser plume. The nanosized silicon carbonitride particles that are produced can be rapidly condensed onto a laser-heated sub-

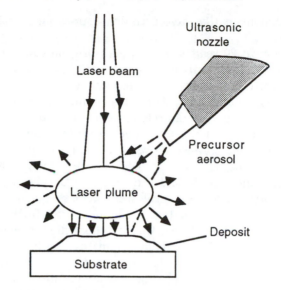

Figure 1. Technique to form ceramic coatings by rapid condensation of fine particles from laser plume onto laser-heated substrate. (Reproduced from reference 76. Copyright 1990 American Chemical Society.)

strate and sintered to form a deposited layer (Figure 1). In an alternate configuration, the nanoparticles of the ceramic may be collected for processing into bulk material. This process is capable of being scaled up and has great promise for the preparation of ceramic coatings and parts.

Ceramic Fibers via Polymer Pyrolysis. The research direction with which preceramic polymer chemistry first started, the preparation of continuous ceramic fibers, also is a current research direction and no doubt still will be one in the future. The fabrication of ceramic fibers that are useful at high temperatures from a preceramic polymer is the most challenging and difficult task in the preceramic polymer field. The best organic fibers are prepared from linear, high-molecular weight polymers. However, polymers of this type, in general, are not useful ceramic precursors because they fragment into volatile products, either starting monomer or cyclic oligomers, when pyrolyzed, leaving little or no solid residue. The good preceramic polymers that have been prepared to date are cross-linked, network-type polymers, generally of low molecular weight ($\overline{M}_n$ ~500–3000), hence still soluble. They can be converted to fibers only with difficulty, and the green fibers that result are rather fragile. Their conversion to ceramic fibers also is fraught with difficulties. If melt-spun, they must be cured prior to pyrolysis, and if the cure chemistry involves hydrolysis or oxidation, oxygen is introduced into the ceramic. As noted for the Nicalon fibers, the oxygen content causes fatal problems at high temperatures.

Today, almost 20 years since Yajima's original research, the Nicalon fibers of the Nippon Carbon Company still are the only preceramic polymer-derived ceramic fibers that are commercially available in greater than developmental quantities. Other companies have silicon carbide-, silicon nitride-, or silicon carbonitride-based fibers under development, but their future is as yet uncertain. The Dow Corning Corporation has developed preceramic polymer-derived ceramic fibers that are nanocrystalline, near-stoichiometric silicon carbide (>95% β-SiC) with only a small amount of graphitic carbon and little if any amorphous phase content. Densities of >97% of theoretical were achieved, and average tensile strengths of up to 2–6 GPa and elastic moduli of >420 GPa (77).

A commercial ceramic fiber with excellent high-temperature properties for use in composite technology remains a prime goal of the defense–aerospace establishments in the United States and other countries. It is, perhaps, ironic that one of the recommendations of a study of the U.S. high-performance fiber industry by the National Materials Advisory Board, published last year, was that research on ceramic fibers should be undertaken to develop fibers with enhanced stability for use at temperatures above 1200 °C. Exactly the same recommendation was being made more than 10 years ago! Such slow progress is not for want of trying!

Oxidative and hydrolytic cures of melt-spun fibers bring major problems when the ceramic fiber produced is heated to high temperatures, and therefore alternative cures have been examined in more recent work. For example, the Dow Corning HPZ polysilazane fiber precursor, of gross composition $[(SiH_{39.7})][(CH_2)_3Si]_{24.2}(NH)_{37.3}$-$(N)_{22.6}]$, is cured by exposure to a trichlorosilane, $RSiCl_3$ (preferably $R = H$). This exposure results in a chemical reaction in which $(CH_3)_3SiCl$ is eliminated and further cross-linking takes place (78). Thionyl chloride may be used in place of $RSiCl_3$ in this cure step (79). Alternatively, radiation curing of the green fiber is possible. Electron beam or γ-irradiation of uncured Nicalon polycarbosilane fibers of very low oxygen content effected the required cure. The resulting, essentially oxygen-free, ceramic fibers had much greater tensile strength and Young's modulus compared to those of fibers that had been subjected to the usual air cure (80, 81a).

Ceramic fibers of this type (cured by electron-beam irradiation) will be commercialized by the Nippon Carbon Company in 1995 under the trade name Hi-Nicalon (81b). These SiC fibers contain only ~0.5 wt% oxygen but do contain about 8.5 wt% free carbon. In contrast to the original Nicalon fibers, they retain a tensile strength of more than 2 GPa after being heated in argon for 10 h at 1500 °C, and they retain their fibrous form and flexibility after being heated in argon for 1 h at 2000 °C (81c, 81d). The problem of excess free carbon has been handled by carrying out the pyrolysis in part in a stream of hydrogen, which serves to remove free carbon. The resulting near-stoichiometric SiC fibers are still in the developmental stage (*81b*).

A completely different approach to spinning of SiC fibers that also uses a preceramic polymer is the slurry spinning method developed by workers at DuPont (*82*). In this procedure, an ultrasonicated slurry containing polycrystalline α- or β-SiC, B_4C as sintering aid, a dispersant, and 18.5% by weight of Nicalon polycarbosilane (as a binder) in xylene was evaporated to a toothpaste consistency, homogenized, and then spun by extrusion through 4-mil holes and attenuated to a final diameter of 50 μm. Pyrolysis in argon then gave SiC fibers of ~30-μm diameter that retain their room-temperature properties after being heated for 120 h at 1500 °C. Apparently, with such SiC–polycarbosilane systems a cure step is not necessary.

Reference 83 gives an excellent review of the structure and properties of ceramic fibers prepared by pyrolysis of organosilicon polymers.

Ceramics Containing Heteroelements Other Than Silicon via Polymer Pyrolysis.

Most of this discussion has focused on silicon-based preceramic polymers. However, other ceramics have some of the superior properties of the type that materials scientists are seeking: boron nitride and carbide; aluminum nitride (more of interest for electronic applications because of its high thermal conductivity); and the early transition metal carbides, nitrides, borides, and silicides. Research activity directed toward useful polymeric precursors for these materials has been growing. Precursors for boron nitride have received the greatest attention, most work being devoted to borazine-derived polymers (reviewed in ref. 84). However, higher boron hydrides also have served as the boron source in polymeric precursors for BN. Thus pyrolysis of poly(2-vinylpentaborane(9)) in a stream of ammonia to 1000 °C gave BN in high yield (*85*). Mentioned earlier was a similar ammonia pyrolysis of $[B_{10}H_{12}$ · diamine]$_n$ polymers, which also gave BN in high yield (*50*). Pyrolysis of poly(2-vinylpentaborane(9)) in argon to 1000 °C resulted in formation of boron carbide in high yield (*86*).

A melt-spinnable, thermoplastic polymer of composition $[(EtAl-NH)_a(Et_2AlNH_2)_b \cdot (AlEt_3)_c]_n$ ($a + b/c \sim 50$) was prepared by Bolt, Tebbe and co-workers (*87–89*) by reaction of triethylaluminum with ammonia. Aluminum nitride fibers could be prepared by melt-spinning the polymer and pyrolyzing initially in ammonia and, at higher temperatures, in nitrogen. Heating to 1600–1800 °C resulted in formation of polycrystalline AlN.

A novel electrochemical preparation of a polymeric precursor for aluminum nitride was reported by German workers (*90, 91*). In this process, metallic aluminum was dissolved anodically in a mixture of a primary amine and acetonitrile that contained a tetraalkylammonium salt as the supporting electrolyte. A polymeric iminoalane gel was formed on removing the solvent and heating the residue to 150 °C. Pyrolysis to 1100 °C in

ammonia removed RN substituents, giving crystalline AlN. Such electrochemically generated precursor solutions could be used to prepare AlN
coatings on SiC fibers (92). The same type of poly(iminoalane) precursor,
[HAlN-iso-Pr]$_6$, was prepared by reaction of isopropylamine with LiAlH$_4$
in refluxing heptane (93). Pyrolysis to 1000 °C in argon gave an amorphous, Al- and N-containing ceramic. Heating of this material to 1600 °C
caused crystallization of AlN.

A similar electrolytic procedure served in the synthesis of a polymeric
titanium nitride precursor of empirical formula TiC$_{3.94}$H$_{8.24}$N$_{1.14}$ and of
unknown structure when n-propylamine was used (94). Pyrolysis in dry nitrogen gave titanium carbonitride in 20% yield. Pyrolysis in a stream of
ammonia to 1100 °C resulted in gold-colored residues, typical of TiN.

Titanazane polymers prepared by reaction of [(CH$_3$)$_2$N]$_4$Ti with n-
butylamine, (6, R = n-C$_4$H$_9$) (95) also were useful TiN precursors (51).
Pyrolysis to 1000 °C in a stream of ammonia gave golden-yellow TiN containing only 0.26% by weight of C in 32.6% ceramic yield. Also of interest
is the polymeric solid obtained by hydrolysis of a Ti(O-n-Bu)$_4$–furfuryl alcohol mixture whose pyrolysis in argon to 1150 °C gave TiC, and pyrolysis
in ammonia produced TiN (96).

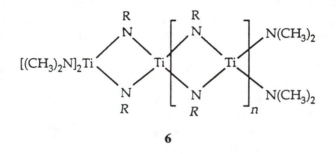

6

Investigation of the Polymer Pyrolysis Process. Also to be
noted among the current research directions in preceramic polymer chemistry are the many studies aimed at gaining a better understanding of the
polymer pyrolysis process and of the ceramic that is formed. Techniques
such as thermogravimetric analysis (TGA) combined with gas chromatography (GC) and mass spectrometry (MS) or Fourier transform IR (FTIR)
spectroscopy have been found useful in the study of the evolution of volatile products in the pyrolysis as a function of temperature. Solid-state
NMR spectroscopy (^{29}Si, ^{11}B, ^{13}C, ^{1}H, ^{27}Al, etc.) and diffuse reflectance
FTIR spectroscopy have been used to study the developing amorphous
solid pyrolysis residue, and powder X-ray diffraction to study the developing and final crystalline phases that are formed at higher temperatures.

Also applied to investigation of the solids produced in the pyrolysis
have been scanning electron microscopy (SEM), transmission electron mi-

croscopy (TEM), electron spectroscopy for chemical analysis (ESCA), X-ray photoelectron spectroscopy (XPS), extended X-ray absorption fine structure spectroscopy (EXAFS), small-angle X-ray scattering (SAXS), and wide-angle X-ray diffractometry. From such studies, a good picture of the pyrolysis chemistry of the organosilicon preceramic polymers is emerging. Furthermore, various laboratories are carrying out detailed studies of the microstructure development as a function of the starting polymer and the pyrolysis process variables. A selection of papers that report studies of the type mentioned in this section is given in references 97–121. The understanding gained from such studies is vital to the further progress of the field.

Other Recent Work on the Polymer Pyrolysis Route to Ceramics. In the years 1985 to the present, numerous research and development studies of preceramic polymer chemistry and applications have been done throughout the world. Some of these have already been discussed, and a bibliography of selected references to papers that have appeared in this period follows.

1. silicon carbide and silicon oxycarbide precursors

    - polysilanes *(122–133)*
    - polycarbosilanes *(134–146)*
    - polysiloxanes *(147–155)*
    - transition metal containing organosilicon polymers *(156–158)*
    - Al-containing organosilicon polymers *(159–161)*

2. silicon nitride, silicon carbonitride and silicon oxynitride precursors *(162–187)*

The focus of this list is on the synthesis and characterization of mostly silicon-containing preceramic polymers and to their pyrolytic conversion to ceramic materials. No attempt is made to assess the practical value of the polymer systems reported in these papers. Many more references report detailed studies of some of these polymers and of the ceramics obtained from them, especially of the microstructure and the physical, chemical, and mechanical properties of the ceramics formed, of specific applications in the fabrication of fibers and coatings, of their use as binders and matrices, and of their conversions to ceramic powders. Such references are not included.

The Future of Preceramic Polymer Chemistry

These then are some of the current research directions. But where, one might ask, is the field of preceramic polymer chemistry and technology heading? Indeed, is it a viable field that has a future? The answer to these questions is not at all clear at the present time.

As noted already, preceramic polymer chemistry is an area of *applied* polymer science–ceramic science. It is market-driven: The end-goal is a product. Preceramic polymer chemistry has been supported in the research laboratory and developed by industry because there were important potential commercial applications in sight. In the middle-to-late 1980s there was general enthusiasm for and ever-increasing research and development (R&D) activity in this new field of polymer science and, more generally, in advanced ceramics, first in Japan, then in the United States, and more recently in Europe.

In the meantime, the picture has changed. Most of the immediate applications of preceramic polymers mentioned earlier were in the area of aerospace, and a large fraction of the R&D activities in the United States received major support from the Departments of Defense and Energy, the National Aeronautic and Space Administration, and the National Science Foundation. In the time of big defense budgets that supported the development of ever-more sophisticated and demanding (in terms of stresses to be tolerated) jet aircraft, space vehicles, missiles, etc., the future of advanced structural ceramics, and thus of the preceramic polymer field, seemed secure. Preceramic polymers were urgently needed, and the activity in the field—chemistry and ceramics—grew rapidly; many chemical companies initiated research in this new area.

Now, however, with the end of the Cold War, defense budgets worldwide are getting smaller. Budget deficits lead to reductions in space-related activities. The National Aerospace Plane, to which much of the advanced structural ceramics technology would be applied, is in jeopardy. In short, the perceived market for preceramic polymers is shrinking. Add to this the fact that the time required for development of a final product in the preceramic polymer technology area has turned out to be much longer than had been expected, so that R&D costs are very high. As a result, those companies whose vision is restricted to short-to-moderate term profitability that had ventured into the preceramic polymer–composite technology area have dropped out again. Such departures from the scene of advanced structural ceramics have become a worldwide phenomenon during the past 2 years. The companies that still are active in this area now are very much fewer in number, and the future of the field is in their hands.

In the United States, the Dow Corning Corporation is the major player remaining, and this company has developed very promising polymeric precursors for silicon carbide and silicon nitride fibers and parts. Support of preceramic polymer *chemistry* by U.S. government agencies is now much diminished and still decreasing, although some development programs aimed at ceramic fiber development and fabrication of ceramic coatings and especially of ceramic and metal matrix composites still are being funded.

The preceramic polymer—advanced ceramics technology is a vital one that is needed in continuing development of defense and aerospace programs in the 21st century. It should be applicable also in other sectors such as surface transportation and energy. However, penetrating those markets will require imagination and ingenuity, as well as further technical advances.

I have focused on the *chemistry* of preceramic polymers. The success of a preceramic polymer will be measured by its utility in the production of useful materials—ceramics in most cases, but other materials as well (e.g., strengthened metal parts by means of ceramic fiber-reinforced metal matrix composites). If the field of preceramic polymer chemistry is to flourish, good chemistry must be followed by good ceramics and good ceramics must ultimately give a good final product. To make this happen, the chemist must interact and collaborate with the ceramist—materials scientist, and both must interact with the engineers of the end-use industry. Preceramic polymer chemistry presents interesting chemical challenges to the synthetic chemists and, because of these essential interactions with materials scientists and engineers, considerably widens the horizons.

Acknowledgments

I am grateful for generous support of the MIT research mentioned in this review by the Office of Naval Research, the Air Force Office of Scientific Research, the National Science Foundation, Akzo Corporate Research America Inc., Alcoa, and the Rhône-Poulenc Company. I also acknowledge a useful discussion with W. H. Atwell of the Dow Corning Corporation.

References

1. Aylett, B. J. In *Special Ceramics 1964;* Popper, P., Ed.; Academic: London, 1965; pp 105–113.
2. Chantrell, P. G.; Popper, P. In *Special Ceramics 1964;* Popper, P., Ed.; Academic: London, 1965; pp 87–103.
3. *Carbon Fibres and Their Composites;* Fitzer, E., Ed.; Springer: Berlin, Germany, 1985.
4. Sheppard, L. M. *Am. Ceram. Soc. Bull.* **1988** *67,* 1897.
5. *Ceramic Matrix Composites;* Naslain, R.; Harris, B., Eds.; Elsevier: Amsterdam, Netherlands, 1990.
6. *Gmelin Handbook of Inorganic Chemistry,* 8th ed.; Silicon Supplement, Vol B3, System No. 15; Springer: Berlin, Germany, 1986.

7. Smoak, R. H.; Korzekwa, T. M.; Kunz, S. M.; Howell, E. D. "Silicon Carbide" In *Kirk-Othmer Encyclopedia of Chemical Technology*, 3rd ed.; Wiley: New York, 1978; pp 520–535.
8. Weiss, J. *Annu. Rev. Mater. Sci.* **1981**, *11*, 381.
9. Messier, D. R.; Croft, W. J. In *Preparation and Properties of Solid State Materials;* Wilcox, W. R., Ed.; Dekker: New York, 1982; pp 131–212.
10. Wynne, K. J.; Rice, R. W. *Annu. Rev. Mater. Sci.* **1984**, *14*, 297.
11. Verbeek, W. U.S. Patent 3 853 567 (1974).
12. Tansjö, L. *Acta Chem. Scand.* **1960**, *14*, 2097.
13. Pearce, C. A.; Lloyd, N. C. U.S. Patent 3 580 941 (1971).
14. Penn, B. G.; Ledbetter, F. E., III; Clemons, J. M.; Daniels, J. G. *J. Appl. Polym. Sci.* **1982**, *27*, 3751.
15. Wills, R. R.; Markle, R. A.; Mukherjee, S. P. *Am. Ceram. Soc. Bull.* **1983**, *62*, 904.
16. Winter, G.; Verbeek, W.; Mansmann, M. U.S. Patent 3 892 583 (1975).
17. Verbeek, W.; Winter, G. Ger. Offen. 2 236 078 (March 21, 1974).
18. Fritz, G.; Matern, E. *Carbosilanes. Synthesis and Reactions;* Springer: Berlin, Germany, 1986.
19. Yajima, S.; Hayashi, J.; Omori, M. *Chem. Lett.* **1975**, 931.
20. Yajima, S.; Okamura, K.; Hayashi, J. *Chem. Lett.* **1975**, 1209.
21. Yajima, S.; Hayashi, J.; Omori, M.; Okamura, K. *Nature (London)* **1976**, *260*, 683.
22. Yajima, S.; Okamura, K.; Hayashi, J.; Omori, M. *J. Am. Ceram.* **1976**, *59*, 324.
23. Yajima, S.; Hasegawa, Y.; Hayashi, J.; Iimura, M. *J. Mater. Sci.* **1978**, *13*, 2569.
24. Hasegawa, Y.; Okamura, K. *J. Mater. Sci.* **1983**, *18*, 3633.
25. Yajima, S. *Am. Ceram. Soc. Bull.* **1983**, *62*, 893.
26. Burkhard, C. A. *J. Am. Chem. Soc.* **1949**, *71*, 963.
27. Yajima, S.; Iwai, T.; Yamamura, T.; Okamura, K.; Hasegawa, Y. *J. Mater. Sci.* **1981**, *16*, 1349.
28. Yamamura, T.; Ishikawa, T.; Shibuya, M.; Hisayuki, T.; Okamura, K. *J. Mater. Sci.* **1988**, *23*, 258.
29. Seyferth, D.; Wiseman, G. H.; Schwark, J. M.; Yu, Y.-F.; Poutasse, C. A. In *Inorganic and Organometallics Polymers;* Zeldin, M.; Wynne, K. J.; Allcock, H. R., Eds.; ACS Symposium Series 360; American Chemical Society: Washington, DC, 1988; pp 143–155
30. Seyferth, D.; Yu, Y.-F. In *Design of New Materials;* Cocke, D. L.; Clearfield, A., Eds.; Plenum: New York, 1987; pp 79–94.
31. Seyferth, D. In *Transformations of Organometallics into Common and Exotic Materials: Design and Activation* (NATO, ASI Ser. E, No. 141); Laine, R. M., Ed.; Nijhoff: Dordrecht, Netherlands, 1988; 133–154.
32. Seyferth, D. In *Silicon-Based Polymer Science: A Comprehensive Resource;* Zeigler, J. M.; Fearon, F. W. G., Eds.; Advances in Chemistry 224; American Chemical Society: Washington, DC, 1990; pp 565–591.
33. Rice, R. *Am. Ceram. Soc. Bull.* **1983**, *62*, 889.
34. Peuckert, M.; Vaahs, T.; Brück, M. *Adv. Mater.* **1990**, 398.
35. Toreki, W. *Polym. News* **1991**, *16*(1), 6.

36. Pouskouleli, G. *Ceram. Int.* **1989,** *15,* 213.
37. Mutsuddy, B. C. *Ceram. Int.* **1987,** *13,* 41.
38. Riedel, R. In *Concise Encyclopedia of Advanced Ceramic Materials;* Brook, R. J., Ed.; Pergamon: Oxford, England, 1991; p 299.
39. Wu, H.-J.; Interrante, L. V. *Macromolecules* **1992,** *25,* 1840.
40. Brown-Wensley, K. A.; Sinclair, R. A. U.S. Patent 4 537 942 (1985).
41. Wood, T. G. Ph.D. Dissertation, Massachusetts Institute of Technology, 1984.
42. Seyferth, D.; Wood, T. G.; Tracy, H. J.; Robison, J. L. *J. Am. Ceram. Soc.* **1992,** *75,* 1300.
43. Harrod, J. F. In *Inorganic and Organometallic Polymers;* Zeldin, M.; Wynne, K. J.; Allcock, H. R., Eds.; ACS Symposium Series 360; American Chemical Society: Washington, DC, 1988; pp 89–100.
44. Isoda, T.; Arai, M. Jpn. Kokai Tokkyo Koho, JP 60 145, 903 [85 145, 903], August 1, 1985; *Chem. Abstr.* *104,* 36340r.
45. Isoda, T.; Kaya, H.; Nishii, H.; Funayama, O.; Suzuki, T. *J. Inorg. Organomet. Polym.* **1992,** *2,* 151.
46. Seyferth, D.; Wiseman, G. H.; Prudhomme, C. *J. Am. Ceram. Soc.* **1984,** *66,* C-13.
47. Seyferth, D.; Wiseman, G. H. *J. Am. Ceram. Soc.* **1984,** *67,* C-132.
48. Seyferth, D.; Wiseman, G. H. U.S. Patent 4 482 669 (1984).
49. Han, H. N.; Lindquist, D. A.; Haggerty, J. S.; Seyferth, D. *Chem. Mater.* **1992,** *4,* 705.
50. Seyferth, D.; Rees, W. S., Jr. *Chem. Mater.* **1991,** *3,* 1106.
51. Seyferth, D.; Mignani, G. *J. Mater. Sci. Lett.* **1988,** *7,* 487.
52. Okamura, K.; Sato, M.; Hasegawa, Y. *Ceram. Int.* **1987,** *13,* 55.
53. Seyferth, D.; Yu, Y.-F.; Koppetsch, G. E. U.S. Patent 4 719 272 (1988).
54. Endo, H.; Ueki, M.; Kubo, H. *J. Mater. Sci.* **1990,** *25,* 2503.
55. Jiang, D. L.; Wang, J. H.; Li, Y. L.; Ma, T. *Mater. Sci. Eng. Part A,* **1989,** *109,* 401.
56. Seyferth, D.; Lang, H.; Sobon, C. A.; Borm, J.; Tracy, H. J.; Bryson, H. *J. Inorg. Organomet. Polym.* **1992,** *2,* 59.
57. Seyferth, D.; Plenio, H. *J. Am. Ceram. Soc.* **1990,** *75,* 2131.
58. Mazdiyasni, K. S.; Ruh, R. *J. Am. Ceram. Soc.* **1981,** *64,* 415.
59. Isomura, K.; Fukuda, T.; Ogasahara, K.; Fashi, T.; Uchimura, R. *Tetsu to Hagane* **1989,** *75,* 1612; *Chem. Abstr.* *111,* 218290b.
60. Nakamura, Y.; Nakajima, M. Jpn. Kokai Tokkyo Koho, JP 01 83, 506 [89 83, 506] (March 29, 1989); *Chem. Abstr.* *111,* 44274g.
61. Nakamura, Y.; Nakajima, M. Jpn. Kokai Tokkyo Koho, JP 01 83, 507 [89 83, 507] (March 29, 1989); *Chem. Abstr.* *111,* 44275h.
62. Funayama, O.; Arai, M.; Aoki, H.; Tashiro, Y.; Katahata, T.; Sato, K.; Isoda, T.; Suzuki, T.; Kohshi, I. *Eur. Pat. Appl.* EP 404,503 (1990); *Chem. Abstr.* *114,* 191120u.
63. Czekaj, C. L.; Hackney, M. L. J.; Hurley, W. J., Jr.; Interrante, L. V.; Sigel, G. A. *J. Am. Ceram. Soc.* **1990,** *73,* 352.
64. Interrante, L. V.; Hurley, W. J., Jr.; Schmidt, W. R.; Kwon, D.; Doremus, R. H.; Marchetti, P. S.; Maciel, G. E. *Ceram. Trans.* **1991,** *19,* (Adv. Compos. Mater.), 3; *Chem. Abstr.* *117,* 13070g.

65. Schmidt, W. R.; Hurley, W. J., Jr.; Doremus, R. H.; Interrante, L. V.; Marchetti, P. S. *Ceram. Trans.* **1991,** *19* (Adv. Compos. Mater.), 19; *Chem. Abstr. 117,* 13071h.

66. Semen, J.; Loop, J. G. *Ceram. Eng. Sci. Proc.* **1991,** *12,* 1967.

67. Atwell, W. H.; Burns, G. T.; Zank, G. A. In *Inorganic and Organometallic Polymers;* Harrod, J. F.; Laine, R. M., Eds.; Kluwer: Dordrecht, Netherlands, 1991; pp 147–159.

68. Riedel, R.; Seher, M.; Becker, G. *J. Eur. Ceram. Soc.* **1989,** *5,* 113.

69. Riedel, R.; Passing, G.; Schönfelder, H.; Brook, R. J. *Nature (London)* **1992,** *355,* 714.

70. Kodama, H.; Miyoshi, T. *Adv. Ceram. Mater.* **1988,** *3,* 177.

71. Seibold, M.; Greil, P. *Advances in Materials Processing;* Exner, H. E.; Schuhmacher, V., Eds.; DGM Inform. Ges., Oberursel (Germany), 1990, Vol. 1, p 641.

72. Greil, P.; Seibold, M. *Chem. Trans.* **1991,** *19* (Adv. Compos. Mater.), 43; *Chem. Abstr. 117,* 13074m.

73. Seyferth, D.; Bryson, N.; Workman, D. P.; Sobon, C. A. *J. Am. Ceram. Soc.* **1991,** *74,* 2687.

74. Magee, A. P.; Strutt, P. R.; Gonsalves, K. E. *Chem. Mater.* **1990,** *2,* 232.

75. Gonsalves, K. E.; Strutt, P. R.; Xiao, T. D.; Klemens, P. G. *J. Mater. Sci.* **1992,** *27,* 3231.

76. *Chem. Eng. News, Sept. 10, 1990,* p 23.

77. Lipowitz, J.; Rabe, J. A.; Zank, G. A. *Ceram. Eng. Sci. Proc.* **1991,** *12,* 1819.

78. LeGrow, G. E.; Lim, T. F.; Lipowitz, J.; Reaoch, R. S. *Am. Ceram. Soc. Bull.* **1987,** *66,* 363.

79. Foley, P. U.S. Patent 4 693 914 (1987).

80. Okamura, K.; Matsuzawa, T.; Hasegawa, Y. *J. Mater. Sci. Lett.* **1985,** *4,* 55.

81a. Okamura, K.; Sato, M.; Seguchi, T.; Kawanishi, S. *Proceedings of the First Japan International SAMPE Symposium* (Nov. 28–Dec. 1, 1989); p 929.

81b. Ishikawa, H. Joint U.S.–Japan Symposium on Inorganic–Organic Hybrid Materials, Pacific Grove, CA, May 15–19, 1994.

81c. Takeda, M.; Imai, Y.; Ichikawa, H.; Ishikawa, T.; Seguchi, T.; Okamura, K. *Ceram. Eng. Sci. Proc.* **1991,** *12,* 1007.

81d. Takeda, M.; Imai, Y.; Ichikawa, H.; Ishikawa, T.; Seguchi, T.; Okamura, K. *Ceram. Eng. Sci. Proc.* **1992,** *13,* 209.

82. Silverman, L. A.; Hewett, W. D., Jr.; Blatchford, T. P.; Beller, A. J. *J. Appl. Polym. Sci. Appl. Polym. Symp.* **1991,** *47,* 99.

83. Lipowitz, J. *J. Inorg. Organomet. Polym.* **1991,** *1,* 277.

84. Paine, R. T.; Narula, C. K. *Chem. Rev.* **1990,** *90,* 73.

85. Mirabelli, M. G. L.; Sneddon, L. G. *Inorg. Chem.* **1988,** *27,* 3271.

86. Mirabelli, M. G. L.; Sneddon, L. G. *J. Am. Chem. Soc.* **1988,** *110,* 3305.

87. Bolt, J. D.; Tebbe, F. N. *Mater. Res. Soc. Symp. Proc.* **1988,** *108,* 337.

88. Baker, R. T.; Bolt, J. D.; Reddy, G. S.; Roe, C.; Staley, R. S.; Tebbe, F. N.; Vega, A. J. *Mater. Res. Soc. Symp. Proc.* **1988,** *121,* 471.

89. Tebbe, F. N.; Bolt, J. D.; Young, R. J., Jr.; Van Buskirk, O. R.; Mahler, W.; Reddy, G. S.; Chowdhry, U. *Adv. Ceram.* **1989,** *26* (Ceram. Substrates Packages Electron. Appl.), 63; *Chem. Abstr. 112,* 239441n.

90. Seibold, M.; Rüssel, C. *Mater. Res. Soc. Symp. Proc.* **1988**, *121*, 477; *J. Am. Ceram. Soc.* **1989**, *72*, 1503.

91. Distler, P.; Rüssel, C. *J. Mater. Sci.* **1992**, *27*, 133.

92. Teusel, I.; Rüssel, C. *J. Mater. Sci.* **1990**, *25*, 3531.

93. Sugahara, Y.; Onuma, T.; Tanegashima, O.; Kuroda, K.; Kato, C. *J. Ceram. Soc. Jpn.* **1992**, *100*, 101.

94. Rüssel, C. *Chem. Mater.* **1990**, *2*, 241.

95. Bradley, D. C.; Torrible, E. G. *Can. J. Chem.* **1963**, *41*, 134.

96. Jiang, Z.; Rhine, W. E. *Chem. Mater.* **1991**, *3*, 1132.

97. Lipowitz, J.; Turner, G. L. *Polym. Prepr. (Am. Chem. Soc. Div. Polym. Chem.)* **1988**, *29*, 74.

98. Lipowitz, J.; Turner, G. L. *Solid State NMR Polymers* (Proc. 3rd. Annu. Chem. Conf. North Am. Solid State NMR Polym.); Mathias, L. J., Ed.; Plenum: New York, 1988; p 305; *Chem. Abstr.* *116*, 179443k.

99. Lipowitz, J.; LeGrow, G.; Lim, T.; Langley, N. *Ceram. Trans.* **1989**, *2* (Silicon Carbide 87), 421.

100. Chang, Y. W.; Zangvil, A.; Lipowitz, J. *Ceram. Trans.* **1989**, *2* (Silicon Carbide 87), 435.

101. Lipowitz, J.; Rabe, J. A.; Frevel, L. K.; Miller, R. L. *J. Mater. Sci.* **1990**, *25*, 2118.

102. Freeman, H. A.; Rabe, J. A. *Microbeam Anal.* **1990**, *25*, 212; *Chem. Abstr.* **114**, 233488c.

103. Schmidt, W. R.; Interrante, L. V.; Doremus, R. H.; Trout, T. K.; Marchetti, P. S.; Maciel, G. E. *Chem. Mater.* **1991**, *3*, 257.

104. Schmidt, W. R.; Marchetti, P. S.; Interrante, L. V.; Hurley, W. J., Jr.; Lewis, R. H.; Doremus, R. H.; Maciel, G. E. *Chem. Mater.* **1992**, *4*, 937.

105. Sirieix, F.; Goursat, P. *Rev. Int. Hautes Temp. Refract.* **1990**, *26*, 75.

106. Penot, C.; Fabre, A.; Goursat, P.; Bahloul, D.; Lecomte, A.; Danger, A.; Lespade, P. *Ann. Chim. (Paris)* **1992**, *17*, 155.

107. Delverdier, O.; Monthioux, M.; Oberlin, A.; Lavedrine, A.; Bahloul, D.; Goursat, P. *J. High Temp. Chem. Processes* **1992**, *1*, 139.

108. Sorarù, G. D.; Babonneau, F.; Mackenzie, J. D. *J. Non-Cryst. Solids* **1988**, *106*, 256.

109. Gerardin, C.; Taulelle, F.; Livage, J.; Birot, M.; Dunoguès, J. *Bull. Magn. Reson.* **1990**, *12*, 84.

110. Babonneau, F.; Barre, P.; Livage, J.; Verdaguer, M. *Mater. Res. Soc. Symp. Proc.* **1990**, *180* (Better Ceram. Chem. 4), 1035.

111. Sorarù, G. D.; Glisenti, A.; Granozzi, G.; Babonneau, F.; Mackenzie, J. D. *J. Mater. Res.* **1990**, *5*, 1958.

112. Babonneau, F.; Sorarù, G. D.; Mackenzie, J. D. *J. Mater. Sci.* **1990**, *25*, 3664.

113. Babonneau, F.; Livage, J.; Laine, R. M. *Polym. Prepr. (Am. Chem. Soc. Div. Polym. Chem.)* **1991**, *32*, 579.

114. Gerardin, C.; Taulelle, F.; Livage, J. *J. Chim. Phys. Phys.-Chim. Biol.* **1992**, *89*, 461.

115. Hommel, H.; Miguel, J. L.; Legrand, A. *Ind. Ceram. (Paris)* **1990**, No. 849, 344.

116. Laffon, C.; Flank, A. M.; Lagarde, P.; Laridjani, M.; Hagege, R.; Olry, P.; Cotteret, J.; Dixmier, J.; Miguel, J. L. *J. Mater. Sci.* **1989,** *24,* 1503.
117. Corriu, R. J.; Leclerq, D.; Mutin, P. H.; Vioux, A. *Chem. Mater.* **1992,** *4,* 711.
118. Choong, N. S.; Yive, K.; Corriu, R. J.; Leclerq, D.; Mutin, P. H.; Vioux, A. *Chem. Mater.* **1992,** *4,* 1263.
119. Morrone, A. A.; Toreki, W.; Batich, C. D. *Mater. Lett.* **1991,** *11,* 19.
120. Chaim, R.; Heuer, A. H.; Chen, R. T. *J. Am. Ceram. Soc.* **1988,** *71,* 960.
121. Poupeau, J. J.; Abbe, D.; Jamet, J. *Mater. Sci. Res.* **1984,** *17* (Emergent Process Methods High Technol. Ceram.), 287.
122. Baney, R. H., Gaul, J. H., Jr.; Hilty, T. K. *Organometallics* **1983,** *2,* 859.
123. Lipowitz, J.; LeGrow, G. E.; Lim, T. F.; Langley, N. *Ceram. Eng. Sci. Proc.* **1988,** *9,* 931; *Ceram. Trans.* **1989,** *2,* 421.
124. Atwell, W. H. In *Silicon-Based Polymer Science: A Comprehensive Resource;* Zeigler, J. M.; Fearon, F. W. G., Eds.; Advances in Chemistry 224; American Chemical Society: Washington, DC, 1990; pp 593–606.
125. West, R.; David, L. D.; Djurovich, P. I.; Yu, H. *Am. Ceram. Soc. Bull.* **1983,** *62,* 899.
126. Schilling, C. L. *Brit. Polym. J.* **1986,** *18,* 355.
127. Schilling, C. L., Jr.; Wesson, J.; Williams, T. C. *J. Polym. Sci. Polym. Symp.* **1983,** *70,* 121.
128. Zhang, Z.-F.; Babonneau, F.; Laine, R. M.; Mu, Y.; Harrod, J. F.; Ralin, J. A. *J. Am. Ceram. Soc.* **1991,** *74,* 670.
129. Carlsson, D. J.; Cooney, J. D.; Gauthier, S.; Worsfold, D. J. *J. Am. Ceram. Soc.* **1990,** *73,* 237.
130. Qui, H.; Du, Z. *J. Polym. Sci. Part A Polym Chem.* **1989,** *27,* 2849, 2861.
131. Kumar, K.; Litt, M. H. *J. Polym. Sci. Part C Polym Lett.* **1988,** *26,* 25.
132. Kalchauer, W.; Pachaly, B. *Ger. Offen. DE* 4 013 059 (1991); *Chem. Abstr.* *116,* 60244f.
133. Hengge, E.; Winterberger, M. *J. Organomet. Chem.* **1992,** *433,* 21.
134. Seyferth, D. In *Inorganic and Organometallic Polymers;* Zeldin, M.; Wynne, K. J.; Allcock, H. R., Eds.; ACS Symposium Series 360; American Chemical Society: Washington, DC, 1988; pp 21–24 (Review).
135. Schilling, C. L., Jr.; Wesson, J.; Williams, T. C. *Am. Ceram. Soc. Bull.* **1983,** *62,* 912.
136. Hurwitz, F. I.; Heimann, P. J.; Gyekenyesi, J. Z.; Masnovi, J.; Bu, X. Y. *Ceram. Eng. Sci. Proc.* **1991,** *12,* 1292.
137. Wu, H.-J.; Interrante, L. V. *Chem. Mater.* **1989,** *1,* 564.
138. Whitmarsh, C. K.; Interrante, L. V. *Organometallics* **1991,** *10,* 1336.
139. Boury, B.; Carpenter, L.; Corriu, R. J.; Mutin, P. H. *New J. Chem.* **1990,** *14,* 535.
140. Boury, B.; Corriu, R. J.; Douglas, W. E. *Chem. Mater.* **1991,** *3,* 487.
141. Boury, B.; Corriu, R. J.; Leclerq, D.; Mutin, P. H.; Planeix, J.-M.; Vioux, A. *Organometallics* **1991,** *10,* 1457.
142. Bouillon, E.; Pailler, R.; Naslain, R.; Bacqué, E.; Pillot, J.-P.; Birot, M.; Dunoguès, J. *Chem. Mater.* **1991,** *3,* 356.
143. Sartori, P.; Habel, W.; Aefferden, B. van; Hurtado, A. M.; Dose, H. R.; Alkan, Z. *Eur. J. Solid State Inorg. Chem.* **1992,** *29,* 127.

144. Ijadi-Maghsoodi, S.; Pang, Y.; Barton, T. J. *J. Polym. Sci. Part A Polym. Chem.* **1990**, *28*, 955.
145. Meyer, M. K.; Akinc, M.; Ijadi-Maghsoodi, S.; Zhang, X. Barton, T. J. *Ceram. Eng. Sci. Proc.* **1991**, *12*, 1019.
146. Seyferth, D.; Lang, H. *Organometallics* **1991**, *10*, 551.
147. White, D. A.; Oleff, S. M.; Boyer, R. D.; Budinger, P. A.; Fox, J. R. *Adv. Ceram. Mater.* **1987**, *2*, 45.
148. White, D. A.; Oleff, S. A.; Fox, J. R. *Adv. Ceram. Mater.* **1987**, *2*, 53.
149. Chorley, R. W.; Lednor, P. W. *Adv. Mater.* **1991**, *3*, 474.
150. Hurwitz, F. I.; Hyatt, L.; Gorecki, J.; D'Amore, L. *Ceram. Eng. Sci. Proc.* **1987**, *8*, 732.
151. Ishida, H.; Shick, R.; Hurwitz, F. *Polym. Mater. Sci. Eng.* **1990**, *63*, 877.
152. Hurwitz, F. I.; Farmer, S. C.; Terenka, F. M.; Leonhardt, T. A. *J. Mater. Sci.* **1991**, *26*, 1247.
153. Heimann, P. J.; Hurwitz, F. I.; Rivera, A. L. *Ceram. Trans.* **1991**, *19* (Adv. Compos. Mater.), 27.
154. Burns, G. T.; Taylor, R. B.; Xu, Y.; Zangvil, A.; Zank, G. A. *Chem. Mater.* **1992**, *4*, 1313.
155. Renlund, G. M.; Prochazka, S.; Doremus, R. H. *J. Mater. Res.* **1991**, *6*, 2716, 2723.
156. Babonneau, F.; Livage, J.; Sorarù, G. D.; Carturan, G.; Mackenzie, J. D. *New J. Chem.* **1990**, *14*, 539.
157. Thorne, K.; Liimatta, E.; Mackenzie, J. D. *J. Mater. Res.* **1991**, *6*, 2199.
158. Babonneau, F.; Sorarù, G. D. *J. Eur. Ceram. Soc.* **1991**, *8*, 29.
159. Czekaj, C. L.; Hackney, M. L. J.; Hurley, W. J., Jr.; Interrante, L. V.; Sigel, G. A. *J. Am. Ceram. Soc.* **1990**, *73*, 352.
160. Schmidt, W. R.; Hurley, W. J., Jr.; Doremus, R. H.; Interrante, L. V.; Marchetti, P. S. *Ceram. Trans.* **1991**, *19*, (Adv. Compos. Mater.) 19.
161. Babonneau, F.; Sorarù, G. D.; Thorne, K. J.; Mackenzie, J. D. *J. Am. Ceram. Soc.* **1991**, *74*, 1725.
162. Blum, Y. D.; Laine, R. M. *Organometallics* **1986**, *5*, 2081.
163. Blum, Y. D.; Laine, R. M.; Schwartz, K. B.; Rowcliffe, D. J.; Bening, R. C.; Cotts, D. B. *Mater. Res. Soc. Symp. Proc.* **1986**, *73* (Better Ceram. Chem. 2), 389.
164. Biran, C.; Blum, Y. D.; Glaser, R.; Tse, D. S.; Youngdahl, K. A.; Laine, R. M. *J. Mol. Catal.* **1988**, *48*, 183.
165. Laine, R. M.; Blum, Y. D.; Tas, D.; Glaser, R. In *Inorganic and Organometallic Polymers;* Zeldin, M.; Wynne, K. J.; Allcock, H. R., Eds.; ACS Symposium Series 360; American Chemical Society: Washington, DC, 1988; p 124.
166. Blum, Y. D.; Schwartz, K. B.; Laine, R. M. *J. Mater. Sci.* **1989**, *24*, 1707.
167. Arkles, B. *J. Electrochem. Soc.* **1986**, *133*, 233.
168. Nakaido, Y.; Otani, Y.; Kozakai, N.; Otani, S. *Chem. Lett.* **1987**, 705.
169. Lebrun, J.-J.; Porte, H. U.S. Patents 4 689 252 (1987); 4 689 382 (1987); 4 694 060 (1987).
170. Colombier, C. *Proc. Eur. Ceram. Soc. Conf. 1st* **1989**, 43.
171. Yu, Y.-F.; Mah, T. I. *Mater. Res. Soc. Symp. Proc.* **1986**, *73* (Better Ceram. Chem. 2), 559.

172. Burns, G. T.; Angelotti, T.; Hannemann, L. F.; Chandra, G.; Moore, J. A. *J. Mater. Sci.* **1987,** *22,* 2608.

173. Burns, G. T.; Chandra, G. *J. Am. Ceram. Soc.* **1989,** *72,* 333.

174. Burns, G. T.; Saha, C. K.; Zank, G. A.; Freeman, H. A. *J. Mater. Sci.* **1992,** *27,* 2140.

175. Kanner, B.; King, R. E., III. In *Silicon-Based Polymer Science: A Comprehensive Resource;* Zeigler, J. M.; Fearon, F. W. G., Eds.; Advances in Chemistry 224; American Chemical Society: Washington, DC, 1990; pp 607–618.

176. Yive, N. S. C. K.; Corriu, R. J.; Leclerq, D.; Mutin, P. H.; Vioux, A. *New J. Chem.* **1991,** *15,* 85.

177. Vaahs, T.; Brück, M.; Böcker, E. D. G. *Adv. Mater.* **1992,** *4,* 224.

178. Toreki, W.; Creed, N. A.; Batich, C. D. *Polym. Prepr. (Am. Chem. Soc. Div. Polym. Chem.)* **1990,** *31,* 611.

179. Schwark, J. M. U.S. Patents 4 929 704 (1990); 5 001 090 (1991); 5 021 533 (1991); 5 032 649 (1991).

180. Schwark, J. M. *Polym. Prepr. (Am. Chem. Soc. Div. Polym. Chem.)* **1991,** *32,* 567.

181. Saha, C. K.; Freeman, H. A. *J. Mater. Sci.* **1992,** *27,* 4651.

182. Qui, H.; Yu, S.; Ren, H.; Du, Z. *Polym. Bull.* **1992,** 607.

183. Abe, Y.; Ozai, T.; Kuno, Y.; Nagao, Y.; Misono, T. *J. Inorg. Organomet. Polym.* **1992,** *2,* 143.

184. Mizutani, N.; Liu, T. Q. *Ceram. Trans.* **1990,** *12* (Ceram. Powder Sci. 3), 59.

185. Narsavage, D. M.; Interrante, L. V.; Marchetti, P. S.; Maciel, G. E. *Chem. Mater.* **1991,** *3,* 721.

186. Dippel, K.; Werner, E.; Klingebiel, U. *Phosphorus Sulfur Silicon Relat. Elem.* **1992,** *64,* 15.

187. Pillot, J.; Richard, C.; Dunoguès, J.; Birot, M.; Pailler, R.; Mocaer, D.; Naslain, R.; Olry, P.; Chassagneux, E. *Ind. Ceram. (Paris)* **1992,** No. 867, 48.

RECEIVED for review November 9, 1992. ACCEPTED revised manuscript March 26, 1993.

Molecular Magnets: An Emerging Area of Materials Chemistry

Joel S. Miller[1,3] and Arthur J. Epstein[2]

[1]DuPont Science and Engineering Laboratories, Experimental Station E328, Wilmington, DE 19880–0328

[2]Department of Physics, Ohio State University, Columbus, OH 43210–1108

The anticipated attributes of molecule-based organic–polymeric magnetic materials may enable their use in future generations of electronic, magnetic and photonic–photronic devices. Some organometallic solids comprising linear chains of alternating metallocenium donors, D, and cyanocarbon acceptors, A, for example, $\cdots D^{\cdot+}A^{\cdot-}D^{\cdot+}A^{\cdot-}\cdots$, exhibit cooperative magnetic phenomena, that is, ferro-, antiferro-, ferri-, and metamagnetism. For $[Fe^{III}(C_5Me_5)_2]^{\cdot+}[TCNE]^{\cdot-}$ (Me is methyl; TCNE is tetracyanoethylene), bulk ferromagnetic behavior is observed below the critical (Curie) temperature, T_c, of 4.8 K. Replacement of Fe^{III} with Mn^{III} leads to a ferromagnet with a T_c of 8.8 K, in agreement with mean-field models developed for this class of materials. Extension to the reaction of a vanadium(0) complex with TCNE leads to the isolation of a magnet with a $T_c \sim 400$ K, which exceeds the decomposition temperature of the material. This result demonstrates that a magnetic material with a T_c substantially above room temperature is achievable in a molecule-based organic–polymeric material.

SOCIETY HAS BENEFITED FROM MAGNETS for centuries and increasingly continues to do so. In an early, if not the earliest, example of tech-

[3]Current address: Department of Chemistry, University of Utah, Salt Lake City, UT 84112

0065–2393/95/0245–0161$14.00/0

nology transfer involving materials science, the compass was invented (*1*). Today, with annual western-world sales of $200 million (*2*), magnets are indispensable to our high-technology society. Magnets are used in a wide variety of technologies ranging from magneto-mechanical machines (frictionless bearings, medical implants, magnetic separators, etc.), to acoustic devices (loudspeakers, microphones, etc.), to telecommunicator information technology (switches, sensors, magnetic resonance imaging, magnetic and optical disks, etc.), and to motors and generators (*3–9*). In the future, so-called "smart" materials and systems will undoubtedly rely upon "smart" switches, sensors, and transducers that will be composed in part of magnetic materials. Magnets consequently are a key focus of modern materials research programs.

Magnetic materials share the attributes of being d- or f-orbital transition or lanthanide atom–ion based with an extended network bonding in at least two dimensions. They are prepared by high-temperature metallurgy. A molecule-based material is composed of molecular or molecular ion entities and not atoms or atomic ions. A molecular magnet is a magnet that is molecule-based. A molecule-based organic–polymeric material should enable the deliberate modulation of the magnetic properties by established methods in organic chemistry; the combining of magnetic properties with mechanical, electrical, or optical properties; and the simplifying of the fabrication through low-temperature processing (Table I). These attributes as well as those of conductivity and superconductivity in molecule-based organic–polymeric materials may enable the use of organic materials in future generations of electronic, magnetic, and photonic–photronic devices.

Molecule-based ferromagnetic compounds, although postulated in the 1960s, have been realized only within the past decade (*10–23*). Herein we will very briefly review the magnetic phenomena necessary to appreciate the physical properties of this class of materials and describe some of our research results.

Magnetic Behavior: A Synopsis

Magnetism arises from the intrinsic spin of an electron and the manner in which the electron spins on adjacent atoms (or molecules) couple with each other. Magnetism is detected by a materials attraction (or repulsion) to a magnet. Every electron has a minute magnetic moment associated with the quantum mechanical spin. The electrons reside in orbitals, and each orbital may have two electrons—one spin-up ($\uparrow$) and one spin-down ($\downarrow$)—such that the spins cancel. A radical has one (or more) unpaired electrons and a net spin. [An unpaired electron is designated by a dot

Table I. Representative Examples of Magnetic Materials

Building Unit	Example[a]	Active Spin Site	Processing Method
Atom-based[b]	Fe, Fe_3O_4, Co_5Sm, CrO_2	Fe, Co, Sm, Cr	metallurgy
	C	C^c	pyrolysis–metallurgy
Molecule-based	$[FeCp^*_2][TCNE]$	Fe^{III}, $TCNE^d$	growth from solution
	$Mn(hfac)_2(NITEt)$	Mn^{II}, nitroxide[d]	growth from solution
	$Cu^{II}Mn^{II}(obbz)$	Mn^{II}, Cu^{II}	growth from solution
	$p\text{-}NO_2C_6H_4NIT$	nitroxide[d]	growth from solution
	$V[TCNE]_x \cdot y(solvent)$	V, $TCNE^d$	growth from solution

[a]Abbreviations: TCNE is tetracyanoethylene, hfac is hexafluoroacetylacetonate, NIT is nitroxide, NITEt is ethylnitronyl nitroxide, and obbz is oxamidobis(benzoato).
[b]Commercial products.
[c]Proposed.
[d]p-orbital based.

(·).] Usually radicals are sufficiently far enough apart that their coupling energy is small compared to the coupling-breaking (thermal) energy. Such spins do not couple (i.e., align cooperatively), but form instead a weak magnet (paramagnet) (Figure 1a). When the radicals become close to each other, the coupling energies increase ultimately to influence the alignment of the neighboring spins. If this alignment occurs in such a manner that the spins oppose each other, it leads to antiferromagnetic behavior (Figure 1b).

Bulk ferromagnetic behavior, albeit rare, occurs when the spins in a solid align in the same direction (Figure 1c) and result in a net magnetic moment. Thus, ferromagnetism requires that the individual unpaired spins interact collectively with each other to align themselves. Ferrimagnets have different moments (essentially different numbers of spins) on neighboring sites, which, although they align antiferromagnetically, cannot cancel; this condition leads to a net magnetism for the solid (Figure 1d). A vitally important concept is that this commercially useful highly magnetic behavior is not a property of a molecule, atom, or ion; it is a cooperative solid-state (bulk) property.

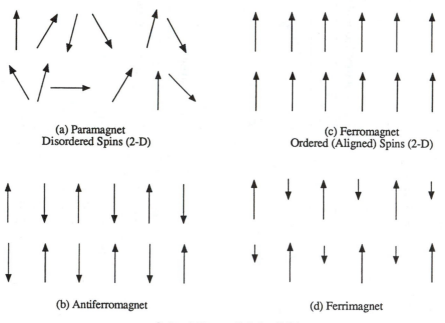

(a) Paramagnet
Disordered Spins (2-D)

(c) Ferromagnet
Ordered (Aligned) Spins (2-D)

(b) Antiferromagnet

(d) Ferrimagnet

Ordered (Opposed) Spins (2-D)

Figure 1. Schematic illustration of spin alignment for a 2-D paramagnet (a), antiferromagnet (b), ferromagnet (c), and ferrimagnet (d).

Structure and Magnetic Properties of Electron-Transfer Donor–Acceptor Salts

In 1979 $[Fe^{III}Cp^*_2]^{\cdot+}[TCNQ]^{\cdot-}$ (Cp* is pentamethylcyclopentadienide, C_5Me_5; TCNQ is 7,7,8,8-tetracyano-*p*-quinodimethane) was characterized as a metamagnet (24); that is, below a 1.6-kG applied field the magnetization is characteristic of an antiferromagnet, whereas above 1.6 kG a sharp rise and approach to magnetization saturation characteristic of a ferromagnet is observed. These materials are composed of alternating cation donors, D, and anion acceptors, A, for example, $\cdots D^{\cdot+}A^{\cdot-}D^{\cdot+}A^{\cdot-}\cdots$ (20–26) (Figure 2).

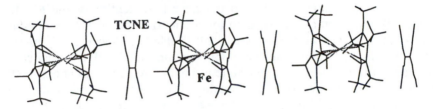

Figure 2. Alternating donor–acceptor, $\cdots D^{\cdot+}A^{\cdot-}D^{\cdot+}A^{\cdot-}\cdots$, linear chain structure of $[Fe^{III}Cp^*_2]^{\cdot+}[A]^{\cdot-}$ [A is TCNQ, TCNE, DDQ, $C_4(CN)_6$, $[Fe^{II}Cp_2][TCNE]$, and $[Fe^{III}Cp^*_2]^{\cdot+}[C_3(CN)_5]^-$. The structure shows views of adjacent out-of-registry chains for A = TCNE.

Although several metamagnets, for example, $FeCl_2$, had been reported (27), $[FeCp^*_2]^{\cdot+}[TCNQ]^{\cdot-}$ was the first example in which a one-, two-, or three-dimensional (1-D, 2-D, or 3-D) covalently bonded network structure is not present in the structure. Subsequently, we sought to elucidate the structure–function relationship via the systematic modification of the acceptor A, the C_5Me_5-ring substituent groups, and the metal ion to understand the steric–electronic features necessary to stabilize magnetism and ultimately design a molecular-based magnet. On the basis that a smaller radical anion would have a greater spin density that could lead to increased spin–spin interactions, we sought to identify stable radical ions smaller than $[TCNQ]^{\cdot-}$, **1**, and selected $[TCNE]^{\cdot-}$, **2** (TCNE is tetracyanoethylene).

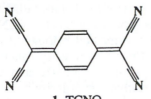

1, TCNQ

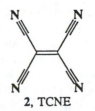

2, TCNE

$[Fe^{III}Cp*_2]^{\bullet+}[TCNE]^{\bullet-}$ was prepared and also found to possess the $\cdots D \cdot^+ A \cdot^- D \cdot^+ A \cdot^- \cdots$ motif (Figure 2) (28). With application of only the earth's magnetic field ($\sim$0.3 G) a spontaneous magnetization is observed (29). The saturation magnetization for single crystals aligned parallel to the $\cdots D \cdot^+ A \cdot^- D \cdot^+ A \cdot^- \cdots$ stacking axis is 36% greater than iron on a gram-atom basis and is in agreement with the calculated saturation moment for ferromagnetic alignment of the donor and the acceptor. The critical (Curie) temperature, T_c, was determined to be 4.8 K. The magnetization versus applied field data for $[Fe^{III}Cp*_2]^{\bullet+}[TCNE]^{\bullet-}$ exhibit hysteresis loops (Figure 3) with a large coercive field of 1 kG at 2 K (29).

Fitting the susceptibility data to different physical models aids in the understanding of the microscopic spin interactions. The Curie–Weiss law, $\chi \propto 1/(T - \theta)$, where χ is the molar magnetic susceptibility and θ is the Curie–Weiss constant, can be used to parametrize the higher temperature ($T > 130$ K) data with $\theta = 30$ K (28, 29) (Table II). Above 16 K, interactions among the nearest neighbor spins within individual 1-D chains are sufficient to understand the magnetic behavior. A 1-D Heisenberg model with a ferromagnetic exchange coupling, J, of 19 cm^{-1} models the data down to 16 K (29). These models, however, are insufficient to explain the magnetic susceptibility below 16 K as long-range spin correlations and 3-D spin interactions become increasingly important, until permanent long-range (3-D) ferromagnetic order occurs at a T_c of 4.8 K. Variation of the low-field magnetic susceptibility with temperature above T_c, magnetization with temperature below T_c, and the magnetization with magnetic field at T_c enabled a measurement of three critical exponents (29) for the magnetic field parallel to the chain axis. The observed values are in accord with a transition to a 3-D ordered magnetic state.

To identify the structure–function relationship with the goal of preparing a magnet with a higher critical temperature, the properties of $\cdots$DADA$\cdots$ structured compounds based on $[M(C_5R_5)_2]^+$ were studied. Three modifiable entities are replacement of the Me groups with H and Et, use of alternate open- and closed-shell anions, and replacement of Fe^{III} with other metal ions.

Substituted C_5-rings may maintain the fivefold symmetry necessary to lead to a cation with a degenerate partially occupied molecular orbital and a Kramers doublet as established for ferrocenium and decamethylferrocenium. Three such ferrocenes have been studied: ferrocene, 1,2,3,4,5-pentamethylferrocene, and decaethylferrocene. Ferrocene is more difficult to oxidize than decamethylferrocene and is not oxidized by TCNE. Nonetheless, diamagnetic $[Fe^{II}Cp_2][TCNE]$ (35) forms and belongs to the same structure type (35–41) (Figure 2). FeCpCp* (Cp is cyclopentadienide, C_5H_5) is a sufficiently strong donor to reduce TCNE, and the simple (1:1) 1-D salt was prepared (42). It exhibits weak ferromagnetic coupling as evidenced from the fit of its susceptibility to the Curie–Weiss

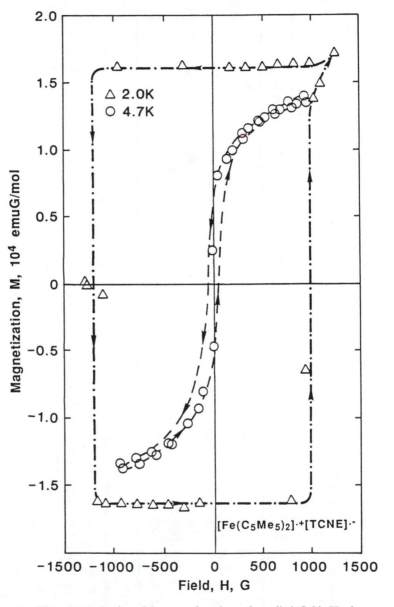

Figure 3. The magnetization, M, *as a function of applied field,* H, *for [FeCp**$_2$*]*$^{\bullet +}$*[TCNE]*$^{\bullet -}$ *shows hysteresis loops at 2 K.*

Table II. Summary of the Critical Temperatures and Coercive Fields for $[MCp^*_2]\cdot^+[A]\cdot^-$

| | $[TCNE]\cdot^-$ | | | | $[TCNQ]\cdot^-$ | |
	$[FeCp^*_2]\cdot^+$	$[MnCp^*_2]\cdot^+$	$[CrCp^*_2]\cdot^+$	$[FeCp^*_2]\cdot^+$	$[MnCp^*_2]\cdot^+$	$[CrCp^*_2]\cdot^+$
$S_D{}^+$	1/2	1	3/2	1/2	1	3/2
$S_A{}^-$	1/2	1/2	1/2	1/2	1/2	1/2
T_c, K	4.8	8.8	3.65[a]	2.55[b]	6.3[c]	3.3[d]
θ, K	+16.8[e]	+22.6	+22.2	+12.3	+10.5[f]	+11.6[g]
H_c, G (K)	1000(2)	1200 (4.2)	—[h]		3600(3)	—[h]
Reference	29	30	31	24	32	33

[a]0.15 G.
[b]Metamagnetic with a 16-kG critical field.
[c]50 G, T_c from the maximum slope of dM/dT is reported as 6.2 K.
[d]15 G, T_c from the maximum slope of dM/dT is reported as 3.1 K.
[e]$\theta_\parallel = +30$ K, $\theta_\perp = +10$ K; 0.15 G.
[f]9.0 K (34).
[g]12.8 K (34).
[h]Not observed.

expression with $\theta = +3.2$ K, but cooperative 3-D magnetic behavior is not observed to the lowest temperature studied ($\sim$2 K) (*42*). Likewise a linear chain structure is proposed for $[Fe^{III}(C_5Et_5)_2]^{\bullet+}[TCNE]^{\bullet-}$, which also exhibits ferromagnetic coupling as evidenced from the fit of its susceptibility data to the Curie–Weiss expression with $\theta = +7.5$ K, but again a cooperative 3-D magnetic state was not observed (*43*).

To test the necessity of a ^2E ground state, the TCNE electron-transfer salt with the lower symmetry $Fe(C_5Me_4H)_2$ donor was prepared (*44*). The magnetic susceptibility can be fit by the Curie–Weiss expression with $\theta \sim 0$ K. The absence of 3-D ferro- or antiferromagnetic ordering above 2.2 K is observed for $[Fe(C_5Me_4H)_2]^{\bullet+}[TCNE]^{\bullet-}$, in contrast with that reported for $[FeCp^*_2]^{\bullet+}[TCNE]^{\bullet-}$. The lack of magnetic ordering apparently arises from poorer intra- and intermolecular overlap within and between the chains, leading to substantially weaker magnetic coupling for $[Fe(C_5Me_4H)_2]^{\bullet+}[TCNE]^{\bullet-}$. This weaker coupling suppresses the spin ordering temperature. Alternatively, because of the overall C_{2v}, symmetry, the $[Fe(C_5Me_4H)_2]^{2+}$ charge-transfer excited state may be a singlet, not a triplet as expected for $[FeCp^*_2]^{2+}$. However, the admixture of a singlet, not a triplet charge-transfer excited state should lead to antiferromagnetic, not ferromagnetic, coupling. Significant antiferromagnetic coupling was also absent for $[Fe(C_5Me_4H)_2]^{\bullet+}[TCNE]^{\bullet-}$, and thus reduced overlap with neighboring radicals is likely the more important effect of modification of the cation.

Substitution of TCNE with, for example, TCNQ (*26*), $C_4(CN)_6$ (*45*), and DDQ (2,3-dichloro-5,6-dicyanobenzoquinone) (*46, 47*), or $[M(S_2C_2(CF_3)_2)_2]_n^{\bullet-}$ [$n = 2$, M = Ni (*48*) or Pt (*49*)] acceptors leads to $\cdots D \cdot^+ A \cdot^- D \cdot^+ A \cdot^- \cdots$ structured complexes with dominant ferromagnetic coupling (Table III). Replacement of $[TCNE]^{\bullet-}$ with diamagnetic $[C_3(CN)_5]^-$ also leads to the formation of a $\cdots D \cdot^+ A \cdot^- D \cdot^+ A \cdot^- \cdots$ structured phase. It exhibits only Curie susceptibility ($\theta = -1$ K) (*28*).

The $[Co^{III}Cp^*_2]^+[TCNE]^{\bullet-}$ complex with a diamagnetic donor was prepared and exhibits essentially the Curie susceptibility anticipated for $[TCNE]^{\bullet-}$ (*28*). Attempts to prepare $[M^{III}Cp^*_2]^+$ (M = Ru or Os) salts of $[TCNE]^{\bullet-}$ have yet to lead to compounds suitable for comparison with the highly magnetic Fe^{III} phase (*50*). Replacement of Fe^{III} in $[Fe^{III}Cp^*_2]^+[A]^{\bullet-}$ (A is TCNE or TCNQ) with doublet Ni^{III} or quartet Cr^{III} leads to compounds exhibiting antiferromagnetic and ferromagnetic (Table III) magnetic properties.

Although a magnetic ground state (ferro-, ferri-, antiferromagnetic, etc.) is not observed for bis(dithiolato)metallate salts of decamethylferrocenium, these salts exhibit ferromagnetic coupling, as evidenced by Curie–Weiss θ constants that range from 0 to +27 K (Table IV). This behavior provides an insight into a structure–function relationship. Of the compounds studied, the $[MCp^*_2]\{M'[S_2C_2(CF_3)_2]_2\}$ [M' = Ni or Pt;

Table III. Curie–Weiss θs and Critical Temperatures
for $\cdots D\cdot^{+}A\cdot^{-}D\cdot^{+}A\cdot^{-}\cdots$ Structures

Salt with $\cdots D\cdot^{+}A\cdot^{-}D\cdot^{+}A\cdot^{-}\cdots$ Structure	θ $(K)^{a}$	T_c $(K)^{b}$	Ref.
$[FeCp^{*}_2]\cdot^{+}[TCNE]\cdot^{-}$	+16.9	4.8	28, 29
$[FeCp^{*}_{20.955}\cdot^{+}[CoCp^{*}_2]_{0.045}{}^{+}[TCNE]\cdot^{-}$		4.4	51, 52
$[FeCp^{*}_2]_{0.923}\cdot^{+}[CoCp^{*}_2]_{0.0777}{}^{+}[TCNE]\cdot^{-}$		3.8	51, 52
$[FeCp^{*}_2]_{0.915}\cdot^{+}[CoCp^{*}_2]_{0.085}{}^{+}[TCNE]\cdot^{-}$		2.75	51, 52
$[FeCp^{*}_2]_{0.855}\cdot^{+}[CoCp^{*}_2]_{0.145}{}^{+}[TCNE]\cdot^{-}$		0.75^{c}	51, 52
$[FeCp^{*}_2]\cdot^{+}\{Pt[S_2C_2(CF_3)_2]_2\}\cdot^{-}$	+27		49
$[MnCp^{*}_2]\!:\!^{+}[TCNE]\cdot^{-}$	+22.6	8.8	30
$[CrCp^{*}_2]\!:\!^{+}[TCNE]\cdot^{-}$	+22.2	3.65	31
$[FeCp^{*}_2]\cdot^{+}\{Ni[S_2C_2(CF_3)_2]_2\}\cdot^{-}$	+15		48
$[CrCp^{*}_2]\!:\!^{+}[TCNQ]\cdot^{-}$	+11.6	3.5^{d}	33
$[FeCp^{*}_2]\cdot^{+}[TCNQ]\cdot^{-}$	+12.3	2.55^{e}	6, 34
$[MnCp^{*}_2]\!:\!^{+}[TCNQ]\cdot^{-}$	+10.5	6.5^{f}	32
$[Fe(C_5Et_5)_2]\cdot^{+}[TCNE]\cdot^{-}$	+7.5		43
$[Fe(C_5Et_5)_2]\cdot^{+}[TCNQ]\cdot^{-}$	6.1		43
$[FeCpCp^{*}]\cdot^{+}[TCNE]\cdot^{-}$	+3.3		35
$[Fe(C_5Me_4H)_2]\cdot^{+}[TCNQ]\cdot^{-}$	+0.8		44
$[Fe(C_5Me_4H)_2]\cdot^{+}[TCNE]\cdot^{-}$	−0.3		44
$[CoCp^{*}_2]^{+}[TCNE]\cdot^{-}$	−1		28
$[FeCp^{*}_2]\cdot^{+}[C_3(CN)_5]^{-}$	−1.2		28
$[NiCp^{*}_2]\cdot^{+}[TCNE]\cdot^{-}$	−11.5		34

^{a}For polycrystalline samples.

$^{b}T_c$ was determined from a linear extrapolation of the steepest slope of the $M(T)$ data to the temperature at which $M = 0$.

^{c}Alternating current (ac) (100-Hz measurement).

^{d}15 G, T_c from the maximum slope of dM/dT is reported as 3.1 K.

^{e}Metamagnetic.

^{f}50 G, T_c from the maximum slope of dM/dT is reported as 6.2 K.

M = Fe (48) or Mn (53)] series have a 1-D chain structure. For M = Fe they also have the greatest θ values, greatest effective moments, and the most pronounced field dependence of the susceptibility. The Pt analogue with θ = +27 K possesses 1-D $\cdots D\cdot^{+}A\cdot^{-}D\cdot^{+}A\cdot^{-}\cdots$ chains, whereas the Ni analogue with only a θ of 15 K possesses zig-zag 1-D $\cdots D\cdot^{+}A\cdot^{-}D\cdot^{+}A\cdot^{-}\cdots$ chains and longer M$\cdots$M separations (11.19 vs. 10.94 Å for the Pt ana-

Table IV. Curie–Weiss θs and μ_{eff} for FeCp*$_2$]{M[S$_2$C$_2$(CN)$_2$]$_2$}$_n$}

Anion	Structural Arrangement	Spin Repeat Unit	Susceptibility	
			θ (K)	μ_{eff} μ_B[a]
{Ni[S$_2$C$_2$(CN)$_2$]$_2$}·$^-$	D·$^+$[A]$_2^{2-}$D·$^+$ dimer[b]	D·$^+$	0	2.83
α-{Pt[S$_2$C$_2$(CN)$_2$]$_2$}·$^-$	··D·$^+$A·$^-$D·$^+$·· sheets 1-D, ··D·$^+$[A]$_2^{2-}$·· chains[b]	D·$^+$ + ⅓A·$^-$	+6.6	3.05
β-{Pt[S$_2$C$_2$(CN)$_2$]$_2$}·$^-$	1-D, ··D·$^+$A·$^-$·· chains D·$^+$[A]$_2^{2-}$D·$^+$ dimers[b]	D·$^+$ + ⅓A·$^-$	+9.8	3.10
{Ni[S$_2$C$_2$(CF$_3$)$_2$]$_2$}·$^-$	1-D, ··D·$^+$A·$^-$··[c]	D·$^+$ + A·$^-$	+15	3.73
{Pt[S$_2$C$_2$(CF$_3$)$_2$]$_2$}·$^-$	1-D, ··D·$^+$A·$^-$·· chains	D·$^+$ + A·$^-$	+27	3.76

[a] μ_{eff} is effective magnetic moment, and μ_B is Bohr magneton; $\mu = (8\chi T)^{1/2}$.
[b] [A]$_2^{2-}$ = isolated $S = 0$ {M[S$_2$C$_2$(CN)$_2$]$_2$}$_2^{2-}$ dimer.
[c] Zig-zag chains.

logue). Thus, the enhanced magnetic coupling arises from the stronger intrachain coupling. In contrast, for $[MnCp^*_2]\{M'[S_2C_2(CF_3)_2]_2\}$ (M' = Ni, Pd, or Pt) (53), weak ferromagnetic coupling ($\theta = +2.82 \pm 0.8$ K) and metamagnetic behavior a with T_c of 2.5 $\pm$ 0.3 K are observed. This observation is comparable to the T_c reported for $[FeCp^*_2]^{\cdot+}[TCNQ]^{\cdot-}$ (24).

Replacement of $S = \frac{1}{2}[FeCp^*_2]^{\cdot+}$ in $[FeCp^*_2]^{\cdot+}[TCNE]^{\cdot-}$ with $S = 1$ $[MnCp^*_2]^{\cdot+}$ increases T_c nominally in accord with theory [i.e., $T_c \propto S(S + 1)$], and replacement of $[FeCp^*_2]^{\cdot+}$ in $[Fe(Cp^*_2]^{\cdot+}[TCNQ]^{\cdot-}$ with $[MnCp^*_2]^{\cdot+}$ destabilizes the metamagnetic behavior; replacement of $[FeCp^*_2]^{\cdot+}$ in $[MnCp^*_2]\{M'[S_2C_2(CF_3)_2]_2\}$ with $[MnCp^*_2]^{\cdot+}$ stabilizes metamagnetic behavior (53).

In contrast, $[FeCp^*_2]\{Ni[S_2C_2(CN)_2]_2\}$ possesses isolated $D^{\cdot+}A_2^{2-}D^{\cdot+}$ dimers with only one spin per repeat unit and has a zero θ and no field-dependence of the susceptibility (48). Between these limiting cases are the α- and β-$[FeCp^*_2]\{Pt[S_2C_2(CN)_2]_2\}$, which have 1-D $\cdots D^{\cdot+}A^{\cdot-}D^{\cdot+}A^{\cdot-}\cdots$ strands in one direction and $\cdots DAAD\cdots$ units in another direction and their θ and effective magnetic moment (μ_{eff}) values are intermediate in value (Table IV). This result is consistent with the presence of one-third of the anions having a singlet ground state. Thus, a correlation exists between the presence of 1-D $\cdots D^{\cdot+}A^{\cdot-}D^{\cdot+}A^{\cdot-}\cdots$ chains and the existence and magnitude of ferromagnetic coupling, as evidenced by a θ value.

These results support the necessity of 1-D $\cdots D^{\cdot+}A^{\cdot-}D^{\cdot+}A^{\cdot-}\cdots$ chain structure for achieving ferromagnetic coupling and ultimately bulk ferromagnetic behavior as observed for $[FeCp^*_2]^{\cdot+}[TCNE]^{\cdot-}$ (28).

Models for Molecule-Based Magnetic Materials

Several mechanisms for obtaining the spin alignment throughout the solid necessary for ferromagnetism in a molecule-based organic–polymeric material have evolved since the early 1970s. A fact worthy of emphasis is that the mechanisms that have been proposed are for the pairwise stabilization of ferromagnetic coupling among spins, and this stabilization is insufficient to account for three-dimensional ferromagnetic behavior, for which interactions in all three directions are crucial. The proposed models include the following

- configuration mixing of a virtual triplet excited state with the ground state for a $\cdots[D]^{\cdot+}[A]^{\cdot-}[D]^{\cdot+}[A]^{\cdot-}\cdots$ chain

- Heider–London spin-exchange between positive spin density on one radical and negative spin density on another

- very high spin multiplicity radicals based on strongly coupled unpaired electrons in orthogonal orbitals

- the antiferromagnetic coupling of alternating spin sites with a different number of spins per site; this property leads to ferrimagnetic behavior.

The model of configuration mixing of a virtual triplet charge-transfer excited state with the ground state for a $\cdots[D]\cdot^+[A]\cdot^-[D]\cdot^+[A]\cdot^-\cdots$ chain to stabilize ferromagnetic coupling was originally introduced by McConnell (*54, 55*). The large separations and energy differences between ions on adjacent sites lead to small overlaps of the frontier orbitals and prevent the formation of metallic energy bands for mixed-stack electron-transfer salts. Instead, for a $[D]\cdot^+[A]\cdot^-$ pair with half-occupied nondegenerate highest occupied molecular orbitals (HOMOs) the spins couple antiferromagnetically (Figure 4-Ia). (This conclusion assumes that the virtual charge transfer involves only the highest energy partially occupied molecular orbital. Circumstances in which virtual excitation from a lower lying filled (or to a high lying filled) orbital dominate the admixing exciting state are conceivable, and the orbital degeneracy and symmetry restrictions are relaxed.)

Admixture of the higher energy charge-transfer states with the ground state lowers the total electronic energy and stabilizes antiferromagnetic coupling ('↑↓'). Figures 4-IIb and 4-IIc illustrate an electron being delocalized onto an adjacent site. This energy reduction and delocalization does not occur when the two electron spins are parallel ('↑↓', ferromagnetically aligned) in accord with the Pauli exclusion principle. Thus, antiferromagnetic coupling is achieved. For other electron configurations involving partially occupied degenerate orbitals, ferromagnetic coupling can be achieved along and between chains. The magnetic properties of several electron-transfer salts based on metallocenes and cyanocarbons were discussed within this model (*9*).

Figure 5 illustrates how the energy of the systems depicted in Figure 4 is lowered upon virtual charge-transfer admixture of the ground state, E_{gs}, with one intrachain, two interchain, and with a third interchain excited states E_{es}. The relevance of this model is the subject of current controversy.

Heider–London spin-exchange was also reported by McConnell (*56*) in 1963 to be a mechanism for ferromagnetic exchange. He claimed that radicals with "large positive and negative atomic π-spin densities ... [that] pancake ... so that atoms of positive spin density are exchanged coupled ... to atoms of negative spin density in neighboring molecules ... gives a ferromagnetic exchange interaction." Here ferromagnetic exchange results from the incomplete cancellation of antiferromagnetic coupled spin components. This mechanism describes pairwise ferromagnetic exchange, not bulk ferromagnetism. Ferromagnetic exchange in three dimensions is necessary for bulk ferromagnetic behavior. To achieve ferromagnetic exchange via this mechanism, routes to spin pairing, for example, bond formation, must be avoided.

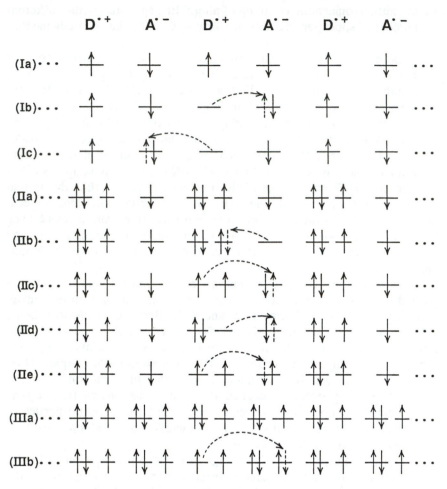

Figure 4. Schematic illustration of stabilization of antiferromagnetic or ferromagnetic coupling. If both the D and A have a half-filled nondegenerate partially occupied molecular orbital (POMO) (s¹) (Ia) then the A ← D (or D ← A) charge-transfer excited state (Ib or equivalently Ic) stabilizes antiferromagnetic coupling. If the D (or A) have a non-half-filled degenerate POMO (e.g., d³, assumed here to be the D) (IIa), then the D ← A charge-transfer excited state formed (IIb) or A ← D charge-transfer excited states formed via excitation of a "spin-up" electron (IIc or IId) will stabilize antiferromagnetic coupling. In contrast, the excited state formed via the D ← A charge-transfer (IIe) stabilizes ferromagnetic coupling. Hund's rule predicts this to be the dominant excited state that admixes with the ground state. If the D (or A) has a half-filled degenerate POMO (e.g., d³) (IIIa), then the A ← D (or D ← A) charge-transfer (disproportionation) excited state (IIIb) stabilizes ferromagnetic coupling.

workers (*13*) focused on the same mechanism through utilization of metal ions and nitroxide radicals with extended network structures. Similar to that already discussed, adjacent spin sites couple antiferromagnetically ('↑↓') whereas next-adjacent spin sites couple ferromagnetically ('↑↑'). Thus, via alternating sites with a larger number of spins (e.g., five for $S = \frac{5}{2}Mn^{II}$) with sites with fewer spins (e.g., one for $S = \frac{1}{2}Cu^{II}$ or nitroxide ligands, i.e., '↑↓'), the Ss cannot cancel, and this property thereby leads to a ferrimagnetic system (*see* Figure 1d). Kahn and his group (*73*) reported several examples of highly magnetic systems developed by this strategy using Mn^{II} and Cu^{II}; for example, $Cu^{II}Mn^{II}(obbz)\cdot H_2O$ [obbz is oxamido-bis(benzoato)] has a 14 K T_c. The Gatteschi and Rey groups (*74, 75*) focused on Mn^{II}–nitroxide systems, for example, $Mn^{II}(hfac)_2NITEt$ (hfac is hexafluoroacetylacetonate; NITEt is ethylnitronyl nitroxide), which is a ferrimagnet with a T_c of 8.1 K.

Calculation of the Curie Temperature, T_c

Heisenberg, along with Dirac (*76*), solved expressions that relate T_c to exchange integrals between sites i and j, J, for the general spin case assuming that only the number of equivalent nearest neighbor sites, z, is important. This simplest mean-field model leads to the following expression for the general spin case (*77*):

$$T_c = \frac{2JzS(S + 1)}{3k_B} \tag{1}$$

where k_B is the Boltzmann constant.

This model is constrained to have only one spin site S and a unique J. These are poor assumptions for the $[MCp*_2]^+[TCNE]^{\cdot-}$ system, as the distinct spin sites may have different values of S and the crystal structure implies four distinctly different nearest neighbor J values, that is, J_z, J_{AA}, J_{DD}, and J_{OR}^{DA} (Figure 6). A mean-field expression for T_c was developed under the assumption that these J values are all identical (*78*), that is,

$$T_c = \frac{64J\left[S_A S_D + S_A^2 S_D + S_A S_D^2 + S_A^2 + S_D^2\right]}{3\left[\left[S_A^2 + 2S_A^3 + S_A^4 + S_D^2 + 2S_D^3 + S_D^4 + 34(S_A S_D + S_A^2 S_D + S_A S_D^2 + S_A^2 S_D^2)\right]^{\frac{1}{2}} - (S_A + S_A^2 + S_D + S_D^2)\right]}$$

Equation 2

Thus, assuming the intra- and interchain interactions remain unchanged as the donor is varied, the effective exchange integral (J_{eff}) will

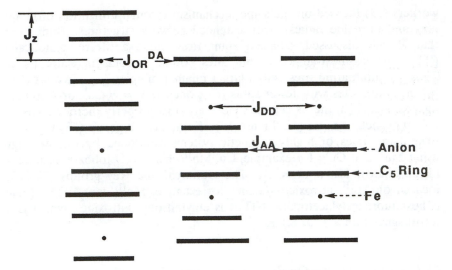

Out-of-Registry   In-Registry

Figure 6. Schematic illustration of the structure of orthorhombic [FeCp$_2$]$^{\cdot+}$[TCNE]$^{\cdot-}$ showing intrachain pairwise out-of-registry and in-registry interactions.*

remain constant, and consequently the relative T_c can be calculated for different values of S_A and S_D (Table V). For [Fe^{III}Cp*$_2$]$^{\cdot+}$[TCNE]$^{\cdot-}$ T_c is $4J_{eff}$ within this mean-field model. Thus, because T_c is observed to be 4.8 K, then J_{eff} is 1.2 K (0.83 cm^{-1}). This result points out that the value of J_{eff} is indeed small, but sufficient to enable a T_c at experimentally accessible temperatures.

The key advantage of equation 2 is that T_c can be scaled for different S_D values (Table V). Thus, T_c for an $S_D = 1$ donor (with an $S_A = \frac{1}{2}$ acceptor) is 6.8 J or 1.7 times greater than that for a $S_D = \frac{1}{2}$ donor (with an $S_A = \frac{1}{2}$ acceptor) system. T_c is 4.8 K for [Fe^{III}Cp*$_2$]$^{\cdot+}$[TCNE]$^{\cdot-}$, then,

Table V. Calculated Mean-Field T_c as a Function of S_D for $S_A = \frac{1}{2}$ Systems

S_D	T_c (J)	Scaled $T_c{}^a$
$\frac{1}{2}$	4	1.00
1	6.8	1.70
$\frac{3}{2}$	10	2.50
2	13.7	3.43

$^a T_c / T_c (S_D = \frac{1}{2})$.

all else being equal T_c is expected to be 8.2 K for the isostructural $[Mn^{III}Cp^*_2]^{\cdot-}[TCNE]^{\cdot-}$. This value is in good agreement with the reported value of 8.8 K for $[Mn^{III}Cp^*_2]^{\cdot-}[TCNE]^{\cdot-}$ *(30)*. (More complex equations using four different J values lead to an improved scaling of T_c as a function of S_D *(30, 78)*. The observed T_c is in excellent agreement with the experimental values and suggests that similar exchange interactions operate for both $[FeCp^*_2]^{\cdot+}[TCNE]^{\cdot-}$ and $[MnCp^*_2]^{\cdot+}[TCNE]^{\cdot-}$.

Although the observed T_c is in excellent agreement with this mean-field model prediction and is suggestive of similar exchange interactions operating for both the $[Fe^{III}Cp^*_2]^{\cdot+}[TCNE]^{\cdot-}$ and $[MnCp^*_2]^{\cdot+}[TCNE]^{\cdot-}$ systems, the Curie–Weiss θ observed for the $[MnCp^*_2]^{\cdot+}[TCNE]^{\cdot-}$ system in the high-temperature regime is 22.6 K and does not scale with the increased magnitude of the cation radical spin, S. Instead it is slightly less than the value of θ recorded for the $[FeCp^*_2]^{\cdot+}[TCNE]^{\cdot-}$; hence these mean-field scaling models must be used with caution.

We also treated theoretically the $S_D = \frac{3}{2}/S_A = \frac{1}{2}$ case (i.e., $[CrCp^*_2]^{\cdot+}[TCNE]^{\cdot-}$) and obtained a predicted scaled value for T_c of 12 K. Experimental study of the magnetic susceptibility of the $[CrCp^*_2]^{\cdot+}[TCNE]^{\cdot-}$ shows a ferromagnetic transition at 3.65 K *(31)*, which is substantially lower than that of either $[FeCp^*_2]^{\cdot+}[TCNE]^{\cdot-}$ and $[MnCp^*_2]^{\cdot+}[TCNE]^{\cdot-}$ (Table III). This trend is also observed for $[MnCp^*_2]^{\cdot+}[TCNQ]^{\cdot-}$ ($T_c = 6.5$ K) *(32)* and $[[CrCp^*_2]^{\cdot+}[TCNQ]^{\cdot-}$ ($T_c = 3.5$ K) *(53)*. The anomalously low value of T_c is further compounded by the ferromagnetic–magnetic exchange, in contrast to the prediction of antiferromagnetic exchange and a ferrimagnetic ground state for the configuration mixing model described earlier (this model successfully yields the sign of the magnetic exchange of all other linear-chain metallocene cases studied so far). Experimental study of the $[CrCp^*_2]^{\cdot+}[TCNE]^{\cdot-}$ is complicated by its extreme sensitivity to the presence of oxygen, which in turn modifies the observed exchange interactions *(31)*.

To test the critical importance of the one-dimensionality, spinless $S = 0$, $[CoCp^*_2]^+$ cations were randomly substituted for the cation in $[FeCp^*_2]^{\cdot+}[TCNE]^{\cdot-}$ structure. This substitution resulted in the formation of random finite magnetic chain segments embedded onto the linear chains *(51, 52)*. A precipitous reduction of T_c occurred with increasing $[CoCp^*_2]^+$ content, in excellent agreement with theoretical concepts *(79)* (Table III). With 14.5% replacement of Co^{III} for Fe^{III}, the T_c plummets from 4.8 to 0.75 K. The extreme sensitivity of the 3-D ordering temperature T_c stands as a cautionary note in attempts to observe a high T_c for solids composed of high-spin oligomers. For such systems the ratio of the intra- to interoligomer exchange may be very large. Given the finite length of oligomers, the results of the aforementioned doping experiments suggest a significant suppression of 3-D ordering.

In summary, cooperative magnetic properties of $[MCp^*_2]^{\cdot+}[A]^{\cdot-}$ (M is Cr, Mn, or Fe; A is TCNE or TCNQ) are reported here. The important magnetic properties are summarized in Table II.

Room-Temperature Polymeric Magnet

With observation of bulk ferromagnetic behavior for $[MnCp*_2]^{\cdot+}[TCNE]^{\cdot-}$, the preparation of the electron-transfer salt of $V(C_6H_6)_2$ and TCNE was identified for study (80–83). $V^I(C_6H_6)_2]^{\cdot+}$, like $[Mn^{III}Cp*_2]^{\cdot+}$, is an $S = 1$ cation possessing a $3E_{2g}$ ground state $(a_{1g}^{1}e_{2g}^{3})$ (84, 85). Addition of $V(C_6H_6)_2\cdot$ to an excess of TCNE in dichloromethane resulted in an amorphous black precipitate with a nominal composition of $V(TCNE)_x \cdot yCH_2Cl_2$ ($x \sim 2; y \sim \frac{1}{2}$). However, because of the extreme insolubility of the precipitate and reactivity of the solvent and extreme air and water sensitivities, variations in composition as a function of preparation conditions were observed. The first step in the reaction is electron transfer from $V(C_6H_6)_2\cdot$ to TCNE followed by loss of the benzene ligands. The material exhibits strong, broad absorptions at 2099 and 2188 cm^{-1}, which are assigned to $\nu_{C\equiv N}$ or perhaps $\nu_{N=C=C}$. The breadth of the $\nu_{C\equiv N}$ absorptions and their relatively low frequency are consistent with the presence of reduced TCNE, with nitrogens coordinated to vanadium.

The $V(TCNE)_{2x} \cdot y(CH_2Cl_2)$ has a field-dependent magnetization, M, between 1.4 and 350 K and moment, μ_{eff} ($\mu_{eff} = [8\chi T]^{1/2}$ or $[8MT/H]^{1/2}$ where H is applied field) (Figure 7). The nearly linear increase of M with decreasing temperature (82) is unusual and may reflect the contribution of the two spin sublattices (V and TCNE) and the effects of disorder. Hysteresis with a coercive field of 60 Oe is observed at room temperature (Figure 8). The strong magnetic behavior is readily observed by its being attracted to a permanent magnet at room temperature (Figure 9). This system is the first and only example of molecule-based organic–polymeric material with a critical temperature above room temperature. The critical temperature exceeds 350 K, the thermal decomposition temperature of the sample. A linear extrapolation of the magnetization to a temperature at which it would vanish leads to an estimate of a T_c of ~400 K. An independent estimate of T_c was obtained (83) using the empirical correlation between the spin-wave coefficient and T_c that exists for amorphous magnets (86).

Because of the structural disorder and variable composition of the magnetic material, the structure has yet to be elucidated. TCNE may bond to metals in many ways. For early transition metals, linear or bent V–N σ-bonds are anticipated. Presently we formulate each vanadium as being surrounded by up to six ligands, which are primarily Ns from different TCNEs. Cl from the weak CH_2Cl_2 ligand or from oxidative addition of CH_2Cl_2 may also coordinate (Figure 10). Any trace oxygen that is

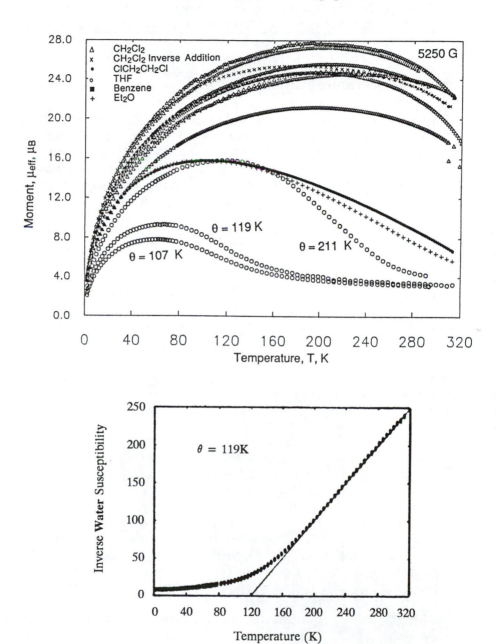

Figure 7. Effective moment, μ_{eff} as a function of temperature, T, for several preparations of $V(TCNE)_x \cdot y(solvent)$.

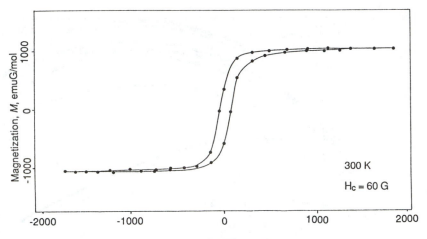

Figure 8. Hysteresis, M(H), of V(TCNE)$_x$ · y(CH$_2$Cl$_2$) at room temperature. (The data were taken on a vibrating sample magnetometer). (Reproduced with permission from reference 18. Copyright 1991 American Association for the Advancement of Science.)

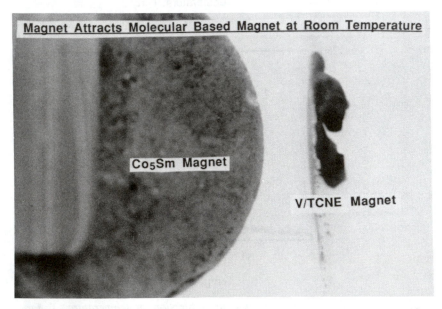

Figure 9. Photograph of a powdered sample of the magnet being attracted to a Co$_5$Sm magnet.

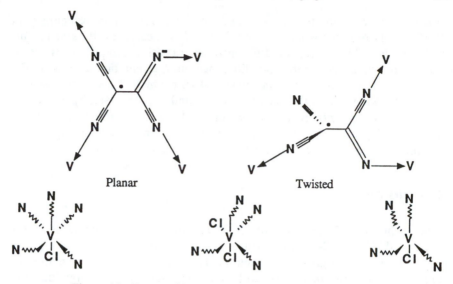

Figure 10. Proposed local bonding about each TCNE and V.

present will strongly bond to the vanadium. The TCNEs may bind up to four different Vs via σ-N bonds. This fragment may be planar or twisted (Figure 10); nonetheless, its ability to bind to more than one vanadium enables the construction of a 3-D network structure that supports strong 3-D spin–spin coupling necessary for a 400-K T_c. By using tetrahydrofuran (THF) or MeCN as alternative solvents, materials with $T_c < 350$ K can be isolated and characterized. Given the geometrical constraints imposed by the coordination of TCNE to a metal ion, an open structure with unfilled vanadium coordination sites is not unexpected.

Although the structure is unknown, it is interesting to speculate as to the type of magnetic coupling present from this system. On the basis of the IR data and elemental analysis, the precipitate appears to be best formulated as $V^{II}(TCNE)_2 \cdot \frac{1}{2}(CH_2Cl_2)$ with $S = \frac{3}{2}$ V^{II} and two $S = \frac{1}{2}[TCNE]^{\bullet-}$s. For ferromagnetic coupling S_{total} is $\frac{5}{2}$, and assuming g is 2, the saturation magnetization, M_s, is expected to be 28×10^3 emuG/mol. For antiferromagnetic coupling between the V^{II} and the two $[TCNE]^{\bullet-}$s and hence ferrimagnetic behavior, the S_{total} is $\frac{1}{2}$ and M_s is expected to be 5.6×10^3 emuG/mol. This value is in good agreement with 6.0×10^3 emuG/mol observed at 2 K at 19.5 kG (*80–83*).

The results of the physical studies emphasize the importance of 3-D coupling in the V–TCNE–solvent system. As a 3-D network structure present, equation 1 may be used to estimate the effective exchange interaction, J_{eff}, operative in the room-temperature magnet system (solvent is CH_2Cl_2). Assuming $T_c = 400$ K, S being the root-mean square of S_A ($\frac{1}{2}$) and S_D ($\frac{3}{2}$), and $z = 5$, then $J_{eff} = 70$ K (70 cm^{-1}). This result is only 2.6

times greater than the J obtained for the intrachain exchange in [FeCp*$_2$][TCNE] for which T_c= 4.8 K, yet T_c is nearly 2 orders of magnitude greater. Therefore, the differences in transition temperatures are in large part concerned with the difference in J alone but also with the change in dimensionality from quasi-1-D (for which T_c is suppressed to $\sim[J_\parallel/J_\perp]^{1/2}$, where $J_\parallel$ is the (large) in-chain and $J_\perp$ is the (small) interchain exchange) to essentially a 3-D system. Hence, we conclude that for achieving high T_c in molecular–polymeric material, it is a significant advantage to increase the dimensionality of the spin network.

Conclusion

Since the initial reports of 3-D cooperative metamagnetism for [FeCp*$_2$][TCNQ] (24) and ferromagnetism in [FeCp*$_2$][TCNE] (25, 28, 29) very rapid progress has been made in the synthesis of materials, models for the origin and control of the spin-exchange interaction, and theories for the origin and details of the observed magnetic behavior such as T_c and saturation magnetization. The chronology of increases in T_c is compared to that of the organic and ceramic superconductors in Figure 11. From these exemplary materials exhibiting cooperative magnetic ordering, the field of molecule-based organic–polymeric magnets has evolved to include a wide range of phenomena including magnetism in the V–TCNE molecule-based material significantly above room temperature. As this research area has evolved, several chemical features have emerged as important considerations in designing new magnetic materials. Clearly, to prepare a molecule-based magnet, both the donor and acceptor, if present, must be radicals. Such radicals need only have one spin per site; however, a greater number of spins per site is expected to lead to higher T_c materials.

For donor–acceptor based systems, the competition between ferromagnetic and antiferromagnetic interactions arises from the $\cdots D \cdot^+ A \cdot^- D \cdot^+ A \cdot^- \cdots$, as well as $A \cdots A$ interactions. Subtle changes in the orbital overlaps presumably lead to subtle changes in magnetic coupling. Thus, akin to proteins, the primary, secondary, and tertiary structures are crucial for achieving the desired cooperative magnetic properties.

Current limitations to the discovery of new molecule-based organic–polymeric magnets requires the rational design of solid-state structures, which remains an art. This situation is due to the formation of numerous polymorphs, complex and solvated compositions, as well as undesired structure types. The growth of crystals enabling the study of the single-crystal structure and magnetic properties is also an important limitation. With the dramatic rapid evolution of the field, major advances are expected to occur in this new arena of solid-state science.

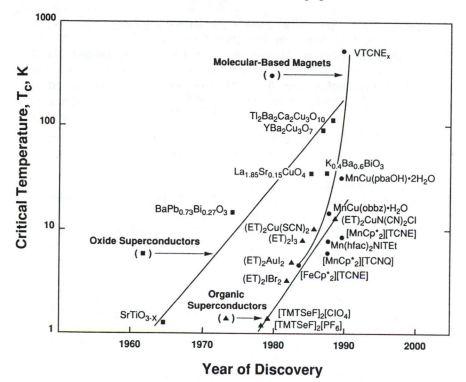

Figure 11. Time evolution of the discovery of increasing critical tempera-tures, T_c, for molecule-based magnetic materials as well as organic and oxide superconductors. (T_c on the order 10^3 K has been reported for some metallurgically prepared magnets.)

Acknowledgments

We have benefited greatly from the stimulating interactions and continued collaborations with our colleagues, postdoctoral fellows, and students, for they have made important contributions and enabled the rapid progress of the work reported herein. We furthermore gratefully acknowledge the support from the Department of Energy, Division of Materials Sciences (Grant DE–FG02–86BR45271).

References

1. Hellemans A.; Bunch, B. *Timetables of Science;* Simon and Schuster: New York, 1988; p 19.
2. *Magnets in Your Future* **1991**, *5*(6), 22. Hart, W. G. Intertech Conference on Polymer-Bonded Magnets '92, Rosemont, IL, and private communication.

3. Robinson, A. L. *Science (Washington, D.C.)* **1984**, *223,* 920.
4. Buschow, K. H. J. *Mater. Sci. Rep.* **1986**, *1,* 1.
5. Cohen, M. In *Advancing Materials Research;* Psaras, P. A.; Langford, H. D.,
    Eds.; National Academy Press: Washington, DC, 1987; p 91.
6. Croat, J. J.; Herbst, J. F. *MRS Bull.* **1988**, *13 (June),* 37.
7. White, R. M. *Science (Washington, D.C.)* **1985**, *229,* 4807.
8. Wallace, W. E. *J. Less Comm. Met.* **1984**, *100,* 85.
9. *Materials Science and Engineering for the 1990s;* National Research Council;
    National Academy Press: Washington, DC, 1989; p 94.
10. Buchachenko, A. L. *Russ. Chem. Rev.* **1990**, *59,* 307. *Usp. Khim.* **1990**, *59,*
    529.
11. Kahn, O. *Struct. Bonding (Berlin)* **1987**, *68,* 89.
12. Kahn, O.; Journaux, Y. In *Inorganic Materials;* Bruce, D. W.; O'Hare, D.,
    Eds.; John Wiley & Sons: New York, 1993; p 59.
13. Caneschi, A.; Gatteschi, D.; Sessoli, R.; Rey, P. *Acc. Chem. Res.* **1989**, *22,*
    392.
14. Dulog, L. *Nachr. Chem. Tech. Lab.* **1990**, *38,* 448.
15. Ishida, H. *Encyc. Poly. Sci. Eng.* (Supp. Vol.) **1989**, S446.
16. Sugawara, T. *Yuki Gos. Kag.* **1989**, *47,* 306.
17. Miller, J. S.; Epstein, A. J.; Reiff, W. M. *Acc. Chem. Res.* **1988**, *21,* 114.
18. Miller, J. S.; Epstein, A. J.; Reiff, W. M. *Science (Washington, D.C.)* **1988**,
    240, 40.
19. Miller, J. S.; Epstein, A. J. In *New Aspects of Organic Chemistry;* Yoshida,
    Z.; Shiba, T.; Ohsiro, Y., Eds.; VCH Publishers: New York, 1989; p 237.
20. Miller, J. S.; Epstein, A. J. *Chemtech* **1991**, *21,* 168.
21. Miller, J. S.; Epstein, A. J.; Reiff, W. M. *Chem. Rev.* **1988**, *88,* 201.
22. "Ferromagnetic and High Spin Molecular-Based Materials" (Proceedings of
    the Conference); Miller, J. S.; Dougherty, D. A., Eds.; *Mol. Cryst. Liq. Cryst.*
    1989, *176.*
23. "Proceedings of the Conference on Molecular Magnetic Materials;" Kahn,
    O.; Gatteschi, D.; Miller, J. S.; Palacio, F., Eds.; *NATO ARW Molecular
    Magnetic Materials* **1991**, *E198.*
24. Candela, G. A.; Swartzendruber, L.; Miller, J. S.; Rice, M. J. *J. Am. Chem.
    Soc.* **1979**, *101,* 2755.
25. Miller, J. S.; Epstein, A. J.; Reiff, W. M. *Mol. Cryst. Liq. Cryst.* **1986**, *120,*
    27.
26. Miller, J. S.; Zhang, J. H.; Reiff, W. M.; Preston, L. D.; Reis, A. H., Jr.;
    Gerbert, E.; Extine, M.; Troup, J.; Ward, M. D. *J. Phys. Chem.* **1987**, *91,*
    4344.
27. Stryjewski, E.; Giordano, N. *Adv. Phys.* **1977**, *26,* 487.
28. Miller, J. S.; Calabrese, J. C.; Rommelmann, H.; Chittapeddi, S.; Zhang, J.
    H.; Reiff, W. M.; Epstein, A. J. *J. Am. Chem. Soc.* **1987**, *109,* 769.
29. Chittapeddi, S.; Cromack, K. R.; Miller, J. S.; Epstein, A. J. *Phys. Rev. Lett.*
    1987, *58,* 2695.
30. Yee, G. T.; Manriquez, J. M.; Dixon, D. A.; McLean, R. S.; Groski, D. M.;
    Flippen, R. B.; Narayan, K. S.; Epstein, A. J.; Miller, J. S. *Adv. Mater.* **1991**,
    3, 309.
31. Zuo, F.; Zane, S.; Zhou, P.; Epstein, A. J.; McLean, R. S.; Miller, J. S. *J.
    Appl. Phys.* **1993**, *3,* 215.

32. Broderick, W. E.; Thompson, J. A.; Day, E. P.; Hoffman, B. M. *Science (Washington, D.C.)* **1990,** *249,* 410.
33. Broderick, W. E.; Hoffman, B. M. *J. Am. Chem. Soc.* **1991,** *113,* 6334.
34. McLean, R. S.; Miller, J. S., unpublished data.
35. Miller, J. S.; Calabrese, J. C.; Reiff, W. M.; Glatzhofer, D. T., unpublished data.
36. Webster, O. W.; Mahler, W.; Benson, R. E. *J. Am. Chem. Soc.* **1962,** *84,* 3678.
37. Rosenblum, M.; Fish, R. W.; Bennett, C. *J. Am. Chem. Soc.* **1964,** *86,* 5166.
38. Brandon, R. L.; Osipcki, J. H.; Ottenberg, A. *J. Org. Chem.* **1966,** *31,* 1214.
39. Adman, E.; Rosenblum, M.; Sullivan, S.; Margulis, T. N. *J. Am. Chem. Soc.* **1967,** *89,* 4540.
40. Foxman, B., private communication.
41. Sullivan, B. W.; Foxman, B. *Organometallics* **1983,** *2,* 187.
42. Miller, J. S.; Calabrese, J. C.; Glatzhofer, D. T., unpublished data.
43. Chi, K.-M.; Calabrese, J. C.; Reiff, W. M.; Miller, J. S. *Organometallics* **1991,** *10,* 688.
44. Miller, J. S.; Glatzhofer, D. T.; O'Hare, D. M.; Reiff, W. M.; Chackraborty, A.; Epstein, A. J. *Inorg. Chem.* **1989,** *27,* 2930.
45. Chi, K.-M.; Zhang, J. H.; Reiff, W. M. *J. Am. Chem. Soc.* **1987,** *109,* 4584.
46. Gerbert, E.; Reis, A. H.; Miller, J. S.; Rommelmann, H.; Epstein, A. J. *J. Am. Chem. Soc.* **1982,** *104,* 4403.
47. Miller, J. S.; Krusic, P. J.; Dixon, D. A.; Reiff, W. M.; Zhang, W. M.; Anderson, E. C.; Epstein, A. J. *J. Am. Chem. Soc.* **1986,** *108,* 4459.
48. Miller, J. S.; Calabrese, J. C.; Epstein, A. J. *Inorg. Chem.* **1989,** *27,* 4230.
49. Miller, J. S.; Calabrese, J. C.; Epstein, A. J., unpublished data.
50. O'Hare, D.; Miller, J. S.; Green, J. C.; Chadwick, J. C. *Organometallics* **1988,** *7,* 1335.
51. Narayan, K. S.; Kai, K. M.; Epstein, A. J.; Miller, J. S. *J. Appl. Phys.* **1991,** *69,* 5953.
52. Narayan, K. S.; Morin, B. G.; Miller, J. S.; Epstein, A. J. *Phys. Rev. B* **1992,** *46,* 6195.
53. Broderick, W. E.; Thompson, J. A.; Hoffman, B. M. *Inorg. Chem.* **1991,** *30,* 2960.
54. McConnell, H. M. *Proc. R. A. Welch Found. Chem. Res.* **1967,** *11,* 144.
55. Miller, J. S.; Epstein, A. J. *J. Am. Chem. Soc.* **1987,** *109,* 3850.
56. McConnell, H. M. *J. Chem. Phys.* **1963,** *39,* 1910.
57. Izoka, A.; Murata, S.; Sugawara, T.; Iwamura, H. *J. Am. Chem. Soc.* **1985,** *107,* 1786.
58. Izoka, A.; Murata, S.; Sugawara, T.; Iwamura, H. *J. Am. Chem. Soc.* **1987,** *109,* 2631.
59. Buchachenko, A. L. *Mol. Cryst. Liq. Cryst.* **1989,** *307,* 176.
60. Soos, Z. G.; McWilliams, P. C. M. *Mol. Cryst. Liq. Cryst.* **1989,** *369,* 176.
61. Kollmar, C.; Kahn, O. *J. Am. Chem. Soc.* **1991,** *113,* 7987.
62. Kollmar, C.; Couty, M.; Kahn, O. *J. Am. Chem. Soc.* **1991,** *113,* 7994.
63. Tchougreff, A. L.; Missurkin, I. A. *Chem. Phys.* **1991,** *153,* 371.
64. Kollmar, C.; Kahn, O. *J. Chem. Phys.* **1992,** *96,* 2988.
65. Morimoto, S.; Itoh, K.; Tanaka, F.; Mataga, N. *Preprints of the Symposium on Molecular Structure;* Tokyo, 1968; p 6.

66. Mataga, N. *Theor. Chim. Acta* **1968**, *10*, 372.
67. Ovchinnikov, A. A. *Theor. Chim. Acta* **1978**, *47*, 297.
68. Iwamura, H. *Pure Appl. Chem.* **1986**, *58*, 187.
69. Sugawara, T.; Murata, S.; Kimura, K.; Iwamura, H. *J. Am. Chem. Soc.* **1985**, *107*, 5293.
70. Sugawara, T.; Murata, S.; Kimura, K.; Iwamura, H. *J. Am. Chem. Soc.* **1984**, *106*, 6449.
71. Sugawara, T.; Bandow, S.; Kimura, K.; Iwamura, H.; Itoh, K. *J. Am. Chem. Soc.* **1986**, *108*, 368.
72. Nakamura, N.; Inoue, K.; Iwamura, H.; Fujioka, T.; Sawaki, Y. *J. Am. Chem. Soc.* **1992**, *114*, 1484.
73. Nakatani, F.; Carriat, J. Y.; Journaux, Y.; Kahn, O.; Lloret, F.; Renard, J. P.; Pei, Y.; Sletten, J.; Verdaguer, M. *J. Am. Chem. Soc.* **1989**, *111*, 5739.
74. Caneschi, A.; Gatteschi, D.; Renard, J. P.; Rey, P.; Sessoli, R. *Inorg. Chem.* **1989**, *28*, 3314.
75. Caneschi, A.; Gatteschi, D.; Renard, J. P.; Rey, P.; Sessoli, R. *J. Am. Chem. Soc.* **1989**, *111*, 785.
76. Dirac, P. A. M. *The Principles of Quantum Mechanics*, 2nd ed.; Oxford University Press: New York, 1935. Dirac, P. A. M. *Proc. Roy. Soc.* **1926**, *112A*, 661.
77. Van Vleck, J. H. *The Theory of Electric and Magnetic Susceptibilities;* Oxford University Press: London, 1932.
78. Dixon, D. A.; Suna, A.; Miller J. S.; Epstein, A. J. In *NATO ARW Molecular Magnetic Materials;* Kahn, O.; Gatteschi, D.; Miller, J. S.; Palacio, F., Eds.; **1991**, *E198*, 171.
79. Stinchcombe, R. B. In *Phase Transitions and Critical Phenomena;* Domb, C.; Lebowitz, J. L., Eds.; Academic: London, 1983, Vol. 7; p 152.
80. Manriquez, J. M.; Yee, G. T.; McLean, R. S.; Epstein, A. J.; Miller J. S. *Science (Washington, D.C.)* **1991**, *252*, 1415.
81. Epstein, A. J.; Miller J. S. In *Conjugated Polymers and Related Materials: The Interconnection of Chemical and Electronic Structure* (Proceedings of Nobel Symposium NS–81); Oxford University Press: New York, 1993; p 475. *La Chimica & La Industria* **1993**, *75*, 185.
82. Miller J. S.; Yee, G. T.; Manriquez, J. M.; Epstein, A. J. In *Conjugated Polymers and Related Materials: The Interconnection of Chemical and Electronic Structure* (Proceedings of Nobel Symposium NS–81); Oxford University Press: New York, 1993; p 461. *La Chimica & La Industria* **1992**, *74*, 845.
83. Epstein, A. J.; Miller J. S. *Mol. Cryst. Liq. Cryst.* **1993**, *228*, 99.
84. Robbins, J. L.; Edelstein, N.; Spencer, B.; Smart, J. C. *J. Am. Chem. Soc.* **1982**, *104*, 1882.
85. Cloak, F. G. N.; Dix, A. N.; Green, J. C.; Perutz, R. N.; Seddon, E. A. *Organometallics* **1983**, *2*, 1150.
86. Luborsky, F. E. In *Ferromagnetic Materials;* Wohlfarth, E. P., Ed.; North-Holland Publishing Co.: Amsterdam, Netherlands, 1980, Vol. 1; p 452.

RECEIVED for review November 9, 1992. ACCEPTED revised manuscript May 12, 1993.

Optimization of Microscopic and Macroscopic Second-Order Optical Nonlinearities

Seth R. Marder

Jet Propulsion Laboratory, California Institute of Technology, Pasadena, CA 91109, and Molecular Materials Resource Center, Beckman Institute, California Institute of Technology, Pasadena, CA 91125

Basic concepts of linear and nonlinear polarizability are briefly reviewed. Various approaches for optimizing the first hyperpolarizability (β) are discussed. It is shown, computationally, that an optimal degree of bond-length alternation in polyenic molecules is required to maximize β. The aromaticity of the π-electron system is an important factor in determining bond-length alternation. An example of a molecule whose π-electron system is not strongly aromatic in the neutral canonical resonance form and has large β is described. The second hyperpolarizability (γ) is correlated with bond-length alternation and exhibits both positive and negative peaks as a function of this parameter. Finally, the organic salt, 4-N,N-dimethylamino-4'-N'-methylstilbazolium toluene-p-sulfonate (DAST), which has an extremely large macroscopic second-order optical nonlinearity, is discussed.

Linear and Nonlinear Polarization

In 1893, Pockels demonstrated that the polarization state of light passing through a crystal was modified by the application of a static electric field to the crystal. In 1961, shortly after the invention of lasers, Franken et al. (*1*) observed that a focused ruby laser beam passing through a quartz crystal produced a faint beam at the ruby laser's second harmonic. These are two classic examples of nonlinear optical effects whereby the phase or po-

0065–2393/95/0245–0189$12.50/0

larization state of light may be modulated, or where new optical frequencies may be generated. The linear electro-optic or Pockels effect and second harmonic generation (SHG) are two technologically important second-order nonlinear optical (NLO) effects. The Pockels effect can be used to modulate the amplitude, phase, or polarization state of an optical beam. This effect can be used to impress information on an optical carrier signal or to route optical signals between fiber-optic channels, and therefore has applications in telecommunications and in integrated optics. SHG and other second-order optical mixing processes allow the extension of the useful frequency range of lasers, which is important for laser-based remote sensing and for optical communication. In optical disk devices, a fourfold enhancement in bit density can be achieved by halving the wavelength of the output of an 820-nm GaAs laser used to write the bits to 410 nm, and thereby reducing the diffraction-limited spot size to which the beam can be focused.

The origin of these NLO effects is related to the modification of the refractive indices of the material by applied electric fields and the modulation of light beams by the field-dependent indices. A more detailed discussion of nonlinear polarization can be found elsewhere (2). The refractive index of a material is related to the ease with which the charges in it can be displaced by the electric field of the light, that is, its polarizability. To gain some insight into how induced polarization behaves as a function of applied electric field, E, (an electric field is a vector quantity; however, for the purposes of this discussion we will only be concerned with its magnitude), it is instructive to consider the expansion of induced polarization, or dipole moment (μ) as a Taylor series, where

$$\mu = \mu_0 + E \left(\frac{\partial \mu}{\partial E} \right)_{E \to 0} + \frac{1}{2} E^2 \left(\frac{\partial^2 \mu}{\partial E^2} \right)_{E \to 0} + \frac{1}{6} E^3 \left(\frac{\partial^3 \mu}{\partial E^3} \right)_{E \to 0} \quad (1)$$

or,

$$\mu = \mu_0 + \alpha E + \frac{\beta}{2} E^2 + \frac{\gamma}{6} E^3 + \cdots \quad (2)$$

where μ_0 is the static dipole of the molecule, α is the linear polarizability; and the higher order terms β and γ (equation 2) are called the first and second hyperpolarizabilities, respectively. The terms beyond αE are not linear in E and are therefore referred to as the nonlinear polarization and give rise to NLO effects. Nonlinear polarization becomes more important with increasing field strength, because it scales with higher powers of the field. Under normal conditions, $\alpha E > (\beta/2)E^2 > (\gamma/6)E^3$. Thus, few observations of NLO effects were made before the invention of the laser with its associated large electric fields. The observed bulk polarization density (P) is given by an expression analogous to equation 2:

$$P = P_0 + \chi^{(1)}E + \chi^{(2)}E^2 + \chi^{(3)}E^3 + \cdots \tag{3}$$

where the $\chi^{(n)}$ susceptibility coefficients are tensors of order $n + 1$ and P_0 is the intrinsic static dipole moment density of the sample.

Second-Order Effects

The electronic charge displacement (polarization) induced by an oscillating electric field (e.g., light) can be viewed as a classical oscillating dipole that itself emits radiation at the oscillation frequency. For linear first-order polarization, the radiation has the same frequency as the incident light. What is the frequency of the re-emitted light for a NLO material? The electric field of a plane light wave can be expressed as

$$E = E_0 \cos(\omega t) \tag{4}$$

where ω is the frequency of the electric field and t is time; equation 3 can then be rewritten as

$$P = P_0 + \chi^{(1)}E_0 \cos(\omega t) + \chi^{(2)}E_0{}^2 \cos^2(\omega t) \\ + \chi^{(3)}E_0{}^3 \cos^3(\omega t) + \cdots \tag{5}$$

Because $\cos^2(\omega t)$ equals $\frac{1}{2} + \frac{1}{2}\cos(2\omega t)$, the first three terms of equation 5 become

$$P = (P_0 + \frac{1}{2}\chi^{(2)}E_0{}^2) + \chi^{(1)}E_0 \cos(\omega t) \\ + \frac{1}{2}\chi^{(2)}E_0{}^2 \cos(2\omega t) + \cdots \tag{6}$$

Physically, equation 6 states that the polarization consists of a second-order direct-current (DC) field contribution to the static polarization (first term), a frequency component ω corresponding to the incident light frequency (second term), and a new frequency-doubled component, 2ω (third term). Thus, if an intense light beam passes through a second-order NLO material, light at twice the input frequency will be produced, as well as a static electric field. The first process is called second harmonic generation (SHG), and the second is called optical rectification. SHG is a form of three-wave mixing, because two photons with frequency ω combine to generate a single photon with frequency 2ω. The oscillating dipole re-emits at all of its polarization frequencies, so light is observed at both ω and 2ω. This analysis can be extended to third-order and higher order terms. By analogy, third-order processes involve four-wave mixing.

As written, equation 5 is a simplified picture in which a single field of frequency ω acts on the material. The general picture of second-order NLO effects involves the interaction of two distinct waves of frequencies ω_1 and ω_2 in an NLO material. In this case, polarization occurs at sum $(\omega_1 + \omega_2)$ and difference $(\omega_1 - \omega_2)$ frequencies. This electronic polarization will therefore re-emit radiation at these frequencies, with contributions that depend on the magnitude of the NLO coefficient, $\chi^{(2)}$ (which is itself frequency-dependent). The combination of frequencies is called sum (or difference) frequency generation (SFG). SHG is a special case of SFG, where the two frequencies are equal. The sum is the second harmonic, and the difference is the DC component.

An applied electric field can also change a material's linear susceptibility and thus its refractive index. This effect is known as the linear electro-optic (LEO) or Pockels effect, and it can be used to modulate light by changing the voltage applied to a second-order NLO material. At the atomic level, the applied voltage is anisotropically distorting the electron density within the material, and thus, application of a voltage causes the optical beam to "see" a different material with a different polarizability and a different anisotropy of the polarizability. Because the anisotropy is changed upon application of an electric field, a beam of light can (1) have its polarization state (i.e., ellipticity) changed by an amount related to the strength and orientation of the applied voltage and (2) travel at a different speed and possibly in a different direction. Quantitatively, the change in the refractive index as a function of the applied electric field is approximated by the general expression:

$$\frac{1}{N_{ij}{}^2} = \frac{1}{n_{ij}{}^2} + r_{ijk}\mathbf{E_k} + s_{ijkl}\mathbf{E_k}\mathbf{E_l} + \cdots \tag{7}$$

where N_{ij} are the induced refractive indices, n_{ij} are the refractive indices in the absence of the electric field, r_{ijk} are the linear or Pockels coefficients, s_{ijkl} are the quadratic or Kerr coefficients (discussed later), and $\mathbf{E}$ is an applied field as described earlier. The subscripts refer to orientation of the field with respect to a material coordinate system. The optical indicatrix (which characterizes the anisotropy of the refractive index) therefore changes as the electric field within the sample changes. Electro-optic coefficients are frequently defined in terms of r_{ijk}. The r coefficients form a tensor (just as do the coefficients of α). The first subscript refers to the resultant polarization along a defined axis, and the following subscripts refer to the orientations of the applied electric fields. The Pockels effect involves two fields mixing to give rise to a third, and therefore r_{ijk} is a third rank tensor. The Pockels effect has many important technological applications. Light traveling through an electro-optic material can be phase- or polarization-modulated by refractive index changes induced by an applied electric field. Devices exploiting this effect include optical switches, modulators, and wavelength filters.

Third-Order Effects

Just as the interaction of light with a second-order NLO material will create light at its second harmonic, interaction of light with third-order NLO molecules will likewise create a polarization component at its third harmonic. The induced polarization for a bulk material would lead to third harmonic generation through $\chi^{(3)}$, the material susceptibility analogous to γ. In addition, a component at the fundamental will modulate the refractive index in a manner similar to the linear electro-optic effect. Similarly, the application of a voltage will also induce a refractive index change in a third-order NLO material. These effects are known as the optical and the DC Kerr effects, respectively. For materials, the sign of $\chi^{(3)}$ will determine if the third-order contribution to the refractive index is positive or negative. Materials with positive $\chi^{(3)}$ have the property of self-focusing a laser, and those with negative $\chi^{(3)}$ will be self-defocusing.

Another interesting third-order NLO effect is degenerate four-wave mixing. Two beams of light interacting within a material create an interference pattern (Figure 1) that leads to a spatially periodic variation in light intensity across the material. The induced change in refractive index of a third-order NLO material is proportional to the *intensity* of the applied field. Thus, if the beams are interacting with a third-order NLO material, the result will be a refractive index grating. When a third beam is incident on this grating, a fourth beam, called the phase conjugate, is diffracted from the grating. This process is called four-wave mixing; two writing beams and a probe beam result in a fourth phase-conjugate beam.

A potential use of this phenomenon is in phase-conjugate optics. Phase-conjugate optics takes advantage of a special feature of the diffracted beam: Its path exactly retraces the path of one of the writing beams. As a result, a pair of diverging beams impinging on a phase-conjugate mirror will converge after "reflection". In contrast, a pair of diverging beams reflected from an ordinary mirror will continue to diverge (Figure 2).

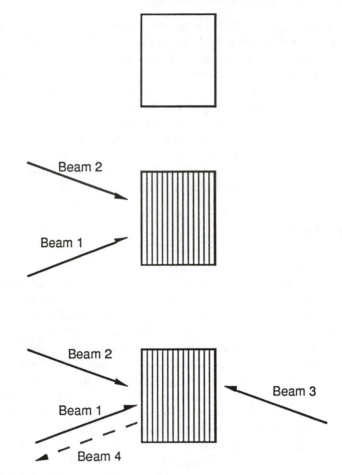

Figure 1. A phase-conjugate material in the absence of an applied field (top); beams 1 and 2 create a refractive index grating (middle); beam 3 interacts with the grating, creating beam 4, which is the phase conjugate of beam 1 (bottom).

Thus, distorted optical wavefronts can be reconstructed by using phase-conjugate optical systems (Figure 3).

The Need for Nonlinear Optical Materials

Since the advent of lasers, there has been an explosion in concepts for using light to carry and process information. The current push to exploit photonic technologies has created a need for high-performance NLO materials whose properties are very sensitive to applied fields. Physicists and

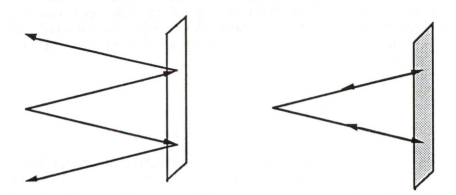

Figure 2. Left: A diverging set of beams reflected off of a normal mirror continue to diverge. Right: A diverging set of beams reflected off of a phase-conjugate mirror exactly retrace their original path and are therefore recombined at their point of origin.

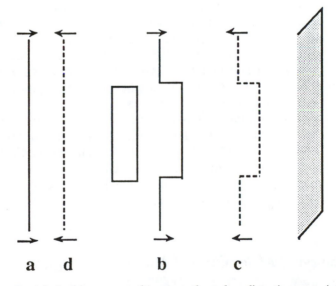

a   d   b   c

Figure 3. (a) A planar wave (b) passes through a distorting material that introduces an aberration, and (c) the light interacts with a phase-conjugate mirror to create the phase-conjugate wavefront. (d) A phase-conjugate wave passes through the distorting material on the reverse path, and thus cancels the original aberration and produces an undistorted wavefront.

theoreticians have developed rather complex models to describe the origin of NLO effects that are often couched in complicated equations involving descriptions of tensor components in terms of transition dipole moments and energy differences between states. For example, the individual tensorial components of β derived from perturbation theory lead to a sum-over-states formulation, for β *(3, 4)*:

$$\beta \propto \sum_{n,n'} \frac{\mu_{gn}\mu_{nn'}\mu_{n'g}}{E_{ng}E_{n'g}} - \mu_{gg} \sum_n \frac{\mu_g\mu_{n'g}}{E_{ng}^2} \tag{8}$$

where μ_{gn} and $\mu_{n'g}$ are the transition dipole moments between the ground-state (g) and excited-states (n or n'), $\mu_{nn'}$ is the transition dipole moment between two excited-states, and E_{ng} and $E_{n'g}$ are energy gaps between electronic states *(3–7)*. The role of the materials chemist is to translate these models into optimized molecules and then incorporate these molecules into optimized materials. Regarding the question of what is optimized, the magnitude of the desired optical nonlinearity will be only one of many criteria that will ultimately dictate the material of choice.

In addition to high nonlinearity, materials must also satisfy stringent requirements on optical absorption; scattering loss; ease of fabrication; and mechanical, thermal, photochemical, and environmental stability. Our current understanding of NLO materials suggests that these variables are frequently interrelated and that there is usually no ideal NLO material. The material of preference for a given application typically possesses the best compromise among a variety of variables.

Recently, organic molecules have received much attention for NLO applications because they exhibit large nonlinear polarizabilities and are amenable to optimization of their properties by rational modification of their structures. The remainder of this chapter focuses on our research at the Jet Propulsion Laboratory and the Beckman Institute to understand, optimize, and ultimately use organic materials for second-order NLO applications.

Optimization of Molecular and Macroscopic Optical Nonlinearities

Molecular Nonlinearities: Bond-Length Alternation, Aromaticity, β, and γ. The ability to build devices that exploit the NLO properties of organic materials relies on a fundamental understanding of relationships between chemical structure and molecular nonlinearities *(2, 8–10)*. Large second-order optical nonlinearities are associated with structures that are very polarizable in an asymmetric manner (e.g., the ease of polarization in one direction is different than in the opposite direction). As previously noted, polarization is a *nonlinear* function of the applied field, and the efficacy of a molecule to be polarized in this non-linear manner is called its first hyperpolarizability (β).

Organic molecules with delocalized π-electron systems capped with end groups of disparate electron affinities (electron donors and acceptors) to induce asymmetry, can fulfill these requirements. Thus, as the electrons interact with the oscillating electric field in light, they will show a preference to shift from donor toward acceptor and a reluctance to shift in the opposite direction. Two prototypical examples of NLO chromophores are *p*-nitroaniline (*p*-NA) and 4-*N,N*-dimethylamino-4'-nitrostilbene (DANS) (Figure 4). In *p*-NA, the benzene ring π-system provides the polarizable electrons. The amine acts as the donor, the nitro acts as the acceptor, and these groups induce the directional asymmetry into the polarizability. In DANS, the two benzene rings and the double bond are all conjugated, providing a longer π-system. This increased conjugation length, in general, leads to increased linear and nonlinear polarizability. A study of numerous classes of conventional organic molecules has appeared (*11–13*).

From a quantum mechanical viewpoint, the nonlinear polarization arises from an electric-field-induced mixing of the ground state with a charge-transfer (CT) state of a molecule wherein the electron density of the molecule has been preferentially shifted toward one end. We showed (*14*) that an optimal combination of donor and acceptor strengths leads to the correct degree of mixing needed to maximize β. We next sought to

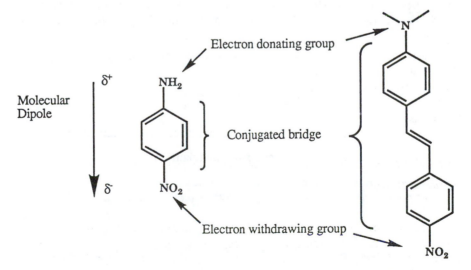

Electron donating group

Molecular Dipole

δ^+

NH_2

Conjugated bridge

δ^-

NO_2

Electron withdrawing group

Figure 4. Typical π-electron conjugated organic chromophores: left, p-ni-troaniline (p-NA); and right, 4-N,N-dimethylamino-4'-nitrostilbene (DANS).

correlate hyperpolarizabilities, with a relevant molecular parameter, bond-length alternation, that we believe can be tuned systematically (15, 16). Bond-length alternation refers to the difference in length between adjacent carbon–carbon bonds in a polymethine $[(CH)_n]$ chain. Polyene molecules have alternating double (short, 1.34 Å) and single (long, 1.45 Å) bonds. Thus, polyenes show a high degree of bond-length alternation.

In donor–acceptor polyenes, this parameter is related to the degree of ground-state polarization in the molecule. To better understand this correlation, we discuss the wave functions of the ground state in terms of a linear combination of the two limiting charge-transfer resonance structures. For polyenes with weak donors and acceptors, the neutral resonance form dominates the ground-state wave function and the molecule exhibits a high degree of bond-length alternation (15, 16). As the donor and acceptor strengths increase, so will the contribution of the charge-separated resonance form to the ground-state wave function (Figure 5A) (15, 16).

When the two resonance structures contribute equally, as in a symmetrical cyanine (Figure 6), the molecule will exhibit essentially no bond-length alternation. The charge-separated form may even dominate the ground-state wave function. In this case the molecule will exhibit bond-length alternation, but it will have the opposite sense from that of the neutral donor–acceptor polyene (Figure 5A, right side).

Molecules for which the neutral resonance form contains aromatic rings will have a diminished contribution of the charge-separated form to the ground-state wave function because of the energetic price associated with the loss of aromaticity in that form. As a result, molecules with aromatic ground states will tend to be more bond-length alternated for a

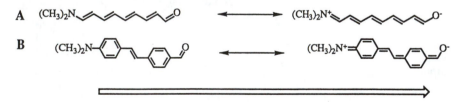

Increasing polarization

Figure 5. Neutral and charge-transfer resonance forms: **A,** *donor–acceptor polyene; and* **B,** *a donor–acceptor stilbene (containing aromatic rings).*

(CH₃)₂N⌇⌇⌇⌇⌇⁺N(CH₃)₂ ⟷ (CH₃)₂N⁺⌇⌇⌇⌇⌇N(CH₃)₂

Figure 6. Two resonance forms for a symmetric cyanine. Their degeneracy results in essentially no bond-length alternation.

given donor and acceptor pair than for a polyene of comparable length (Figure 5B) (*15, 16*).

Bond-length alternation is thus a measurable parameter that is related to the *mixing* of the two charge-transfer resonance descriptions in the actual ground-state form of the molecule. For aromatic groups, the neutral form will exhibit minimal bond-length alternation, whereas the charge-separated form will exhibit significant bond-length alternation (Figure 5B, right side). Thus, to the extent that we are using bond-length alternation as a measure of the mixing between the two forms, it is important to account for the fact that different molecular building blocks ($-CH=CH-$, $-C\equiv C-$, and $-C_6H_4-$) will have different bond-length alternations at a point where the mixing between the two charge-transfer forms is equivalent. Nonetheless, if these caveats are kept in mind, considerable insight can be gained from examining the correlation between hyperpolarizability and bond-length alternation (for now, with bond-length alternation as defined for polymethines).

Computational Studies on the Correlation of β and Bond-Length Alternation. We sought to develop a computational procedure to vary systematically the ground-state polarization of a molecule and observe the resulting changes in bond-length alternation, in α, β, and γ. Ideally, we would describe accurately the geometry and electronic structure of various donor–acceptor molecules in solvent environments of varying polarity. Currently, several research groups are developing procedures to simulate the effects of solvent on polarizable molecules (*17*). When these techniques have been optimized, it will be of great interest to couple them with computational analyses of hyperpolarizabilities.

At this time, one simple way to polarize a molecule is to apply an external electric field and permit the molecule to assume a new equilibrium geometry and electronic configuration. Although the polarizations in a molecule created by donors and acceptors or by solvent stabilization of charge separation are not strictly analogous to polarization created by an external electric field, this method qualitatively reproduces experimental trends in geometry and polarizabilities as a function of increasing ground-state polarization. We therefore used the AM1 Hamiltonian in the MOPAC package (*18, 19*) to examine the donor–acceptor substituted polyene, $(CH_3)_2N-(CH=CH)_4-CHO$ (Figure 5A), under the influence of an external perturbation designed to vary the ground-state polarization and geometry (*15, 16*). This perturbation was two positive and two negative point charges (Sparkles) moved in steps from 40 to 5 Å toward each end of the molecule. At each distance, the geometry was optimized, and the dipole moment (μ), α, β, and γ were calculated by a finite field subroutine. A bond-length alternation parameter, defined here as the differ-

ence between the double- and single-bond lengths (Ångstroms) in the neutral canonical resonance structure, was calculated for each geometry-
optimized structure. A plot of β (Figure 7) versus bond-length alternation
is consistent with our previous prediction based on a four-orbital calculation (14).

As can be seen in Figure 7, β peaks at a value of -0.04 Å of bond-
length alternation. Thus, the peak is closer to the cyanine limit than to
the bond-length alternated polyene limit. Previously, this optimal degree
of bond-length alternation was not reached because researchers had used
molecules in which the aromatic bridge itself strongly biased the molecule
to resist charge-transfer polarization and thus prevented sufficient
ground-state polarization from being achieved.

Experimental Efforts To Optimize β. To avoid paying the large energetic price of losing aromatic bridge stabilization in the charge-transfer
form, we synthesized molecules with thiobarbituric acid acceptors such as
that shown in bottom of Figure 8 (Marder, S. R. et al., unpublished
results). In the ground state of these molecules, one ring is aromatic and
the other is somewhat quinoidal, and in the charge-transfer state these
roles are effectively reversed. Thus, the π-electron bridge no longer has a
strong energetic bias for one form; now the necessary mixing is dictated by
the electron affinities of the end groups. Previously, Cahill and Singer
(20) examined compounds with thiobarbituric acids for anomalous-phased
matched SHG.

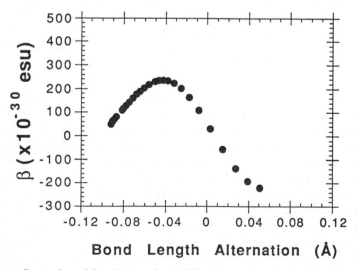

*Figure 7. A plot of first hyperpolarizability, β, as a function of bond-length
alternation for $(CH_3)_2N-(CH=CH)_4-CHO$ with peaks at -0.04 Å.*

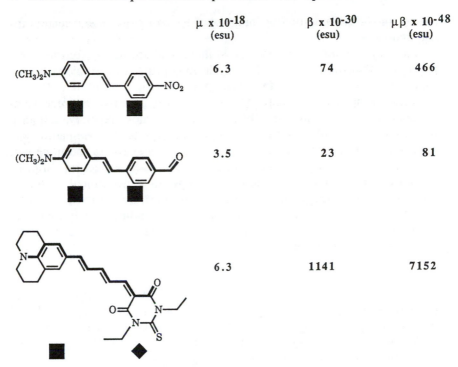

	$\mu \times 10^{-18}$ (esu)	$\beta \times 10^{-30}$ (esu)	$\mu\beta \times 10^{-48}$ (esu)
	6.3	74	466
	3.5	23	81
	6.3	1141	7152

Figure 8. Dipole moment (μ); first hyperpolarizability (β), and their dot product, μβ (which is a useful figure of merit for polymer applications, for DANS (top), dimethylaminoformylstilbene (middle), and a thiobarbituric acid derivative (bottom). The middle and bottom compounds have the same end groups and number of atoms (this pathway is highlighted in bold). Rings that lose aromaticity upon polarization have a square below them, and the one that gains aromaticity upon polarization has a diamond below it.

We showed that molecules containing thiobarbituric acid acceptors have greatly enhanced nonlinearities in comparison to conventional molecules of the same size (Marder, S. R. et al., unpublished results). For example, the thiobarbituric acid chromophore (Figure 8, bottom) that we developed has a large molecular hyperpolarizability ($\beta = 1141 \times 10^{-30}$ esu) compared to commonly used DANS (Figure 8, top), or 4-*N,N*-di-methylamino-4'-formylstilbene (Figure 8, middle), as measured by elec-tric-field-induced second harmonic generation, EFISH, (21–23) in chloro-form, at 1.907 μm. (Each of these molecules has the same number of atoms from the donor to the acceptor, and the last two even have the same end-group atoms; however, the low-energy absorption band for the thiobarbituric acid compound centered at 680 nm leads to roughly a factor of 2.2 dispersive enhancement, using a two-level model correction (5–7), whereas for the stilbene there is only about a factor of 1.3 enhancement in measured value).

The key difference between the stilbene and the thiobarbituric acid derivative is that the topology of the π-electron bridge in the thiobarbituric acid has been designed not to impede polarization. Dirk et al. (*24*) also observed enhanced β relative to DANS when the acceptor benzene ring was replaced by a less aromatic thiazole ring, and more recently, Jen and co-workers (*25, 26*) observed further enhancements in stilbene-based systems when both benzene rings were replaced with less aromatic thiophene rings. Thus, by reducing the aromaticity of the π-electron bridge, molecules with nonlinearities many times those of conventional chromophores have now been realized. If these molecules are sufficiently stable with respect to the processing and operating conditions required for specific device applications, they will be the active elements in the next generation of nonlinear materials.

Computational Studies on the Correlation of γ and Bond-Length Alternation. The detailed dependence of γ as a function of ground-state polarization and bond-length alternation and its behavior as compared to α and β has not been previously described. Figure 9 shows results of an electric-field-dependent calculation of bond-length alternation and γ for $(CH_3)_2N-(CH=CH)_4-CHO$.

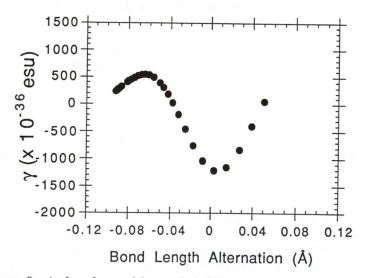

Figure 9. A plot of second hyperpolarizability, γ, as a function of bond-length alternation for $(CH_3)_2N-(CH=CH)_4-CHO$, has a positive peak at ~-0.07 Å and a negative peak at ~0 Å.

Several important structure–property predictions result from our analysis. First, the γ curve for $(CH_3)_2N–(CH=CH)_4–CHO$ exhibits a negative peak at zero bond-length alternation and a positive peak at negative bond-length alternation. Second, the negative peak value is somewhat larger than the positive peak value. Third, the positive peak has a two- to fourfold enhancement of γ relative to that of a nonpolarized (unsubstituted) polyene of similar length (γ = 1,078,410 au for **2** vs. 241,012 au for a 10-carbon polyene and 489,040 au for a 12-carbon polyene) (*27*). Fourth, γ peaks at a larger absolute value of bond-length alternation than β. In fact, when β is peaked, γ is roughly zero. Qualitatively, the behavior of γ predicted here is consistent with that predicted by a three-term model previously derived from perturbation theory (*28–30*) in which:

$$\gamma \propto -\frac{\mu_{gn}^{4}}{E_{gn}^{3}} + \frac{\mu_{gn}^{2}\mu_{nn'}^{2}}{E_{gn}^{2}E_{gn}} + \frac{\mu_{gn}^{2}(\mu_{nn} - \mu_{gg})^{2}}{E_{gn}^{3}} \tag{9}$$

where g, n, and n' are as defined for the sum-over-states expression (equation 8). When the first term is dominant, negative γ results, which is so for cyanines (*31*) and related squarylium dyes (*32*). For centrosymmetric polyenes, the second term dominates, the third term is zero, and the result is a positive γ (*33*). Thus, even when the contribution of the third term is not accounted for, at some point between the polyene and the cyanine limit, the magnitude of the first term will overtake that of the second term. Enhanced γ in donor–acceptor diphenylpolyenes relative to their centrosymmetric parent compounds has also been observed (*12, 13*). For these types of polarized polyenes, the third term will enhance γ. However, because $(\mu_{nn} - \mu_{gg})$ exhibits a peak as function of decreasing bond-length alternation (*14*) and goes through zero at roughly zero bond-length alternation, its contribution will be small near the cyanine limit. Furthermore, μ_{gn}^{4}/E_{gn}^{3} also increases with decreasing ground-state bond-length alternation (*14*) and is of opposite sign to the third term, and therefore this simple analysis of the perturbation theory expression can explain the peaked γ curves that we predict using MOPAC.

Optimization of Macroscopic Second-Order Optical Nonlinearities: Organic Salts.

Second-order NLO effects occur only in molecules lacking an inversion center (*2, 8–10*). Likewise, a bulk material composed of nonlinear molecules must also lack an inversion center if the molecular β is to lead to a macroscopic susceptibility $\chi^{(2)}$ (*2, 8–10*). Unfortunately, 75% of all known, achiral, organic molecules crystallize in centrosymmetric space groups, and thus, have no net $\chi^{(2)}$ (*34*). In 1983 Meredith (*35*) suggested that in organic salts, Coulombic interactions could override the deleterious dipolar interactions that provide a strong driving force for centrosymmetric crystallization. To test this hypothesis, salts of the form $(CH_3)_2NC_6H_4-CH=CH-C_5H_4N(CH_3)^+X^-$ were examined, and most of the counterions gave materials that exhibited relatively large powder SHG efficiencies (35). In particular, $(CH_3)_2NC_6H_4-CH=CH-C_5H_4N(CH_3)^+CH_3OSO_3^-$ (DASM) had a large SHG efficiency roughly 220 times that of urea (fundamental at $\lambda = 1.907$ μm) (35).

Subsequent studies by Marder and co-workers (36–38) and Nakanishi and co-workers (39–41) provided compelling evidence that cationic chromophores with large β, when crystallized with various anions, led to desirable noncentrosymmetric packing arrangements in many cases. In particular, Marder and co-workers found that for a particular class of compounds, the "salt methodology" provides a higher probability of obtaining $\chi^{(2)}$ active crystals, as compared to crystals of conventional neutral dipolar organic compounds (36). Roughly half of the 70 4-methylstilbazolium salts examined exhibited sizeable SHG efficiencies, indicative of a high incidence of noncentrosymmetric packing. More significantly, for most of the nonlinear cations, an anion could be found that led to efficient SHG. One compound, 4-N,N-dimethylamino-4'-N-methylstilbazolium toluene-p-sulfonate, DAST, (Figure 10, top) has a powder SHG efficiency larger than any other material to date, some 20 times greater than LiNbO$_3$ (fundamental at $\lambda = 1.907$ μm).

For electro-optic applications the largest nonlinear coefficients will occur when all of the molecules are aligned in the same direction. In DAST crystals, the only deviation from a completely aligned system is the 20° angle between the long axis of the cations and the polar **a** axis of the crystal (Figure 10, bottom) (36). In DAST, the cations form sheets, with

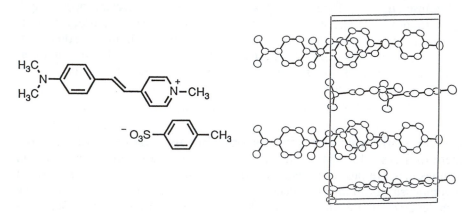

Figure 10. Left: Chemical structure of DAST. Right: Packing diagram for DAST viewed approximately along the b axis. Hydrogen atoms are omitted for clarity.

π−π stacking between the donor and acceptor rings of the cations that lie directly below one another in the sheet. The interplane cation spacing is about 3.4 Å. The sheets of cationic chromophores are interleaved with sheets of toluene-*p*-sulfonate anions. The planes of the toluene-*p*-sulfonate rings lie in the plane of anion sheets and are roughly perpendicular to the molecular planes of the cations. This general structural motif is common to seven of the compounds that have been structurally characterized (*36, 38–42*).

We are currently trying to understand the intermolecular interactions that are responsible for the unusually high incidence of noncentrosymmetric packing in these salt crystals so that we can couple our ability to design molecules where β is optimized with a rational strategy for optimizing packing. We believe that the formation of alternating cationic and anionic sheets can facilitate the formation of macroscopically polar structures as follows (*38*). Donor–acceptor, hydrogen bonding, and π−π stacking interactions often yield polar sheet structures. In neutral dipolar molecules dipole–dipole or steric interactions (indicated by +B, Figure 11, lower left) can occur between neighboring sheets when they are aligned in parallel, and thus an antiparallel orientation often results (Figure 11, lower right) (*43*).

In several of the noncentrosymmetric salt crystals, polar cationic sheets are separated by sheets of anions that are roughly planar or have approximate mirror symmetry, with respect to the plane of the sheet. As a result, the interactions between the cationic sheets may be screened by the anion sheet. The anions can be polar, and so a combination of steric and polar interactions (indicated by −A, Figure 11, upper left) will define the most favorable orientation of the anions with respect to the cations. Because of the rough mirror symmetry of the counterion sheet, the interactions that defined the preferred orientation between the anion sheet and the first cation sheet could provide a driving force to align the subsequent cation sheet roughly parallel to the first. Thus, the intermediate anion sheet could provide a driving force *favoring* a net polar alignment of cation sheets (Figure 11, upper left) and *disfavoring antiparallel alignment* (Figure 11, upper right).

The concept proposed here is in some ways analogous that involved in magnetic interactions in materials. Magnetic dipole–dipole interactions between spins favor antiferromagnetic rather than ferromagnetic alignment of spins. As a result, some research groups that are attempting to synthesize materials with bulk magnetic moments have focused on *ferrimagnetic materials*, that is, materials composed of alternating *antiferromagnetically coupled* sets of two spin systems, wherein the alternating spins are of different magnitude. The symmetry or pseudosymmetry of interactions in two-component polar, layered lattices could give rise to a higher incidence of polar three-dimensional packing arrangements than for

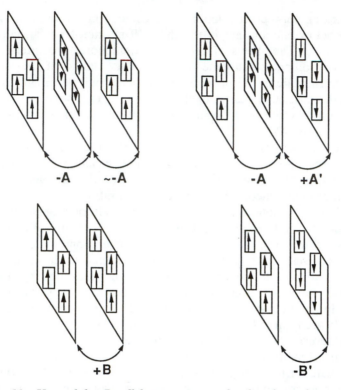

Figure 11. Upper left: Parallel arrangement of polar sheets for a salt. Upper right: Anti-parallel arrangement of polar sheets for a salt. Lower left: parallel arrangement of polar sheets for a neutral dipolar compound. Lower right: Parallel arrangement of polar sheets for a neutral dipolar compound. The letters A, A', B, and B' generically label a possible overall interaction energy between the adjacent sheets; the minus and plus signs indicate whether this interaction is net stabilizing or destabilizing, respectively. (Reprinted with permission from reference 38. Copyright 1992 Royal Society of Chemistry.)

single-component systems, despite the fact that all of the interactions may not necessarily be controlled.

The exceptional powder SHG efficiencies and the excellent alignment of the chromophore in the crystal led to the conclusion that crystals of DAST could have electro-optic properties that could enable the realization of new devices unattainable with state-of-the-art materials. Accordingly, we have grown good-quality specimens of DAST with dimensions up to about $5 \times 5 \times 1$ mm^3 from solution, such that the second harmonic and electro-optic properties of DAST single crystals could be characterized. A large SHG nonlinear coefficient, d_{11}, 620 ± 200 pm/V (roughly 20 times LiNbO$_3$), was measured (fundamental at $\lambda = 1.907$ μm) *(42)*. The electro-optic coefficient, r_{11}, of DAST is 400 ± 150 pm/V, and $n^3 r_{11}$ is 4200 pm/V at 820 nm for frequencies up to at least 100 kHz (*42, 44*). This value is one of the largest for $n^3 r$ ever measured on any material and compares favorably with the value of 1300 pm/V determined at 633 nm by Yoshimura (*45*) for a thin-film crystal of DASM (first examined by Meredith (*35*), as well as the values for 2-methyl-4-nitroaniline (MNA) (640 pm/V), the ceramic lead lanthanum zirconium titanate (PLZT) (2300 pm/V), and LiNbO$_3$ (400 pm/V).

Modulator array devices, such as high-speed spatial light modulators, require large nonlinearity and low power consumption. Therefore, it is desirable for the nonlinear material to have a low dielectric constant, ϵ, to minimize the charge required per switching operation. A useful power figure of merit for an electro-optic modulator is $n^3 r_{11}/\epsilon$, where n is the refractive index. DAST's relatively large n and low ϵ leads to $n^3 r_{11}/\epsilon \sim 800$ pm/V, one of the highest power merit values for any material. For example, LiNbO$_3$ has $n^3 r_{11}/\epsilon = 11$ pm/V. Additionally, thermal studies indicate that DAST is stable to temperatures higher than 250 °C, a condition suggesting that the material could be compatible with silicon device fabrication technology.

The extremely large second harmonic generation, d, and electro-optic, r, coefficients measured for DAST and the high incidence of noncentrosymmetric crystal packing demonstrate the tremendous potential of this material class for use in applications such as optical signal processing, communications, and interconnect applications requiring highly nonlinear, fast, thermally stable materials. Currently, our collaborators are attempting to develop DAST electro-optic modulators for magnetic resonance imaging and phased-array radar applications.

Conclusions, Summary, and Outlook

We have shown that by control of the topology of the π-electron system in organic donor–acceptor molecules, large molecular hyperpolarizabilities can be achieved. We then demonstrated how variation of the counterion in organic salts can give rise to nearly optimized crystal structures. Although DAST has among the highest, if not the highest, electro-optic figure of merit for any known material, its molecular hyperpolarizability has been estimated to be only comparable to that of DANS. This observation suggests that if we are successful in developing a new generation of organic salts with highly optimized chromophores and crystal structures, then nonlinearities 1–2 orders of magnitude greater than DAST could be realized. These materials would have a profound impact on optoelectronics if their secondary properties such as processability, stability, and transparency will allow them to be successfully incorporated into device architectures. In conclusion, the work described emphasizes how the development of a fundamental understanding of the science of nonlinear optics can lead to a rational approach to molecules and materials with optimized properties.

Acknowledgments

This work was performed in part at the Center for Space Microelectronics Technology, Jet Propulsion Laboratory (JPL), California Institute of Technology, under contract with the National Aeronautics and Space Administration (NASA). The work was sponsored by the Defense Advanced Research Projects Agency through a contract administered by the Air Force Office of Scientific Research and by the Strategic Defense Initiative Organization, Innovative Science and Technology Office. Support at the Beckman Institute from the National Science Foundation (Grant CHE–9106689) and the Air Force Office of Scientific Research (Grant F49620–92–J–0177) is also gratefully acknowledged.

The invaluable contributions of Joseph Perry (JPL), David Beratan (JPL and University of Pittsburgh), Bruce Tiemann (JPL and the Beckman Institute), Christopher Gorman (JPL and the Beckman Institute), Andrienne Friedli (the Beckman Institute), Lap-Tak Cheng (DuPont), and Christopher Yakymyshyn (General Electric) are gratefully acknowledged.

References

1. Franken, P. A.; Hill, A. E.; Peters, C. W.; Weinreich, G. *Phys. Rev. Lett.* **1961,** *7,* 118.
2. *Materials for Nonlinear Optics: Chemical Perspectives;* Marder, S. R.; Sohn, J. E.; Stucky, G. D., Eds.; ACS Symposium Series 455; American Chemical Society: Washington, 1991.
3. Ward, J. *Rev. Mod. Phys.* **1965,** *37,* 1.
4. Orr, B. J.; Ward, J. F. *Mol. Phys* **1971,** *20,* 513.
5. Oudar, J. L.; Chemla, D. S. *J. Chem. Phys.* **1977,** *66,* 2664–2668.
6. Levine, B. F.; Bethea, C. G. *J. Chem. Phys.* **1977,** *66,* 1070.
7. Lalama, S. J.; Garito, A. F. *Phys. Rev. A* **1979,** *20,* 1179.
8. *Nonlinear Optical Properties of Organic and Polymeric Materials;* Williams, D. J., Ed.; ACS Symposium Series 233; American Chemical Society: Washington, DC, 1983.
9. Williams, D. J. *Angew. Chem. Int. Ed. Engl.* **1984,** *23,* 690–703.
10. *Nonlinear Optical Properties of Organic Molecules and Crystals;* Chemla, D. S.; Zyss, J., Eds.; Academic: San Diego, CA, 1987, Vols. 1 and 2.
11. Cheng, L.-T.; Tam, W.; Meredith, G. R.; Rikken, G. L. J. A.; Meijer, E. W. *Proc. SPIE* **1989,** *1147,* 61–72.

12. Cheng, L.-T.; Tam, W.; Stevenson, S. H.; Meredith, G. R.; Rikken, G.; Marder, S. R. *J. Phys. Chem.* **1991**, *95*, 10631–10643.
13. Cheng, L.-T.; Tam, W.; Marder, S. R.; Steigman, A. E.; Rikken, G.; Spangler, C. W. *J. Phys. Chem.* **1991**, *95*, 10643–10652.
14. Marder, S. R.; Beratan, D. N.; Cheng, L.-T. *Science (Washington, D.C.)* **1991**, *252*, 103–106.
15. Marder, S. R.; Gorman, C. B.; Cheng, L.; Tiemann, B. G. *Proc. SPIE* **1993**, *1775*, 19–31.
16. Gorman, C. B.; Marder, S. R. *Proc. Natl. Acad. Sci. U.S.A.* **1993**, *90*, 11297.
17. Luzhkov, V.; Warshel, A. *J. Am. Chem. Soc.* **1991**, *113*, 4491–4499.
18. Stewart, J. J. P. *J. Comput. Chem.* **1989**, *10*, 209–220.
19. Stewart, J. J. P. *J. Comput. Chem.* **1989**, *10*, 221–264.
20. Cahill, P. A.; Singer, K. D. In *Materials for Nonlinear Optics: Chemical Perspectives;* Marder, S. R.; Sohn, J. S.; Stucky, G. D., Eds.; ACS Symposium Series 455; American Chemical Society: Washington, DC, 1991; pp 200–213.
21. Singer, K. D.; Garito, A. F. *J. Chem. Phys.* **1981**, *75*, 3572–3580.
22. Levine, B. F.; Bethea, C. G. *Appl. Phys. Lett.* **1974**, *24*, 445.
23. Oudar, J. L.; Person, H. L. *Opt. Commun.* **1975**, *15*, 258.
24. Dirk, C. W.; Katz, H. E.; Schilling, M. L.; King, L. A. *Chem. Mater.* **1990**, *2*, 700–705.
25. Rao, V. P.; Jen, A. K.; Wong, K.; Mininni, R. M. *Proc. SPIE* **1993**, *1775*, 32–41.
26. Wong, K.; Jen, A. K.; Rao, V. P.; Drost, K.; Mininni, R. M. *Proc. SPIE* **1993**, *1775*, 74–84.
27. Garito, A. F.; Heflin, J. R.; Wong, K. Y.; Zamani-Khamiri, O. In *Organic Materials for Non-linear Optics: Royal Society of Chemistry Special Publication No. 69;* Hann, R. A.; Bloor, D., Eds.; Royal Society of Chemistry, Burlington House: London, 1989; pp 16–27.
28. Dirk, C. W.; Kuzyk, M. G. In *Materials for Nonlinear Optics: Chemical Perspectives;* Marder, S. R.; Sohn, J. E.; Stucky, G. D., Eds.; ACS Symposium Series 455; American Chemical Society: Washington, DC, 1991; pp 687–703.
29. Kuzyk, M. G.; Dirk, C. W. *Phys. Rev. A* **1990**, *41*, 5098–5109.
30. Pierce, B. M. *Proc. SPIE* **1991**, *1560*, 148–161.
31. Stevenson, S. H.; Donald, D. S.; Meredith, G. R. In *Nonlinear Optical Properties of Polymers;* Heeger, A. J.; Orenstein, J.; Ulrich, D. R., Eds.; Materials Research Society Symposium Proceedings Vol. 109; Materials Research Society: Pittsburgh, PA, 1988; pp 103–108.
32. Dirk, C. W.; Cheng, L.-T.; Kuzyk, M. G. *Int. J. Quantum Chem.* **1992**, *43*, 27–36.
33. Perry, J. W.; Steigman, A. E.; Marder, S. R.; Coulter, D. R.; Beratan, D. N.; Brinza, D. E.; Klavetter, F. L.; Grubbs, R. H. *Proc. SPIE* **1988**, *971*, 17–24.
34. Nicoud, J. F.; Twieg, R. J. In *Nonlinear Optical Properties of Organic Molecules and Crystal;* Chemla, D. S.; Zyss, J., Eds.; Academic: San Diego, CA, 1987, Vol. 1; pp 227–296.
35. Meredith, G. R. In *Nonlinear Optical Properties of Organic and Polymeric Materials;* Williams, D. J., Ed.; ACS Symposium Series 233; American Chemical Society: Washington, DC, 1983; pp 27–56.

36. Marder, S. R.; Perry, J. W.; Schaefer, W. P. *Science (Washington, D.C.)* **1989,** *245,* 626–628.
37. Marder, S. R.; Perry, J. W.; Schaefer, W. P.; Tiemann, B. G.; Groves, P. C.; Perry, K. J. *Proc. SPIE* **1989,** *1147,* 108–115.
38. Marder, S. R.; Perry, J. W.; Schaefer, W. P. *J. Mater. Chem.* **1992,** *2,* 985–986.
39. Okada, S.; Masaki, A.; Matsuda, H.; Nakanishi, H.; Kato, M.; Muramatsu, R.; Otsuka, M. *Jpn. J. Appl. Phys.* **1990,** *29,* 1112–1115.
40. Koike, T.; Ohmi, T.; Umegaki, S.; Okada, S.; Masaki, A.; Matsuda, H.; Nakanishi, H. *CLEO Technical Digest* **1990,** *7,* 402.
41. Okada, S.; Matsuda, H.; Nakanishi, H.; Kato, M.; Muramatsu, R. Japanese Patent 63–348265, 1988.
42. Perry, J. W.; Marder, S. R.; Perry, K. J.; Sleva, E. T.; Yakymyshyn, C.; Stewart, K. R.; Boden, E. P. *Proc. SPIE* **1991,** *1560,* 302–309.
43. Wright, J. D. *Molecular Crystals;* Cambridge University Press: Cambridge, United Kingdom, 1987; pp 178.
44. Yakymyshyn, C. P.; Marder, S. R.; Stewart, K. R.; Boden, E. P.; Perry, J. W.; Schaefer, W. P. In *Organic Materials for Non-linear Optics II: Royal Society of Chemistry Special Publication No. 91;* Hann, R. A.; Bloor, D., Eds.; Royal Society of Chemistry, Burlington House: London, 1991; pp 108–114.

RECEIVED for review December 22, 1992. ACCEPTED revised manuscript May 3, 1993.

Materials Chemistry of Organic Monolayer and Multilayer Thin Films

Christine M. Bell, Huey C. Yang, and Thomas E. Mallouk*

Department of Chemistry and Biochemistry, University of Texas at Austin, Austin, TX 78712

The materials chemistry of organic monolayer and multilayer films is reviewed. Monolayer thin films prepared by Langmuir–Blodgett (LB) techniques or by self-assembly of silane- or thiol-containing amphiphiles were studied in several applications including chemical sensing, control of surface properties such as wettability and friction, and semiconductor surface passivation. By using modifications of these techniques, organic multilayer films can be grown, and these three-dimensional assemblies lend themselves to more complex functions including optical second harmonic generation (SHG). Synthetic approaches to monolayer and multilayer films and prospects for other future applications are discussed.

MONOMOLECULAR AND MULTILAYER FILMS make up one of the fastest-growing branches of materials chemistry (*1, 2*). The ability to grow ordered, monomolecular and multilayer films on solid surfaces in a controlled fashion has many present and potential applications, which range from fields such as microelectronics, nonlinear optics, and lubrication to studies of protein adsorption and biosensors. It is, like most of materials chemistry, a multidisciplinary effort, involving scientists in chemistry, physics, biology, and engineering.

*Corresponding author

0065–2393/95/0245–0211$12.00/0

Study of such films has been progressing for decades, but the recent explosion of research can be traced to significant improvements in the field of surface analytical chemistry. Reflection infrared spectroscopy, ellipsometry, low-energy electron diffraction, X-ray photoelectron spectroscopy, and contact-angle measurements have been joined by newer methods such as scanning tunneling microscopy (3), atomic force microscopy (4), surface Raman spectroscopy (5), and surface near-edge X-ray absorption fine structure (NEXAFS) spectroscopy (6) to provide more detailed pictures of these molecular films. Friction force microscopy is another new method used to discriminate between two different species in a phase-separated thin film with a lateral resolution of 5 Å (7). The availability of these new analytical tools has allowed scientists to develop, within the past decade, a variety of new synthetic strategies and applications for monolayer and multilayer films on solid surfaces. This review focuses on the techniques and materials used to prepare molecular thin films and on some of their current and future applications.

Monolayer Thin Films

It is appropriate to begin a discussion of monomolecular films with the decades-old, yet still effective and versatile, method of Langmuir and Blodgett (8). In their technique, a film is grown by the coherent transfer of a compressed monolayer that has preassembled at an air–liquid interface. This preassembly is accomplished by using an amphiphilic molecule. The polar end of the molecule is soluble in the polar liquid, usually water, and the nonpolar end prefers to orient away from the water toward the "nonpolar" air. The monolayer that forms is mechanically delivered onto a solid substrate such as glass or a metal. Hundreds of layers can be grown sequentially by this technique, by repetition of the transfer of new monolayers formed at the interface (Figure 1).

Langmuir–Blodgett (LB) monolayers are interesting for many reasons (9–11), one of the most important being their structural resemblance to lipid bilayer membranes. An intriguing application of this similarity, which shows a functional as well as structural analogy to biological membranes, is an ion-channel sensor developed by Umezawa and co-workers (12). In their device (Figure 2), an LB film that terminates in a phosphate group is formed on an electrode. When an ion such as Ca^{2+} is introduced, it bridges two molecules of the film, thus opening a channel through which marker ions in solution, such as $Fe(CN)_6^{4-}$, can penetrate and be sensed at the electrode surface. The sensor can be shut off by a quencher, such as ethylenediaminetetraacetic acid (EDTA), which binds the cation and closes the channel. The sensor is dependent upon concentration and has the advantage that amplification of the analyte ion can be

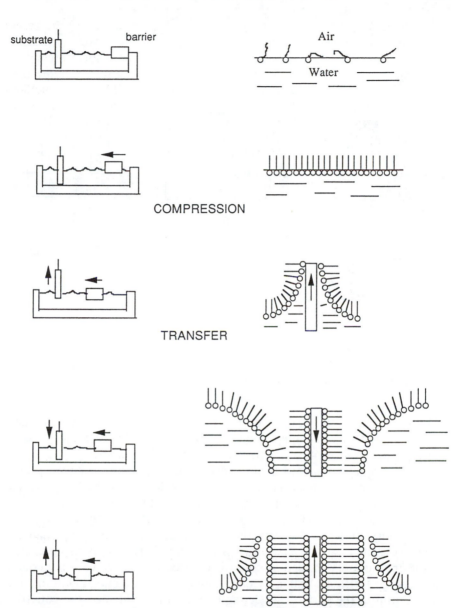

Figure 1. Formation of Langmuir–Blodgett monolayers at an air–water interface, and transfer to a solid substrate to form a Y-type multilayer film. The compression of a monolayer in the Langmuir trough is shown schematically on the left, and a schematic microscopic representation of the air–water interface during compression and transfer steps is shown on the right.

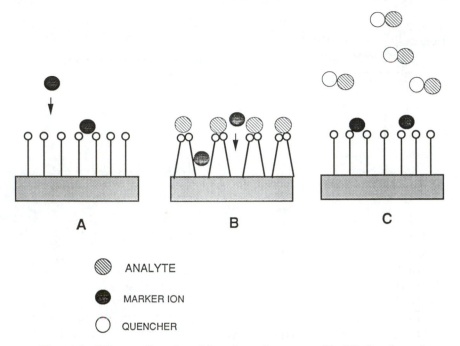

Figure 2. LB monolayer-based ion-channel sensors with (A) the channel closed, (B) the channel opened and marker ions sensed, and (C) the channel closed by binding of analyte with quencher (12).

achieved because one analyte ion can induce the sensing of many marker ions. In addition to its applications for sensing, this device could be used for probing the mechanism of membrane systems in biology.

Many applications of LB films have been reported, but the technique suffers drawbacks such as the necessity for mechanical manipulations, the low stability of the films, and their requirement for planar surfaces. For these reasons, a second strategy, the so-called self-assembly process, has been studied intensively over the past decade. In this approach, a molecule with a surface-active group spontaneously adsorbs from either vapor or solution. The self-assembly is induced by the decrease in surface free energy ($\simeq$10–40 kcal/mol) that results from covalent binding of the head group to the surface. Ordering of the film is aided by the cohesive packing forces of other elements of the molecule, such as the spacer (usually an alkyl chain) and terminal functional groups. Because these monolayers are self-assembling, that is, there is a thermodynamic driving force for their formation, they can displace many species that physisorb to the surface, such as water or exogenous amphiphilic substances (contaminants from the laboratory environment that make for "dirty" surfaces under ambient pressure conditions). This means that few special conditions, such as a rigorously clean environment, or ultra-high vacuum, are needed as they may be in other thin-film preparations.

In 1946, Bigelow et al. (*13*) first reported the phenomenon of self-assembled molecular films on solid surfaces, when they noticed, quite accidentally, that a flask was not wetted by a dilute solution of eicosyl alcohol ($C_{20}H_{41}OH$) in *n*-hexadecane. Other surfaces that were similarly affected included borosilicate glass (Pyrex), soda glass, nickel, iron, steel, aluminum, and chromium. They concluded that an oleophobic and hydrophobic monolayer of the fatty acid was responsible for the lack of wetting. In a later study (*14*), measuring contact potential differences of fatty acids adsorbed on Pt and NiO, Timmons and Zisman showed that, on Pt, a chain length of 14 or greater gave the highest ΔV and highest contact angle with methyl iodide. With $n > 13$, intermolecular attractive forces between neighboring alkyl chains become large enough to force the molecules into a close-packed array. The terminal surface is thought to consist of close-packed methyl groups, a feature that explains the methyl iodide contact angle of 68–71°, a value also obtained for a bulk long-chain alkane crystal known to terminate in close-packed methyl groups. Levine and Zisman (*15*) used these monolayers to lower the coefficient of friction (defined as friction force divided by load force) of glass–steel from 1.1 to 0.05. In addition, there was reduced wear on the glass substrate as a result of protection by the film.

More recent studies focused on the self-assembly of carboxylic acids on aluminum and glass, alkyltrichlorosilanes on oxidized surfaces such as silicon–silicon dioxide, and alkylthiols and organic disulfides on gold. Carboxylic acids are adsorbed on $Al–Al_2O_3$ by proton dissociation to give a carboxylate species that participates in coordination chemistry at the oxide surface (*16, 17*). Close-packed monolayers, with the alkyl chains tilted about 10° from the surface normal, are produced if the alkyl chain contains at least 12 carbons. However, these are dynamic monolayers in that they undergo facile exchange with materials in solution. Thus, although these layers are interesting, they suffer from reversible binding, which makes them susceptible to decomposition through chemically, mechanically, and thermally induced molecular desorption.

In contrast, alkyltrichlorosilanes adsorbed on oxidized surfaces, first reported by Polymeropoulos and Sagiv (*18, 19*) are irreversibly bound. These highly stable, well-ordered monolayers are grown through the covalent binding of active silane molecules to the surface (*20–24*). This growth is achieved by the formation of a covalent Si–O–Si bond (Figure 3). As before, these monolayers are stabilized by van der Waals interactions between adjacent alkyl chains and are further stabilized by various degrees of planar polymerization between the head groups. Contact angle and Fourier transform infrared (FTIR) measurements, along with molecular modeling, indicate that the chains are tilted 15° from the surface nor-

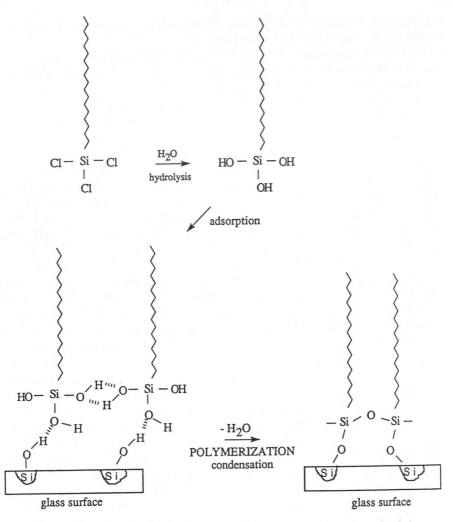

Figure 3. Chemisorption and lateral polymerization of an n-octadecyltri-chlorosilane *(OTS) monolayer on glass (19).*

mal and that the CH_2 groups are in an all-trans conformation. The silane is soluble in chloroform, but the condensed film is completely resistant to chloroform treatment, as well as acid and thermal treatment, and this behavior is indicative of its irreversible adsorption.

In an early experiment, Sagiv (25) was able to exploit this irreversibility in order to make templated skeletal monolayers. Mixed layers were formed by the simultaneous adsorption of octadecyltrichlorosilane (OTS) with dye molecules. Because the dye molecules were merely physisorbed, they could be removed with solvent to leave the OTS intact. The resulting skeletal monolayers were able to readsorb the same and other dye molecules into the holes with geometric discrimination. Such layers hold promise for use in adsorption chromatography (especially in light of the fact that self-assembled monolayers can be grown on nonplanar, that is, surfaces with high surface areas), in sensors, and for controlling the orientation of molecules in some chemical reactions.

Because controlled growth of molecular layers presents possibilities in the field of optics and nonlinear optics, studies have been performed to assess the consequences of inserting aromatic and polar aromatic groups into silane monolayers. Tillman, Ulman, and co-workers (26, 27) introduced phenoxy- and sulfonyl-substituted phenyl groups into long-chain alkylsilane molecules. They found that if the alkyl chains on either side of the aromatic group were sufficiently long, then ordered, high-quality monolayers, similar to OTS, albeit with greater vertical tilt, could be formed. Shortening the length of the chain between the phenoxyl group and the surface gave increasingly less ordered films. Ulman and Scaringe (28) suggested that, for mixed aromatic–paraffin chain molecules, a "commensurability of intra-assembly planes" is necessary, but not sufficient, for good packing. That is, the close-packing of tilted alkyl chains should complement and encourage the good packing of another molecular fragment, for example, the herringbone arrangement of straight phenyl rings, in order to get a good match.

Another practical application of alkylsilane monolayers is in the field of tribology, which is the study of friction and its related aspects, such as lubrication and adhesion (29). Chemisorbed or covalently bound monolayers are attractive because, unlike liquid lubricants, their strong binding to the surface prevents them from migrating or transferring from one surface to another. This property is important in magnetic and optical disc technologies, as well as in other areas where friction between solid surfaces must be minimized. DePalma and Tillman (30) reported that the coefficient of friction for boundary lubrication with an OTS monolayer on glass was 0.07, and no transfer or wear was evident even after repeated traverses by the glass slider. The minimum coefficient corresponded to one monolayer, in agreement with other studies that suggested that one monolayer was sufficient to reduce friction considerably. The reduction in friction coefficient was comparable to that found previously for noncovalently bound hydrophobic monolayers (15). Alkylsilanes should represent a better choice as durable wear-resistant monolayers, because the covalent Si–O–Si bond, along with the in-plane polymerization, holds them more strongly to the surface.

A related property is wettability. As noted, the contact angle of a liquid drop on a surface can be used to deduce the surface free energies for the various interfaces, for example, solid–vapor, solid–liquid, and liquid–vapor. Manipulation of these interactions can have interesting effects. Chaudhury and Whitesides (31) recently reported a simple, yet elegant, example of this manipulation of surface energies. They prepared a silicon wafer with a gradient of surface energies by creating a concentration gradient of decyltrichlorosilane during a vapor-phase deposition. At the high-concentration end, a close-packed, hydrophobic monolayer was formed. As the concentration over the wafer decreased, the surface became more hydrophilic, presumably due to less order of the resulting film. When the wafer was held at a 15° angle and a drop of water was placed on the lower, hydrophobic end, the drop moved uphill, against gravity, in response to the surface gradient. The motion was effected by the decrease in the area of the liquid–solid interface having the larger free energy (hydrophobic end) and the increase in that having lower free energy. The overall free energy was decreased enough to overcome the force of gravity. This behavior demonstrates nicely the ability to control surface properties by means organic thin films. It may find useful applications in water repellency (e.g., in steam condensers), in adhesion (e.g., as wear-resistant coatings), and in fundamental studies of liquid–surface dynamics.

Monolayers of alkylsilanes are very attractive for many applications. They do, however, require an oxidized surface, a condition that may not be ideal in some circumstances. Alkylthiols and organic disulfides on gold offer an attractive alternative (32–39). These monolayers feature relatively strong attachment involving covalent, slightly polar sulfur–gold bonds, high surface coverages approaching bulk packing, and orientational ordering with chain tilts of about 20–30° from the surface normal. The initial adsorption process is very fast, as it is with silane monolayers, but physical observables such as contact angles can change slowly, sometimes over periods of days. The slowness of the self-assembly process for thiols may signal long-range surface ordering that cannot be attained in silane monolayers. This inability results because the siloxane linkage forms quickly and irreversibly, with associated polymerization, whereas thiol monolayers have more freedom to reorganize and hence anneal away a larger proportion of surface defects.

Alkylthiol monolayers on gold electrodes are excellent systems for studies of fundamental electrochemical phenomena, heterogeneous electron transfer, ion transport, and double-layer properties. Porter et al. (34) used electrochemical capacitance and heterogeneous electron-transfer measurements to examine the integrity of nanoscale assemblies. They showed that alkylthiol monolayers on gold provided high-quality barriers for both electron and ion-transfer processes. This ability to passivate surfaces could lead to improved insulator layers for some semiconductors, especially ones such as gallium arsenide, whose lack of a stable insulating oxide prevents its use in many devices. Alkylthiol monolayers on etched GaAs (100) impart some stability toward oxidation as well as significantly reducing electron–hole pair recombination velocities (40, 41).

Miller and co-workers (*42, 43*) were able to form extremely compact and defect-free monolayers using ω-hydroxyalkanethiols, and used them to probe the dynamics of electron transfer between solution-phase redox couples, such as $Fe(CN)_6^{4-/3-}$, and electrode surfaces. The logarithmic dependence of the electron-transfer rate constant on the thickness of the film and the lack of temperature dependence at high overpotentials confirmed a tunneling mechanism, in contrast to earlier studies in which the current was probably defect-mediated. They were able to use these well-characterized tunneling barriers to perform kinetic measurements over a wide range of overpotentials, even on very facile redox couples. Because of the low current densities, mass-transfer effects were mostly eliminated and they were able to obtain the density of states by using the approach of Bennett (*44*). These compared well with the predictions of Marcus theory (*45*) and demonstrate the use of organic monolayers for fundamental studies.

Others have employed organic films for similar fundamental studies. Li and Weaver (*46, 47*) showed that the rate of electron transfer of redox centers, attached through an aliphatic spacer chain to an electrode surface, decreased exponentially as the chain length was increased. Chidsey, et al. (*48*) used mixed monolayers of unsubstituted and ferrocene-terminated alkanethiols on gold to probe the electron-transfer kinetics of identical, isolated electroactive sites. Because the redox center was bound, the importance of defect sites on electron transfer was minimized. Again, longer chain lengths gave slower rates. These assemblies, with their potential for systematic changing of the structure, could be employed to probe the mechanism of electron transfer and tunneling, including the importance of factors such as solvation and chemical structure (*49, 50*).

Because the gold–sulfur bond is so strong and specific, bifunctional molecules, with a thiol at one end, can be used to form self-assembled monolayers with the confidence that the terminal functional group that is not thiol will be oriented away from the gold surface. The chemical properties of this functional group can then be exploited. Two interesting applications of this idea were recently reported. Rubinstein and co-workers (*51, 52*) prepared thin films capable of ion-selective sensing. Their two-component films consisted of 2,2′-thiobis(ethylacetoacetate) (TBEA) and octadecylmercaptan (OM). The TBEA binds Cu^{2+} and Pb^{2+}, which can then be sensed electrochemically as a surface-bound species by the underlying gold electrode. Other ions, such as Fe^{2+} and Fe^{3+}, do not have access to the rest of the electrode because it is well blocked by the ordered OM layer.

In another study (*53*), biotin-terminated alkylthiols were bound to gold and were able to recognize and bind streptavidin. If the packing density was too high, there was lower efficiency of the coupling because of steric problems with the large protein. This example illustrates a significant advantage in the use of monolayers on solid surfaces for sensors, particularly biosensors. This advantage arises from the ease of transduction of the chemical signal, in this case molecular binding, to an electrical signal. A recent study in which this advantage was exploited was a surface acoustic wave sensor developed by Ricco, Crooks, and co-workers (54). They formed monolayers of 11-mercaptoundecanoic acid. The terminal carboxylate groups bound Cu^{2+}, to which diisopropyl methylphosphonate, a nerve agent stimulant, was absorbed, inducing a frequency drop in the surface acoustic wave (SAW) device. Their sensor was concentration-dependent, fast, durable, reversible, and selective. Water and several common organic solvents gave no response, although other ligands that might also be expected to bind Cu^{2+} were not investigated. This study showed that simple molecules could be self-assembled to form surfaces sensitive to important analytes. Other examples in which monolayer-modified surfaces can be used for sensing include the quartz crystal microbalance (55) and optical devices.

Multilayer Thin Films

It could be argued that, after the "flatland" chemistry of two-dimensional monolayers is exhausted (if ever), there is no place to go but up. That is, three-dimensional structures composed of individual layers stacked on top of each other provide, in principle, many more avenues for research and practical applications than do simple monolayers. Multilayer films offer the promise of amplification of the properties of a single layer, for example, in optical second harmonic generation (SHG). Also, if one prepares films in which chemically different layers can be juxtaposed to form heterostructures, then in principle rather intricate device functions, which are needed in areas such as artificial photosynthesis, molecular electronics, chemical sensing, and separations, might be attained. The problem is that the synthesis and characterization of these assemblies is generally more challenging than that of single monolayers, and consequently the science of such systems is at present less highly developed.

The Langmuir–Blodgett technique, illustrated in Figure 1, does provide a convenient and versatile route to organic multilayer films as well as monolayers. In the process shown in Figure 1, the stacking of layers is tail-to-tail and head-to-head, forming a so-called "Y-type" LB film. An alternative procedure is to grow multilayers in head-to-tail fashion ("Z-type"), which is the necessary sequence to create polar films appropriate for SHG. Popovitz-Biro et al. (56) showed that the type of LB film generated, Y or Z, may be controlled quite nicely by simply adjusting the polarity of the terminal groups of the amphiphile. With sufficiently polar tail groups, the aqueous subphase will wet the monolayer-coated substrate;

that is, the contact angle of the liquid with the substrate will be less than 90° (Figure 4). Under these conditions polar (Z-type) multilayer films result. If the individual molecules composing these layers have nonzero hyperpolarizability along the film-stacking axis, the resulting assembly shows SHG effects under sufficiently intense illumination.

As noted, LB films suffer from mechanical and chemical instability because they are metastable phases assembled at the air–water interface by means of applying surface pressure. The poor stability of these systems is exacerbated in multilayer films, although certain tricks, such as intralayer cross-linking of olefinic groups, have been developed to improve their durability (*57*). The problem of growing them, in practical applications, on nonplanar or indeed on tortuous surfaces (such as supports of high surface area) nevertheless remains. For this reason, several multilayer film-growth procedures based on spontaneous adsorption of molecules from solution and concomitant self-assembly, analogous to monolayer-forming techniques, have been developed.

An important conceptual development in this regard occurred when Netzer, Sagiv, and co-workers (*58, 59*) showed that self-assembling silane

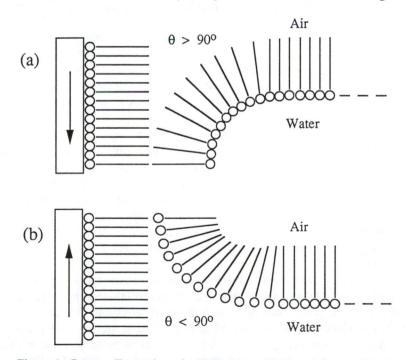

Figure 4. Part a: Formation of a Y-type LB multilayer, using amphiphiles with hydrophobic tail groups. The aqueous subphase does not wet the exposed monolayer (θ > 90°). Part b: Formation of a Z-type multilayer film, using polar tail groups wetted by the subphase (θ < 90°) (56).

layers could be grown, one upon the next, using a series of sequential chemical activation–adsorption steps. Alkyltrichlorosilane molecules bearing a terminal olefin group were adsorbed onto the surface of silicon, and then the exposed olefin was oxidized to a terminal alcohol. The next layer was then adsorbed and activated in the same way, and so on. Although the crystallinity of the uppermost monolayer decayed as more layers were added, Ulman and co-workers (60) later showed that a modification of this technique, using carboxylate tail groups, could be used to prepare highly ordered, polar multilayer films that were chemically and mechanically extremely stable. Recognizing that this sequential adsorption procedure gave rise naturally to polar films (analogous to Z-type LB multilayers), Li et al.(61) prepared multilayer films containing hyperpolarizable stilbazole derivatives aligned with their long axes roughly parallel to the stacking axis. These multilayer films gave relatively strong SHG signals. The square root of the SHG intensity was linear with number of layers, as expected if the same degree of polar order is maintained from one layer to the next. The film-growth procedure in this case involved several sequential adsorption–reaction steps, including steps in which poly(vinyl alcohol) was introduced between silane layers to provide structural reinforcement (Figure 5).

Stable organic monolayers can be formed by using this sequential adsorption–reaction technique, but relatively little is known about the detailed three-dimensional structure of the surface phases formed; that is, no bulk analogue of these materials exists, and so any structural characterization comes from surface analytical techniques such as reflectance IR spectroscopy, ellipsometry, contact-angle measurements, and surface-sensitive spectroscopies such as X-ray photoelectron spectroscopy (XPS). One potential solution to this problem is to attempt to form multilayer films with the same composition as known bulk layered phases. The same problems attend thin-film characterization, but layer thicknesses and composition can be compared with crystallographic data from bulk material that is fully characterized. An additional advantage of this approach derives from the fact that the lamellar structure of such materials represents their lowest free energy form, and so it is likely that multilayers formed on surfaces will be relatively stable.

This idea was first demonstrated by Lee and co-workers (62–64), who prepared thin films of zirconium alkanebisphosphonates by sequentially adsorbing their Zr^{4+} and phosphonic acid components on suitably prepared surfaces, as shown in Figure 6. Ellipsometric measurements show that these $Zr(O_3PC_nH_{2n}PO_3)$ films grow in stepwise fashion, with an increase in thickness per layer corresponding closely to the crystallographically determined layer spacing in well-annealed bulk solids for $n = 6$–10. The sequential dipping procedure may be automated, and as many as a hundred layers can be grown, although this procedure is rather tedious be-

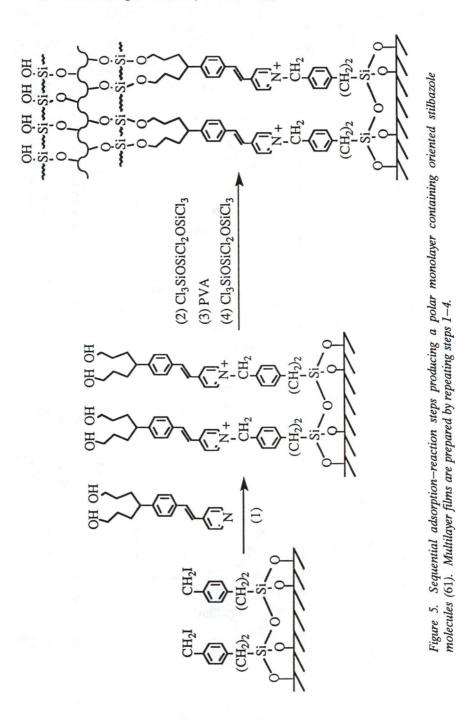

Figure 5. Sequential adsorption–reaction steps producing a polar monolayer containing oriented stilbazole molecules (61). Multilayer films are prepared by repeating steps 1–4.

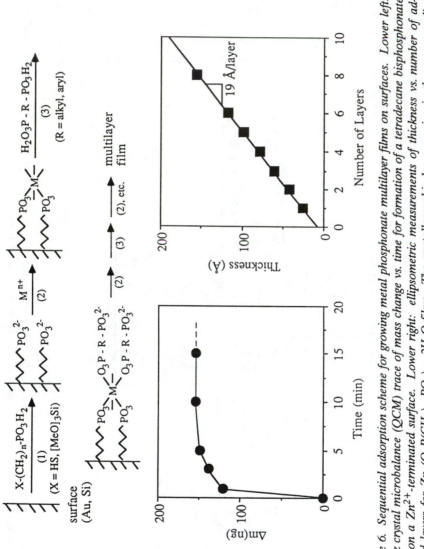

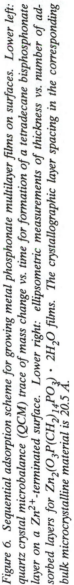

Figure 6. Sequential adsorption scheme for growing metal phosphonate multilayer films on surfaces. Lower left: quartz crystal microbalance (QCM) trace of mass change vs. time for formation of a tetradecane bisphosphonate layer on a Zn^{2+}-terminated surface. Lower right: ellipsometric measurements of thickness vs. number of adsorbed layers for $Zn_2(O_3P(CH_2)_{14}PO_3) \cdot 2H_2O$ films. The crystallographic layer spacing in the corresponding bulk microcrystalline material is 20.5 Å.

cause the complete assembly of each layer requires on the order of 4 h with tetravalent or trivalent (65) metal ions. The growth of divalent metal ion (Zn^{2+} and Cu^{2+}) films from methanol or ethanol solutions containing about 5% water is considerably faster, each adsorption step requiring on the order of 5–10 min, as illustrated in the quartz crystal microbalance (QCM) traces shown in Figure 6 (66, 67).

These materials are currently being evaluated as active components of small-molecule sensing devices, based on shape-selective intercalation reactions of analytes such as NH_3 and CO_2 (68–70), and as insulating layers for use in metal–insulator–semiconductor (MIS) and metal–insulator–metal thin-film devices (71). Katz et al. (72) also showed that this layer growth procedure, which normally gives rise to structures analogous to Y-type LB films, can be modified to produce polar, SHG-active multilayers. Using a three-step adsorption sequence, they incorporated oriented azo dyes into growing multilayer films. These polar multilayers show second-order nonlinear optical effects comparable to $LiNbO_3$, one of the most widely used inorganic nonlinear materials. The inorganic metal phosphate superstructure is chemically and thermally very durable and imparts orientational stability to the dye molecules up to at least 150 °C (at which temperature the azo dye itself begins to decompose). Other polar organic media, such as Z-type LB multilayers and poled polymers, undergo orientational randomization at much lower temperatures.

Future Directions: New Techniques and Applications

Several recent developments have been described that, in principle, could significantly expand the scope of synthetic techniques available for preparing organic multilayer heterostructures. In the adsorption techniques described here, binding of one layer to the next involves strong covalent or ionic bonds. This requirement limits the number of functional group interactions available for constructing multilayers to a relatively small number: siloxane bonds, metal phosphonate bonds, and ionic copper–oxygen and copper–sulfur bonds in carboxylate–thiolate multilayers (73). Multilayers based on polycation–polyanion ionic interactions (74) and on transition metal–ligand coordinate covalent interactions also were described (75, 76), although in neither case are the details of film structure known. Perhaps the most interesting new development in multilayer film-growth techniques involves sequential adsorption reactions from the gas phase to give either covalent (77) or noncovalent (78) interlayer links in the absence of solvent.

The gas-phase technique offers two very significant advantages: First, it is possible to eliminate solution-phase contaminants, such as amphiphiles and metal ions, which can present serious problems even at very low levels in adsorption reactions. Second, multilayers can be assembled via relatively weak interactions, such as hydrogen bonding, because solvolysis of these weak bonds is eliminated. This second feature may open the door to preparation of multilayers through a rather wide variety of terminal functional groups such as alcohols, carboxylic acids, primary amines, and various Lewis acid and base groups. Of course, the technique does suffer the disadvantage that the layer-forming molecules must be sufficiently volatile; for these reasons only relatively small molecules have been studied by this technique.

One of the potential applications of organic monolayer and multilayer films is in microelectronics or "molecular electronics". Because their thickness can be controlled quite precisely, insulating films (containing long alkyl chains) have been studied as insulators in semiconductor MIS devices (40, 57, 71), and monolayers containing oriented electron-donor–acceptor amphiphiles also were shown to act as electrochemical rectifiers (79). Micrometer-thick "all-organic" transistors have been made from polymeric materials (80), but the preparation of such devices on a truly thin-film scale (using, for example, a trilayer film of electroactive molecules) remains a synthetic challenge for multilayer chemistry. The gas-phase techniques described by Crooks and co-workers (77, 78) may lend themselves most easily to these goals, as they are compatible with various molecular functionalities, as well as with vapor-phase microelectronic fabrication techniques that involve masking and dry-etching. Ultimately, one might hope to develop an organic analogue of molecular beam epitaxy using covalent or noncovalent interactions to induce registry of one layer on the next.

Finally, one new multilayer film-growth technique has not yet been applied to organic molecules, or in fact even to anisotropic materials, but nevertheless could be quite interesting in this regard. This electrochemical technique has been used by Switzer et al. (81) to prepare well-ordered superlattices of Tl_2O_3 and PbO_2, in which the layers alternate on a length scale of a few nanometers. They are grown from alkaline solutions containing Tl^+ and Pb^{2+}, the latter at much higher concentration. PbO_2 is deposited at more positive potentials than Tl_2O_3, and so at the PbO_2 deposition potential a mixed film is formed; at relatively high overpotentials, however, the film is mostly PbO_2, because both ions are oxidized at a rate controlled by diffusion to the electrode surface, and the concentration of Pb^{2+} is much higher. At more negative potentials, only Tl_2O_3 is deposited. Therefore, by switching back and forth between more negative and positive potentials, a superlattice of alternating composition is created. The amount of charge passed in each half-cycle can be controlled quite precisely, so a very regularly spaced superlattice can be seen in X-ray diffraction experiments.

The beauty of this electrochemical technique is that it produces each layer very rapidly, with precise control of thickness, and it can be computer-controlled to create very large superlattices. One might imagine preparing similar superlattices from electroactive organic compounds, for example, metallocenes or derivatives of molecules such as tetrathiafulvalene (TTF), tetramethylphenylenediamine (TMPD), and tetracyanoquinodimethane (TCNQ). The oxidation of metal ions at an electrode might also be a means of creating in situ controlled amounts of insoluble organic salts, such as phosphonates or thiolates, which could then be deposited as multilayers or even as superlattices.

Acknowledgments

We thank the Texas Advanced Research Program and the National Science Foundation (Grant CHE–9217719) for financial support. T. E. Mallouk also thanks the Camille and Henry Dreyfus Foundation for support in the form of a Teacher–Scholar Award.

References

1. Swalen, J. D.; Allara, D. L.; Andrade, J. D.; Chandross, E. A.; Garoff, S.; Isreaelachvili, J.; McCarthy, T.; Murray, R.; Pease, R. F.; Rabolt, J. F.; Wynne, K. J.; Yu, H. *Langmuir* **1987,** *3,* 932.
2. Ulman, A. *An Introduction to Ultrathin Organic Films: From Langmuir–Blodgett to Self-Assembly;* Harcourt Brace Jovanovich: Boston, MA, 1991.
3. Kim Y. T.; Bard, A. J. *Langmuir* **1992,** *8,* 1096.
4. Binnig, G.; Quate, C. F.: Gerber, C. *Phys. Rev. Lett.* **1986,** *56,* 930.
5. Campion, A. *Annu. Rev. Phys. Chem.* **1985,** *36,* 549.
6. Outka, D. A.; Stohr, J.; Rabe, J.; Swalen, J. D. *J. Phys. Chem.* **1988,** *88,* 4076.
7. Overney, R. M.; Meyer, E.; Frommer, J.; Brodbeck, D.; Lüthi, R.; Howald, L.; Güntherodt, H.-J.; Fujihira, M.; Takano, H.; Gotoh, Y. *Nature (London)* **1992,** *359,* 133.
8. Blodgett, K. B.; Langmuir, I. *Phys. Rev.* **1937,** *51,* 964.
9. Gaines, G. L., Jr. *Insoluble Monolayers at Liquid–Gas Interfaces;* Wiley: New York, 1966.
10. *Molecular Engineering in Ultrathin Polymeric Films;* Stroeve, P.; Franses, E., Eds.; Elsevier: New York, 1987.
11. Agarwal, V. K. *Phys. Today* **1988,** *June,* 40.
12. Sugawara, M.; Kojima, K.; Sazawa, H.; Umezawa, Y. *Anal. Chem.* **1987,** *59,* 2842.
13. Bigelow, W. C.; Pickett, D. L.; Zisman, W. A. *J. Colloid Sci.* **1946,** *1,* 513.

14. Timmons, C. O.; Zisman, W. A. *J. Phys. Chem.* **1965**, *69*, 984.
15. Levine, O.; Zisman, W. A. *J. Phys. Chem.* **1957**, *61*, 1068.
16. Allara, D. L.; Nuzzo, R. G. *Langmuir* **1985**, *1*, 45.
17. Allara, D. L.; Nuzzo, R. G. *Langmuir* **1985**, *1*, 52.
18. Polymeropoulos, E. E.; Sagiv, J. *J. Chem. Phys.* **1978**, *69*, 1836.
19. Sagiv, J. *J. Am. Chem. Soc.* **1980**, *102*, 92.
20. Sagiv, J. *Isr. J. Chem.* **1979**, *18*, 339.
21. Maoz, R.; Sagiv, J. *J. Colloid Interface Sci.* **1984**, *100*, 465.
22. Gun, J.; Iscovici, R.; Sagiv, J. *J. Colloid Interface Sci.* **1984**, *101*, 201.
23. Gun, J.; Sagiv, J. *J. Colloid Interface Sci.* **1986**, *112*, 457.
24. Wasserman, S. R.; Tao, Y. T.; Whitesides. G. M. *Langmuir* **1989**, *5*, 1074.
25. Sagiv, J. *Isr. J. Chem.* **1979**, *18*, 346.
26. Tillman, N.; Ulman, A.; Schildkraut, J. S.; Penner, T. L. *J. Am. Chem. Soc.* **1988**, *110*, 6136.
27. Tillman, N.; Ulman, A.; Elman, J. F. *Langmuir* **1990**, *6*, 1512.
28. Ulman, A.; Scaringe, R. *Langmuir* **1992**, *8*, 894.
29. Adamson, A. W. *Physical Chemistry of Surfaces*, 4th ed.; John Wiley and Sons: New York, 1982.
30. DePalma, V.; Tillman, N. *Langmuir* **1989**, *5*, 868.
31. Chaudhury, M. K.; Whitesides, G. M. *Science (Washington, D.C.)* **1992**, *256*, 1539.
32. (a) Blackman, L. C. F.; Dewar, M. J. S. *J. Chem. Soc.* **1957**, 171. (b) Blackman, L. C. F.; Dewar, M. J. S.; Hampson, H. *J. Appl. Chem.* **1957**, *7*, 160.
33. (a) Nuzzo, R. G.; Allara, D. L. *J. Am. Chem. Soc.* **1983**, *105*, 4481. (b) Nuzzo, R. G.; Fusco, F. A.; Allara, D. L. *J. Am. Chem. Soc.* **1987**, *109*, 2358.
34. Porter, M. D.; Bright, T. B.; Allara, D. L.; Chidsey. C. E. D. *J. Am. Chem. Soc.* **1987**, *109*, 3559.
35. Bain, C. D.; Whitesides, G. M. *J. Am. Chem. Soc.* **1988**, *110*, 5897.
36. Troughton, E. B.; Bain, C. D.; Whitesides, G. M.; Nuzzo, R. G.; Allara, D. L.; Porter, M. D. *Langmuir* **1988**, *4*, 365.
37. Whitesides, G. M.; Laibinis, P. E. *Langmuir* **1990**, *6*, 87.
38. Thomas, R. C.; Sun, L.; Crooks, R. M.; Ricco. A. J. *Langmuir* **1991**, *7*, 620.
39. Folkers, J.; Laibinis, P. E.; Whitesides, G. M. *Langmuir* **1992**, *8*, 1330.
40. Sheen, C. W.; Shi, J. X.; Martensson, J.; Parikh, A. N.; Allara, D. L. *J. Am. Chem. Soc.* **1992**, *114*, 1514.
41. Lunt, S. R.; Santangelo, P. G.; Lewis, N. S. *J. Vac. Sci. Technol. B.* **1991**, *9*, 2333.
42. Miller, C.; Cuendet, P.; Gratzel, M. *J. Phys. Chem.* **1991**, *95*, 877.
43. Miller, C.; Gratzel, M. *J. Phys. Chem.* **1991**, *95*, 5225.
44. Bennett, A. J. *J. Electroanal. Chem.* **1975**, *60*, 125.
45. Marcus, R. A. *J. Phys. Chem.* **1965**, *43*, 679.
46. Li, T. T.-T.; Weaver, M. *J. Am. Chem. Soc.* **1984**, *106*, 6108.
47. Li, T. T.-T.; Weaver, M. *J. Electroanal. Chem.* **1985**, *188*, 121.
48. Chidsey, C. E. D.; Bertozzi, C. R.; Putvinski, T. M.; Mujsce, A. M. *J. Am. Chem. Soc.* **1990**, *112*, 4301.
49. Chidsey, C. E. D. *Science (Washington, D.C.)* **1991**, *251*, 919.
50. Finklea, H. O.; Hanshew, D. D. *J. Am. Chem. Soc.* **1992**, *114*, 3173.
51. Rubinstein, I.; Steinberg, S.; Tor, I.; Shanzer, A.; Sagiv, J. *Nature (London)* **1988**, *332*, 426.

52. Steinberg, S.; Rubinstein, I. *Langmuir* **1992**, *8*, 1183.
53. Haussling, L.; Ringsdorf, H.; Schmitt, T. J.; Knoll, W. *Langmuir* **1991**, *7*, 1837.
54. Ricco, A. J.; Kepley, L. J.; Thomas, R. C.; Sun. L.; Crooks, R. M. *Tech. Dig.* (IEEE Solid-State Sensor and Actuator Workshop) Hilton Head, SC, 1992.
55. Wang, J.; Frostman, L. M.; Ward, M. D. *J. Phys. Chem.* **1992**, *96*, 5224.
56. Popovitz-Biro, R.; Hill, K.; Hung, D. J.; Lahav, M.; Leiserowitz, L.; Sagiv, J.; Hsiung, H.; Meredith, G. R.; Vanherzeele, H. *J. Am. Chem. Soc.* **1990**, *112*, 2498.
57. Roberts, G. G. *Adv. Phys.* **1985**, *34*, 475.
58. Netzer, L; Sagiv, J. *J. Am. Chem. Soc.* **1983**, *105*, 674.
59. Maoz, R.; Netzer, L.; Gun, J.; Sagiv. J. *J. Chim. Phys.* **1988**, *85*, 1059.
60. Tillman, A.; Ulman, A.; Penner, T. L. *Langmuir* **1989**, *5*, 101.
61. Li, D.; Ratner, M. A.; Marks, T. J.; Zhang, C. H.; Yang, J.; Wong, G. K. *J. Am. Chem. Soc.* **1990**, *112*, 7389.
62. Lee, H.; Kepley, L. J.; Hong, H.-G.; Mallouk, T. E. *J. Am. Chem. Soc.* **1988**, *110*, 618.
63. Lee, H.; Kepley, L. J.; Hong, H.-G.; Akhter, S.; Mallouk, T. E. *J. Phys. Chem.* **1988**, *92*, 2597.
64. Cao, G.; Hong, H.-G.; Mallouk, T. E. *Acc. Chem. Res.* **1992**, *25*, 420.
65. Akhter, S.; Lee, H.; Hong, H.-G.; Mallouk, T. E.; White, J. M. *J. Vac. Sci. Technol. A* **1989**, *7*, 1608.
66. Yang, H. C.; Aoki, K.; Hong, H.-G.; Sackett, D. D.; Arendt, M. F.; Yau, S.-L.; Bell, C. M.; Mallouk, T. E. *J. Am. Chem. Soc.* **1993**, *115*, 11855.
67. Aoki, K.; Brousseau, L. C.; Mallouk, T. E. *Sens. Actuators* **1993**, *14*, 703.
68. Brousseau, L. C.; Aoki, K.; Cao, G.; Garcia, M. E.; Mallouk, T. E. In *Multifunctional Mesoporous Solids;* NATO Adv. Study Inst. Ser.; Sequeira, C. A. C., Ed.; Vol. 400, 1993, pp 225–236.
69. Cao, G.; Mallouk, T. E. *Inorg. Chem.* **1991**, *30*, 1434.
70. Frink, K. J.; Wang, R.-C.; Colon, J. L.; Clearfield, A. *Inorg. Chem.* **1991**, *30*, 1439.
71. Kepley, L. J.; Sackett, D. D.; Bell, C. M.; Mallouk, T. E. *Thin Solid Films* **1992**, *208*, 132.
72. Katz, H. E.; Scheller, G.; Putvinski, T. M.; Schilling, M. L.; Wilson, W. L.; Chidsey, C. E. D. *Science (Washington, D.C.)* **1991**, *254*, 1485.
73. Evans, S. D.; Ulman, A.; Goppert-Berarducci, K. E.; Gerenser, L. J. *J. Am. Chem. Soc.* **1991**, *113*, 5866.
74. Decher, G.; Hong, J. D.; Schmitt, J. *Thin Solid Films* **1992**, *210–211*, 831.
75. Li, D.; Huckett, S. C.; Frankcom, T.; Paffett, M. T.; Farr, J. D.; Hawley, M. E.; Gottesfeld, S.; Thompson, J. D.; Burns, C. J.; Swanson, G. I. In *Supramolecular Architecture;* ACS Symposium Series 499; Bein, T., Ed.; American Chemical Society: Washington, DC, 1992; pp 33–45.
76. Li, D.; Smith, D. C.; Swanson, B. I.; Farr, J. D.; Paffett, M. T.; Hawley, M. E. *Chem. Mater.* **1992**, *4*, 1047.
77. Sun, L.; Thomas, R. C.; Crooks, R. M.; Ricco, A. J. *J. Am. Chem. Soc.* **1991**, *113*, 8550.
78. Sun, L.; Kepley, L.; Crooks, R. M. *Langmuir* **1992**, *8*, 2101.
79. Geddes, N. J.; Sambles, J. R.; Jarvis, D. J.; Parker, W. G.; Sandman, D. J. *Appl. Phys. Lett.* **1990**, *56*, 1916.

80.   Garnier, F.; Horowitz, G.; Peng, X; Fichou, D. *Adv. Mater.* **1990,** *2,* 592.
81.   Switzer, J. A.; Shane, M. J.; Phillips, R. J. *Science (Washington, D.C.)* **1990,** *247,* 444.

RECEIVED for review November 9, 1992. ACCEPTED revised manuscript April 13, 1993.

Orientation-Dependent NMR Spectroscopy as a Structural Tool for Layered Materials

Mark E. Thompson, David A. Burwell, Charlotte F. Lee, Lori K. Myers, and Kathleen G. Valentine

Department of Chemistry, Princeton University, Princeton, NJ 08544

This chapter discusses the use of NMR spectroscopy of aligned thin films to determine the structures of two layered inorganic compounds. The change in the observed NMR spectrum of an aligned sample as it is reoriented in the magnetic field is usually significant. This orientation-dependent change in the spectra can be used to determine the structure of selectively labeled groups within a solid. For layered materials, the formation of an aligned sample is straightforward. The crystals tend to have a platelike morphology, and deposition of these materials on planar substrates leads to films with the majority of the crystals lying parallel to the surface of the substrate. The chapter covers the background theory needed to understand the orientation-dependent NMR experiment and reports on studies carried out for two different systems. The structures of a layered phosphonate and a layered intercalation compound were determined by the NMR techniques.

THE INTERCALATION OF ORGANIC MOLECULES into layered inorganic compounds has been studied extensively over the past 25 years (*1–3*). In these topochemical reactions, inorganic host layers separate to accommodate interaction with organic guest molecules. A fully intercalated material consists of regularly alternating organic and inorganic layers.

0065–2393/95/0245–0231$13.75/0

Intercalation can significantly alter both the physical nature and chemical reactivity of a host. For example, intercalation has been used to modify a host's optical properties, superconducting critical temperature, and interlayer magnetic coupling (4–5). Applications of intercalation compounds include materials design (6, 7), ion exchange (8–10), and catalysis (11, 12). Some recent and very interesting applications involve the use of these compounds to modify electrode surfaces (13, 14), prepare low-dimensional conducting polymers (15, 16), and assemble molecular multilayers at solid–liquid interfaces (17).

A typical problem with layered materials such as those described is the determination of structure. The particle size is large enough for powder X-ray diffraction to be carried out, but it is typically too small for single-crystal X-ray diffraction. Moreover, the microcrystals themselves usually possess long-range order only along an axis normal to the layers. The crystallographic registry of adjacent layers in the directions parallel to the lamellae is typically poor. The orientation-dependent NMR techniques described in this chapter were developed to determine structure in these microcrystalline layered materials. These NMR experiments depend on being able to fabricate an aligned thin film of crystals, which is fairly straightforward for the platelike intercalation compounds. The NMR spectrum of the ordered sample is then examined as a function of its orientation relative to the applied magnetic field.

The two systems that will be used to illustrate the orientation-dependent NMR technique will be described in detail, but a number of other systems may also be examined by these techniques. A great deal of work has been done with liquid-crystalline materials (18, 19), lipid bilayers (20, 21), and proteins (22–25). In all of these systems the entire sample is ordered along a single axis. All that is required of a crystalline material for it to be suitable for this experiment is that it exist in a platelike morphology, so that an oriented thin film can be formed. Layered intercalation compounds are ideally suited for these studies. The pure host solids are crystalline and can typically be grown into single crystals large enough for X-ray crystallographic analysis. Upon intercalation the crystals become disordered, with good order observed only along the axis parallel to the layer normal. This disorder makes single-crystal X-ray diffraction study worthless for determining structure.

NMR spectroscopy has been used to study oriented thin films of microcrystalline materials by us for α-Zr(O$_3$POH)$_2 \cdot$ H$_2$O intercalation compounds (vide infra) and by Resing and co-workers (26–28) for highly oriented pyrolytic graphite compounds. A large number of other host compounds could be examined by these NMR techniques. In particular, a large number of intercalation compounds have been prepared with clay minerals, metal dichalcogenides, metal phosphorus trichalcogenides, metal

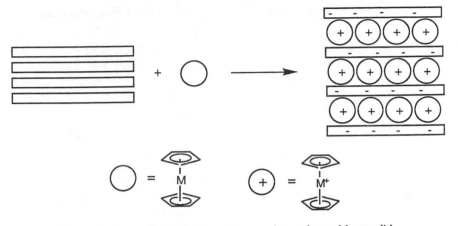

Figure 1. Intercalation of a metallocene into a layered host solid.

oxyhalides, and metal oxides, many of which would be amenable to structural study by orientation-dependent NMR spectroscopy (*29*).

A good example of the sort structural detail that can be obtained from this technique, which is not readily available from other analytical or spectroscopic techniques, involves the structures of metallocene intercalation compounds. Cobaltocene and ferrocene intercalate into layered hosts, as shown in Figure 1. The orientation of the metallocene relative to the host layers cannot be determined from X-ray diffraction data because the dimensions of the metallocenes parallel and perpendicular to the C_5 axis are nearly identical. These two different orientations give very different NMR spectra, however, and can readily be differentiated, as will be described.

The methods used to promote intercalation of organic guest molecules into inorganic hosts can generally be divided into three types: redox, ion-exchange, and acid–base reactions (*1–3*). Intercalation reactions can also be driven by the reaction of an amine or alcohol with a layered acyl chloride compound. This acyl chloride host is a zirconium phosphonate, which is prepared from the layered carboxylic acid, $Zr(O_3PCH_2CH_2COOH)_2$. The preparation of the acyl chloride is done in two steps: The acid is first deprotonated with NH_3 and then treated with $SOCl_2$ (*30*):

$$Zr(O_3PCH_2CH_2COOH)_2 \xrightarrow{NH_3} Zr(O_3PCH_2CH_2COO^-NH_4^+)_2 \quad (1a)$$
$$\xrightarrow{SOCl_2} Zr(O_3PCH_2CH_2COCl)_2 \quad (1b)$$

The acyl chloride derivative then reacts with amines or alcohols to give the corresponding amides and esters (*31*):

$$Zr(O_3PCH_2CH_2COCl)_2 \xrightarrow{\text{RNH}_2} Zr(O_3PCH_2CH_2CONHR)_2 \quad (2)$$

Our initial studies in the amide and ester chemistry focused on the propionic acid derivative, $Zr(O_3PCH_2CH_2COOH)_2$. When we tried to carry out the same reactions for the acetic or butyric acid derivatives (i.e., $Zr(O_3P(CH_2)_nCOOH)_2$, $n = 1$ or 3), we found only partial conversion of these compounds to the acyl chloride derivatives. The carboxylic acid could be converted to the ammonium salt quantitatively, but the subsequent reaction with $SOCl_2$ gave only a 40% conversion to the acyl chloride (as evidenced by IR spectroscopy). A significant difference in reactivity was observed between compounds that have an odd versus an even number of methylenes between the phosphorus and the carboxylic acid group. Most likely, this variation in reactivity is related to structural differences in the pendant organic group that alter the host's ability to interact with guest molecules. A detailed investigation of the proton conductivity of these same carboxy-terminated zirconium phosphonates was reported by Alberti et al. (32). Their results showed that compounds with $n = 1$, 3, and 5 exhibit much higher ionic conductivity than the compounds with $n = 2$ and 4. They too proposed that structural changes associated with an odd–even n effect are responsible for this behavior.

We believed that it was important to better understand these changes and any effect they might have on the chemical and physical properties of $Zr(O_3P(CH_2)_nCOOH)_2$. Unfortunately, zirconium phosphonate crystals of sufficient size for structural determination by conventional X-ray diffraction (XRD) techniques have yet to be grown. Next, we present orientation-dependent ^{31}P and ^{13}C solid-state NMR studies that were used to resolve the structure of the organic moiety in $Zr(O_3PCH_2COOH)_2$ (33).

Our group is also interested in studying the structure and reactivity of a variety of organometallic intercalation compounds. α-Zirconium hydrogen phosphate, α-$Zr(O_3POH)_2 \cdot H_2O$, (ZrP) is a versatile host for organometallic guests, intercalating both redox-active (34) and amine-substituted compounds (35–37). One of the organometallic intercalation compounds that we have prepared with ZrP is the ferrocenylethylamine intercalation compound, $[CpFe(C_5H_4CH_2CH_2NH_2)]_{0.6} \cdot ZrP$. One question we have about the structure in this material is the orientation of the organometallic guest molecule with respect to the layers of the host solid. The observed interlayer expansion of 14.5 Å is consistent with a bilayer of organometallic species between adjacent host layers. But because the dimensions of Cp_2Fe parallel and perpendicular to the C_5 axis are nearly identical, the interlayer expansion does not indicate the orientation of the ferrocenyl group relative to the host lamellae (38–43). A similar situation was observed in other metallocene intercalation compounds. To determine the orientation of the metallocenes in the **ZrP** intercalation compound, we carried out an orientation-dependent ^2H solid-state NMR study with the layered intercalation compound 2-(d_9-ferrocenyl)ethylamine · **ZrP** (*44*), as described later.

NMR Background

Nuclei with $I = \frac{1}{2}$, Such as ^{31}P and ^{13}C. A number of interactions affect the NMR linewidths of dipolar nuclei in solids; the two most important are chemical-shift anisotropy (CSA) and dipolar broadening (*45, 46*). Large CSA is observed in powdered samples because the chemical shift observed for a given particle is dependent on the particle's orientation relative to the magnetic field (vide infra). The powder sample consists of a random distribution of particle orientations, leading to a large dispersion of chemical shifts. Broadening due to CSA can be eliminated by rapidly spinning the sample; this spinning acts to average all of the particles to a single orientation. For dilute nuclear spins, such as ^{13}C (natural abundance 1.1%) the homonuclear dipolar effects are negligible, and only heteronuclear dipolar broadening needs to be considered. The heteronuclear dipolar interactions depend on the orientation of the bond vector in the magnetic field, being a minimum when $(3\cos^2\theta - 1)$ goes to 0 ($\theta = 54.74°$, referred to as the magic angle). Additional narrowing in hydrocarbon spectra is accomplished by proton decoupling during the ^{13}C acquisition.

The large magnetic moment of the proton can also be used to increase sensitivity by transferring magnetization to the ^{13}C nuclei in a cross-polarization (CP) experiment. To achieve high-resolution and high-sensitivity ^{13}C spectra in the solid state, several techniques are used together. Spinning the sample at the magic angle (MAS), heteronuclear decoupling, and cross-polarization used together can lead to narrow lines. The CP–MAS ^{13}C NMR spectrum of $Zr(O_3PCH_2COOH)_2$ is shown in Figure 2C. For comparison the ^1H-decoupled static spectrum of the same powdered sample is shown in Figure 2B. The isotropic chemical shift (σ_{iso}) observed in a CP–MAS experiment is the mean of the three principal components of the chemical-shielding tensor $[(\sigma_{11} + \sigma_{22} + \sigma_{33})/3]$ (vide infra).

The resonance frequency for a given nucleus in a solid is dependent on both the chemical environment around the nucleus and the orientation of the crystal relative to the magnetic field. The effect of the chemical environment on the NMR resonance frequency for a given nucleus is described by the chemical-shielding tensor. In the diagonalized form, the

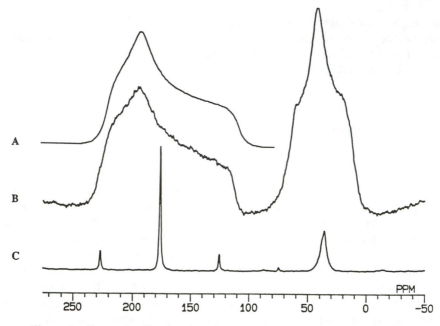

Figure 2. Curve A: Simulated static pattern for the carboxyl resonance of $Zr(O_3PCH_2COOH)_2$. *Curve B: ^{1}H-decoupled static powder spectrum of* $Zr(O_3PCH_2COOH)_2$. *Curve C: ^{13}C CP–MAS NMR spectrum of* $Zr(O_3$-$PCH_2COOH)_2$. *(Reproduced with permission from reference 56. Copyright 1986.)*

three principal components of this tensor are referred to as the principal axis system (PAS). The PAS is an orthogonal axis system, the three components being σ_{11}, σ_{22}, and σ_{33}. The PAS can be thought of as the *xyz* coordinates of the chemical shift. Each different nucleus in the solid has a unique PAS. To use an NMR experiment for structural studies one must know how the PAS is related to the molecular framework that makes up the solid. The mapping of PAS onto the molecular frame can be determined from either single-crystal rotation studies (*47, 48*) or selective labeling in powdered samples (*49–55*). We have determined the mapping of the PAS onto the molecular frames for the ^{13}C and ^{31}P of $Zr(O_3PCH_2{}^{13}COOH)_2$ (Figure 3) using selectively labeled powder samples (*56*). The orientations of the PASs relative to the molecular frames for both the ^{31}P and ^{13}C are similar to analogous phosphonates and carboxylic acids, respectively, reported by others (*57–60*).

The chemical shift for an arbitrary orientation of a crystal in an applied magnetic field is given by

$$\sigma(\theta,\phi) = \sigma_{11} \sin^2\theta \cos^2\phi + \sigma_{22} \sin^2\theta \sin^2\phi + \sigma_{33} \cos^2\theta \qquad (3)$$

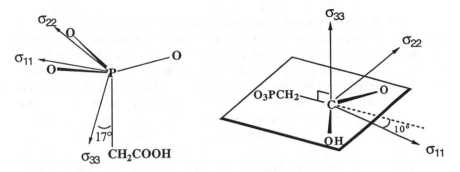

Figure 3. Orientation of the ^{31}P (top) and ^{13}C (bottom) chemical-shielding tensors in the molecular frame of $Zr(O_3PCH_2COOH)_2$.

where θ is the angle between σ_{33} and the applied magnetic field (B_0) and ϕ is the angle between the projection of B_0 onto the $\sigma_{11} - \sigma_{22}$ plane and σ_{11} (Figure 4a). On the basis of equation 3, if B_0 lies along the σ_{33} axis, the chemical shift will be equal to σ_{33} ($\theta = 0°$). Similarly, if the field lies along either of the other axes, the chemical shift will be equal to that PAS value. A simple way to think about the relationship between the PAS and the observed resonance frequency is to consider the ellipsoid defined by the PAS, with B_0 originating at the center of the ellipsoid (Figure 4b). The resonance frequency is given by the distance from the center of the ellipsoid to the point where B_0 intersects the surface; this distance is exactly what is given by equation 3 for particular values of θ and ϕ. As the crystal (and thus the ellipsoid) is rotated, the point of intersection of B_0 with the surface of the ellipsoid changes, and so does the resonance frequency.

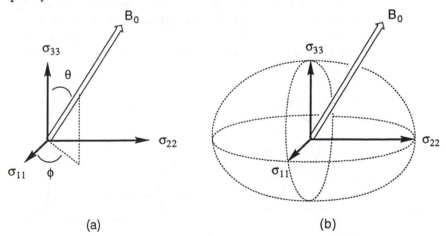

(a)   (b)

Figure 4. Definition of angles θ and ϕ (a), and PAS ellipsoid (b) (see text).

Both the large linewidth and the asymmetry of a static powder spectrum (e.g., Figure 2B) are a result of the random orientation of the individual microcrystals. The signal observed in the static spectrum can be thought of as a superposition of the resonances for each of the particles. The particles have a random distribution of θ and ϕ values. Solving equation 3 for all values of θ and ϕ for the carboxyl carbon of $Zr(O_3P\text{-}CH_2COOH)_2$ gives the line shape shown in Figure 2A (*61, 62*). An important point is that an individual particle can have a range of different chemical shifts, depending on its orientation, but the maximum and minimum values for the chemical shift are equal to the maximum and minimum PAS values. The values for the principal components can be obtained directly from the powder spectra. σ_{22} is always the feature closest to the center of the powder pattern, and σ_{11} is always the feature that lies closest to σ_{22}. The remaining feature in the powder pattern is σ_{33}. The principal components are indicated in Figure 5 (left) for a general powder pattern. If the PAS is axially symmetric (two of the principal components are equal), the two equal values are termed $\sigma_{\perp}$, and the unique component is $\sigma_{\parallel}$. A typical powder pattern for an axially symmetric tensor is shown in Figure 5 (right).

Nuclei with $I > \frac{1}{2}$, Such as ^2H.

Thus far we have dealt exclusively with dipolar nuclei. A great deal of important information can be obtained from quadrupolar nuclei as well (*45, 63–65*). A great many quadrupolar nuclei have been studied, and the most heavily studied nucleus is ^2H. This chapter will be restricted to $I = 1$ quadrupolar nuclei, particularly ^2H.

In a magnetic field the Zeeman effect breaks the degeneracy of the three levels of an $I = 1$ nucleus to give three different energy levels. In a spherically symmetric electric field the two allowed transitions ($-1 \rightarrow 0$ and $0 \rightarrow +1$) are degenerate. In the presence of a nonspherically symmetric electric field, which is common for molecular solids, the quadrupo-

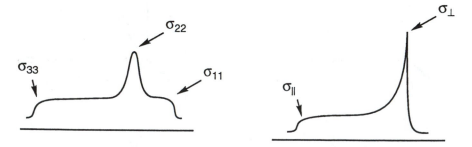

Figure 5. Powder pattern and principal components for a general powder pattern (left) and an axially symmetric powder pattern (right).

lar moment of the nucleus acts to break the degeneracy of these transitions. The net result of this action is a significant broadening of the lines in the ^{2}H NMR spectrum relative to the lines observed for dipolar nuclei. A typical linewidth for a solid-state ^{2}H NMR spectrum is 250–300 kHz (ca. 6000 ppm at 41 MHz).

In the same way that the chemical-shielding tensor could be used to describe the NMR spectrum of a dipolar nucleus, a tensor can be used to describe the NMR spectra of quadrupolar nuclei. This electric-field gradient (or EFG) tensor for quadrupolar nuclei has three mutually perpendicular components, termed V_{11}, V_{22}, and V_{33} (*45, 46*). For most deuterocarbons the electric-field gradient is axially symmetric, because it is dominated by the C–D bonding electrons. In an axially symmetric system the singular component is V_{33}, which is termed the principal component of the EFG tensor, and it is spatially represented by a vector lying along the C–D bond. Nonaxially symmetric systems are characterized by an asymmetry parameter, η, (*21, 63–65*), but for the purposes this chapter we will assume that all ^{2}H nuclei are acted on by an axially symmetric electric field.

The energy difference between the two allowed transitions for an $I = 1$ nucleus is related to the orientation of the sample relative to the magnetic field. In the same way equation 3 describes the orientation-dependent NMR spectra of dipolar nuclei, an equation can be developed for quadrupolar nuclei. Because the system here is axially symmetric, only a single angle needs to be considered, where θ is the angle between V_{33} and the magnetic field (B_0). Rather than calculating the energy of a single transition, the energy difference between the two transitions, $\Delta\nu$, is typically calculated. $\Delta\nu$ is given by equation 4:

$$\Delta\nu \;=\; \frac{3}{4}\left[\frac{e^2qQ}{h}\right](3\cos^2\theta - 1) \qquad (4)$$

where eQ is the quadrupole moment, q is the electric field gradient, and h is Planck's constant; e^2qQ/h is the quadrupolar coupling constant. For a single crystal with a single magnetically inequivalent ^{2}H atom in the unit cell, the ^{2}H NMR spectrum would consist of two peaks, with $\Delta\nu$ dependent on the orientation of V_{33} relative to B_0. For a randomly oriented powder sample a more complicated signal is observed (Figure 6b). This powder spectrum, a Pake doublet, (*45, 46*) can be thought of as the superposition of two axially symmetric powder signals, one from each of the lines observed for a single-crystalline sample (Figure 6a). The two powder signals have the same isotropic chemical shift [$(V_{11} + V_{11}, V_{22} + V_{11}, V_{22}$ $V_{33})/3$], but different signs. The peaks in the powder spectrum arise from

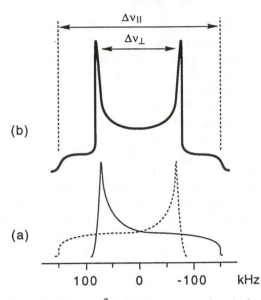

Figure 6. General solid-state ^2H NMR spectrum for single-crystalline (a) and randomly oriented powder (b) samples.

particles with θ close to 90°, and their energy difference is $\Delta\nu_\perp$ (Figure 6b). The shoulders of the spectrum come from a θ close to 0 contribution, and their energy difference is termed $\Delta\nu_\parallel$.

Because of the tremendous linewidths, ^2H atoms with different chemical shifts cannot generally be resolved. Fortunately, ^2H atoms undergoing different types of motional averaging can be resolved (45, 63, 66). The affect of motional averaging on the observed line shapes of ^2H NMR spectra can be dramatic. For this chapter we will restrict our attention to systems that are undergoing a simple rotation about a three-or-higher fold axis, for example, the rotation of a metal-bound d_5-cyclopentadienyl ligand (d_5-Cp) about its C_5 axis. The effect that this rotation has on the spectrum is directly dependent on the angle between the axis of rotation and V_{33} of the static system, γ. $\Delta\nu$ can be calculated by an equation similar to equation 4, with the inclusion of a term to compensate for the motional averaging:

$$\Delta\nu = \frac{3}{4}\left[\frac{e^2qQ}{h}\right](3\cos^2\theta - 1)\left[\frac{3\cos^2\gamma - 1}{2}\right] \qquad (5)$$

Motional averaging always leads to a decrease in $\Delta\nu$, with the magnitude of the decrease dependent on γ. For the d_5-Cp ligand the angle between the C_5 axis and V_{33} is 90°. On the basis of equation 5, the expected $\Delta\nu$ is half that of a static system, which is exactly what is observed for crystalline

d_{10}-Cp$_2$Fe at room temperature (*67*) or d_{10}-Cp$_2$Fe in solid matrices (*68–71*).

In addition to decreasing the linewidth, the motional averaging has a very important effect. The principal component of the EFG tensor (V_{33}) for the motionally averaged system is parallel to the axis of rotation. Thus for the d_5-Cp example, in the static case each of the five ^{2}H atoms will have their V_{33} oriented in a different direction, giving rise to five different values of $\Delta\nu$ based on equation 4. In a rapidly rotating d_5-Cp all five ^{2}H atoms will have an effective V_{33} parallel to the C_5 axis, giving rise to a single value of $\Delta\nu$ for all five ^{2}H atoms. Thus in a motionally averaged sample, $\Delta\nu$ is not dictated by the angle between the C–D bond vector and B_0, but by the angle between the C_5 axis and B_0. Another important effect that motional averaging has is to shorten spin relaxation times relative to static nuclei. For example, in crystalline d_{40}-nonadecane at 165 K the –CD$_3$ end groups are rapidly rotating about their C_3 axes and have a spin–lattice relaxation time (T^1) of 3 ms, and the static –CD$_2$– groups have T^1 values of ~100 s (*72*).

Nature of the Sample

The preceding discussion illustrates how to use NMR techniques to determine the orientations of various functional groups relative to the axes of large crystals. If one had crystals large enough for such an experiment, and all that was sought was structural information, a single-crystal X-ray diffraction study would probably be more appropriate than an NMR experiment. However, the NMR experiment is very well suited to microcrystalline or poorly crystalline materials.

Two of the advantages that the NMR experiment has over X-ray diffraction are (1) fairly good structural data can be obtained from a uniaxially aligned powder sample, and (2) the NMR experiment of a selectively labeled compound gives signals from the labeled sites only. The second advantage can be very useful for looking at the structural or dynamical properties of different parts of a given complex. By moving the label throughout the sample, each site can be examined independently of the other sites.

The preparation of a uniaxially aligned thin film for these NMR studies is rather straightforward (*33, 44*). Layered materials tend to have platelike morphologies. When slurries of these materials are allowed to settle, the crystals usually lie flat, parallel to the surface onto which they are deposited. This self-ordering is similar to what is seen when a deck of cards is dropped on the floor. The procedure for preparing NMR samples of microcrystalline materials involves first sonicating the powdered sample in ethanol for a short period of time to disperse the crystals. The slurry is then cast onto pieces of glass coverslip (6 × 20 mm) and allowed to dry. The resulting films consist predominantly of regions that are two to four crystals thick, the crystals lying parallel to the glass substrate (Figure 7). Small regions of the film can be observed; these regions are much thicker and have a random orientation of the crystals. Twenty-five of these coverslips are then stacked up and placed into an NMR sample holder (Figure 7). The vast majority of the crystals in this sample have their layer normals lying parallel to each other and perpendicular to the glass plates. The sample is inserted into the NMR probe and, the spectrum is recorded with the layer normals parallel and perpendicular to the applied field.

Structure of $Zr(O_3PCH_2COOH)_2$

Phosphorus-31 NMR Spectra of $Zr(O_3PCH_2{}^{13}COOH)$. Ori-

ented ^{31}P CP NMR spectra of ^{13}C-labeled $Zr(O_3PCH_2{}^{13}COOH)_2$ were taken from reference 33 and are presented in Figure 8. On the left side of Figure 8 are shown the experimental CSA powder pattern (d), the experimental spectrum with the layer normal perpendicular to B_0 (c), and the difference spectrum (b). The analogous spectra obtained with the layer normal parallel to B_0 are shown on the right side of Figure 8. Difference spectra were used to correct for the small portions of the film that are not aligned. The effect of uniaxial sample orientation is readily seen by comparing the two difference spectra. The resonance frequencies seen in the perpendicular difference spectrum fall exclusively in the $\sigma_{11} - \sigma_{22}$ region of the CSA powder pattern, whereas frequencies in the σ_{33} region were

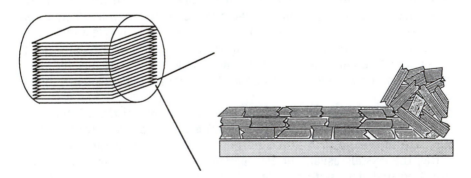

Figure 7. Samples for orientation-dependent NMR studies. A side view of the proposed thin film of crystals on the glass slide is shown on the bottom, and a packed NMR rotor is shown on the top.

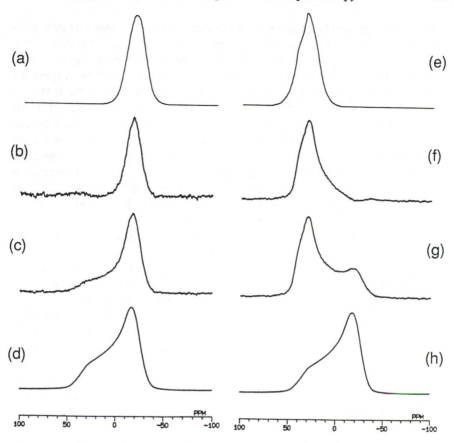

Figure 8. 109.4-MHz ^{31}P CP NMR spectra of oriented $Zr(O_3\text{-}PCH_2{}^{13}COOH)_2$: (a) simulated perpendicular spectrum, (b) perpendicular difference spectrum, (c) perpendicular experimental spectrum, (d) experimental powder spectrum, (e) simulated parallel spectrum, (f) parallel difference spectrum, (g) parallel experimental spectrum, and (h) experimental powder spectrum.

observed in the parallel difference spectrum. These spectra indicate, in a qualitative sense, that σ_{33} was nearly coincident with B_0 when the $Zr(O_3P)_2$ layer normal was parallel to B_0, and that B_0 lies in the $\sigma_{11} - \sigma_{22}$ plane (θ is close to 90°) when the layer normal is perpendicular to B_0.

The oriented spectra were quantitatively interpreted using equation 3 and the predetermined orientation of the ^{31}P-phosphonate chemical-shielding PAS in the molecular frame of $Zr(O_3PCH_2COOH)_2$ (33). When the layer normal was positioned parallel to B_0, the resulting difference spectrum (Figure 8) exhibited an asymmetric band of resonance frequencies centered at 30 ppm. To simplify analysis of the line shape, the observed range of chemical shifts was taken as the full width at half maximum intensity. Thus, the range of chemical shift in the parallel difference spectrum is 30 ± 10 ppm. The θ/ϕ combinations resulting in chemical shifts within the 30 ± 10-ppm range were then calculated by using equation 3 and the principal elements. The θ/ϕ combinations consistent with the chemical shifts exhibited in the parallel difference spectrum are represented by the shaded region in Figure 9 (33). On the basis of this plot, θ was limited to $0-34°$. Only angles from $0-90°$ need be considered, because angles greater than $90°$ are related by symmetry. The range of chemical shifts exhibited in the perpendicular difference spectrum (Figure 9) is 16 ± 10 ppm, corresponding to the θ/ϕ combinations shown in Figure 9. From this plot, θ was limited to $59-90°$. The θ/ϕ plots derived from the two difference spectra cannot be directly superimposed (after a $90°$ phase shift in θ) to further limit ϕ, because the uniaxial orientation does not allow correlation of ϕ from one plot to the other.

Each of these difference spectra independently suggests that the P–C bond is perpendicular to the $Zr(O_3P)_2$ layers. This conclusion was arrived at by considering the orientation of σ_{33} in the molecular frame. To get an upper limit on the degree of disorder (crystal and molecular combined) in the alignment of the sample on the glass slides, the linewidths of the oriented spectra were examined. The mapping of the ^{31}P-phosphonate chemical-shielding tensor in $Zr(O_3PCH_2COOH)_2$ has σ_{33} $17°$ off the P–C bond (Figure 3a). The effect of ideal uniaxial alignment in the thin-film sample would therefore sweep out a $\theta = 17°$ cone when the layer normal is positioned parallel to B_0. With the layer normal perpendicular to B_0, a $17°$ cone about $\theta = 90°$ would be swept out. The θ/ϕ plots for both the parallel and perpendicular orientations indicate a range for θ of $30-35°$. If these ranges are corrected for the $17°$ cone angle expected on the basis of the mapping of the PAS onto the molecular frame, the upper limit for the disorder in alignment is $\sim15°$. Figures 8a and 8e show simulated oriented spectra for the perpendicular and parallel orientations assuming that the P–C bond is perpendicular to the $Zr(O_3P)_2$ layers and that the microcrystalline platelets are disordered by $\pm15°$ with respect to the glass slides.

Carbon-13 NMR Spectra of $Zr(O_3PCH_2{}^{13}COOH)$.

Oriented ^{13}C CP NMR spectra of ^{13}C-labeled $Zr(O_3PCH_2{}^{13}COOH)_2$ were taken from reference 33 and are shown in Figure 10. The left side of Figure 10 contains the experimental CSA powder pattern (d), the experimental spectrum with the layer normal perpendicular to B_0 (c), and the difference spectrum (b), and the right side shows the analogous spectra obtained with the layer normal parallel to B_0. Chemical-shift frequency ranges of the difference spectra and the corresponding θ/ϕ plots were calculated as

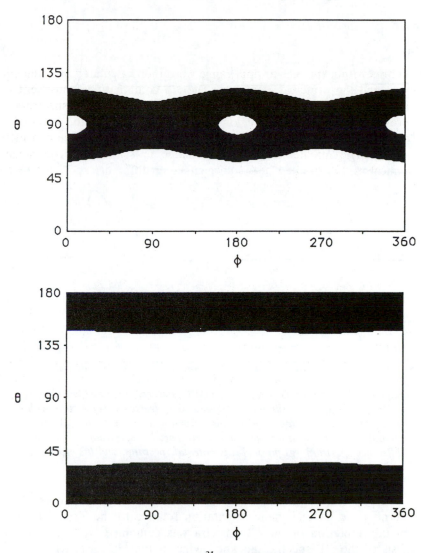

Figure 9. θ/ϕ plots for the oriented ^{31}P CP NMR spectra with layer normal perpendicular (top) and parallel (bottom) to B_0. The shaded regions indicate θ/ϕ combinations that are consistent with the experimental data.

described for the ^{31}P spectra using the chemical-shielding principal elements extracted from the powder pattern of $Zr(O_3PCH_2{}^{13}COOH)_2$. Chemical shifts range from 207 to 173 ppm in the perpendicular difference spectrum and from 184 to 108 ppm in the parallel difference spectrum. In the perpendicular case, θ was found to range from 50 to 90°; whereas in the parallel case, θ ranged from 0 to 70°.

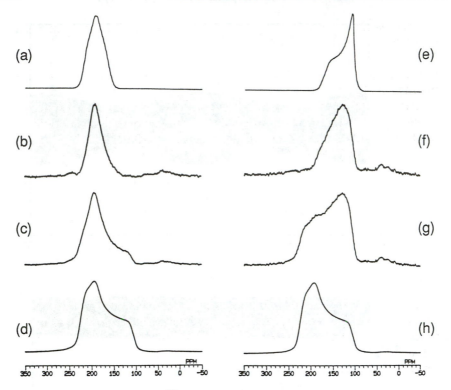

Figure 10. 67.9-MHz ^{13}C CP NMR spectra of oriented $Zr(O_3$-$PCH_2{}^{13}COOH)_2$: (a) simulated perpendicular spectrum, (b) perpendicular difference spectrum, (c) perpendicular experimental spectrum, (d) experimental powder spectrum, (e) simulated parallel spectrum, (f) parallel difference spectrum, (g) parallel experimental spectrum, and (h) experimental powder spectrum.

Despite the large ranges for θ, certain conclusions could be drawn concerning the COOH plane orientation relative to the $Zr(O_3P)_2$ layers when interpretation of the ^{13}C spectra was combined with results from analysis of the ^{31}P spectra and knowledge of the ^{13}C carboxyl chemical-shielding tensor in the molecular frame. For this analysis, it was assumed that the P–C bond was perpendicular to the $Zr(O_3P)_2$ layers and that the microcrystalline platelets of $Zr(O_3PCH_2{}^{13}COOH)_2$ were disordered $\pm15°$ relative to the glass slides. The orientation of the COOH plane relative to the $Zr(O_3P)_2$ layers was defined in terms of the P–C–C–O dihedral angle, referred to here as ρ.

To determine the range of values for ρ, the following method was used. First, the $\pm15°$ disorder was subtracted from the θ ranges determined from the ^{13}C perpendicular and parallel difference spectra to obtain the ranges expected for an ideally ordered sample. This subtraction resulted in new ranges of $\theta_\perp = 65–90°$ for the perpendicular case and $\theta_\| = 0–55°$ for the parallel case. In this idealized situation, the minimum value of $\theta_\perp$ corresponds to 90° minus the maximum possible value of $\theta_\|$. Thus, the maximum allowable value for $\theta_\|$ is 25° for this sampling of ideally oriented platelets. From previous analyses, the P–C–C bond angle was determined to be ~110°, which restricts θ in this idealized system to 20–90°, giving a minimum $\theta_\|$ of 20°. Through geometric considerations it can be established that for the ideally ordered crystallites:

$$\rho = \sin^{-1}\sqrt{\cos^2(\theta_\|)(1 + \tan^2 20°)} \tag{6}$$

$$\rho = \sin^{-1}\sqrt{\cos^2(\theta_\perp)(1 + \tan^2 70°)} \tag{7}$$

From equations 6 and 7, ρ can be calculated to range from 75° to 90°. Calculated perpendicular and parallel ^{13}C NMR spectra that reflect this range of ρ and the ±15° orientational disorder are shown in Figures 10a and 10e. The range of dihedral angles, as shown in Figure 11, may be a reflection of the microcrystalline nature of zirconium phosphonates. A ball-and-stick model of the expected structure for the organic portion of $Zr(O_3PCH_2COOH)_2$ is shown in Figure 11.

Reactivity of $Zr[O_3P(CH_2)_nCOOH]_2$

For the reaction of $Zr[O_3P(CH_2)_nCOO^-]_2$ with $SOCl_2$ to give $Zr[O_3P-(CH_2)_nCOOCl]_2$, n = odd compounds showed significantly lower conver-

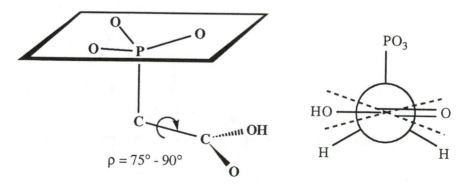

Figure 11. Newman projection and model of the calculated structure of $Zr(O_3PCH_2COOH)_2$ (left) and Newman projection (right). The dashed lines show the maximum and minimum dihedral angles.

sion to the acyl chloride than the n = even compounds (vide supra). The structure of $Zr(O_3PCH_2COOH)_2$ suggests an explanation for this odd–even difference. The mechanism for the $SOCl_2$ reaction is outlined in Scheme I (73). In the first step of the reaction, chloride ion is liberated and acyl intermediate, X, is formed. X is subsequently attacked by chloride ion to give the tetrahedral intermediate, Y, which collapses to give SO_2Cl^- and the acyl chloride. The expected structure of X, formed from $Zr(O_3PCH_2COO^-)_2$, is shown in Figure 12a. Because of steric interactions, the C(O)OSOCl group in X must rotate out of the preferred conformation found for the carboxylic acid derivative. A space-filling model of this intermediate is shown in Figure 12b.

The next step in the conversion to the acyl chloride involves attack of chloride ion at the carbonyl carbon of X to give the tetrahedral intermediate, Y. For the zirconium phosphonate, the approach of Cl^- to the carbonyl carbon of X is hindered by the adjacent carboxyl groups. In the early stages of the reaction, the steric congestion is low enough that Cl^- attack gives Y, which collapses to give the acyl chloride. As more of the carboxyls are converted to acyl chloride groups, the steric demands of the surface become greater, and the rate of formation of the tetrahedral intermediate drops. The rate of formation of Y ultimately drops close to zero when 40% of the carboxyl groups have been converted to acyl chlorides. Similar steric effects would be expected for all of the n = odd compounds, because their structures should be similar to $Zr(O_3PCH_2COOH)_2$ (Figure 13).

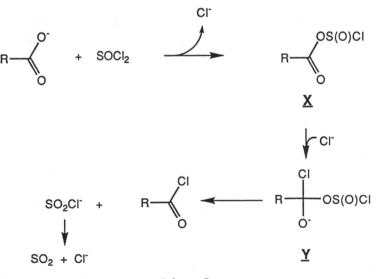

Scheme I.

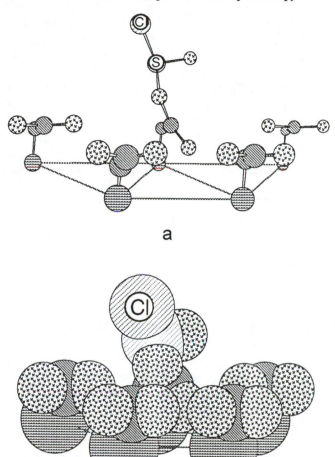

Figure 12. Ball-and-stick (a) and space-filling (b) models of the intermediate X in the conversion of $Zr(O_3PCH_2COO^-)_2$ to an acyl chloride.

For the $Zr[O_3P(CH_2)_nCOO^-NH_4^+]_2$ compounds with an even number of methylenes (n = even) the orientation of the carboxyl groups relative to the host lamellae should be quite different (Figure 13). Here the plane of the carboxylate group will be perpendicular to the lamellar plane, and X can be formed without any change in conformation. The Cl attack should proceed smoothly to give Y, because the steric demands of the adjacent carboxyl groups are much lower for the carboxyls in this conformation (relative to the n = odd case described). The result is that the conversion to the acyl chloride is complete.

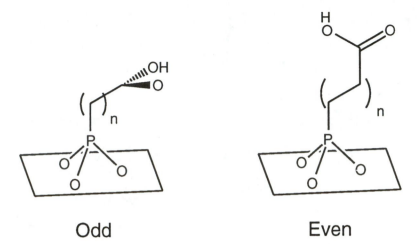

Odd   Even

Figure 13. Odd vs. even structures for $Zr[O_3P(CH_2)_nCOOH]_2$ compounds.

Structure of Ferrocenylalkylamino Intercalation Compounds

Equations 4 and 5 can be used to calculate the expected splittings for ferrocene, which has a quadrupolar coupling constant of 198 kHz *(74)*. In the d_9-ferrocenyl-$CH_2CH_2NH_2$ intercalation compound, the unsubstituted d_5-Cp ring of ferrocene is expected to be rotating rapidly about the C_5 axis at room temperature, while the substituted ring should be frozen because it is bound to the hydroxyl groups on the lamellae of **ZrP** through the amino appendage. Rapid rotation of the d_5-Cp rings in $(d_{10}\text{-}Cp_2Co)_{0.25}TaS_2$ has been observed at room temperature *(68–71)*. For a static C–D bond in the nonrotating d_4-Cp ring of the intercalation compound, the expected splitting for the inner doublet, $\Delta\nu_\perp$, is 148 kHz and for the wings, $\Delta\nu_\parallel$, 297 kHz. The splitting for a motionally averaged C–D bond in the rapidly rotating d_5-Cp ring would be narrowed according to equation 5. Rotation of the C–D bond about the C_5 axis creates an averaged EFG tensor whose principal component is directed along the C_5 axis. The splitting of the inner doublet is calculated to be 74 kHz, and the wings, 148 kHz. The powder pattern for the ferrocenyl **ZrP** intercalation compound should be a superposition of two powder patterns—that for the frozen d_4-Cp ring and that for the freely rotating ring. The inner doublet of the powder pattern of the frozen ring (148 kHz) overlaps the wings in the powder pattern of the freely rotating ring (148 kHz).

The room-temperature ^2H NMR powder pattern for the intercalation compound is shown in Figure 14, *(75, 76)* $\Delta\nu_\perp$ = 71 kHz and $\Delta\nu_\parallel$ = 145 kHz. To simplify the spectra, all of the ^2H spectra were collected with a re-

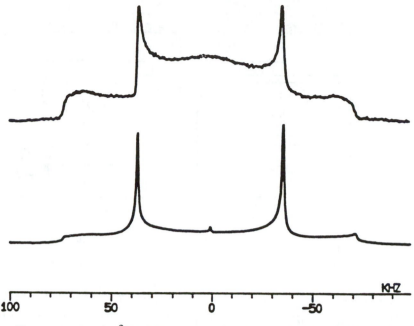

*Figure 14. Powder ^2H NMR spectra of d_9-ferrocenyl-$CH_2CH_2NH_2$ · **ZrP** (top) and d_{10}-ferrocene (bottom).*

cycle time of 2 s. With this recycle time the rotating d_5-Cp ring has ample time to relax, but the static ring will be saturated because of its longer T^1 (44). Under these conditions the spectra will be dominated by the rotating d_5-Cp. The powder pattern of d_{10}-ferrocene is shown for comparison to the intercalation compound in Figure 14. The inner doublet is also at $\Delta\nu = 71$ kHz and the wings at $\Delta\nu = 145$ kHz. The inner doublets in both spectra are assigned to a rapidly rotating d_5-Cp ring and is consistent with the calculated splitting for a rotating d_5-Cp ring. The striking difference in the powder patterns of the intercalation compound and ferrocene is the intensity of the shoulders. The greater intensity in the shoulders of the powder patterns of the intercalation compounds suggests that some of the intensity is due to the inner doublet ($\Delta\nu_1$) of the static ring, even with the short recycle time. Because of the limitations of the spectral window of the NMR spectrometer, the shoulders in the powder pattern of the static d_4-Cp ring could not be observed. Therefore, analysis of the oriented spectra (vide infra) is based only on the rotating ring.

The ^2H NMR spectra of the oriented thin films can provide information on the orientation of ferrocene between the **ZrP** layers. Three possible orientations of ferrocene in the layered matrix are shown in Figure 15: C_5 axis perpendicular to the layers, C_5 axis parallel to the layers, and random.

C_5 axis ⊥ to layers C_5 axis ‖ to layers

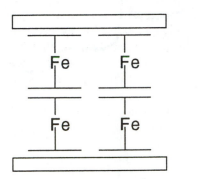
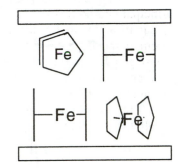

$\theta = 0°$ $\theta = 90°$
$\Delta\nu = \text{max.}$ $\Delta\nu = 1/2\ \text{max.}$

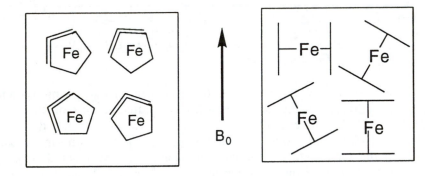

$\theta = 90°$ $\theta = \text{random}$
$\Delta\nu = 1/2\ \text{max.}$ $\Delta\nu = \text{powder}$

Random

Ferrocenes will be randomly oriented
to B_0 at all sample orientations
∴ $\Delta\nu = \text{powder}$ for both orientations.

Figure 15. Possible orientations of ferrocenyl group in zirconium hydrogen phosphate. θ is the angle between the average EFG tensor and the applied magnetic field. Pictures on the left refer to samples with the layer normal parallel to B_0, and pictures to the right represent samples with the layer normal parallel to B_0.

If the C_5 axis is perpendicular to the layers, oriented spectra are expected for both parallel and perpendicular orientations of the layer normal to the magnetic field. When the layer normal is parallel to B_0, the principal component of the averaged EFG tensor is parallel to B_0 ($\theta = 0$ in equation 5), so the maximum splitting (145 kHz) is expected for the oriented spectrum. When the layer normal is turned 90° so that it is perpendicular to B_0, the principal component of the averaged EFG tensor is now perpendicular to B_0 and a splitting of 71 kHz is expected for the oriented spectrum.

The second possible orientation of ferrocene is with the C_5 axis parallel to the layers. In this orientation all of the C_5 axes are parallel to the lamellae of **ZrP**. When the layer normal is parallel to B_0, the principal component of the averaged EFG tensor is perpendicular to B_0 and a splitting of 71 kHz is expected. However, when the layer normal is perpendicular to B_0, the averaged EFG tensor is now found at all angles with respect to B_0 because of the disorder of the guest molecules in the plane parallel to the host lamellae. In addition, the crystals are disordered in the plane of the glass coverslips, which predicts a random orientation of the individual EFG tensors. Therefore, a powder pattern is expected when layer normal is perpendicular to the magnetic field.

The last possibility is with the ferrocene guest randomly oriented between the layers. Here a powder pattern will be observed for both parallel and perpendicular orientations of the layer normal to B_0.

The ^{2}H NMR spectra of a thin film of 2-(d_9-ferrocenyl)ethylamine · **ZrP** aligned perpendicular and parallel to B_0 are shown in Figure 16. The powder patterns for this compound are also included for comparison. To account for those particles that are misaligned on the glass coverslips, a portion of the powder pattern is subtracted from the raw oriented spectra (vide supra). The corrected spectra are shown on the top of each figure. The corrected spectrum shows that when the layer normal is parallel to B_0, there is intensity at $\Delta\nu_1 = 71$ kHz and no intensity at the wings. The slight broadening seen in the parallel spectrum is presumably due to the fact that the aligned crystals in the thin film are not exactly parallel to the coverslips, but lie within a narrow range of angles close to parallel. The corrected spectrum for the layer normal oriented perpendicular to B_0 shows a null signal indicating that the raw spectrum is a simple powder pattern.

The ^{2}H NMR spectrum of the oriented thin film indicates that ferrocene lies with its C_5 axis parallel to the layers of the host (Figure 16b). We completed analogous NMR experiments with two other deuteroferrocenyl intercalation compounds. The orientation-dependent NMR spectra of $[d_9$-ferrocenyl-$(CH_2)_n NH_2]_m$ · **ZrP** ($n = 1$ or 3) in aligned thin films are very similar to those described for the ferrocenylethylamino **ZrP** compound. Thus the ferrocenyl groups lie with their C_5 axes parallel to the

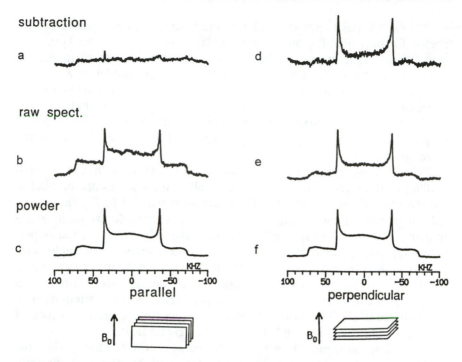

Figure 16. ^2H *NMR spectra of 2-(d_9-ferrocenyl)ethylamine · ZrP: (a) difference spectra (parallel minus powder), (b) spectrum of oriented sample with layer normal parallel to the applied magnetic field, (c) powder spectrum, (d) difference spectra (perpendicular minus powder), and (e) spectrum of oriented sample with layer normal perpendicular to the applied magnetic field. (Reproduced with permission from reference 44. Copyright 1988.)*

host lamellae for all three compounds. The orientation determined for the intercalated ferrocenylethylamine is the opposite of the orientation of cobaltocene intercalated into TaS_2, which has been inferred from other ^2H NMR experiments (*68, 71*). X-ray diffraction studies (*77*) indicate that ferrocenium ion intercalated into FeOCl lies with its C_5 parallel to the surface, similar to what is observed here. A similar structure is suggested for ferrocene intercalated into SnS_2 by O'Hare et al. (*43*) based on single-crystal NMR studies.

Acknowledgment

The purchase of the Jeol 270 NMR spectrometer was made possible by National Science Foundation Grant CHE–89–09857. Acknowledgment is also made to the National Science Foundation (Grant DMR–9113002), Dow Chemical Company, and to the Air Force Office of Scientific Research (Grant AFOSR–90–0122) for the financial support of this work.

References

1. *Intercalation Chemistry;* Whittingham, M. S.; Jacobson, A. J., Eds.; Academic: New York, 1982.
2. Schölhorn, R. In *Inclusion Compounds;* Atwood, J. A., Ed.; Academic: London, 1984; and references therein.
3. Clearfield, A. *Comments Inorg. Chem.* **1990,** *10,* 89.
4. Dresselhaus, M. S. *Mater. Sci. Eng. B1* **1988,** 259–277; and references therein.
5. Formstone, C. A.; Fitzgerald, E. T.; O'Hare, D.; Cox, P. A.; Kurmoo, M.; Hodby, J. W.; Lillicrap, D.; Goss-Custard, M. *J. Chem. Soc. Chem. Commun.* **1990,** 501–503.
6. Clearfield, A. In *Design of New Materials;* Cocke, D. L.; Clearfield, A., Eds.; Plenum: New York, 1988.
7. Giannelis, E. P.; Mehrota, V.; Russell, M. W. *Mater. Res. Soc. Proc.* **1990,** *180.*
8. Alberti, G.; Constantino, U. *Intercalation Chemistry;* Whittingham, M. S.; Jacobson, A. J., Eds.; Academic: New York, 1982; Chapter 5.
9. Alberti, G.; Constantino, U.; Marmottini, F. In *Recent Developments in Ion Exchange;* Williams, P. A.; Hudson, M. J., Eds.; Elsevier Applied Science: New York, 1987.
10. Clearfield, A. *Chem. Rev.* **1988,** *88,* 125–148.
11. Alberti, G.; Constantino, U. *J. Mol. Catal.* **1984,** *27,* 235–250.
12. Clearfield, A. *J. Mol. Catal.* **1984,** *27,* 251–262.
13. Li, Z.; Lai, C.; Mallouk, T. E. *Inorg. Chem.* **1989,** *28,* 178–182.
14. Rong, D.; Kim, Y. I.; Mallouk, T. E. *Inorg. Chem.* **1990,** *29,* 1531–1535.
15. Kanatzidis, M. G.; Wu, C.; Marcy, H. O.; DeGroot, D. C.; Kannewurf, C. R. *Chem. Mater.* **1990,** *2,* 222–224.
16. Kanatzidis, M. G.; Hubbard, M.; Tonge, L. M.; Marks, T. J.; Marcy, H. O.; Kannewurf, C. R. *Synth. Met.* **1989,** *28,* C89–C95.
17. Lee, H.; Kepley, J.; Hong, H.; Mallouk, T. E. *J. Am. Chem. Soc.* **1988,** *110,* 618–620.
18. Wiesner, U.; Schmidt-Rohr, K.; Boeffel, C.; Pawelzik, U.; Spiess, H. W. *Adv. Mater.* **1990,** *2,* 484.
19. Oulyadi, H. Lauprêtre, F.; Monnerie, L;, Mauzac, M.; Richard, H.; Gasparoux, H. *Macromolecules* **1990,** *23,* 1965.
20. Cornell, B. A.; Separovic, F.; Baldassi, A. J.; Smith, R. *Biophys. J.* **1988,** *53,* 67.
21. Nicholson, L. K.; Cross, T. A. *Biochemistry* **1989,** *28,* 9379.
22. Cross, T. A.; Opella, S. J. *J. Am. Chem. Soc.* **1983,** *105,* 306.
23. Cross, T. A.; Opella, S. J. *J. Mol. Biol.* **1985,** *182,* 367.
24. Cross, T. A. *Biophys. J.* **1986,** *49,* 124.

25. Opella, S. J.; Stewart, P. L.; Valentine, K. G. *Q. Rev. Biophys.* **1987,** *19,* 7.
26. Resing, H. A.; Slotfedt-Ellingsen, D. J. *J. Magn. Reson.* **1980,** *38,* 401.
27. Miller, G. R.; Poranski, C. F.; Resing, H. A. *J. Chem. Phys.* **1984,** *80,* 1708.
28. Miller, G. R.; Moran, M. J.; Resing, H. A.; Tsang, T. *Langmuir* **1986,** *2,* 194.
29. *Intercalation Chemistry;* Whittingham, M. S.; Jacobson, A. J., Eds.; Academic: New York, 1982.
30. Burwell, D. A.; Thompson, M. E. *Chem. Mater.* **1991,** *3,* 14.
31. Burwell, D. A.; Thompson, M. E. *Chem. Mater.* **1991,** *3,* 730.
32. Alberti, G.; Constantino, U.; Casciola, M.; Vivani R.; Peraio, A. *Solid State Ionics* **1991,** *46,* 61.
33. Burwell, D. A.; Valentine, K. G.; Timmermans, J. H.; Thompson, M. E. *J. Am. Chem. Soc.* **1992,** *114,* 4144.
34. Johnson, J. W. *J. Chem. Soc. Chem. Commun.* **1980,** 263.
35. Lee, C. F.; Thompson, M. E. *Inorg. Chem.* 1991, *30,* 3.
36. Chatakondu, K.; Green, M. L. H.; Qin, J.; Thompson, M. E.; Wiseman, P. J. *J. Chem. Soc. Chem. Commun.* **1988,** 223.
37. Chatakondu, K.; Formstone, C.; Green, M. L. H.; O'Hare, D.; Twyman, J. M.; Wiseman, P. J. *J. Mater. Chem.* **1991,** *1,* 205.
38. Jacobson, A. J. *Intercalation Chemistry;* Whittingham, M. S.; Jacobson, A. J., Eds.; Academic: New York, 1982; Chapter 7.
39. Dines, M. B. *Science (Washington, D.C.)* **1975,** *188,* 1210.
40. Clement, R; Davies, W. B.; Ford, K. A.; Green, M. L. H.; Jacobson, A. J. *Inorg. Chem.* **1978,** *17,* 2754.
41. Nazar, L. F.; Jacobson, A. J. *J. Chem. Soc. Chem. Commun.* **1986,** 570.
42. O'Hare, D.; Evans, J. S. O.; Wiseman, P. J.; Prout, K. *Angew. Chem. Int. Ed. Engl.* **1991,** *30,* 1156.
43. Grey, C.; Evans, J. S. O.; O'Hare, D.; Heyes, S. J. *J. Chem. Soc. Chem. Commun.* **1991,** 1380.
44. Lee, C. F.; Myers, L. K.; Valentine, K. G.; Thompson, M. E. *J. Chem. Soc. Chem. Commun.* **1992,** 201.
45. Fyfe, C. A. *Solid State NMR for Chemists;* C. F. C. Press: Ontario, Canada, 1983.
46. Sanders, J. K. M.; Hunter, B. K. *Modern NMR Spectroscopy, a Guide for Chemists;* Oxford University Press: Oxford, United Kingdom, 1987.
47. Haberkorn, R. A.; Stark, R. E.; van Willigen, H.; Griffin, R. G. *J. Am. Chem. Soc.* **1981,** *103,* 2534.
48. Sherwood, M. H.; Facelli, J. C.; Alderman, D. W.; Grant, D. M. *J. Am. Chem. Soc.* **1991,** *113,* 750.
49. K. G. Valentine, A. L. Rockwell, L. M. Gierasch, S. J. Opella, *J. Magn. Reson.* **1987,** *73,* 519.
50. Oas, T. G.; Hartzell, C. J.; McMahon, T. J.; Drobny, G. P.; Dahlquist, F. W. *J. Am. Chem. Soc.* **1987,** *109,* 5956.
51. Oas, T. G.; Hartzell, C. J.; Dahlquist, F. W.; Drobny, G. P. *J. Am. Chem. Soc.* **1987,** *109,* 5962.
52. Hartzell, C. J.; Whitfield, M.; Oas, T. G.; Drobny, G. P. *J. Am. Chem. Soc.* **1987,** *109,* 5966.
53. Teng, Q.; Cross, T. A. *J. Magn. Reson.* **1989,** *85,* 439.
54. Wasylishen, R. E.; Penner, G. H.; Power, W. P.; Curtis, R. D. *J. Am. Chem. Soc.* **1989,** *111,* 6082.

55. Separovic, F.; Smith, R.; Yannoni, C. S.; Cornell, B. A. *J. Am. Chem. Soc.* **1990,** *112,* 8324.
56. Burwell, D. A.; Valentine, K. G.; Thompson, M. E. *J. Magn. Reson.* **1992,** *97,* 498.
57. Veeman, W. S. *Prog. NMR Spectrosc.* **1984,** *16,* 193, and references therein.
58. Griffin, R. G.; Ruben, D. J. *J. Chem. Phys.* **1975,** *63,* 1272.
59. Nagoaka, S.; Terao, T.; Imashiro, F.; Saika, A.; Hirota, N. *Chem. Phys. Lett.* **1981,** *80,* 580.
60. Van Calsteren, M.; Birnbaum, G.; Smith, I. C. P. *J. Chem. Phys.* **1987,** *86,* 5405.
61. Mehring, M. *Principles of High Resolution NMR in Solids;* Springer-Verlag: Berlin, Germany, 1983.
62. Alderman, D. W.; Solum, M. S.; Grant, D. M. *J. J. Chem. Phys.* **1986,** *84,* 3717.
63. Speiss, H. W. *Colloid Polym. Sci.* **1983,** *261,* 193.
64. Seelig, J.; MacDonald, P. M. *Acc. Chem. Res.* **1987,** *20,* 221.
65. Davis, J. H. *Biochim. Biophys. Acta* **1983,** *737,* 117.
66. Greenfield, M. S.; Ronemus, A. D.; Vold, R. L.; Vold, R. R.; Ellis, P. D.; Raidy, T. E. *J. Magn. Reson.* **1987,** *72,* 89.
67. Olympia, P. L.; Wei, J. Y., Jr.; Fung, B. M. *J. Chem Phys.* **1969,** *51,* 1610.
68. Heyes, S. J.; Clayden, N. J.; Dobson, C. M.; Green, M. L. H.; Wiseman, P. J. *J. Chem. Soc. Chem. Commun.* **1987,** 1560.
69. Clayden, N. J.; Dobson, C. M.; Heyes, S. J.; Wiseman, P. J. *J. Inclusion Phenom.* **1987,** *5,* 65.
70. Lowery, M. D.; Wittebort, R. J.; Sorai, M.; Hendrickson, D. N. *J. Am. Chem. Soc.* **1990,** *112,* 4214.
71. Heyes, S. J.; Clayden, N. J.; Green, M. L. H.; Wiseman, P. J.; Dobson, C. M., submitted to *J. Am. Chem. Soc.*
72. Boden, L.; Clark, L. D.; Hanlon, S. M.; Mortimer, M. *Faraday Symp. Chem. Soc.* **1978,** *3,* 69.
73. Pine, S. H.; Hendrickson, J. B.; Cram, D. J.; Hammond, G. S. *Organic Chemistry,* 4th ed.; McGraw Hill: New York, 1980; Chapter 8.
74. Olympia, P. L., Jr.; Wei, J. Y.; Fung, B. M. *J. Chem Phys.* **1969,** *51,* 1610.
75. Mihailov, M.; Dirlikov, S.; Peeva, N.; Georgieva, Z. *Makromol. Chem.* **1975,** *176,* 789.
76. Morgan, G. T.; Walton, E. *J. Chem. Soc.* **1933,** 1064.
77. Palvadeau, P.; Coic, L.; Rouxel, J.; Ménil, F.; Flournè, L. *Mater. Res. Bull.* **1981,** *16,* 1055.

RECEIVED for review November 9, 1992. ACCEPTED revised manuscript April 13, 1993.

Nanoscale, Two-Dimensional Organic–Inorganic Materials

E. Giannelis

Department of Materials Science and Engineering, Cornell University, Ithaca, NY 14853

Self-assembled organic–inorganic nanostructures were synthesized by intercalation of layered silicates. The materials design and synthesis rely on hydrogen and van der Waals interactions and largely depart from the more conventional synthesis based on covalent bonding. By controlling subtle guest–host interactions, the properties of the assembly can be dramatically altered. Polymer nanocomposites can also be synthesized by intercalative polymerization and direct polymer intercalation using a similar methodology. The physical and mechanical properties of the resulting nanocomposites are attributed to the confinement of the polymer chains in the nanometer-size galleries of the host.

THE DESIGN, SYNTHESIS, AND CHARACTERIZATION of nanophase materials, also called mesoscopic materials, are the subjects of intense research (*1–3*). This activity has been inspired, in part, by the realization that nanophase materials often exhibit new (or crossover) phenomena and that their physical and chemical properties are sometimes dramatically different from the bulk counterparts. Nanophase materials belong to a new family of materials with sizes intermediate to those usually studied by chemists and materials scientists, and therefore they pose new challenges regarding their synthesis and characterization. New design principles and synthetic strategies that allow control at the nanometer level as well as analytical techniques capable of probing nanostructures are required.

Although this chapter focuses mainly on our effort (*4–11*), intercalation of layered solids is emerging as a new strategy with the potential of

0065–2393/95/0245–0259$12.75/0

synthesizing new generations of organic–inorganic, two-dimensional nano-composites (*12–20*). Intercalation, shown schematically in Figure 1, is a process by which guest species are inserted in the nanometer-size gaps (galleries) that exist between the layers of a host lattice (*21*). The materials design and synthesis is based on a modular process by which the final structure is assembled by self-organization of preformed subunits. The self-assembling approach largely departs from the conventional sequential covalent bond formation in that it relies on weaker and less-directional bonding such as hydrogen- and van der Waals bonding (*22*). Because self-assembled structures are at equilibrium, their synthesis involves reversible interactions leading to systems with a high degree of organization and order.

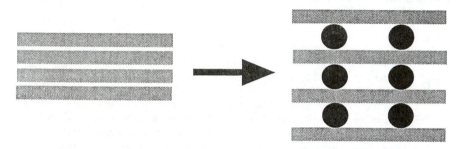

Figure 1. The formation of a multilayer by intercalation of a layered solid.

Although self-assembly is a ubiquitous process in nature (*23*), it has only recently emerged as a viable new strategy for the synthesis of nano-scale materials. Biological systems, for instance, are especially notorious for having mastered the art of assembling complex nanostructures at different spatial levels from relatively simple building components. Examples range from strong and tough composites to highly selective sensing membranes. The characteristic robustness and stability of the self-assembled nanostructures are usually associated with their structures representing thermodynamic minima.

In this chapter we first review some of the essential structural characteristics and intercalation properties of layered silicates. We briefly discuss some of the thermodynamic issues related to the synthesis and stability of self-assembled systems. We then give examples of self-assembled organic–inorganic nanostructures based on intercalation and discuss their properties. We close with a brief summary and a few remarks on future directions.

Structure and Intercalation Properties of Mica-Type Layered Silicates

Crystalline solids with a layered lattice structure can ingest (intercalate) various guest species into the galleries that are formed between the layers to produce well-ordered, molecular multilayers (*21*). Graphite and transition metal chalcogenides are some common examples of layered hosts that have been studied quite extensively over the past several decades. In comparison, little work has been devoted to the layered complex oxides and more specifically to the mica-type layered silicates (MTSs). Mica-type silicates offer a unique combination of intercalative and ion-exchange properties. These properties can be exploited to introduce new functionalities and thus modify the physical and mechanical properties of the host. In addition, MTSs, when fully exfoliated, possess aspect ratios in the same range as fibers, a feature that makes them very attractive as molecular reinforcements.

MTSs, as well as the better known talc and mica, belong to the general family of 2:1 layered silicates (*24*). Their lattice structure consists of two-dimensional layers that are formed by sandwiching two SiO_2 tetrahedral sheets to an edge-shared octahedral sheet (Figure 2). Depending on the type and population of the cations in the octahedral sheet, different layered silicates are obtained. For instance, when all octahedral sites are occupied by magnesium, the mineral talc is obtained. Alternatively, in pyrophyllite only two-thirds of the octahedral sites are filled with aluminum. In crystals of talc and pyrophyllite, the layers are held together by weak dipolar and van der Waals forces creating interlayers or galleries between the layers. The sum of a single layer (~1 nm) plus the interlayer represents the repeat unit or d-spacing in the multilayer and is conveniently obtained from X-ray diffraction patterns.

In talc and pyrophyllite the silicate layers are electrically neutral. In contrast, micas exhibit a positive charge deficiency, denoted by x, where x equals the charge density per unit cell (equivalent to an Si_8O_{20} unit), due to the isomorphous substitution of silicon by aluminum in the silicate layers. The charge deficiency is generally balanced by potassium cations that reside in the galleries between the silicate layers. In "ideal" micas the layer charge is high enough that all the ditrigonal cavities formed by the oxygen atoms on the basal plane are occupied by potassium cations.

The talc–pyrophyllite group ($x = 0$) and the micas ($x = 2$) represent the end members of a hierarchy defined by the degree of layer charge neutrality. Mica-type layered silicates, also called swelling silicates, with layer charge densities of $0 < x < 2$ define a separate host family with unique

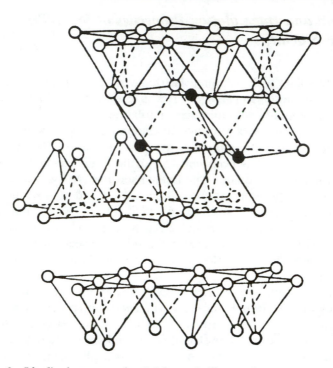

Figure 2. Idealized structure for 2:1 layered silicates showing two tetrahedral sheets sandwiching an octahedral sheet. Open circles represent oxygen atoms; closed circles are hydroxyl groups. In mica the galleries between the layers are usually occupied by potassium cations.

physical and chemical properties not found in the end members. Like micas, cations reside in the galleries of MTSs to balance the charge deficiency in the layers. But unlike the micas, MTSs can intercalate a large number of polar molecules in their galleries. Intercalation of guest species into the host galleries is facilitated by the much weaker interlayer compared to the intralayer binding forces. In addition, the hydrated gallery cations in MTSs can be readily exchanged with almost any other cation (mono- or polynuclear, organic or inorganic) by a simple ion-exchange reaction. Thus, a variety of neutral or positively charged molecules of virtually any size can be introduced in the host galleries by using direct intercalation and ion-exchange reactions, respectively.

Thermodynamic Issues in Self-Assembled Materials

An important aspect of self-assembled structures is that their synthesis involves a modular process. Stable covalent units are assembled into the final structure by weaker nondirectional bonding. As Whitesides et al. (*22*) pointed out, for the final assembly to be stable the relatively weak noncovalent interactions must be collectively stable. Relative to typical covalent bonding with energies ranging from 100 to 300 kcal/mol, hydrogen and van der Waals bonds are very weak with bonding energies of 0.1–5 kcal/mol. Furthermore, bonding energies in H and van der Waals bonds are comparable to thermal energies at room temperature. These considerations suggest that, in order for the assembly to be stable, many weaker interactions that collectively overcome the thermal energy of the system are required.

One approach to maximize the interactions is to increase the interaction area between subunits in the final assembly. With polymers, this step might involve stretching the polymer chains to increase favorable interactions. The average configuration will then be one that is more stretched out than in the random coil. Moreover, these interactions must be able to overcome the entropic penalty of ordering. Entropic favorable release of water molecules might be an important contribution to overcoming the unfavorable loss of translational energy upon intercalation. In addition, the H and van der Waals interactions between subunits in the assembly must be more favorable than the competing interactions with solvent molecules.

In the following sections we present a few examples of self-assembled nanostructures. The structure and properties of the assembly are controlled by weak noncovalent guest–host interactions. Understanding the thermodynamic issues, and more specifically the interplay between enthalpy and entropy, is critical in designing new structures with controlled properties.

Highly Organized Molecular Assemblies

The ability to control the orientation and alignment of guest molecules and thus the properties of the assembly by mediating subtle guest–host interactions is already well documented. An example from our laboratory involves the synthesis of well-organized assemblies of the Cu-metallated form of tetrakis(1-methyl-4-pyridyl)porphyrin (CuTMPyP) in the galleries of Na hectorite (with a charge density per unit cell of $x = 0.7$) and Li fluorohectorite (with $x = 1.6$) (*4*). X-ray diffraction, electronic absorption, and electron spin resonance (ESR) spectra of the intercalates are consistent with the porphyrin guest molecules being oriented parallel with respect to the hectorite layers. In contrast, the porphyrin molecules are oriented near 45° to the host layers in fluorohectorite (Figure 3).

ESR spectra of oriented film samples at room temperature are shown in Figures 4 and 5. The spectra were recorded at different sample orientations corresponding to the magnetic field direction oriented parallel or

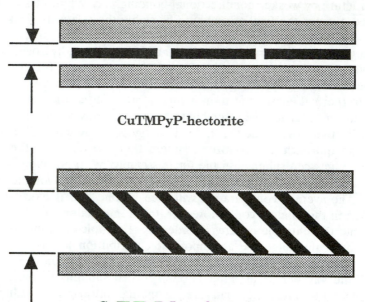

Figure 3. Orientation of porphyrin guest molecules in hectorite and fluorohectorite.

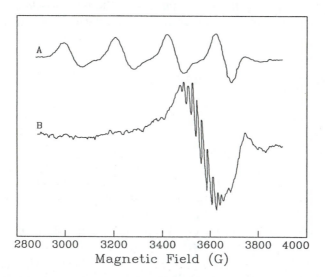

Figure 4. ESR spectra of oriented films of CuTMPyP–hectorite with the silicate layers positioned perpendicular (A) and parallel (B) to the magnetic field. (Reproduced from reference 4. Copyright 1990 American Chemical Society.)

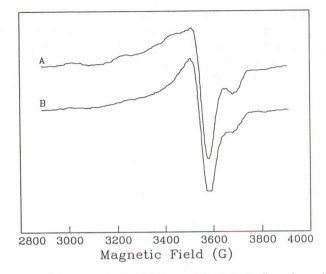

Figure 5. ESR spectra of oriented films of CuTMPyP–fluorohectorite with the silicate layers positioned parallel (A) and perpendicular (B) to the magnetic field. (Reproduced from reference 4. Copyright 1990 American Chemical Society.)

perpendicular to the plane of the host layers. When the magnetic field is oriented parallel to the hectorite layers, only the $g_\perp$ is observed, and the $g_\parallel$ component is observed when the magnetic field is perpendicular to the host layers. Because $g_\parallel$ and $g_\perp$ are observed for the perpendicular and parallel orientations, respectively (Figure 4), the molecular plane of the porphyrin complex must be oriented parallel to the silicate layers. In contrast, the ESR spectrum of the fluorohectorite intercalate is orientation-independent with both parallel and perpendicular components present regardless of field orientation (Figure 5). The lack of orientation dependence indicates that the porphyrin molecules are oriented near 45° to the host layers so that the magnetic field "sees" the same molecular environment regardless of film orientation. This argument is supported by X-ray diffraction data that show gallery heights of 4.4 and 10.5 Å for the hectorite and fluorohectorite intercalates, respectively, consistent with the dimensions of the porphyrin molecule and the proposed molecular orientation.

What is the underlying mechanism that controls the orientation of the guest molecules and why does hectorite behave differently from fluorohectorite? The layer charge density in fluorohectorite is more than double that in hectorite. If the porphyrin molecules are taken to have a square shape, a monolayer with the porphyrin rings oriented parallel to the silicate layers is sufficient to spatially balance the charge of the host layers in hectorite. A parallel orientation is preferred, because the interaction between the layers and the π-electrons in the porphyrin molecules is thus maximized. In contrast, an inclined arrangement of the porphyrin molecules is predicted and observed for fluorohectorite because of its higher charge density. The porphyrin molecules are incapable of balancing the host layer charge when oriented parallel to the layers. Charge balance is accomplished by forcing the porphyrin molecules to attain a tilted configuration that can pack more guest molecules per surface area. Therefore, the orientation of the guest molecules is controlled by the area available per exchange site, and it depends on the arrangement that optimizes charge balance with the area required to accommodate each guest molecule.

If intercalation was effective only in controlling the structural features without any effect on the properties of the assembly, its utility as a means of synthesizing new materials would have been limited. The real advantage is that in addition to the structural control, intercalation offers the means to control the properties of the assembly. By carefully selecting the host and guest species and by controlling subtle guest–host interactions, the properties of the assembly can be dramatically altered. Examples include modifying the ground- and excited-state properties of the guest molecules as well as the properties of the host lattice, as we previously reported (*4, 25, 26*). In addition, intercalation enhances the thermal and oxidative stability of the guest molecules because of the molecular confinement in the host galleries. An example from our work involves intercalation of ethylenediamine- (en-) functionalized buckminsterfullerene, $C_{60}(en)_6$ (*9*). Li fluorohectorite readily undergoes an intercalative ion-exchange reaction when added to a solution of $C_{60}(en)_6$. The X-ray diffraction pattern of oriented films show that a relatively well-ordered multilayer is obtained with (001) diffractions corresponding to a monolayer of guest molecules in the host galleries.

The thermal and oxidative stability of $C_{60}(en)_6$-intercalated fluorohectorite was studied by high-temperature X-ray diffraction as well as by thermogravimetric analysis (TGA) and differential scanning calorimetry (DSC). Figure 6 shows the TGA traces for the intercalated and chloride salt of $C_{60}(en)_6$. The onset of decomposition for the intercalated carbon clusters does not commence until 800 °C in air, which is ~350 °C higher than that for the unintercalated molecules. That the C_{60} clusters remain intact at high temperatures is also evident from X-ray diffraction data. X-ray diffraction patterns of the $C_{60}(en)_6$ intercalate after being heated in air to 750 °C still show the presence of a multilayer with a gallery height of 0.7 nm, in excellent agreement with the size of C_{60}. The remarkable enhancement of the thermal and oxidative stability of the intercalated carbon clusters is attributed to their confinement in the silicate galleries, which significantly modifies the reaction pathways that lead to decomposition.

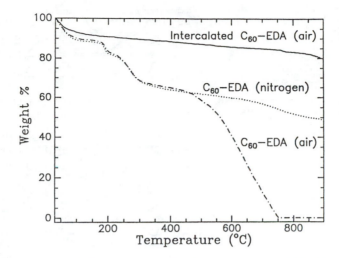

Figure 6. TGA of $C_{60}(en)_6$-intercalated fluorohectorite (solid line), and $C_{60}(en)_6$ chloride salt in air (dashed–dotted line) and nitrogen (dotted line). The samples were heated at 10 °C/min to 900 °C. (Reproduced from reference 9. Copyright 1992 American Chemical Society.)

Polymer Nanocomposites

We also used intercalation of layered solids as a means to synthesize new polymer nanocomposites. Our approach is divided into two major synthetic strategies: (1) intercalation of single polymer chains either by in situ polymerization or direct intercalation and (2) dispersion of single silicate layers (1 nm thick) in a continuous polymer matrix (Figure 7). The distinguishing feature of these nanocomposites, in contrast to those by more conventional approaches, is their self-assembling nature. This feature leads to highly-organized, two-dimensional structures with nanometer dimensions.

The whole thrust of producing novel composites by intercalation of layered solids, as opposed to those with the properties of a simple mixture of the components (i.e., polymer and ceramic) is that the molecular features and the apparent synergism between subunits could impart unique properties not necessarily limited by the rule of mixtures. Therefore, the new composites can be designed to exhibit the overriding strength and dimensional and thermal stability of ceramics with the fracture properties, processibility, and dielectric properties of polymers.

Intercalation of Single Polymer Chains. Intercalative (in situ) polymerization of various monomers in the silicate galleries yields highly

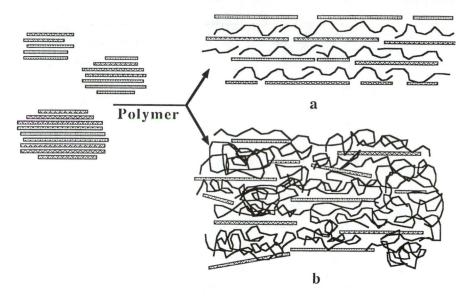

Figure 7. Schematic composite structures obtained using layered ceramics. The rectangular bars represent the ceramic layers. Key: a, single polymer layers intercalated in the ceramic galleries; and b, composites obtained by dispersion of the ceramic layers in a continuous polymer matrix. (Reproduced with permission from reference 11. Copyright 1992 Minerals, Metals, and Materials Society.)

oriented multilayers consisting of single chains of the polymer alternately stacked with the layers of the host. Aniline and pyrrole, for example, diffuse readily in the silicate galleries, and they form polyaniline (PANI) and polypyrrole, respectively, by an oxidative polymerization mechanism catalyzed by Cu^{2+} cations (5, 8). The Cu^{2+} cations are introduced by a simple ion-exchange reaction prior to exposing the host to the monomer. The reaction for aniline is represented by the following equation

$$\overline{Cu^{2+}} + n\,C_6H_5NH_2 \longrightarrow \overline{Cu^{2+}(PANI)}$$

where the horizontal lines identify the layered structure. The polyaniline–silicate hybrid contains 18 wt% polymer as determined by chemical analysis. X-ray diffraction patterns of oriented films (Figure 8) confirm that a highly ordered multilayer is obtained with a gallery height corresponding to the thickness of a single polymer chain, as shown in Figure 9.

Polyaniline is an electrical conductor when appropriately doped. Doping can be accomplished by protonation, which results in the forma-

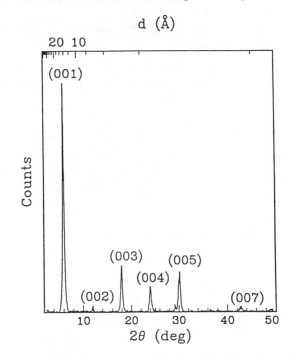

Figure 8. X-ray diffraction pattern of polyaniline–fluorohectorite hybrid. (Reproduced with permission from reference 5. Copyright 1990 Pergamon Press.)

tion of a metallic polaron lattice and phase segregation between metallic and insulating islands. Electrical conduction results from limited tunneling between the metallic islands. DC conductivity measurements of the PANI hybrid show a substantial increase in conductivity after exposure to HCl vapors. The in-plane electrical conductivity, $\sigma_{\parallel}$, measured at room temperature by the four-probe technique is 0.05 S/cm. The transverse conductivity is about 10^{-7} S/cm, which corresponds to an electrical anisotropy $\sigma_{\parallel}/\sigma_{\perp} \approx 5 \times 10^{5}$. The observed anisotropy is in accord with the film microstructure, which consists of single conducting layers separated by the 1-nm-thick layers of the insulating host.

The conductivity, even when corrected for the presence of the insulating host, is lower than that usually reported for PANI. In addition, the electronic absorption spectra of the composite in both conducting and insulating forms are blue-shifted by about 70 nm with respect to PANI. The energy shifts are, most likely, due to the effect of molecular confinement in the intercalated state. The two-dimensional inorganic matrix not only enforces a particular conformation on the chains but also minimizes the extensive electronic interaction that normally occurs in the polymer

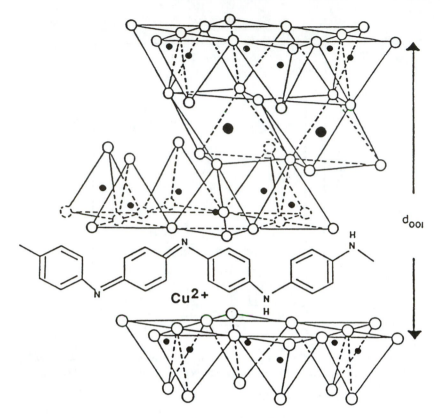

Figure 9. Intercalation of single polyaniline chains in the silicate galleries.

chains. This increased localization of the charge carriers leads to a decrease in the electrical conductivity of the composite.

Fracture toughness tests with the crack front oriented perpendicular to the layers yield a fracture toughness of 1.2 MPa/m for the hybrid; films of pristine silicate could not be tested because of their brittleness. Thus, at least qualitatively, the fracture toughness of the hybrid is superior to the ceramic matrix. The fracture toughness with the crack front parallel to the layers is about 0.3 MPa/m, in agreement with the anisotropic character of the hybrid and the rather weak bonding of the polymer—silicate interface. Figures 10 and 11 show the fracture surfaces of the pristine and polymer-intercalated silicate, respectively. The enhanced ordering of the hybrid is probably due to reorganization of the silicate layers after intercalation to maximize the hydrogen-bond interactions between the guest molecules and the host. The pristine host is less oriented than the hybrid, so a larger energy dissipation during fracture due to the increased surface area is expected for the hybrid, in agreement with our data.

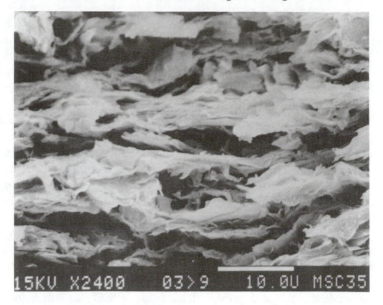

Figure 10. SEM of a cross section of pristine silicate. (Reproduced with permission from reference 11. Copyright 1992 Minerals, Metals, and Materials Society.)

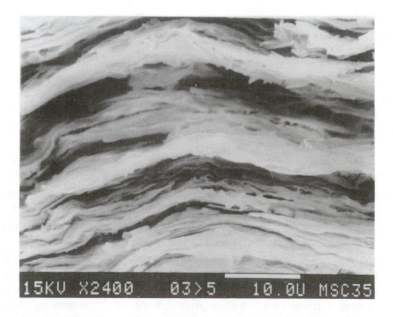

Figure 11. SEM of a cross section of polyaniline–silicate hybrid. (Reproduced with permission from reference 11. Copyright 1992 Minerals, Metals, and Materials Society.)

The in-plane storage modulus for the hybrid is 13.0 ± 2.7 GPa at room temperature. For comparison, the storage modulus of pristine silicate is 5.5 ± 1.2 GPa, and that for polyaniline films is approximately 3 GPa. The rather low value for the in-plane modulus of silicate (approximately 1 order of magnitude lower than mica) implies that the measured value represents an effective modulus due to the sliding of the silicate platelets and not that due to the in-plane atomic bonding. The sliding of the layers is probably facilitated by the presence of intercalated molecules that act as a lubricant. We attribute the enhanced in-plane modulus of the hybrid to the better ordering of the composite compared to the pristine silicate host. This better ordering increases the interaction area between the polymer and the silicate layers and increases the stability of the assembly.

Two-dimensional nanocomposites can also be synthesized by direct polymer intercalation. One advantage compared to the intercalative polymerization is that well-characterized polymers of different molecular weights can be used, and the effect on the properties of the composites can be studied. We use polyethylene oxide (PEO) as an example, although various other polymers have been intercalated as well. The reaction is schematically shown as follows; the horizontal lines represent the layered structure.

$$\text{Na}^+ + \text{PEO} \longrightarrow \text{Na}^+(\text{PEO})$$

Synthesis involves mixing an aqueous suspension of montmorillonite (a natural silicate) with a water solution of the polymer, although intercalation can also take place from the melt in the absence of any solvent. X-ray diffraction patterns of oriented films (Figure 12) show that PEO intercalation results in a highly ordered multilayered structure with a repeat unit corresponding to single polymer chains in a helical form. Comparing the X-ray diffraction patterns of the PEO-intercalated and pristine silicate (Figure 12) reveals a dramatic enhancement of the structural ordering upon intercalation, similar to that for polyaniline. The enhanced ordering is probably due to the tendency of the system to maximize the interaction area between the polymer chains and the host layers.

Polymers, in general, can be amorphous or semicrystalline, depending on the regularity of the polymer chains. The degree of crystallinity and the size of the crystallites have a profound effect upon the physical and mechanical properties of the polymer. Segmental motion of the disordered chains in the amorphous regime takes place above the glass-transition temperature, T_g, while the polymer crystals melt at the melting temperature, T_m. Both glass transition and melting are accompanied by

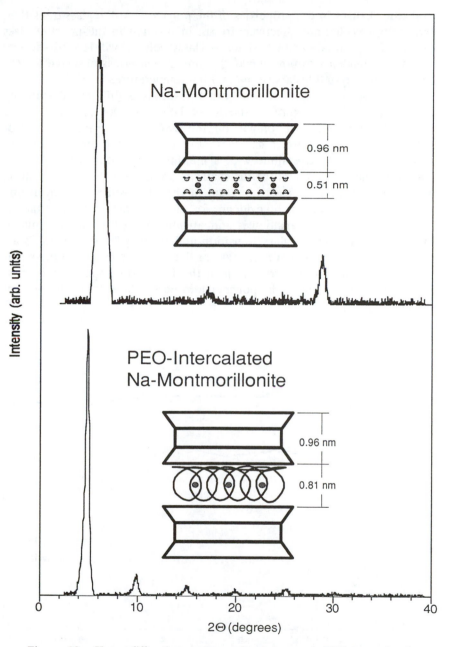

Figure 12. X-ray diffraction pattern of pristine and PEO-intercalated montmorillonite. The 0.51-nm gallery height for pristine montmorillonite is due to the presence of two water layers; the gallery height becomes 0.81 nm upon PEO intercalation.

dramatic changes in the properties of the polymer. At the glass transition, the molecules that are effectively frozen in position in the polymer glass become free to rotate and translate. A glassy polymer will lose its stiffness and have a tendency to flow above T_g. Similar behavior is observed for the chains in the crystallites above the melting temperature.

A convenient way to follow both transitions is differential scanning calorimetry (DSC). Figure 13 shows the DSC trace for the intercalated PEO–silicate hybrid. For comparison, the DSC trace of a PEO–montmorillonite physical mixture is shown in Figure 14. At room temperature PEO is about 85% crystalline with a melting temperature of 65 °C; the glass transition for the amorphous chains is around −50 °C. Both melting and glass transition are clearly present in the PEO–silicate physical mixture. In contrast, there is no evidence for the glass transition while the intensity of the melting transition is dramatically reduced in the intercalated hybrid. The absence of glass transition and melting for the hybrid is attributed to the molecular confinement of the PEO chains due to intercalation. (The small melting transition in the intercalated spectrum is most likely due to a fraction of the polymer located on the external surfaces of the host layers.)

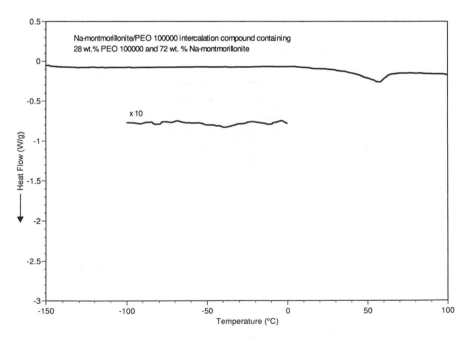

Figure 13. DSC heat flow vs. temperature for montmorillonite-intercalated PEO.

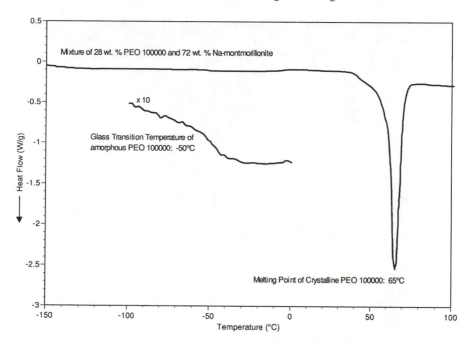

Figure 14. DSC trace of a physical mixture of PEO–montmorillonite. The ratio of silicate to PEO is the same as that for the intercalated hybrid.

Nanocomposites Produced by Molecular Dispersion of the Ceramic.

A second approach, first explored by researchers at the Toyota Research Center (*27*), is based on an alternative design and synthetic strategy. In this case, the nanocomposites consist of single silicate layers (1 nm thick) homogeneously dispersed in a continuous polymer matrix (Figure 7b). To maximize the effect of the silicate, the host must ideally be completely exfoliated into single layers and homogeneously dispersed in the polymer matrix. Because of the poor dispersability of pristine MTSs in nonaqueous solvents, the host layers are functionalized prior to mixing with the polymer. Functionalization involves rendering the surfaces hydrophobic through a simple ion-exchange reaction with alkylammonium cations. The presence of the hydrophobic moieties not only significantly improves dispersability, but it might also affect the chemistry at the polymer–silicate interface.

Figure 15 shows a transmission electron microscopy (TEM) image of a polyimide–silicate composite prepared by mixing a solution of polyamic acid in *N*-methylpyrrolidinone (NMP) with a suspension of silicate also in NMP. After solvent evaporation, the hybrid is cured at 350 °C to transform the polyamic acid into polyimide. The host layers, although not complete-

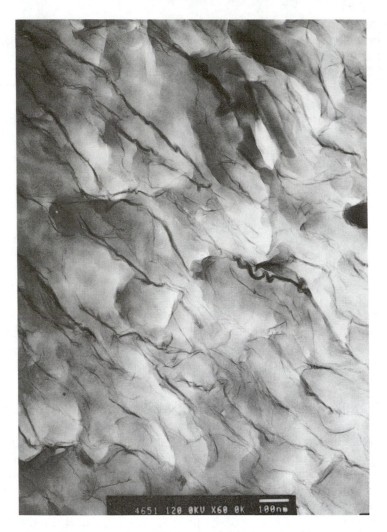

Figure 15. Cross section TEM image of silicate–polyimide composite.

ly exfoliated, tend to orient parallel to each other as a result of the dipole–dipole interactions that usually operate over rather long distances.

Despite the rather low amount of silicate (less than 10 wt%) that is
required in this design, a significant enhancement in the properties of the
composite has been observed. For example, the Toyota group (27)
showed that polyimide composites containing as low as 2 wt% silicate
exhibit a 60% decrease in the permeability of water (in the z direction),
while the thermal expansion coefficient (x,y direction) is reduced by 25%
compared to the bulk polymer. Experiments in our laboratory also
showed that similar composites have an enhanced modulus while main-

taining the dielectric characteristics of polyimide (Figure 16). This combination of properties makes the new composites very attractive candidates for the dielectric layers in electronic packaging applications. In addition, the very low loading of the inorganic phase is crucial in developing lightweight composites.

What are the possible factors that might contribute to the enhanced properties of the nanocomposites? A substantial change in the properties of a composite is observed when the reinforcing phase forms an interconnected network (corresponding to the *percolation threshold*) spanning the entire dimensions of the material. For *anisotropic* percolating objects, such as platelets, the volume fraction occupied at the percolation threshold depends on the aspect ratio, *a*, of the object, and it decreases as *a* increases. The silicate layers possess an unusually high aspect ratio, so this effect might be at least partially responsible for the observed behavior. Another possibility is that the silicate layers might induce ordering of the polymer chains at the polymer–ceramic interface. The role of the silicate surface might be to lower the thermodynamic barrier for crystallization. In the absence of silicate, the chains are randomly oriented, but near a surface they tend to orient parallel to the surface because of the geometric barrier (*28, 29*). This different arrangement persists over rather long distances as the chains become more rigid because in the semi-rigid chains of polyimide, states containing sharply bent conformations have considerably higher energy than those with straight conformations. An alternative ex-

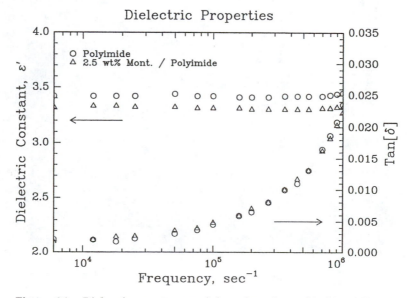

Figure 16. Dielectric constant and loss data for polyimide and polyimide–silicate composite containing 2.5 wt% silicate.

planation is that the slow evaporation of the solvent due to the presence of the silicate layers might be indirectly responsible for the polymer crystallization by decreasing the viscosity and providing a higher mobility for the polymer chains.

Preliminary evidence supports our hypothesis that the silicate surface induces polymer crystallization. Figure 17 is an optical microscope image showing spherulites formed when 1 wt% silicate was added to a polymer solution in NMP. [The polyamic acid (PMDA–MDA) was prepared from pyromellitic dianhydride (PMDA) and methylenedianiline (MDA).] The crystallites are absent when the polymer is processed similarly but without any silicate.

Conclusions and Future Directions

We presented several examples of self-assembled organic–inorganic nanostructures synthesized by intercalation of layered silicates. The materials design and synthesis largely departs from the conventional sequential covalent bond formation, and it relies on weaker and less-directional bonding such as hydrogen and van der Waals bonding. By carefully selecting

Figure 17. Optical micrograph taken in polarized light showing the presence of polymer crystals (spherulites).

the host and guest species and by controlling subtle guest–host interactions, the properties of the assembly can be dramatically altered.

We also showed that polymer nanocomposites can be synthesized by in situ polymerization in the galleries of layered silicates or direct polymer intercalation. The resulting hybrids consist of nanometer-scale, well-ordered multilayers, and they are highly anisotropic. Because of the intercalation and confinement, the polymers show no crystallization or glass transition because crystallization and glass transition are cooperative phenomena, and they are generally characteristic of unconstrained polymers.

Layered silicates can also be used as a means to create molecularly dispersed interfaces in a continuous polymer matrix resulting in lightweight composites with properties dramatically different from bulk polymers. These dramatic changes might be due to the enhanced ordering of the polymer chains near the polymer–silicate interfaces as well as the unusually high aspect ratio of the silicate plates.

Finally, a few comments regarding future directions are in order. Besides studying the effect of confinement on first- and second-order transitions, we have started using these nanostructures to study surface dynamics. The dynamic behavior of molecular liquids and polymers confined in restricted geometries has received increasing theoretical and experimental attention. Studying dynamics at interfaces represents a relatively difficult problem because interfaces are usually buried between two bulk phases, and few available analytical techniques have the required sensitivity. Although porous matrices have been used in the past, because of the pore curvature it is not clear how these studies can relate to liquid dynamics in a planar geometry. In addition, with increasing pore size, a two- to three-dimensional transition is expected. For all these reasons, intercalated nanostructures are ideal model systems for studying dynamics at solid–liquid interfaces because no specialized analytical techniques are required as a result of the dominance of interfaces. Experiments with both simple molecules and polymers are already underway.

Acknowledgments

This work was sponsored by the National Science Foundation (Grant DMR–8818558) through the Materials Science Center at Cornell, and by Corning Inc. and IBM. I thank my coworkers V. Mehrotra, O. Tse, R. A. Vaia, and T.-C. Sung, whose work has made this report possible. I also thank E. J. Kramer, W. Sachse, B. B. Sauer, and J. Smith for experimental assistance and many helpful discussions.

References

1. Brus, L. E.; Brown, W. L.; Andres, R. Averback, R. S.; Goddard, W. A. III; Kaldor, A.; Louie, S. G.; Moskovits, M.; Peercy, P. S.; Riley, S. J.; Siegel, R. W.; Spaepen, F. A.; Wang, Y. *J. Mater. Res.* **1989**, *4,* 704.
2. *Research Opportunities for Materials with Ultrafine Microstructures;* National Academy Press: Alexandria, VA, 1989.
3. *Science (Washington, D.C.)* **1991**, *254,* 1300.
4. Giannelis, E. *Chem. Mater.* **1990**, *2,* 627.
5. Mehrotra, V.; Giannelis, E. *Solid State Commun.* **1990**, *77,* 155.
6. Mehrotra, V.; Giannelis, E. In *Polymer-Based Molecular Composites;* Shaefer, D. W.; Mark, J. E., Eds.; MRS Proceedings; Materials Research Society: Pittsburgh, PA, 1990; p 171.
7. Giannelis, E.; Mehrotra, V.; Russell, M. W. In *Better Ceramics Through Chemistry;* Brinker, C. J.; Clark, D. E., Ulrich, D. R.; Zelinski, B. J. J., Eds.; MRS Proceedings; Materials Research Society: Pittsburgh, PA, 1990.
8. Mehrotra, V.; Giannelis, E. *Solid State Ionics* **1992**, *51,* 115.
9. Mehrotra, V.; Giannelis, E.; Ziolo, R. F.; Rogalskyi, P. *Chem. Mater.* **1992**, *4,* 20.
10. Giannelis, E.; Mehrotra, V.; Tse, O.; Vaia, R. A.; Sung, T.-C. In *Synthesis and Processing of Ceramics: Scientific Issues;* Rhine, W. E.; Shaw, T. M.; Gottshall, R. J.; Chen, Y., Eds.; MRS Proceedings; Materials Research Society: Pittsburgh, PA, 1992.
11. Giannelis, E. *J. Minerals Metals Mater. Soc.* **1992**, *44,* 28.
12. Kanatzidis, M. G.; Marcy, H. O.; McCarthy, W. J.; Kannewurf, C. R.; Marks, T. J. *Solid State Ionics* **1989**, *32/33,* 594.
13. Kanatzidis, M. G.; Wu, C. *J. Am. Chem. Soc.* **1989**, *111,* 4139.
14. Cao, G.; Mallouk, T. E. *J. Solid State Chem.* **1991**, *94,* 59.
15. Pillion, J. E.; Thompson, M. E. *Chem. Mater.* **1991**, *2,* 222.
16. Divigalpitiya, W. M. R.; Frindt, R. F.; Morrison, S. R. *J. Mater. Res.* **1991**, *6,* 1103.
17. Liu, Y.-J.; DeGroot, D. C.; Schindler, J. L.; Kannewurf, C. R.; Kanatzidis, M. G. *Chem. Mater.* **1991**, *3,* 992.
18. Nazar, L. F.; Zhang, Z.; Zinkweg, D. *J. Am. Chem. Soc.* **1992**, *114,* 6239.
19. Messersmith, P. B.; Stupp, S. I. *J. Mater. Res.* **1992**, *7,* 2599.
20. Aranda, P.; Ruiz-Hitzky, E. *Chem. Mater.* **1992**, *4,* 1395.
21. *Intercalation Chemistry;* Whittingham, M. S.; Jacobson, A. J., Eds.; Academic: New York, 1982.
22. Whitesides, G. M.; Mathias, J.; Seto, C. T. *Science (Washington, D.C.)* **1991**, *254,* 1312.
23. *Materials Synthesis Utilizing Biological Processes;* Rieke, P. C.; Calvert, P. D.; Alper, M., Eds.; MRS Proceedings 174; Materials Research Society: Pittsburgh, PA 1990.
24. Pinnavaia, T. J. *Science (Washington, D.C.)* **1983**, *220,* 365.
25. Newsham, M. D.; Giannelis, E.; Pinnavaia, T. J.; Nocera, D. G. *J. Am. Chem. Soc.* **1988**, *110,* 3885.

26. Mehrotra, V.; Lombardo, S.; Thompson, M. O.; Giannelis, E. *Phys. Rev. B.* **1991,** *44,* 5786.
27. Yano, K,; Usuki, A.; Okada, A.; Kurauchi, T.; Kamigaito, O. *Polym. Prepr.* **1991,** 65.
28. Hsiao, B. C.; Chen, E. J. H. In *Controlled Interfaces in Composite Materials;* Ishida, H., Ed.; Elsevier: New York, 1990; p 613.

RECEIVED for review November 9, 1992. ACCEPTED revised manuscript April 6, 1993.

Nanoporous Layered Materials

Thomas J. Pinnavaia

Department of Chemistry and Center for Fundamental Materials Research, East Lansing, MI 48824

Layered compounds can be transformed into pillared nanoporous derivatives in which chemical functionality is designed into the layered host, the pillaring guest, or both. Layered silicate clays and layered double hydroxides represent two complementary structures well-suited for pillaring. In smectite clays, the gallery region between layers can be expanded by intercalative ion-exchange reaction with robust cations. Supergallery clays and tubular silicate-layered silicate heterostructures formed by the intercalation of sol particles is a promising means of forming derivatives in which the gallery height is substantially larger than the thickness of the host layers. Delaminated smectite clays with edge-face layer assembly are a complementary class of nanoporous materials. Layered double hydroxides are pillarable by intercalation of robust polyoxometalate anions and disk-shaped metallo macrocyclic anions. Other layered materials, with functional gallery surfaces, such Group IV metal phosphates, can be pillared by covalently cross-linking adjacent layers with rigid organo groups. In situ formation of gallery pillars by hydrolysis and condensation polymerization of metal alkoxides represents yet another exciting approach to pillared layered structures.

Nanoporous Materials

The economy of the United States, indeed the economy of the entire world, depends critically on nanoporous materials. Consider, for instance, the fact that almost all of the petroleum-derived fuels consumed in the world are processed over catalysts that contain a class of aluminosilicates known as zeolites. These open-framework structures can adsorb a wide variety of organic molecules on their intracrystal surfaces. Once constrained in the nanoporous space of the catalyst, organic reagents can be transformed with unique efficiency into specific reaction products. Thus, zeolites (*1, 2*) and related nanoporous solids currently under development (*3, 4*) are essential for maintaining the low-cost production of the fine chemical feedstocks that support our large-volume manufacturing technologies.

In addition to their use as shape-selective catalysts, nanoporous solids exhibit adsorption and ion-exchange properties useful for a wide variety of advanced technological processes, including environmental pollution control, the design of new structural composites, and novel electronic, optical, and magnetic devices (*5–8*).

In general terms, nanoporous materials are solids with an *accessible* open space in the 1.0–10-nm range. In describing porous materials, the IUPAC has defined three size domains: micropores, <2 nm; mesopores, 2–50 nm; and macropores, >50 nm (*9*). Thus, the nanoporous regime spans the traditional midmicropore to lower-mesopore size range. Meso- and macropores usually are associated with materials that are either finely divided or structurally highly disordered (amorphous). That is, meso- and macroporosity often are consequences of the texture of a material. Figure 1A illustrates the "textural pores" arising, for example, from the random aggregation of platelike particles and from the voids formed within the grains of an amorphous solid such as silica gel.

Micropores can also result from the textural properties of structurally disordered materials (e.g., carbon molecular sieves), but, more commonly, micropores are associated with crystalline materials with open-framework structures. In zeolites and molecular sieves, for example, the oxide framework defines open channels and cavities that can accommodate guest molecules. As illustrated in Figure 1B, these "crystallographic" pores are rigorously regular on an atomic scale. In contrast, textural pores normally exhibit broad size distributions, sometime over hundreds of nanometers.

Until recently, relatively few microporous materials approached the nanoscopic regime. For instance, the faujasitic zeolites used for cracking petroleum are accessed through 12-membered oxygen rings of approximately 0.74-nm diameter (*10*). Molecules with kinetic diameters substantially larger than 0.74 nm are unable to access the intracrystal surfaces of these zeolites. Zeolites VPI-5 and cloverite have 18- and 20-ring apertures, respectively (*11, 12*). In VPI-5, the pore opening is ~1.2 nm in diameter, but in cloverite the ring is not symmetrical.

Crystalline materials with regular pores in the 1–10-nm regime are of considerable current interest because they afford exciting new arenas for molecular assembly and chemical reaction. Early in 1992 Kresge et al. (*13*) disclosed in the patent literature the synthesis of nanoporous zeolites with channel sizes of 6.0 nm or more by using liquid-crystal templates to

A

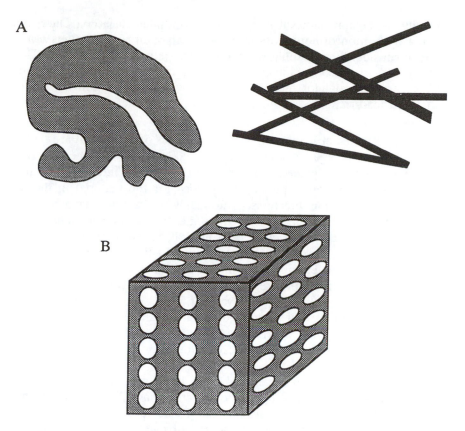

Figure 1. Types of pores found in porous materials: (A) irregular textural pores formed between finely divided platelike crystallites (e.g., kaolin clay) or within the grains of a disordered (amorphous) solid (e.g., silica gel); and (B) crystallographically regular pores formed by an open-framework structure (e.g., zeolites).

direct the crystallization of the aluminosilicate framework. The direct crystalization method promises to be a fruitful approach to the synthesis of nanoporous solids, at least for frameworks based on corner-shared SiO_4 and AlO_4 tetrahedra.

   In addition, a conceptionally complementary synthetic strategy for the design of nanoporous solids is based on the structural modification of layered solids. Owing to the existence of a large number of layered structure types and their ability to intercalate a variety of guests, nanoporous derivatives can be prepared in which chemical functionality is intelligently designed into the layered host, the intercalated guest, or both. This flexibility in synthetic strategy greatly extends the compositional diversity of nanoporous materials for a variety of materials applications. This chapter describes some approaches to the design of nanoporous materials by structural modification of nonporous layered precursors. We begin first with a general consideration of solids with lamellar structures.

Lamellar Solids

Solids with layered structures possess basal planes of atoms that are tightly bonded within the planes but relatively weakly bonded in the direction perpendicular to the planes. The asymmetric bonding interactions translate into greatly different physical properties for the material in the in-plane and out-of-plane directions. The weakly interacting region between the stacked units is usually referred to as the "interlayer" or "gallery" region. When the layers are electrically neutral, as in graphite or FeOCl, the galleries are empty and the basal planes of adjacent layers are in van der Waals contact. Neutral guest molecules often can be incorporated between the host layers to form regularly intercalated derivatives. The incorporation of neutral species into the van der Waals gap typically is accompanied by electron-transfer reaction between the molecular guest and the layered host (14). The free energy change associated with the electron-transfer step provides much of the driving force for the intercalation reaction.

In several classes of lamellar solids, the layered units carry a net electrical charge (15). These include smectite clays, layered double hydroxides, and Group 4 metal phosphates. To achieve an electrically neutral structure, counterions, unusually solvated by water or other polar molecules, occupy the gallery region between layers. Thus, ionic lamellar solids qualitatively resemble the conventional intercalation compounds formed by electron-transfer reactions between neutral guest and layered host precursors. The difference, however, is that in ionic lamellar solids, charge separation between gallery ions and the layers is complete, whereas in conventional intercalates the extent of charge transfer between guest and layered host is seldom complete. Consequently, ionic lamellar compounds can justifiably be described as intercalation compounds, although in practice they are not formed by electron-transfer reactions. Instead, they simply crystallize, complete charge separation between the gallery species and the host layers being a distinguishing feature of their structure.

Owing to their nanoscale periodicity, ionic lamellar solids give rise to very large intracrystalline surface areas of several hundred square meters per gram or more. However, in most cases the gallery surfaces are accessible only to water and other small polar molecules that are capable of solvating the gallery counterions and the charged layer surfaces. Removing the solvating molecules by outgassing at elevated temperatures results in the recollapse of the galleries, especially if the intercalated counterions are small relative to interstices occupied by the ions on the gallery surfaces. If the counterions are relatively large, they can function as molecular props or "pillars" and thereby prevent the galleries from collapsing completely when the solvating medium is removed (*3, 4*). The gallery space might then be accessible to other small molecules the size of H_2O, for example, N_2, CO_2, or NH_3. But simply facilitating the adsorption of small molecules is relatively uninteresting. Ideally, one would like to tailor the gallery structure on a length scale that would allow the accommodation of organic and inorganic molecules for molecular assembly and, perhaps, catalytic chemical conversions. Pillaring reactions of a lamellar host are an important route to achieving these desired structural modifications.

Pillared Lamellar Solids

Pillared lamellar solids are best described as intercalation compounds that meet three important criteria. These are illustrated schematically in Figure 2. First, the gallery species must be sufficiently robust to provide vertical expansion of the galleries and prevent gallery collapse upon dehydration. Second, the pillars must be laterally spaced to allow for interpillar access by molecules as least as large as nitrogen. Simply expanding the galleries to molecular dimensions by intercalation of pillars has no significance for intracrystal adsorption and catalysis if the galleries are effectively stuffed full with the pillars nearly in van der Waals contact. Third, the host layers must be sufficiently rigid to sustain the desired lateral separation of the pillars (*16*). Flexible or "floppy" host layers (e.g., graphite) would be of little use in designing galleries that are nanoporous because the layers would simply fold around the pillars and close off the space between the pillars.

To illustrate the use of pillaring reactions in the design of nanoporous solids, the following two sections consider two layered structure types that currently are under investigation in several laboratories, namely, smectite clays and layered double hydroxides. Then, a few other layered systems of particular interest for pillaring are identified.

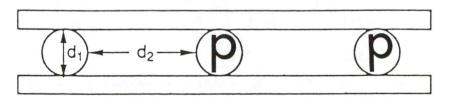

Figure 2. Schematic illustration of a pillared lamellar solid in which the pores are defined by the gallery height (d_1) *and the lateral free separation* (d_2) *between pillaring species P.*

Smectite Clays. Smectite clays are a family of complex layered oxides with 2:1 layer lattice structures analogous to muscovite, phlogopite, and other mica minerals (*17, 18*). Figure 3 illustrates the 2:1 structure in which a central $MO_4(OH)_2$ octahedral sheet is symmetrically cross-linked to two tetrahedral MO_4 sheets. Aluminum, iron, magnesium, and sometimes lithium occupy the octahedral interstices, whereas silicon and, in part, aluminum occupy tetrahedral sites. Various cations, but especially Na^+ and Ca^{2+}, may occupy the gallery surfaces. The nature of the cations filling the tetrahedral and octahedral sites in the 2:1 layers distinguishes the various members of this mineralogical family. Idealized anhydrous unit cell formulas for representative clays are provided in Table I. A major difference between a smectite and a mica is the layer charge density. Smectites typically have charge densities between 0.45 and 1.2 e^- per $O_{20}(OH)_4$ unit cell, whereas the micas are much more highly charged with 2.0 e^- per $O_{20}(OH)_4$ unit. The difference in charge density is in part responsible for the fact that certain metal ion-exchange forms of smectite undergo swelling and ion-exchange in water, whereas micas are not swelled by water and are very difficult to ion-exchange.

The concept of pillaring a lamellar solid as a means of forming microporous derivatives was first demonstrated in 1955 by Barrer and MacLeod (*19*) using smectite clays. In this seminal work, the alkali metal and

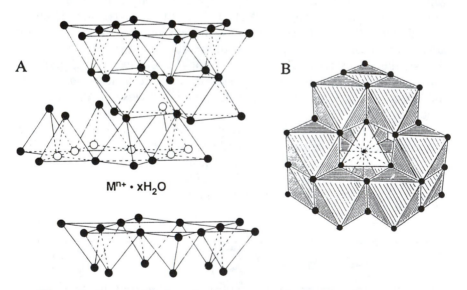

Figure 3. Part A: The layer lattice structure of a smectite clay. The solid circles are oxygen and hydroxyl groups that define two tetrahedral sheets and a central octahedral sheet. The $M^{n+} \cdot nH_2O$ exchange cations occupy positions on the gallery surfaces. Part B: The polyhedral structure of $Al_{13}O_4(OH)_{24}(H_2O)_{12}^{7+}$ used as a pillaring agent for smectite clays.

Table I. Anhydrous Unit Cell Formulas
for Typical Smectite Clay Minerals

Mineral	Typical Unit Cell Formula
Montmorillonite	$Ca_{0.35}[Mg_{0.70}Al_{3.30}](Si_{8.0})O_{20}(OH)_4$
Beidellite	$Na_{0.7}[Al_{4.0}](Si_{7.3}Al_{0.7})O_{20}(OH)_4$
Hectorite	$Na_{0.6}[Li_{0.6}Mg_{5.4}](Si_{8.0})O_{20}(OH)_4$
Saponite	$Na_{0.9}[Mg_{6.0}](Si_{7.1}Al_{0.9})O_{20}(OH)_4$

alkaline earth exchange cations in montmorillonite were replaced by quaternary ammonium ions such as Me_4N^+, among others. Subsequent work, more recently reviewed by Barrer (*20*), led to the pillaring of other smectite clays by a variety of onium ions and metal complex ions of different sizes. By varying the charge on the host layer and hence the lateral separation of the pillars, the pore structure can be tailored to differentiate adsorbates on the basis of molecular size. The molecular sieving properties of a pillared clay can be sufficient to distinguish between molecules differing in size by a few tenths of an angstrom unit.

Metal oxide pillared clays are chemically and thermally more robust than their microporous organo clay counterparts. These derivatives are prepared by first replacing the gallery Na^+ and Ca^{2+} ions in the native mineral with a robust polycation of high charge. For example, aluminum chlorohydrate, $Al_{13}O_4(OH)_{24}(H_2O)_{12}^{7+}$, has been used extensively as a pillaring reagent (*21, 22*). The polyhedral structure of this ion is shown in Figure 3B. Twelve aluminum ions occupy octahedral interstices, and one is positioned at a tetrahedral site at the center of the polyhedron. Water, hydroxyl groups, and bridging oxygens occupy vertices. Thermal dehydration–dehydroxylation of the intercalated polycation converts the ion to a small, metal oxide like particle in the gallery region:

$$Al_{13}O_4(OH)_{24}(H_2O)_{12}^{7+} \longrightarrow \text{``}Al_2O_3\text{''} + 7H^+ \qquad (1)$$

Upon heating, protons are released, and they in part balance the negative charge of the host clay layers. Several review articles (*3, 4, 8, 23*) summarize the synthesis and physical properties of metal oxide pillared clays derived from the intercalation of polyoxocations of aluminum, zirconium, chromium, and many other metals. The Lewis acid sites provided by coordinatively unsaturated metal ions on the pillar and the Brønsted acidity formed upon thermolysis impart novel chemical catalytic properties

(*24–28*). Moreover, the pores between metal oxide pillars often are larger than those found in conventional zeolites.

A schematic illustration of a pillared clay aggregate is provided in Figure 4a. Some defects are introduced as a result of layer folding and irregular platelet sizes. Nevertheless, most of the void volume is in the nanoporous range. If the clay platelet size is very small, a new family of nanoporous materials known as "delaminated clays" can be formed (*29*). These are derivatives in which edge-to-face layer aggregation competes with face-to-face layer aggregation. The edge-to face aggregation of layers is facilitated by very small particle size (<500 Å) or by clays with lathe-like layer morphology. Some face-to-face aggregation allows for the presence of micropores as in conventional pillared clays, but the edge-to-face aggregation mechanism is extensive, and it causes a card-house structure to form. The competitive edge-to-face and face-to-face aggregation mechanisms in a delaminated clay are qualitatively illustrated in Figure 4b. The extent of face-to-face aggregation is apparently limited to domains less than 70 Å, because these materials are typically X-ray amorphous. In contrast, a pillared clay will exhibit one to several orders of 001 X-ray reflections.

Much of the current interest in pillared and delaminated clays focuses on their utility as shape-selective heterogeneous catalysts. Their properties as petroleum cracking catalysts continue to be of major interest (*25, 28, 30*). Figure 5 illustrates the gasoline yields obtained with pillared and delaminated aluminum oxide derivatives (*31*). Included in the figure for comparison purposes are the gasoline yields obtained with an amorphous

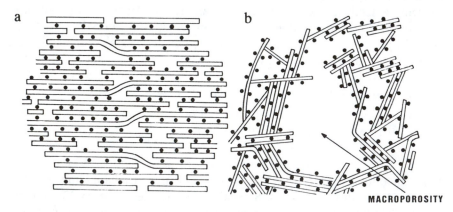

Figure 4. Layer aggregation in pillared clay in which the layers are stacked primarily face-to-face (a) and delaminated clay in which edge-to-face aggregation competes with face-to-face aggregation (b). The slabs represent the 1.0-nm-thick clay layers, and the filled circles represent the pillaring cation.

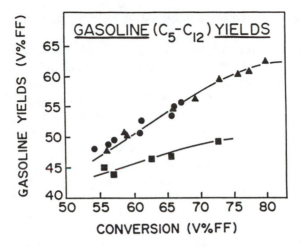

Figure 5. Gasoline selectivity of nanoporous alumina pillared montmoril-lonite (triangles) and delaminated laponite clay (filled circles) catalysts for petroleum gas oil cracking. Included for comparison is the dependence of gasoline yield on percent of fresh feed conversion (%FF) for an amorphous aluminosilicate catalyst (filled squares).

silica alumina catalyst. The yields obtained with the pillared and delaminated clays are substantially larger than those afforded by the amorphous oxide. The shape-selective performance of pillared and delaminated clays is competitive with zeolite catalysts in current commercial use.

In most of the pillared lamellar solids reported to date, the gallery height is comparable to the van der Waals thickness of the host layers. That is, the gallery free space constitutes the *minor* component by volume. It should be possible through the proper choice of pillaring agent to molecularly engineer pillared derivatives in which the gallery height is substantially larger than the thickness of the host layers. The term "supergallery" has been proposed to describe derivatives in which the gallery height is 2 or more times as large as the thickness of the host layers (*32*). If the lateral separation between the pillars can be made comparable to the size of the pillars, then materials with nanopores should be attainable.

Metal oxide pillared clays derived from intercalated polycation precursors contain oxide aggregates of more of less uniform size, as judged by the fact that they exhibit several orders of 001 X-ray reflections. Several investigators (*32–34*) recognized the possibility of preparing pillared clays by direct intercalation of metal oxide sols. This approach could afford supergallery derivatives. Aqueous metal oxide sols are commercially available with average particle sizes in the range of 2.0–4.0 nm and a particle size distribution the order of ±0.5 nm. Such particles could function as pillars for the formation of supergallery pillared clays, provided the particle size was not altered upon intercalation by competitive hydrolysis reactions.

The tubular aluminosilicate imogolite, with the empirical formula $SiAl_2O_3(OH)_4$, is an unusual naturally occurring sol particle with a tunnel-like or tubular structure. This material represents a *molecularly regular* sol particle. The external and internal diameters of the tube are approximately 2.4 and 0.8 nm, respectively, and the tube length can range to several hundred nanometers (*35*). Molecular sieving studies have shown that the intratube channel is indeed available for adsorption of molecules with kinetic diameters smaller than 1.0 nm (*36*).

Imogolite intercalates as a regular monolayer into smectite clays such as Na^+-montmorillonite (*37*). These new tubular-silicate–layered-silicate (TSLS) nanocomposites with a basal spacing of 3.40 ± 0.10 nm can be viewed as a new type of pillared clay in which the pillars themselves are microporous. Figure 6 illustrates the structure of a TSLS complex. Although the tubes are aggregated over limited domains in the lateral direction, the packing of the gallery is incomplete, much like a log-jam struc-

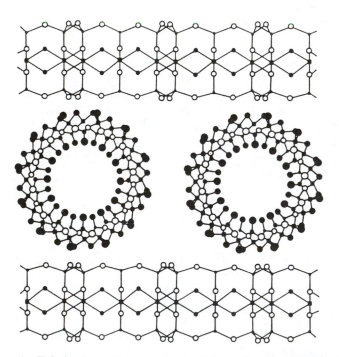

Figure 6. Tubular imogolite monolayers (shown in cross-section) intercalated in the galleries of layered Na^+-montmorillonite. The composition of the imogolite tubes is $(HO)SiO_3Al_2(OH)_3$, as read from the inner to outer tiers of atomic planes constituting the tube walls.

ture. Thus, intratube free space is available, but the intertube pores are accessible only through the intratube channels.

A Brunauer–Emmett–Teller (BET) surface area of 580 m^2/g and a liquid microporous volume of 0.205 mL^3/g was observed for the TSLS species (*38*). A bimodal microporous behavior was indicated by the nitrogen adsorption behavior, suggesting that nitrogen was accessing both intra- and intertube nanopores. Significantly, the intercalated tubes were thermally stable to ~450 °C, whereas the pristine tubes collapse above about 250 °C.

Layered Double Hydroxides. Layered double hydroxides (LDHs) are complementary to smectite clays insofar as the charge on the layers and the gallery ions is reversed; that is, the host layers of an LDH are two-dimensional (2D) polyhydroxy cations, and the gallery species are hydrated anions. The compositions of LDHs are represented by the general formula $[M_{1-x}^{II}M_x^{III}(OH)_2][A^{n-}]_{x/n} \cdot zH_2O$, where A^{n-} is the gallery anion, and M^{II} and M^{III} are divalent and trivalent cations, respectively, that occupy the interstices of edge-shared $M(OH)_6$ octahedral sheets. A large number of compositions are possible (*39–43*), depending on the choice of M^{II}, M^{III}, A^{n-}, and the layer cation stoichiometry, which typically is in the range $x = 0.17–0.33$. As shown in Figure 7a, LDH structures consist of $Mg(OH)_2$-like sheets separated by galleries of hydrated A^{n-} ions.

Despite their complementary structural relationship to smectite clays, LDHs are not easily pillared. Owing to their relatively high layer charge density ($\sim4.0e^+/nm^2$ for LDHs versus $\sim1.0e^-/nm^2$ for smectites), the galleries of LDHs tend to be filled by the pillaring ions themselves. However, polyoxometalate anions (POMs) (*44*) with high charge can be effective reagents for the pillaring of LDHs. These ions have structures consisting of several layers of space-filling oxygen atoms. The first crystalline forms of POM-pillared LDHs were reported for Zn_2Al, Zn_2Cr, and Ni_3Al derivatives using $V_{10}O_{28}^{6-}$ as the pillaring reagent (*45, 46*). The basal spacings of the pillared products (1.19 nm) corresponded to gallery heights of 0.71 nm (three oxygen planes) and to an orientation in which the C_2 axis of the pillaring anion is parallel to the host layers. Adsorption–desorption of N_2 indicated the presence of both small micropores (<1.0 nm) and nanopores with a pore maximum near 2.0 nm.

Keggin ions of the type α-$[XM_{12}O_{40}]^{n-}$, shown in Figure 7b, also are efficient reagents for the pillaring of Zn_2Al–LDH structures (*47*). $[Zn_2Al(OH)_6]NO_3 \cdot 2H_2O$ undergoes facile and complete intercalative ion-exchange reaction with aqueous solutions of $[H_2W_{12}O_{40}]^{6-}$ and $[SiV_3W_9O_{40}]^{7-}$ Keggin ions. Interestingly, no ion exchange occurred with $[PW_{12}O_{40}]^{3-}$ or $[SiW_{12}O_{40}]^{4-}$, and only partial exchange was observed with the Keggin-like species $[PCuW_{11}O_{39}(H_2O)]^{5-}$. These results suggest that the ac-

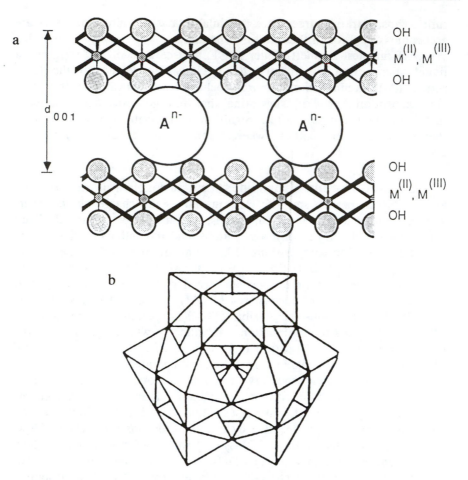

Figure 7. Part a: Structure of layered double hydroxides (LDHs) of the type $[M_{1-x}^{II}M_x^{III}(OH)_2][A^{n-}]_{x/n}$. Gallery water is not shown. Part b: Structure of α-Keggin ion of the type $[XM_{12}O_{40}]^{n-}$.

cessibility of the LDH galleries depends in part on the charge on the POM. The area needed to accommodate a Keggin ion of diameter is 0.98 nm is 0.849 nm^2, and because the area per unit charge in a Zn$_2$Al–LDH is only about 0.166 nm^2, a -5 charge on the Keggin ion is needed to spatially accommodate a monolayer of the ion in the LDH gallery.

The lack of intercalative ion exchange for -3 and -4 Keggin ions is consistent with these simple geometric considerations. The partial replacement of NO$_3^-$ ions by -3 and -4 Keggin ions would require mixing in the same gallery two ions of very different size and charge, and such mixing appears to be thermodynamically unfavorable. Commensurate relationships between the triangular faces of the Keggin ion and those of the gallery surface also may be important in determining gallery access by ion exchange (*48*). For instance, the β-isomer of $[SiV_3W_9O_{40}]^{7-}$ is much slower to intercalate than the α-isomer. This difference in lability suggested that the α-isomer prefers to orient in the galleries with the C_2 axis perpendicular to the basal surfaces.

The pillaring of LDHs by POMs is further complicated by the fact that most LDHs are basic, whereas the POM anions are acidic. Hydrolysis reactions of the LDH and POM can result in products that are poorly ordered or that contain multicrystalline phases. A promising route to pillared forms of basic LDHs is based on the exchange reaction of the desired POM with an expanded LDH precursor intercalated by a large organic anion such as *p*-toluenesulfonate or terephthalate (*46*). The large organic anion is very readily replaced by the POM, and competing side reactions are minimized. Thus, it is possible to prepare pillared forms of highly basic Mg_2Al–LDH hosts with acidic POM pillars such as $[V_{10}O_{28}]^{6-}$ and $[Mo_7O_{24}]^{6-}$.

Well-ordered LDHs interlayered by large organic anions can be relatively difficult to prepare, but more recently a more general and reliable route to such precursors was described (*49*). This method uses as a precursor a hydroxide-exchanged form of the LDH such as synthetic meixnerite, $[Mg_3Al(OH)_8]OH \cdot 2H_2O$. These derivatives swell in polar solvents such as glycerol or glycerol–water mixtures. The swelling greatly enhances the accessibility of the gallery hydroxide ions for reaction with organic acids. A variety of LDH intercalates containing long-chain alkyl carboxylates and α,ω-dicarboxylates have been prepared by this route. Thus, in principle, the gallery height of the organic LDH precursor can be matched to any desired POM pillaring agent.

Disk-shaped, metallo macrocyclic anions such as cobalt(II) phthalocyanine tetrasulfonate $[Co(PcTs)]^{4-}$ are currently being investigated as potential pillaring agents for LDHs (*50*). For $[Co(PcTs)]^{4-}$ intercalation into a $Zn_{1-x}Al_x$–LDH with $x = 0.33$, the anion intercalates with the plane of the macrocycle perpendicular to the LDH layers, as illustrated in Figure 8. At a layer charge density of $x = 0.33$, the gallery anions are too closely spaced to function as true pillaring agents. Lowering the layer charge density may increase the lateral separation of the anions. However, even at $x = 0.33$, the Co(II) centers at the edge of the LDH layers are available for catalytic reaction. LDH-intercalated $[Co(PcTs)]^{4-}$ anions, for instance, catalyze the O_2 oxidation of mercaptides to disulfides under ambient conditions. Such transformations using easily recyclable catalysts are of interest for possible applications in the remediation of polluted groundwaters and industrial effluents. Interestingly, immobilization of the macrocyclic anion on LDH surfaces greatly improves the longevity of the complex. Several hundred catalyst turnovers can be achieved by using the LDH-immobilized catalyst, whereas only ~25 turnovers can be achieved

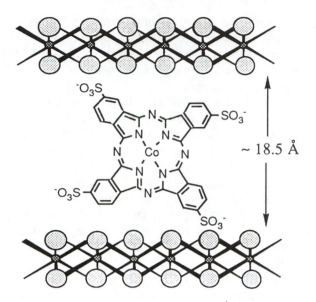

Figure 8. Edge-on intercalation of [Co(PcTs)]⁴⁻ in the galleries of an LDH.

with the homogeneous complex, owing to competing side reactions that lead to its deactivation. Furthermore, Keggin ions intercalated in LDH galleries are effective oxidation catalysts for olefins using peroxide as the oxidizing agent (*51*).

Other Layered Systems

Smectite clays and layered double hydroxides represent only two classes of layered materials that can be chemically modified to form nanoporous derivatives. Several other layered compounds are suitable for pillaring reactions, including layered titanates (*52*), phosphates and phosphonates (*53–59*), silicates (*60*), and niobates (*61*). The Group IV metal phosphates are particularly interesting because, unlike smectite clays, the gallery surfaces are chemically functional. As shown in Figure 9A, the phosphate groups in $Zr(HPO_4)_2$ are monoprotonated. The protons can be replaced by acid–base reaction with metal cations, which in turn can be replaced with pillaring cations such as the $Al_{13}O_4(OH)_{24}(H_2O)_{12}^{7+}$ (*53*). However, this ion completely fills the interlayers, and the stuffed galleries are unaccessible for adsorption.

A much more promising approach, first demonstrated by Dines et al. (*62*), is to replace some of the P–OH groups with rigid difunctional phosphonate groups to form cross-linking bridges between adjacent layers. This

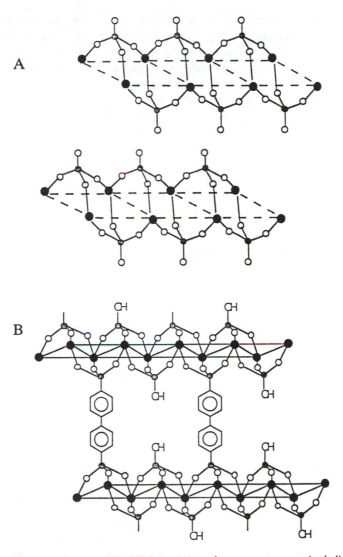

Figure 9. Structures of $Zr(HPO_4)_2$ (A) and a nanoporous mixed diphosphonate−phosphate pillared derivative (B).

approach is illustrated in Figure 9B for a mixed zirconium phosphate−phosphonate containing $P-C_6H_5-C_6H_5-P$ groups as the bridging units. By regulating the relative amounts of phosphate and phosphonate groups in the structure, in principle the average distance between randomly separated cross-linking units can be varied.

Another recent approach to designing nanoporous layered materials is to form the pillaring species directly within the galleries of the host. This concept was first demonstrated for the hydrolysis and polymerization of tetraethyl orthosilicate (TEOS) in the galleries of layered titanates and silicates interlayered by alkylammonium ions (63). The gallery polymerization process under aqueous conditions was sufficiently regular to form crystallographically well-ordered pillared derivatives interlayered by nanoscopic domains of silica. Analogous intragallery polymerization reactions of TEOS can be carried out by using neat amines as the gallery swelling agent (64). This general approach to preparing nanoporous layered materials is only beginning to be explored, and new compositionally diverse derivatives with novel materials properties can be anticipated.

Acknowledgments

Much of the author's research described in this review was supported by National Science Foundation Grant DMR 89–03579 and National Institute of Environmental Health Sciences Grant ESO–4911B.

References

1. Thomas, J. M. *Angew. Chem. Int. Ed. Engl.* **1988,** *27,* 1673.
2. Hlderich, W.; Hesse, M.; Numann, F. *Angew. Chem. Int. Ed. Engl.* **1988,** *27,* 226.
3. Pinnavaia, T. J. *Science (Washington, D.C.)* **1983,** *220,* 365.
4. Vaughan, D. E. W. In *Perspectives in Molecular Sieve Science;* Flank, W. H.; Whyte, T. E., Jr., Eds.; ACS Symposium Series 368; American Chemical Society: Washington, DC, 1988; p 308.
5. Ozin, G. A.; Kuperman, A.; Stein, A. *Angew. Chem. Int. Ed. Engl.* **1989,** *101,* 373.
6. Stucky, G. D.; MacDougell, J. E. *Science (Washington, D.C.)* **1990,** *247,* 669.
7. Bein, T.; Enzel, P.; Beuneu, F.; Zuppiroli, L. In *Electron Transfer in Biology and the Solid State: Inorganic Compounds with Unusual Properties;* Johnson, M. K.; King, R. B.; Kurtz, D. M.; Kutal, C., Jr.; Norton, M. L.; Scott, R. A., Eds.; Advances in Chemistry 226; American Chemical Society: Washington, DC, 1990; p 433.
8. *Pillared Lamellar Structures;* Mitchell, I. V., Ed.; Elsevier: New York, 1990.
9. Gregg, S. J.; Sing, K. S. W. *Adsorption, Surface Area, and Porosity,* 2nd ed.; Academic: New York, 1982; p 25.
10. *Atlas of Zeolite Structure Types,* 3rd ed.; Meier, W. M.; Olson, D. H., Eds.; Butterworth-Heinemann: London, 1992; p 96.
11. Davis, M. E.; Saldarriaga, C.; Montes, C.; Garces, J.; Crowder, C. *Nature (London)* **1988,** *331,* 698.
12. Estermann, M.; McCusker, L. B.; Baerlocker, C.; Merrouche, A.; Kessler, H. *Nature (London)* **1991,** *352,* 320.

13. Kresge, C. T.; Leonowicz, M. E.; Roth, W. J.; Vartuli, J. C. U.S. Patent 5 098 684, 1992.

14. *Intercalation Chemistry;* Whittingham, M. S.; Jacobson, A. J., Eds.; Academic: New York, 1982.

15. Lagaly, G. *Solid State Ionics* **1986**, *22*, 43.

16. Kim, H.; Jin, W.; Lee, S.; Zhou, P.; Pinnavaia, T. J.; Mahanti, S. D.; Solin, S. A. *Phys. Rev. Lett.* **1988**, *60*, 2168.

17. Barrer, R. M. *Zeolites and Clay Minerals as Sorbents and Molecular Sieves;* Academic: New York, 1978; pp 407–483.

18. *Crystal Structures of Clay Minerals and Their X-Ray Identification;* Brindley, G. W.; Brown, G., Eds.; Mineralogical Society: London, 1980.

19. Barrer, R. M.; MacLeod, D. M. *Trans. Faraday Soc.* **1955**, *51*, 1290.

20. Barrer, R. M. *Pure Appl. Chem.* **1989**, *61*, 1903.

21. Johansson, G. *Acta. Chem. Scand.* **1960**, *14*, 771.

22. Bottero, J. Y.; Cases, J. M.; Flessinger, F.; Poirier, J. E. *J. Phys. Chem.* **1980**, *84*, 2933.

23. Burch, R., Ed.; "Pillared Clays", *Catal. Today* **1988**, *2*, 185.

24. Adams, J. M. *Appl. Clay Sci.* **1987**, *2*, 309.

25. Figueras, F. *Catal. Rev. Sci. Eng.* **1988**, *30*, 457.

26. Plee, D.; Schutz, A.; Poncelet, G.; Fripiat, J. J. *Stud. Surf. Sci. Catal.* **1985**, *20*, 343.

27. Laszlo, P. *Science (Washington, D.C.)* **1987**, *235*, 1473.

28. Occelli, M. L. *Stud. Surf. Sci. Catal.* **1988**, *35*, 101.

29. Occelli, M. L.; Landau, S. D.; Pinnavaia, T. J. *J. Catal.* **1987**, *90*, 256.

30. Sterte, J.; Otterstedt, *Appl. Catal.* **1988**, *38*, 131.

31. Occelli, M. L.; Landau, S. D.; Pinnavaia, T. J. *J. Catal.* **1987**, *104*, 331.

32. Moini, A.; Pinnavaia, T. J. *Solid State Ionics,* **1988**, *26*, 119.

33. Lewis, R. M.; Van Santen, R. A. U.S. Patent 4 637 992, 1987.

34. Yamanaka, S.; Tatsoo, N.; Hattori, M. *Mater. Chem. Phys.* **1987**, *17*, 87.

35. Farmer, V. C.; Adams, M. J.; Fraser, A. R.; Palmieri, F. *Clay Miner.* **1983**, *18*, 459.

36. Adams, M. J. *J. Chromatogr.* **1980**, *188*, 97.

37. Johnson, I. D.; Werpy, T. A.; Pinnavaia, T. J. *J. Am. Chem. Soc.* **1988**, *110*, 8545.

38. Werpy, T. A.; Michot, L. J.; Pinnavaia, T. J. In *Novel Materials in Heterogeneous Catalysis;* Baker, R. T. K.; Murrell, L. L., Eds; ACS Symposium Series 437; American Chemical Society: Washington, DC, 1990; 120.

39. Miyata, S. *Clays Clay Miner.* **1980**, *28*, 50.

40. Reichle, W. T. *Chemtech* **1986**, *58*.

41. deRoy, A.; Forano, C.; El Malki, K.; Besse, J.-P. In *Synthesis of Microporous Materials;* Occelli, M. L.; Robson, H., Eds.; Van Nostrand-Reinhold: New York, 1992, Vol. II; pp 108–169.

42. Carrado, K. A.; Kostapapas, A.; Suib, S. L. *Solid State Ionics* **1988**, *26*, 77.

43. Park, I. J.; Kuroda, K.; Kato, C. *Solid State Ionics* **1990**, *42*, 197.

44. Pope, M. T. *Heteropoly and Isopoly Oxometalates;* Spring-Verlag: New York, 1983.

45. Kwon, T.; Tsigdinos, G. A.; Pinnavaia, T. J. *J. Am. Chem. Soc.* **1988**, *110*, 3653.

46. Drezdzon, M. A. *Inorg. Chem.* **1988,** *27,* 4628.
47. Kwon, T.; Pinnavaia, T. J. *Chem. Mater.* **1989,** *1,* 381.
48. Kwon, T.; Pinnavaia, T. J. *J. Molec. Catal.* **1992,** *74,* 23.
49. Dimotakis, E.; Pinnavaia, T. J. *Inorg. Chem.* **1990,** *29,* 2393.
50. Perez, M. E.; Ruano, R.; Pinnavaia, T. J. *Catal. Lett.* **1991,** *11,* 51.
51. Tatsumi, T.; Yamamoto, K.; Tajima, H.; Tominaga, H. *Chem. Lett* **1992,** 815.
52. Anthony, R. G.; Dosch, R. G. *Stud. Surf. Sci. Catal.* **1991,** *63,* 637.
53. Clearfield, A. *Comments Inorg. Chem.* **1990,** *10,* 89.
54. Johnson, J. W.; Jacobson, A. J.; Butler, W. M.; Rosenthal, S. E.; Brody, J. F.; Lewandowski, J. T. *J. Am. Chem. Soc.* **1991,** *111,* 381.
55. Tomlinson, A. A. G. In *Pillared Layered Structures;* Mitchell, I. V., Ed.; Elsevier: New York, 1990; p 91.
56. Alberti, G.; Costantino, V.; Marmottini, F.; Vivani, R.; Zappelli, P. In *Pillared Layered Structures;* Mitchell, I. V., Ed.; Elsevier: New York, 1990; p 119.
57. Cao, G.; Mallouk, T. E. *Inorg. Chem.* **1991,** *30,* 1434.
58. Yamanaka, S.; Hattori, M. *Inorg. Chem.* **1981,** *20,* 1929.
59. Burwell, D. A.; Thompson, M. E. *Chem. Mater.* **1991,** *3,* 730.
60. Schwieger, W.; Heidemann, D.; Bergk, K. H. *Rev. Chim. Miner.* **1985,** *22,* 639.
61. Treacy, M. M.; Rice, S. B.; Jacobson, A. J.; Lewandowski, J. T. *Chem. Mater.* **1990,** *2,* 279.
62. Dines, M. B.; Cooksey, R. E.; Griffith, P. C.; Lane, R. H. *Inorg. Chem.* **1983,** *22,* 1003.
63. Landis, M. E.; Aufdembrink, B. A.; Chu, P.; Johnson, I. D.; Kirker, G. W.; Rubin, M. K. *J. Am. Chem. Soc.* **1991,** *113,* 3189.
64. Dailey, J. M.; Pinnavaia, T. J. *Chem. Mater.* **1992,** *4,* 855.

RECEIVED for review November 9. 1992. ACCEPTED revised manuscript April 5, 1993.

12

Catalytic Materials

Bruce C. Gates

**Department of Chemical Engineering and Materials Science,
University of California, Davis, CA 95616**

Solid catalysts are used in many processes for chemical and fuel conversion and pollution abatement. A typical solid catalyst consists largely of a porous, high-surface-area ceramic material called a support; on its internal surface is a small amount of catalytically active species (e.g., highly dispersed metal) and sometimes other components (promoters) that improve the catalytic properties. This chapter is a summary for those unfamiliar with catalytic materials and includes explanations of why these materials have such properties as porosity, high dispersions of catalytic components, and molecular-sieving character. Examples of catalysts include porous functionalized polymers, zeolites, and supported metals. Complex composite catalytic materials are illustrated by those used for petroleum cracking, ethylene oxidation, and naphtha reforming. Some novel catalytic materials are illustrated, including supported clusters with nearly molecular structures, molecular-sieving carbons, mesoporous aluminosilicates with uniform pores, and membranes.

What Is a Catalyst?

This chapter is a description of materials used as catalysts, written for readers unfamiliar with the subject. A chemist defines a catalyst as a substance that increases the rate of approach to equilibrium of a chemical reaction without being substantially consumed in the process. A materials scientist might describe a catalyst as a device for chemical transformation: Reactant molecules flow into the device and are transformed into product molecules that flow out; energy may be consumed or liberated (Figure 1).

0065–2393/95/0245–0301$12.00/0

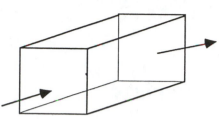

**PRODUCT
MOLECULES
(ENERGY)**

**REACTANT
MOLECULES
(ENERGY)**

Figure 1. Depiction of a catalyst as a device for chemical transformation.

Catalysts are among the most important technological materials, being used in the manufacture of chemicals, fuels, foods, clothing, pharmaceuticals, and materials such as organic polymers (*1–3*). The value of the goods manufactured in the United States in processes that at some stage involve catalysis is about $1 trillion annually; the catalysts used in these processes cost only a few tenths of a percent of the value of the products!

Catalytic technology is central to the abatement of pollutants. In the developed countries, the principal driving force for new catalytic technology is environmental protection.

Most biological reactions are also catalytic; nature's catalysts, which are called enzymes, are proteins, gigantic organic polymers, some of which contain metal ions. As biotechnology emerges, these catalysts are also expected to assume enormous technological importance.

Many important molecular and ionic catalysts work in solution, but these are ignored here. Rather, the focus is on solid catalysts, which are far more important than the others in large-scale processes for conversion of chemicals, fuels, and pollutants (*4*). Many solid elements and compounds, including metals, metal oxides, and metal sulfides, are catalysts. A few industrial catalysts (Table I) are simple in composition, for example, Raney nickel, used for hydrogenation of fats, and γ-Al_2O_3, used for dehydration of ethanol to make ethylene. However, the typical industrial catalyst consists of a variety of components and phases and is so complex that the structure is not well understood.

The *activity* of a catalyst is a measure of how fast it catalyzes a reaction. The *selectivity* is a measure of how well the catalyst directs the conversion to desired products; a highly selective catalyst is much more active for the desired reactions than for undesired side reactions. The *stability* of a catalyst is a measure of how fast it loses activity or selectivity in operation. The *regenerability* is a measure of how effectively a deactivated catalyst can be brought back to a state of high activity and selectivity.

Table I. Some Industrial Catalytic Materials

Catalyst	Reaction Catalyzed	Description of Catalyst
Raney nickel	Fat hydrogenation	Porous particles made by extraction of Al from Al–Ni alloy
γ-Al$_2$O$_3$	Ethanol dehydration	Amorphous particles with micro- and mesopores; surface area ca. hundreds of square meters per gram
Promoted iron	Ammonia synthesis	Iron with alumina as a textural promoter and K and Ca oxides as chemical promoters; microparticle dimension $\sim$30 nm; surface area $\sim$30 m^2/g
Supported silver	Ethylene epoxidation	Silver particles $\sim$1 μm in diameter on α-Al$_2$O$_3$ stabilized with clay binder and promoted with Cs$^+$ and Cl$^-$; surface area $\sim$1 m^2/g
MoS$_2$ supported on γ-Al$_2$O$_3$	Hydrodesulfurization of petroleum	Layers of MoS$_2$ a few atoms thick dispersed on γ-Al$_2$O$_3$ support promoted with Co^{2+}; surface area $\sim$200 m^2/g

Forms of Catalytic Materials

Physical Properties. Catalysts do their work in chemical reactors, which are commonly tubes packed with catalyst particles or tanks with suspended catalyst particles. Catalytic materials in the form of honeycombs (monoliths) are placed in canisters so that a stream of reactant gases flows through the honeycomb channels.

Efficient use of a catalyst requires high rates of catalytic reaction per unit volume. The reaction takes place on the surface of a solid catalyst, and therefore practical applications require catalysts with high surface areas per unit volume. Solids have high surface areas when they consist of very small particles or when they are porous. The typical catalyst is a porous material, with internal surface areas of as much as hundreds of square meters per gram. The typical catalyst is used in the form of particles. Typical particle dimensions are in the range of 30 μm to 1 cm. A porous particle often consists of aggregates of nonporous microparticles, with the void spaces between them constituting a labyrinth that provides the internal pores. A typical pore diameter is about one-fifth of the diameter of a microparticle (Figure 2). The catalytic reaction takes place on the surfaces of the microparticles.

Pores with diameters <2.0 nm are called *micropores*, those with diameters between 2.0 and 5.0 nm are called *mesopores*, and those with diameters >5.0 nm are called *macropores*. Pore volumes and pore size distributions (5) are measured by mercury penetration and by N_2 adsorption or desorption. The uptake of mercury as a function of pressure determines the pore size distribution of the larger pores. In complementary experiments, the sizes of the smallest pores (those 1–20 nm in diameter) are usually determined by measurements characterizing desorption of N_2 from the catalyst. The basis for the measurement is the capillary condensation that occurs in small pores at pressures less than the vapor pressure of the adsorbed nitrogen. The smaller the pore, the greater the lowering of the vapor pressure of the liquid in it, and so a pore size distribution can be found from the amount of N_2 remaining adsorbed as a function of pressure.

Surface areas are determined from measurements of the amount of physically adsorbed (*physisorbed*) nitrogen. Physical adsorption is a process akin to condensation; the adsorbed molecules interact weakly with the

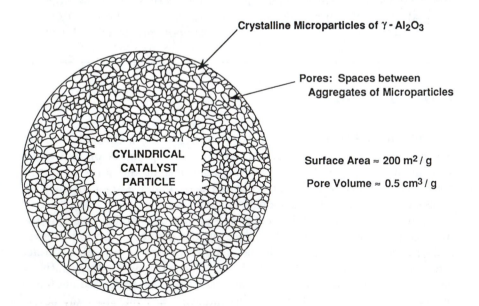

Figure 2. Porous α-Al_2O_3 catalyst particle lacking macropores (not drawn to scale).

surface, and multilayers form. (In contrast, chemical adsorption, or *chemisorption*, involves bonding of the adsorbate with the surface; chemisorption of reactants is the first step in surface catalysis.)

The nature of the bulk of a typical catalytic material is difficult to elucidate because of the complexity of the material. The compositions and structures of catalyst surfaces are even more difficult to elucidate than those of the bulk because of the notorious nonuniformity of solid surfaces and because often only a small fraction of the surface is engaged in catalysis, the rest being a spectator. The term catalytic site is applied to describe the groups on a surface where the catalytic reaction occurs. The nature of catalytic sites is often unknown.

The important physical properties of catalytic materials can be identified by examining how a catalyst is used and what determines its efficiency. The ideal catalyst would work forever, because it would not be consumed in the catalytic process. In reality, however, catalysts lose activity and need to be regenerated and replaced periodically. Thus catalysts must have the robustness to withstand handling and transport and the processes of regeneration and replacement. Catalyst particles used in large reactors must be strong enough to bear the weight of the particles above them, as measured by the crush strength. Some catalyst particles are used in entrained and fluidized bed reactors; thus they are in constant motion and must have resistance to abrasion. Catalysts are often used at high temperatures (ca. 500 °C) and must withstand those temperatures and large swings in temperature.

The physical properties that most affect the performance of a catalyst include the surface area, pore volume, and pore size distribution. The surface-area and pore-volume properties are important because they regulate the tradeoff between the rate of the catalytic reaction and the rate of transport by diffusion of the reactant molecules into the pores and the product molecules out of the pores. The tradeoff is explained as follows: The higher the internal area of the catalytic material, the higher the rate of the reaction per unit volume—up to a limit (6). Beyond the limit, an increase in internal surface area requires such small microparticles (hence such small pores) that the restrictions of the pores limit the rate of transport through them and slow down the reaction so that it is no longer proportional to the internal surface area. And if internal area is gained at the expense of increases in the pore volume, a point is reached at which the material no longer meets the crush strength requirement.

Catalyst particles are usually cylindrical because it is convenient to form them by extrusion. Other shapes (such as extrudates with cloverleaf cross sections) may be dictated by the need to minimize the resistance to diffusion of reactants into the particle interior; thus the goal may be to have a high ratio of external (peripheral) surface area to particle volume and to minimize the average distance from the outside surface to the particle interior without having particles that are so small that the pressure drop of reactants flowing through the reactor will be excessive.

Catalysts are used in the form of monoliths in automobile exhaust converters and other reactors because this form of the material allows rapid transfer of heat and mass (reactants and products) to and from the reactor.

Constituents of Catalyst Particles. Catalyst particles usually consist of a material that is called a support or carrier (which lacks catalytic activity) and other components, including those with catalytic activity and others called promoters. The support is usually the principal component, sometimes being 99% or more of the catalyst. Thus the physical properties of a catalyst are largely determined by the support. Supports are usually ceramic materials, the most common being transition aluminas such as γ-Al_2O_3 and η-Al_2O_3 (2). Others are SiO_2, MgO, mixed oxides such as SiO_2–Al_2O_3, carbon, crystalline aluminosilicates (zeolites), and organic polymers. The advantages of the transition aluminas include the following: They are inexpensive, robust, and stable and can be formed into particles with wide ranges of surface areas and pore size distributions. Much of the research done with transition aluminas and other metal oxides was motivated by catalytic applications, and much of the literature of ceramic materials has been contributed by catalytic scientists.

The typical catalyst incorporates small amounts of catalytically active materials dispersed on the internal surface of the support. Catalytically active components include metals, metal oxides, and metal sulfides (Table I). These components are often expensive, so they are dispersed as small particles (or clusters) on the support surface; clusters of Pt on an Al_2O_3 support may be as small as about 1 nm in dimension, so that almost all of the Pt atoms are exposed at a surface where they can be catalytically engaged. Supported metal catalysts have been applied for decades and are among the first and most important *nanomaterials* to find industrial applications.

Many catalysts incorporate components in addition to the support and the catalytically active components. Components called *promoters* lack significant catalytic activity themselves, but they improve a catalyst by making it more active, selective, or stable. A *chemical promoter* is used in small amounts (e.g., parts per million) and affects the chemistry of the catalytic reaction by influencing and often being part of the catalytic sites. A *textural* (structural) *promoter*, on the other hand, is used in massive amounts and usually plays a role such as stabilization of the material, for instance, by reducing the tendency of the porous material to sinter and lose internal surface area.

Preparation of Catalytic Materials

A typical first step in the preparation of a catalyst (*4, 7*) is the preparation of the support. Ceramic supports are usually prepared by precipitation from aqueous solutions. Nitrates are common anions; alkalis and ammonium are common cations. Supports are often prepared in the form of hydrogels, and mixed oxides such as $SiO_2-Al_2O_3$ are made by cogelation. Careful control of conditions such as pH is important to give uniform materials. There is growing interest in the preparation of catalytic supports by sol–gel methods.

Impurities are removed from supports by washing. Impurities in the preparative solution may be occluded in the solid and difficult to remove by washing. Therefore, ions that might poison the catalyst (e.g., Cl^- or SO_4^{2-}) are avoided. Many metal oxides are cation exchangers, and simple washing does not remove the cations. Instead, the metal ions may be removed by exchange with NH_4^+ ions, and heating of the solid incorporating NH_4^+ ions drives off ammonia and leaves H^+ in surface OH groups.

When metal oxide precipitates and hydrated gels are heated, gases such as steam are evolved and generate microporosity. Porosity can also be created by reduction of a nonporous oxide with hydrogen; porous iron can be made in this way.

In the preparation of γ-Al_2O_3 from $Al(OH)_3$, the material is heated through temperatures of hundreds of degrees Celsius. It decomposes to give a metal oxide consisting of crystalline microparticles and having a surface area of hundreds of square meters per gram. The temperature to which the material is heated determines the principal phases that are formed. As the temperature is raised, several forms of transition aluminas form (including γ-Al_2O_3 at about 500 °C), and the surface area and pore volume decrease. When a temperature of about 1100 °C is reached, the alumina is converted into α-Al_2O_3, a hard, stable, low-surface-area material that finds few catalytic applications.

Macropores can be introduced into transition aluminas by adding particles of a burnable material such as carbon to the $Al(OH)_3$. When the transition alumina is formed, the carbon particles are surrounded by it. When the carbon is burned, macropores are formed. The macropore dimensions are determined by the particle size of the carbon.

The surfaces of transition aluminas and most other metal oxides are covered with polar functional groups including OH groups and O^{2-} ions. Heating the solid drives off water in a reversible process called dehydroxylation. As a result, surface OH groups are lost and Al^{3+} and O^{2-} ions are exposed at the surface. The microparticles of the oxide are bonded strongly to each other through the surface functional groups, for example, by hydrogen bonding. Consequently, a microporous particle made up of such microparticles is a strong material.

On the other hand, some support materials, for example, α-Al_2O_3, have few surface functional groups, and the microparticles are only weakly bonded to each other, and the material lacks strength. To bond the microparticles, *binders* are added. A common binder is a clay mineral such as kaolinite. The clay is added to the mixture of microparticles as they are formed into the desired particle shape, for example, by extrusion. The support is heated to remove water and possibly burnout material and then subjected to a high temperature (possibly about 1500 °C) to cause *vitrification* of the clay; this is a conversion of the clay into a glass-like form that spreads over the microparticles of the support and binds them together. Binders are used in many ceramic materials, not just catalysts.

Catalytically active components are usually added to a support in the form of precursor metal salts in aqueous solutions. Occasionally, organometallic compounds in nonaqueous solutions are used instead. In impregnation, the support may be dried, evacuated, and brought in contact with an excess of solution containing metal salts. The solid is then dried and calcined (brought to a high temperature, usually in air), and it may be treated in hydrogen to reduce a catalytic metal to the zero-valent state. Alternatively, in the incipient wetness method, just enough of the solution is used to fill the pores of the support.

Important parameters that control the adsorption of metal complex precursors from aqueous solution onto the support are the isoelectric point of the metal oxide, the pH of the solution, and the nature of the metal complex. Either cationic or anionic species may be adsorbed. Sometimes these are simple single-metal-atom species, but sometimes they are clusters with more than one metal atom. The nature of the initially adsorbed species may largely determine the structure of the catalytic species that are finally formed. For example, supported catalysts for hydrodesulfurization to produce clean-burning fuels (by reaction of the organosulfur compounds with H_2 to remove the sulfur as H_2S) contain molybdenum sulfide clusters on γ-Al_2O_3 supports; these apparently begin as large anionic clusters containing molybdenum and oxygen that, upon treatment in $H_2S + H_2$, are converted into the sulfide.

Promoters may be added at various stages, for example, as a final step in the preparation or just prior to or during operation.

Examples of Catalytic Materials

Polymer-Supported Catalysts. Some of the simplest and best-understood solid catalysts are organic polymers (8). The one important example in technology consists of cross-linked polystyrene (the support) functionalized with strongly acidic $-SO_3H$ groups (the catalytic sites), as shown in structure **1**. This material, a cation-exchange resin, is used as a catalyst for the conversion of methanol and isobutylene into methyl *tert*-butyl ether (MTBE), a component of gasoline that is clean-burning and has a high octane number.

The practical form of the catalyst is porous beads; the internal surface area may be about 50 m²/g. Methods have been devised to synthesize the polymer in the presence of a liquid such as *n*-heptane, which is a good solvent for the monomers but a poor swelling agent for the polymer; the heptane is removed at the end of the polymerization, and pores are left in its place.

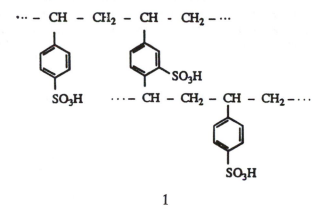

1

Crystalline Aluminosilicates (Zeolites): Shape-Selective Catalysts. Catalytic materials such as the polymers and γ-Al_2O_3 are amorphous, that is, not highly crystalline, as indicted by X-ray diffraction. γ-Al_2O_3 appears to be amorphous even though the microparticles are crystalline, because the microparticles are too small to give sharp X-ray diffraction peaks.

One important class of catalytic materials consists of crystalline solids, namely, zeolites (*9–11*). These materials are unique because they have regular pores that are part of their crystalline structures. The pores are so small (about 1 nm in diameter or smaller) that the materials are *molecular sieves*: Some molecules are small enough to enter the pores, whereas others are too large and are sieved out. Thus zeolites are applied in industrial separations processes, for example, air separation (*see* Chapter 13 of this book).

Zeolites contain Si, Al, and O ions and various other cations. The structures are built up of linked SiO_4 and AlO_4 tetrahedra that share O ions. The tetrahedra are arranged in various ways to give the different zeolites. (Amorphous metal oxides, for example, SiO_2, are also built up of linked SiO_4 tetrahedra.) Some zeolites and their physical properties are listed in Table II.

The zeolites called faujasites (zeolites X and Y) and ZSM-5 are important industrial catalysts. The structure of faujasite is represented in Figure 3 and that of ZSM-5 in Figure 4 (*12*). The points of intersection of the lines represent Si or Al ions; oxygen ions are present at the center of each line. This depiction emphasizes the framework structure of the zeolite and shows the presence of the intracrystalline pore structure. In the center of the faujasite structure is an open space called a supercage, which has a diameter of about 1.2 nm. The pore structure is three-dimensional; the supercages are connected by apertures with diameters of about 0.74 nm. Molecules large enough to fit through these apertures can undergo catalytic reaction in the cages. ZSM-5 also has a three-dimensional structure (Figure 4), with straight parallel pores intersected by zig-zag pores (Figure 4b).

Table II. Pore Dimensions of Some Zeolites

Zeolite	Number of Oxygen Atoms in the Smallest Aperture in the Pores	10× Aperture Dimensions (nm)
Zeolite A	8	4.1
ZSM-5	10	5.1 × 5.5, 5.4 × 5.6
Zeolite X or Y	12	7.4
Zeolite L	12	7.0, 7.1
Mordenite	12	6.7 × 7.0

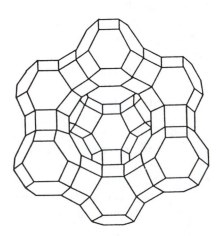

Figure 3. Structure of zeolites X and Y.

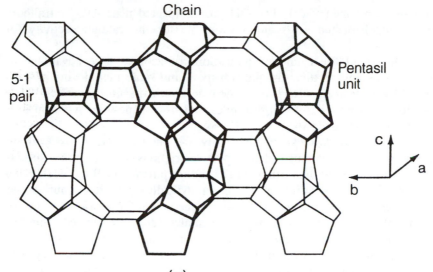

Chain

5-1
pair

Pentasil
unit

(a)

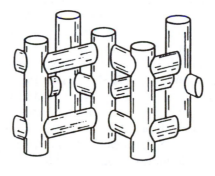

(b)

Figure 4. Structure of zeolite ZSM-5. Part a: structure indicating linkages of SiO₄ and AlO₄ tetrahedra. Part b: schematic representation showing network of straight parallel pores intersecting zig-zag pores.

The zeolite frameworks are built up of SiO_4 tetrahedra, which are neutral, and AlO_4 tetrahedra, which have a charge of -1. The charge of the AlO_4 tetrahedra is balanced by the charges of cations located at various crystallographically defined positions in the zeolite, many of them exposed at the internal surface. The zeolites are thus ion exchangers. The cations may be the catalytically active sites. When the cations are H^+ (in OH groups), the zeolites are acidic. Acidic zeolite Y (called HY) finds enormous industrial application as a component of petroleum cracking catalysts (described later). The OH groups located near AlO_4 tetrahedra are strong Brønsted acids and responsible for the catalytic activity for many reactions.

Almost all the catalytic applications of zeolites take advantage of their acidity and of their unique transport and molecular-sieving properties. The term *shape-selective catalysis* describes the unique effects. The zeolite that has found the most applications as a shape-selective catalyst is the medium-pored ZSM-5.

Different kinds of shape selectivity are known (*13*). *Mass-transport selectivity* is a consequence of transport restrictions whereby some species diffuse more rapidly than others in the zeolite pores. Small molecules in a mixture enter the pores and are catalytically converted, but larger molecules may pass through the reactor unconverted because they do not fit into the pores, where almost all the catalytic sites are located. Similarly, product molecules formed inside a zeolite may be so large that their diffusion out of the zeolite may be slow enough that they are largely converted into other products before they can escape into the product stream.

A different kind of shape selectivity is called *restricted transition-state selectivity*. It is related not to transport restrictions but instead to the size restriction of the catalyst pore, which prevents the formation of transition states that are too large to fit in the pores; thus, reactions proceeding through smaller transition states are favored.

Mass-transport selectivity is illustrated by a process for disproportionation of toluene catalyzed by HZSM-5 (*14*) (Figure 5). The catalytic sites are proton-donating OH groups on the zeolite surfaces in the pores. The desired product is *p*-xylene. The ortho and meta isomers are bulkier than the para isomer and diffuse less readily in the zeolite pores. The transport restriction favors their conversion into the para isomer, which is formed in excess of the equilibrium concentration.

Zeolite-Containing Catalysts for Cracking of Petroleum.

Zeolite Y in an acidic form is the key component of the catalysts used for cracking of petroleum to make motor fuel (*15*). Sometimes the zeolite cations are not only H^+ but also rare earth ions such as La^{3+}, which make the structure more stable. The stability is an important issue because petroleum cracking catalysts are deactivated in a matter of minutes or even seconds by deposition of carbonaceous material (coke) in the pores. To reactivate the catalyst, it is transported by entrainment in a gas stream and flows to a reactor where the carbonaceous deposits are burned off. In the regeneration reactor, the catalyst is subjected to oxygen and steam at temperatures as high as about 800 °C, and it must be robust enough to withstand the rapid temperature cycles, the atmosphere of high-temperature steam, and the abrasion resulting from the transport.

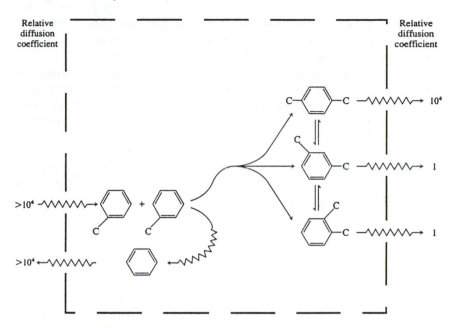

Figure 5. *Diffusion and reaction in HZSM-5-catalyzed disproportionation of toluene. (Reprinted with permission from reference 14. Copyright 1981, Kodansha Scientific, Ltd.)*

The zeolite alone is not a good, practical cracking catalyst. It is so active that it would catalyze the endothermic cracking reaction fast enough to cool the catalyst and slow the reaction to the point that it could not be sustained. Furthermore, the zeolite crystallites are too small and abrasive for use in the entrained flow reactors. The practical cracking catalysts are complex composite materials with optimized properties, consisting of a matrix (e.g., silica–alumina and possibly clay), the catalytically active zeolite, present as crystallites roughly 1 μm in size dispersed in the porous matrix, and other components. The composite catalyst particles have the size (about 50 μm in diameter) and shape to be entrained well by the oil vapors flowing through the cracking reactor. The reactant vapors carry the catalyst particles through the reactor, where they reside for only a few seconds, during which they are largely deactivated by carbonaceous deposits. The particles are separated from the cracked oil, transported continuously to the regenerator (where the carbonaceous material is burned off), and then transported back to the cracking reactor.

Dozens of forms of cracking catalysts are available commercially, with the optimum composition depending on the oil being cracked and the processing goals. A typical cracking catalyst may contain up to about 20 wt% zeolite, but the major component is the matrix. The matrix has a relatively low catalytic activity, but it does catalyze the cracking of molecules that are too large to fit into the zeolite pores; the smaller products are then cracked further in the zeolite pores. The matrix also plays the role of a heat-transfer medium; the cracking reactions are fast and endothermic, and the temperature of the catalyst and the rates of the reactions fall as the catalyst is carried through the reactor. The thermal mass of the matrix keeps the temperature drop from being too large and causing the cracking rate to fall off too quickly. After leaving the reactor, the catalyst particles are heated up in the regenerator where the carbonaceous deposits are burned off, and they return to the reactor at a suitably high temperature.

The catalyst particles present in addition to those consisting of the matrix and the zeolite include small amounts of a supported metal such as platinum on Al_2O_3, which catalyzes CO oxidation in the regenerator, minimizing the emissions of CO into the atmosphere. Metal oxide components in the catalyst mixture minimize the emission of SO_x formed in the regenerator from combustion of organosulfur compounds present in the oil. The metal oxides react with SO_x in the regenerator to make stable metal sulfates. Cycled with the regenerated catalyst to the reducing atmosphere of the cracking reactor, the sulfates are converted into H_2S, which is removed by scrubbing the effluent gas stream.

Some cracking catalysts also contain the zeolite ZSM-5, which is a selective catalyst for conversion of straight-chain alkanes. Straight-chain alkanes are often undesired components in the product because they have low octane numbers. The ZSM-5-catalyzed cracking is selective for straight-chain alkanes because of transition-state selectivity; the branched-chain alkanes are slower to react.

Supported Metal Catalysts. The metals used as catalysts are often expensive and used in a highly dispersed form (*16, 17*). Metals dispersed on supports may be clusters containing only a few atoms, or they may be particles with thousands or more atoms. The larger metal particles have three-dimensional structures and resemble small chunks of metal. The smaller particles (clusters), however, are less like bulk metals and have different catalytic properties.

Ethylene Oxidation to Ethylene Oxide: A Supported Metal Catalyst with an Inert Support. The selective oxidation of ethylene to give ethylene oxide is used on a large scale in the manufacture of ethylene glycol and other chemicals (*18*). The desired reaction is the oxidation of ethylene to give ethylene oxide; undesired reactions are the oxidations of ethylene and of ethylene oxide to give CO_2 and water. The selective oxidation is catalyzed by silver, which is the only selective catalyst.

The catalysts are composite materials containing a support (α-Al_2O_3), catalytically active silver particles, chemical promoters (e.g., Cs^+), and a binder. A catalyst particle is represented in the sketch of Figure 6. Trace amounts of chlorine-containing compounds such as ethylene dichloride are continuously added with the feed to the flow reactor; these compounds, like the alkali metal promoter, increase the selectivity of the catalyst for ethylene oxide.

The support must be virtually inert, and α-Al_2O_3 is an obvious choice; more reactive metal oxides catalyze unselective oxidation. The catalyst has a low surface area, about 1 m^2/g, and large pores to minimize the resistance to diffusion of ethylene oxide product in the pores; greater time of contact of this product with the pores would be undesirable because it would allow a greater opportunity for the conversion of this desired product into CO_2 and water.

Naphtha Reforming: A Catalytic Role of the Support. In some supported metal catalysts, the support is not just an inert platform but plays a catalytic role, as illustrated by catalysts for reforming of naphtha to make high-octane-number gasoline (*19*). The catalysts consist of clusters of metal (originally platinum, but now largely rhenium–platinum) supported on a transition alumina. Platinum is uniquely active for a number of reactions that convert low-octane-number alkanes into higher octane number

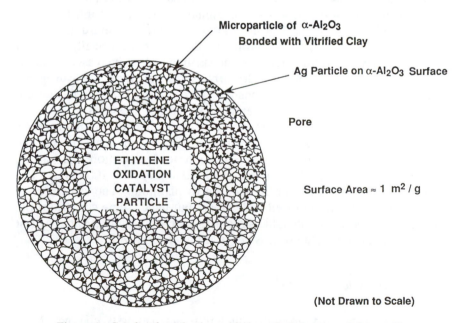

Figure 6. Catalyst for selective oxidation of ethylene to ethylene oxide.

products with nearly unchanged molecular weights. These reactions include dehydrogenation and dehydrocyclization (e.g., n-heptane $\rightarrow$ toluene + H_2).

Another useful reaction is skeletal isomerization, but platinum has only a low activity for this reaction. The isomerization reaction is catalyzed by acids, and, consequently, platinum supported on an amorphous solid acid is a better catalyst than platinum alone. The improvement resulting from this combination is more than at first expected because of an interplay between the two functions of the catalyst. The acidic component alone is not sufficient to generate the carbenium ions that are intermediates in the isomerization reactions because it is too weakly acidic. If it were a strong enough acid to catalyze alkane isomerization, the catalyst would be deactivated rapidly by carbonaceous deposits. However, the metal helps in the isomerization: It catalyzes dehydrogenation of alkanes to give alkenes, which are much stronger bases than alkanes and are easily protonated by the acidic groups of the support surface and thereby converted by carbenium ion routes.

Naphtha reforming is carried out with platinum on Al_2O_3 catalysts at about 500 °C and as much as 50 atm (5000 kPa); the reactant stream contains predominantly H_2. The catalyst is deactivated by carbonaceous deposits, but the deactivation is slow because the high partial pressure of H_2 retards formation of the deposits because aromatic precursors of the deposits and the deposits themselves are hydrogenated. Catalysts are regenerated by burnoff of the deposits, approximately once in 6 months.

The advantage of the high-H_2 partial pressures in retarding catalyst deactivation is offset by the disadvantage that became especially apparent with the removal of tetraethyllead from gasoline and an increased need for high-octane-number aromatic hydrocarbons to replace it. Dehydrogenation of cycloalkanes to give aromatic hydrocarbons is favored thermodynamically at high temperatures and low H_2 partial pressures.

These conditions provide an incentive to operate reforming processes at low pressures. What made low-pressure operation feasible was the discovery of new catalysts that are resistant to deactivation and can be operated economically at low pressures (5–10 atm or 500–1000 kPa). The catalysts incorporate rhenium in addition to platinum. The structures of these catalysts are still not well understood, but under some conditions the two metals form small bimetallic structures that are not deactivated as fast as the monometallic catalyst; it is not well understood why.

Novel Catalytic Materials

As catalytic technology develops, numerous new catalysts emerge. Most are related in structure and composition to earlier catalysts, but occasionally the processing needs lead to discovery and development of new kinds of materials. Several relatively new catalytic materials, not all of them yet important in technology, are mentioned in the following section. The examples have been chosen because of the novelty of the materials, because many of them have been prepared by catalytic scientists and engineers, and because they illustrate the opportunities for materials chemists in catalysis.

Some examples are nitrides of molybdenum and of tungsten, amorphous metals, and supported liquid-phase catalysts (in which the catalyst is dissolved in liquid held in the pores of a solid). Other examples are mentioned in more detail in the next sections.

Molecular-Sieving Carbons. Metal oxides are the common catalyst supports, but sometimes they are impractical, for example, because they are soluble in a basic reactant solution or because they react with components in a reactant stream, such as HF. Carbon is then a likely support of choice, and it is widely used as a support for metal catalysts for hydrogenations. Carbons have been prepared with pores (the slits between layers) having such narrow widths that the materials are molecular sieves. The preparation of the material (e.g., the nature of the organic precursor and the conditions of pyrolysis to generate the carbon) dictate the slit sizes, which can be rather precisely controlled. Thus a variety of catalysts on carbon supports can in prospect be applied as shape-selective catalysts (*20*).

Catalytic Membranes. Shape-selective catalysts, including zeolites and carbons, can be viewed as microscopic devices that simultaneously perform two tasks, conversion (reaction) and separation. Lots of research is being done on macroscopic devices to perform the same tasks: membranes. Early work was done with palladium, a metal that is catalytically active for numerous reactions such as hydrogenations. Hydrogen has a high solubility in palladium and diffuses through it, whereas organic molecules are not transported through palladium unless it is porous. Thus a palladium membrane can be used as a dehydrogenation catalyst that is a separation device, as sketched in Figure 7. The organic reactant flows along one side of a membrane, reacts to give a dehydrogenated product and H_2, and hydrogen diffuses through the membrane catalyst to a product stream on the other side. The reaction is equilibrium-limited, so the removal of the H_2 drives the reaction to a higher conversion than could be achieved if the catalyst were not a selective membrane.

Membrane catalysts still have not found significant applications. Cheaper materials and those with higher permeabilities than palladium are needed if large-scale applications are to emerge, and much of the recent work focuses, for example, on ceramic membranes with controlled pore sizes.

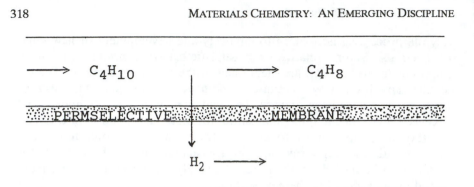

Figure 7. Catalytic membrane reactor for separation of H_2 as it is formed by dehydrogenation of butane. Because the reaction is equilibrium-limited, selective removal of H_2 through the catalytic membrane drives the conversion to butene.

Metal Clusters on Amorphous Supports and in Zeolites.
Metal clusters are comparable to small chunks of metal, but because of their small sizes (some having less than 10 atoms), these nanoparticles have properties different from those of bulk metals. Metal clusters formed by removal of the CO ligands from organometallic precursors exemplified by $[Ir_4(CO)_{12}]$ on amorphous supports have nearly uniform Ir_4 units (*21*) and thus are not metallic and have catalytic properties different from those of conventional supported metal catalysts, in which the catalytic species are usually larger metallic clusters and particles that are highly nonuniform in size and structure.

Metal clusters can also be formed in the pores of zeolites, which may help to stabilize them. Very small and nearly uniform clusters of platinum [about five or six atoms each (*22*)] have been prepared in the pores of zeolite L. This catalyst is highly selective for the formation of benzene from *n*-hexane (*23*) and is finding increasing commercial application.

Metal Oxide Clusters on Supports. Keggin ions (polyoxometallates) consist of two metals and oxygen [e.g., $Nb_2W_4O_{19}$ (*24*)] in a structure that is comparable to a small chunk of metal oxide. Keggin ions on supports are catalytically active for reactions such as selective oxidations. Their uniformity of structure sets them apart from most supported catalysts, and nanomaterials comparable to these might be expected to be selective catalysts because of the uniformity of the surface species.

Mesoporous Molecular Sieves. A large new class of materials with uniform mesopores has been prepared. The pores in these silicates and aluminosilicates can be engineered to be any of a number of sizes in the range of about 15 Å to more than 100 Å (*25*). Surfactant liquid crystals in the synthesis solution may serve as templates around which the porous materials form. These materials are of interest as catalyst supports; they offer opportunities for shape-selective catalysis with molecules larger than those involved in shape-selective zeolite catalysis.

Summary and Assessment

Catalysts are among the most important materials used in technology. Catalytic scientists and engineers have contributed significantly to the understanding of the synthesis, characterization, and application of numerous materials, especially porous ceramics, zeolites, and composites consisting of nanostructures on surfaces of porous materials.

New catalytic materials continue to emerge, and, like most novel materials, they are developed to meet specialized needs. Sometimes, unique properties of materials have been recognized as opening new opportunities in catalysis (e.g., the molecular-sieving properties of carbons), but only rarely does the process of invention and development of new catalytic materials proceed in this direction rather than in the usual direction of recognition of needed properties followed by development of materials that have them. Thus, we foresee that the innovations in catalytic materials will continue to revolve around the standard materials types, namely, sturdy, relatively inexpensive porous materials that incorporate small amounts of expensive components.

The developments are almost certain to be more evolutionary than revolutionary, at least until researchers learn to prepare materials that mimic biological materials as catalysts. The anticipated development in catalytic materials is also almost certain to be relatively rapid, because catalysis will be needed more and more to improve the efficiency of chemical and fuel conversion processes and to minimize the emissions of pollutants into the biosphere.

Acknowledgments

Helpful discussions with H. C. Foley of the University of Delaware and the support of the National Science Foundation (Grant CTS–9300754) are gratefully acknowledged.

References

1. Gates, B. C. In *Kirk-Othmer Encyclopedia of Chemical Technology;* Wiley: New York, 1993, Vol. 5; pp 320–373.

2. Gates, B. C. *Catalytic Chemistry;* Wiley: New York, 1992.
3. *Catalysis—Science and Technology;* Anderson, J. R.; Boudart, M., Eds.; Springer: Berlin, Germany.
4. Satterfield, C. N. *Heterogeneous Catalysis in Industrial Practice;* McGraw-Hill: New York, 1991.
5. Gregg, S. J.; Singh, K. S. W. *Adsorption, Surface Area, and Porosity;* Academic: London, 1967.
6. Weisz, P. B. *Science (Washington, D.C.)* **1973,** *179,* 433.
7. *Catalyst Supports and Supported Catalysts: Theoretical and Applied Concepts;* Stiles, A. B., Ed.; Butterworths: Boston, MA, 1987.
8. Sherrington, D. C.; Hodge, P. *Synthesis and Separations Using Functionalized Polymers;* Wiley: Chichester, England, 1988.
9. Breck, D. W. *Zeolite Molecular Sieves;* Wiley: New York, 1974.
10. Barrer, R. M. *Hydrothermal Chemistry of Zeolites;* Academic: New York, 1982.
11. Barrer, R. M. *Zeolite and Clay Minerals as Sorbents and Molecular Sieves;* Academic: New York, 1978.
12. Olson, D. H.; Kokotailo, G. T.; Lawton, S. L.; Meier, W. M. *J. Phys. Chem.* **1981,** *85,* 2238.
13. Haag, W. O.; Lago, R. M.; Weisz, P. B. *Faraday Discuss. Chem. Soc.* **1981,** *72,* 317.
14. Weisz, P. B. *Proc. Int. Congr. Catal. 7th (Tokyo)* **1981,** *1,* 1.
15. Venuto, P. B.; Habib, E. T., Jr. *Catal. Rev. Sci. Eng.* **1978,** *18,* 1.
16. Anderson, J. R. *Structure of Metallic Catalysts;* Academic: New York, 1975.
17. Boudart, M. *J. Mol. Catal.* **1985,** *30,* 27.
18. Gates, B. C. *Catalytic Chemistry;* Wiley: New York, 1992; pp 392–396.
19. Gates, B. C. *Catalytic Chemistry;* Wiley: New York, 1992; pp 396–403.
20. Bansal, N.; Foley, H. C.; Lafyatis, D. S.; Dybowski, C. *Catal. Today* **1992,** *14,* 305.
21. Gates, B. C.; Koningsberger, D. C. *Chemtech* **1992,** *22,* 300.
22. Vaarkamp, M.; Grondelle, J. V.; Miller, J. T.; Sajkowski, D. J.; Modica, F. S.; Lane, G. S.; Gates, B. C.; Koningsberger, D. C. *Catal. Lett.* **1990,** *6,* 369.
23. Bernard, J. R. *Proc. Int. Zeolite Conf. 5th* Rees, L. V. C., Ed. Heyden: London, 1990, p 686.
24. Day. V. W.; Klemperer, W. G.; Main, D. J. *Inorg. Chem.* **1990,** *29,* 2345.

RECEIVED for review November 9, 1992. ACCEPTED revised manuscript June 2, 1993.

Molecular Sieves for Air Separation

John N. Armor

Air Products and Chemicals, Inc., Allentown, PA 18195–1501

Generally, the commercial separation of air is achieved by cryogenic distillation, pressure-swing adsorption (PSA), or over polymeric membranes. This chapter describes carbon molecular sieves and zeolites that are currently used as adsorbents for the production of nitrogen and oxygen from air by PSA techniques. Carbon molecular sieve (CMS) materials are contrasted with the use of zeolites as molecular sieves for air separation. CMS are amorphous, quasi-graphitic materials that exhibit a high equilibrium capacity for both oxygen and nitrogen, but they are able to discriminate on a kinetic basis for adsorption of these gases. These amorphous carbons are prepared in tonnage quantities with precise control of the pore size distribution in the 3–4-Å region to within 0.2 Å! The preparation and properties of these materials used for the separation of air are described.

MOLECULAR SIEVES ARE USED COMMERCIALLY for the separation of a number of liquid and gaseous mixtures. This chapter focuses on the separation of air into N_2 and O_2. Since the early 1900s the separation of air into N_2 and O_2 has usually been performed by cryogenic distillation (*1*). Recently both solid adsorbents and membranes have taken a large proportion of the production of these commodity chemicals. [N_2 is the second largest chemical (57 billion pounds in 1991), and O_2 is the fourth largest chemical (39 billion pounds).] Solid sorbents can produce either N_2 or O_2 on a customer's site; thus transportation of the gases as liquids by truck is avoided. Liquefied O_2 and N_2 represent a very pure source of these gases and is the best choice for storage or transport of O_2 and N_2. However, when customers can use lower purity N_2 (<99.9%), noncryogenic plants can provide cheaper products at the customer's site. N_2 is used for producing O_2-free environments to prevent fire and explosion, to in-

0065–2393/95/0245–0321$12.00/0

crease storage time for perishable biological products, and to recover oil. O_2 enriches combustion air and thus improves waste incineration, steel production, and nonferrous-metal recovery. It is also used to reduce the amount of chlorine in pulp and paper bleaching and is important for wastewater treatment and bioremediation (2).

In pressure-swing adsorption (PSA), large adsorbent beds are packed with either a zeolite or carbon molecular sieve. Under air pressure either pure O_2 (95%) or N_2 (~99.5%) is produced by cycling multiple beds with periodic pressure swings to adsorb, release, and purge the beds prior to the next cycle. Generally, at least two beds are used in tandem: One is adsorbing O_2 or N_2 while another is being regenerated. The use of a storage tank permits uninterrupted production.

Recently, the use of polymer membranes became another popular means of producing N_2. These membranes are generally polymeric, hollow-fiber tubes that allow O_2, CO_2, and H_2O to permeate the walls and leave a N_2-rich (~95%) feed stream. The selection of the separation technology depends on the customer's application. This chapter focuses on the use of solid sorbents for air separation and is directed at novices in the field of gas separation. Other review articles address cryogenics and membrane separation (3, 4).

Properties of Molecular Sieves

Two major types of molecular sieves are preferred sorbents for the commercial separation of air by PSA processes. Zeolites are equilibrium-selective sorbents that sorb the N_2 from air, and carbon molecular sieves (CMS) are kinetically selective and selectively sorb the O_2.

Zeolites. These are micro crystalline molecular sieves composed of silicate and aluminate units. They are often highly microporous structures consisting of a cavity linked with 8-, 10-, or 12-membered oxygen rings as entrances (built from an infinite three-dimensional array of SiO_4 and $(AlO_4)^-$ tetrahedra linked through common oxygen atoms. Each $(AlO_4)^-$ unit adds a negative charge to the framework, which is counterbalanced by cations. Zeolites are traditionally desired for their shape selectivity or strong acidity in petroleum processing. However, for air separation, shape selectivity at room temperature is not the discriminating feature; rather the discriminating features are the structure type of zeolite and the number, degree of hydroxylation, and location of cations (5) (Figure 1). The location of cations at pore entrances can block access to the cavity and provide some kinetic selectivity. This feature has not seen wide applicability as a principle for air separation.

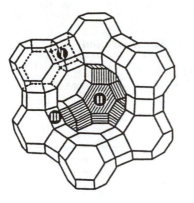

Figure 1. Cation positions and their designations in X zeolite (refer to reference 6).

The cations within the zeolite can exert a substantial electric-field gradient that interacts with the quadrupole moment of N_2. This is a key feature in the separation of N_2 versus O_2. N_2 has a moment of 0.31 Å^3, and O_2 is only 0.1 Å^3 (6). Replacing a monovalent cation by a smaller divalent cation increases the local field gradient and thus increases the interaction between N_2 and the cation (7). These interactions relate to the larger heat of adsorption of N_2 (~7 kcal/mol) versus O_2 (~3 kcal/mol) (8). The heats are also related to surface heterogeneity, which varies with the degree of cation exchange.

Moisture levels within the zeolite can prove crucial to its performance (9). For a CaX zeolite, the highest capacity for N_2 and selectivity for N_2–O_2 is obtained in a thermal treatment step that ensures the cations are in a dehydrated–dehydroxylated state.

Key properties for any air separation sorbent are its equilibrium capacity, equilibrium selectivity, S, for N_2–O_2, and the heat of adsorption. Commercially the capacity of the zeolite is reduced by the use of binders to form pelleted sorbents to reduce pressure drop, permit mass transfer, and minimize attrition. Fast diffusion of gas in and out of the pelleted sorbent is necessary, as well as high attrition resistance, and finally low cost. The sorbent should have high volumetric capacity, good selectivity, and reasonable heats of adsorption. The heat of adsorption is important because as gas sorbs, the sorbent heats up and thus alters the shape of the isotherm. As gas is released, the sorbent cools again and thereby shifts the isotherm and hence its capacity (*refer to* later discussion about Figure 8).

The properties of cation-substituted X zeolites for air separation are described in Table I and Figures 2 and 3. Surprisingly, among the divalent cations Ca^{2+} is preferred over Mg^{2+} or Sr^{2+} for X-type zeolites (references 5 and 8 give details).

Both Union Carbide and Air Products received patents on the use of lithium cations in zeolites (10, 11). For chabazite-type zeolite, the use of

Table I. Properties of Zeolites for Air Separation

Adsorbent	N_2-O_2 Selectivity, 30 °C 1 atm[a]	Capacity [mL(STP)/g], 30 °C, 1 atm		Average N_2 Isosteric Heat (kcal/mol), 0–1 atm[b]
		N_2	O_2	
CaA	4.6	14	4.4	6.5
CaX[c]	8.1	19	5.7	7.1
CaLSX[d]	10.4	26	7.0	7.7
Na mordenite[c]	4.2	15	4.7	6.4
Li chabazite[c]	5.8	25	6.8	7.2

[a]Calculated from pure gas isotherms using ideal solution adsorption theory.
[b]Calculated from isotherm data at 30 and 45 °C.
[c]Reference 12.
[d]Reference 5.

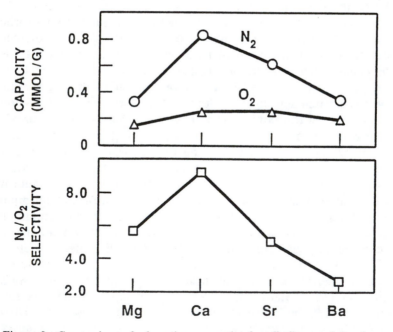

Figure 2. Comparison of adsorption properties for alkaline earth ion forms of X zeolite.

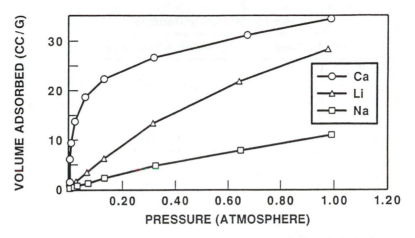

Figure 3. Nitrogen capacities of alkaline cation substituted chabazites.

lithium cations generates a more linear isotherm (Figure 3) and allows the working capacity of the sorbent to be improved for bulk air separation *(12)*.

Carbon Molecular Sieves. These are not traditional activated carbons, but they do represent a subset of a unique type of activated carbons. They are largely bimodal in pore size distribution, consisting of macropores (needed for gas transport) and a high micropore volume at <8 Å. Unlike zeolites they separate air on the basis of their kinetic selectivity of O_2 over N_2 *(11)*. O_2 is a slightly smaller molecule (3.46 Å) versus N_2 (3.64 Å) (Table II). Unlike zeolites, CMS selectively sorb the O_2, which is more economic for the production of high-purity N_2. (There is less O_2 to sorb from air than N_2.)

Activated carbons are often prepared from a natural carbon source (e.g., coal, coconut shells, pitch, and rice hulls) by heating in the absence of O_2 to produce a "char". This char is activated often by treatment with steam at ~900 °C and then with air at <400 °C to produce an amorphous carbon with a great deal of surface functionality. CMS are prepared by making the char at ~600–900 °C to produce a quasi-graphic structure containing a highly disordered array of carbon graphitic planes (Figure 4) *(13)*. These materials are then treated with hydrocarbons to close down the entrances to the micropores to produce a size-selective sieve. Activated carbons often contain a large amount of meso pores, but O_2-selective CMS materials are optimized for micropore capacity below 4 Å. The object is to maximize volumetric capacity for O_2 selectivity with respect to N_2. This maximizing is done with a great deal of precision to reproducibly

Table II. Molecular Dimensions

Molecule	Lennard–Jones, 1σ (Å)
He	2.6
H_2	2.89
Ne	2.75
Ar	3.40
O_2	3.46 (2.8)[a]
N_2	3.64 (3.0)[a]
Kr	3.60
Xe	3.96
CO_2	3.3
CH_4	3.8
n-C_4H_{10}	4.3
iso-C_4H_{10}	5.0
Neopentane	6.2
Benzene	5.85 (3.7 × 7.0 Å)
Cyclohexane	6.0 (4.8 × 6.8 Å)

[a]Van der Waals.

SOURCE: Reference 6.

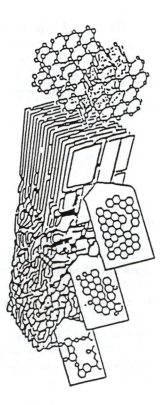

Figure 4. Evolution from a carbonaceous material to graphite.

manufacture materials in tonnage quantities able to distinguish between the ~0.2-Å difference between O_2 and N_2.

The capacity by CMS for O_2 is almost twice that of a 5A zeolite (Tables III and IV); however, allowed to reach equilibrium, CMS materials are slightly O_2-selective. O_2 diffuses faster into the CMS micropores than N_2, so the process runs on very short cycles of air pressurization versus depressurization (~60 s) to sorb much of the O_2 and allow fairly pure N_2 to emerge from the bed (Figure 5).

Table III. Adsorption Capacity of CMS versus Zeolite 5A

Sieve	O_2	N_2	Ar
CMS	8	8	~7.3
Zeolite 5A	~4.3	~12	2.8

NOTE: All values are the amount adsorbed in milliliters per gram at 25 °C and 760 mmHg.

Munzner, Juntgen, and co-workers (*14, 15*) described one preparation of these materials from coal as follows: Mineral coal is activated in 12% O_2 at 230 °C, mixed with 23% soft pitch-water, extruded into ~1/16-inch pellets, heated to 800 °C at ~10 °C/min, and post-treated with benzene (5–10%) vapor at 800 °C for ~20 min to produce hard, granular 2-mm extrudates. These materials can produce 80% enriched O_2 in air or 99% N_2 from air.

Alternatively, coconut shell based CMS are described by Kanebo, Takeda, and Kuraray Chemical Companies. In one U.S. patent (*16*), Ohsaki and Abe describe the treatment of a char derived from coconut shells: treated with coal tar pitch, pelletized, carbonized to 600–900 °C, immersed in mineral acid, washed and dried, impregnated with creosote, heated to 600–900 °C for 10–60 min, and cooled under inert gas. Additional work by Ozaki and Kawabe (*17*) shows that both capacity and selectivity continues to develop between 400 and 700 °C (Table V).

These CMS materials contain no observable level of significant surface functionality. N_2 Brunauer–Emmett–Teller (BET) measurements give rise to almost zero surface area; hence, using BET surface areas to assess the appropriateness of these materials is unwise because the best materials need to have a high degree of porosity below 4 Å. Since N_2 BET is measured at liquid N_2 temperatures, naturally one obtains small values for surface area of CMS materials suitable for air separation. Pore volumes are typically ~0.45 mL/g with micropore volumes of ~0.12 mL/g. It is

Table IV. General Properties of Adsorbents

Property	CMS	Activated Carbon	Zeolite	SiO_2 Gel	Al_2O_3
Particle density (g/cm^3)	0.9–1.1	0.6–1.0	0.9–1.3	0.8–1.3	0.9–1.9
Packed density (g/cm^3)	0.55–0.65	0.35–0.6	0.6–0.75	0.5–0.85	0.5–1.0
Pore volume (mL/g)	~0.5	0.5–1.1	0.4–0.6	0.3–0.8	0.3–0.8
Surface area (m^2/g)	500	750–2000	400–750	200–600	150–350
Mean pore diameter (Å)	4–7	12–20	—[a]	20–120	40–150
Pore shape	slit-cavity ink bottle	veiny	linked ink bottle	—[b]	—[b]

[a]Depends on zeolite.
[b]Irregular.

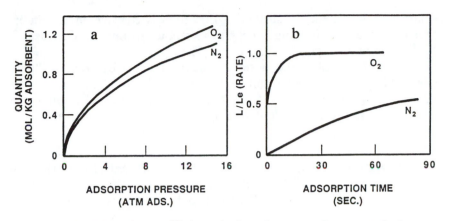

Figure 5. Adsorption equilibria and adsorption rates of oxygen and nitrogen with CMS.

**Table V. O_2 Capacity and Selectivity of CMS
as a Function of Temperature**

Temperature of Heat Treatment (°C)	Equilibrium O_2 Capacity [mL(STP)/g]	Selectivity[a]
Raw material	8.0	1.3
120	3.1	10
400	4.2	16
600	6.9	25
700	7.0	35
Kuraray, 1984 (coal tar)	6.0	26
Use of benzene	7.9	1.3
Xylenol	6.8	35

[a]Volume of O_2 adsorbed in 5 s at 25 °C and 1 atm; then time of adsorption of equivalent volume of $N_2 \div 5$.

desirable to increase the micropore volume at the expense of some of the macropore volume.

A recent family of patents *(18–20)* details critical parameters for the hydrocarbon deposition step to provide the required selectivity. A two-step process of a volatile hydrocarbon deposition ensures precise deposition of carbon at the pore mouths, forming size-selective gates. Size-selective gates ensure that the openings of the micropores are narrowed without filling the micropores themselves. Selective pyrolysis of a molecule that is too large to penetrate the micropores produces a highly selective CMS. First, a larger hydrocarbon is cracked to narrow the mi-

cropores to ~4 Å, then a smaller hydrocarbon is cracked on this intermediate product to fine-tune the pores without a dramatic decrease in
O_2-sorption rates (Figure 6).

Figure 7 describes the pore size distribution (measured by "plug
gauge" adsorption of probe molecules) of O_2-selective (3A-type) versus
nonselective 4A- and 5A-type CMS. Remarkably, the slight size difference
in 3A versus 4A in the 4- to 4.5-Å region is very crucial to their performance.

Production of O_2. The zeolites and CMS described are primarily
used for the production of N_2. PSA cycles can also be used to produce
O_2, but here A- and X-type zeolite sorbents have been used commercially

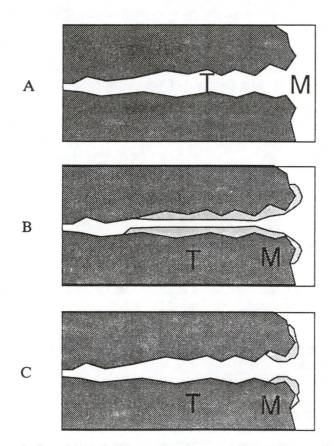

*Figure 6. Deposition of carbon onto–into a pore. Key: A, uncoated mi-
cropore of ~5 Å; B, deposition of carbon via pyrolysis of isobutylene into
the throat (T) of the pore; and C, deposition of carbon via the pyrolysis of
isobutylene into some of the throat (T), but mainly at the pore mouth (M).*

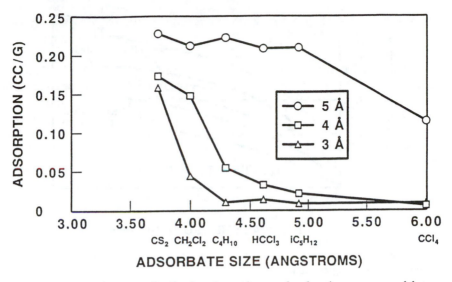

Figure 7. Micropore distribution for carbon molecular sieves measured by the molecular probe method.

(*21*). A process using 5A zeolite operating between 1 and 3 atm can produce 90% O_2 (dry and CO_2-free) at ~1 atm.

PSA Process. The most popular process for gas separation using solid sorbents is the PSA process. For the production of N_2 with a zeolite 5A or 13X bed (*22*), two beds of sorbent are operated as follows:

1. To a regenerated bed, dry feed air is introduced at one end; the other end is closed to bring the column to its highest pressure.
2. At the high pressure, additional feed is introduced, and an O_2-rich product is released at the product end. Just as N_2 breakthrough occurs, the feed is stopped.
3. The bed is countercurrently depressurized to low pressures to release most of the adsorbed N_2.
4. The bed is diluted by some product gas to regenerate the bed.

A number of additional process steps (*22, 23*) can be employed to optimize the recovery and productivity of the process, focusing on the use of as little adsorbent (on a volumetric basis) as possible. The change in pressure (Figure 8) gives rise to a release of gas that is dependent upon the associated temperature change in the bed (*23*). Ideally, this amount of gas should be as large as possible. A more linear isotherm gives rise to a larger working capacity.

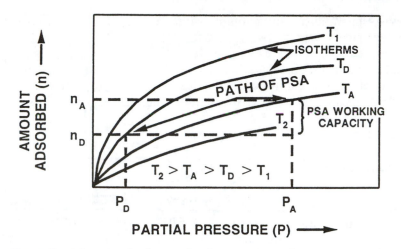

Figure 8. Adsorption isotherms showing pressure and temperature swings and PSA working capacity.

For CMS sorbents, O_2 is the sorbed species and N_2 is collected at the exit of the bed. Desiccants (alumina or silica gel) and activated carbon are often used as guard or pretreatment beds prior to the working sorbent. Typically 99.5% N_2 is produced via a two-bed unit with economics being optimal at ~2000 to 30,000 ft^3/h. Electricity to drive the compressors is 40–50% of the operating costs (4).

Summary

Zeolite molecular sieves are crystalline materials used for the production of O_2 or N_2 by PSA adsorption processes. The high electric fields within the zeolite cavity result in strong adsorption of the N_2 (via its larger quadrupole moment) by the zeolite. On the other hand, the use of CMS for the production of N_2 depends upon a kinetic, size-selective separation of the smaller O_2 molecule from N_2. CMS materials are amorphous compounds formed by the decomposition of hydrocarbons that deposit carbon at the micropore entrances to an activated carbon or char. Both these materials are being actively pursued for the production of new generations of air separation sorbents.

For CMS materials, one key goal is to get as high a volumetric capacity as possible while maintaining selectivity of O_2 versus N_2. There is still ample room for improvement of these materials with regard to future sorbents for air separation. Zeolites have been used for other size-selective separations, but CMS have seen only limited use beyond air separation (15). In this regard, CMS materials may have potential for other size-selective separations.   In addition, with new zeolites and nonzeolitic molecular sieves continually being discovered as well as new techniques [such as chemical vapor deposition (CVD) on zeolites (*24*)], improved molecular sieves for air separation may still be discovered.   The air separation process itself is controlled by the quality of the sieve material.   Better sieves will mean better processes for producing billions of pounds of O_2 and N_2 every year!

Acknowledgments

I acknowledge Charles G. Coe, my teacher and colleague in the field of adsorption science over the past 7 years, for all his help and efforts and for his help in compiling Table I.   In addition, our CMS synthesis work was the result of a substantial team effort by a large group of people including S. R. Auvil, T. A. Braymer, A. L. Cabrera, Charles G. Coe, T. S. Farris, T. R. Gaffney, J. D. Moyer, J. M. Schork, S. R. Srinivasan, and J. E. Zehner.   This manuscript is dedicated to their efforts.

References

1. Jasra, R. V.; Bhat, S. G. T. *Sep. Sci. Technol.* **1988**, *23*, 945–89.
2. Shelley, S. *Chem. Eng.* June 1991, pp 30–39.
3. Thorogood, R. M. *Gas Sep. Purif.* **1991**, *5*, 83–94.
4. Basta, N. *Chem. Eng.* Sept. 26, 1988, 26–31.
5. Coe, C. G. In *Gas Separation Technology;* Vansant, E. F.; Dewolfs, R., Eds.; Elsevier Scientific Publishers: Amsterdam, Netherlands, 1989; p 149.
6. Breck, D. W. *Zeolite Molecular Sieves;* John Wiley and Sons; New York, 1974; p 650.
7. Ruthven, D. M. *Chem. Eng. Prog.* Feb. 1988, 42–50.
8. Siegel, H.; Schollner, R.; Hoffmann, J.; Oehme, W. *Z. Chem.* **1989**, *29*, 77.
9. Coe, C. G.; Parris, G. E.; Srinivasan, R.; Auvil, S. R. Proc. In *Int. Zeolite Conf. 7th (Tokyo);* Iijima, A., Ward, J. W., Murakami, Y., Eds.; Elsevier: New York, 1986; p 1003.
10. Coe, C. G.; Gaffney, T. R.; Srinivasan, R. S. U.S. Patent 4 925 460, 1990.
11. Chao, C. C. U.S. Patent 4 859 217, 1989.
12. Gaffney, T. R.; Coe, C. G.; Srinivasan, R.; Naheiri, T. Presented at the British Zeolite Association Meeting, June 1990.
13. Marsh, H.; Griffiths, J. *Proceedings of the International Symposium on Carbon, New Processes, and New Applications;* Kagaku Gijutsu-sha: Toyohashi, Tokyo, Japan, 1982; p 81.
14. Munzner, H.; Heimbach, H.; Korbacher, W.; Juntgen, H.; Knoblauch, K.; Zundorf, D. U.S. Patent 3 801 513, 1974.
15. Juntgen, H.; Knoblauch, K.; Harder, K. *Fuel* 1981, *60*, 817.

16. Ohsaki, T.; Abe, S. U.S. Patent 4 742 040, 1988.

17. Ozaki, T.; Kawabe, T. Jpn. Kokai Patent No. Sho 62 (1987)–176908.

18. Cabrera, A. L.; Armor, J. N. U.S. Patent 5 071 450, 1991.

19. Gaffney, T. R.; Farris, T. S.; Cabrera, A. L.; Armor, J. N. U.S. Patent 5 098 880, 1992.

20. Armor, J. N.; Braymer, T. A.; Farris, T. S.; Gaffney, T. R. U.S. Patent 5 086 033, 1992.

21. Sircar, S. *Sep. Sci. Technol.* 1988, *23*, 1379.

22. Baron, G. V. In *Gas Separation Technology;* Vansant, E. F.; Dewolfs, R., Eds.; Elsevier Science Publishers: Amsterdam, Netherlands, 1990; pp 137–148.

23. Jasra, R. V.; Choudary, N. V.; Bhat, S. G. T. *Sep. Sci. Technol.* 1991, *26*, 885.

24. Chu, Y. F.; Keweshan, C. F.; Vansant, E. F. In *Zeolites as Catalysts, Sorbents, and Detergent Builders;* Karge, H. G; Weithamp, J., Eds.; Elsevier Science Publishers: Amsterdam, Netherlands, 1989; p 749.

RECEIVED for review November 9, 1992. ACCEPTED revised manuscript March 30, 1993.

14

Nanomaterials: Endosemiconductors and Exosemiconductors

Geoffrey A. Ozin

Advanced Zeolite Materials Science Group, Lash Miller Chemical Laboratories, University of Toronto, Toronto, Ontario, Canada M5S 1A1

Our group has developed inorganic- and organometallic-based synthetic methods that lead to interesting new classes of solid-state nanomaterials, which we refer to as endosemiconductors and exosemiconductors. Here we essentially rearrange the atomic components of well-known bulk semiconductors into periodic assemblies of nanometer-dimension, single-size, and single-shape clusters, housed inside a crystalline nanoporous host material. This method yields a class of nanomaterials considered to be a semiconductor cluster crystal, an expanded semiconductor, or a semiconductor quantum supralattice, which we call an endosemiconductor. If instead the atomic constituents of a bulk semiconductor are juxtaposed into the form of an open-framework nanostructure, the class of material obtained is considered to be a crystalline nanoporous semiconductor or a framework semiconductor, which we call an exosemiconductor.

Nanomaterials: What They Are and Why We Need Them

Nanocomputers built of chips having nanometer-dimension electronic and optical components and interconnects, operating at fantastic speeds with high-mobility electrons or at the speed of light with photons, currently provide the main driving force behind the search for nanomaterials with specific structure–property–function relationships (1). However, a major scientific challenge is to reproducibly produce uniform arrays of single-size

0065–2393/95/0245–0335$16.25/0

and single-shape nanoscale objects in the desirable range of 10–300 Å (2–5). Materials that fall in this size domain, which have dimensions lying between those of molecules (<10 Å) and the bulk (>300 Å), are especially interesting because they are predicted to display novel properties largely as a consequence of their finite small size, dimensionality, organization, and interactions with each other, as well as the surrounding environment (1–5). These properties include enhanced quantum electronic and excitonic optical nonlinearities as a consequence of the spatial, quantum, and dielectric confinement of electrons, holes, and excitons (so called quantum size effects, QSEs) with respect to the nanometer-dimension length scales of the materials (6–7).

The Nanochemistry and Nanophysics Approach to Nanomaterials

In this context, nanochemistry, as opposed to nanophysics, is an emerging subdiscipline of solid-state materials chemistry. It emphasizes the synthesis rather than the engineering aspects of preparing minute pieces of matter with nanometer sizes in one, two, and three dimensions (8). Nano-chemists work toward this objective from the "atom up", whereas nano-physicists operate from the "bulk down". Building and organizing nano-objects under mild and controlled conditions one atom at a time instead of manipulating the bulk should in principle provide the nanochemist with simple, reproducible, and cheap synthetic approaches to materials of perfect atom sizes and shapes, rather than having to use the complex and costly instrumental methods of the nanoengineering physicist. Especially in the size requirement below about 300 Å, nanoengineering physicists usually have intrinsic limitations with respect to the size, dispersion, and the degree of perfection attainable for the final product, as well as the ability for large-scale manufacturing (5).

Nanophysics fabrication methods have improved remarkably over the past few years (9–11). For example, using a combination of sophisticated metal–organic chemical vapor deposition (MOCVD); molecular beam epitaxy (MBE) planar engineering (deposition); and scanned optical, X-ray, ion-, and electron-beam lateral engineering (lithographic) techniques, nanoengineering methods can routinely produce submicrometer-scale objects with essentially any desired architecture. To illustrate this accomplishment, a state-of-the-art quantum dot array, with individual component dots having dimensions of the order of 200 ± 50 Å, is displayed in Figure 1.

To go beyond this practical engineering limit, one must resort to the nanotips of scanning probe microscopes for the simultaneous manipula-

Figure 1. State-of-the-art, reactive ion-beam etching of an organized array of approximately 200-Å diameter quantum dots. (Reproduced with permission from reference 9. Copyright 1989 Academic Press.)

tion and imaging of atomic and nanoscale objects (2). This technique probably represents the ultimate limit in nanoengineering physics and miniaturization and provides access to materials in the size regime at which quantum size effects (QSEs) in superconducting, metallic, semiconducting, and insulating objects can be controlled and exploited for the fabrication of nanoscale electronic and optical devices. But local atomic-scale surface modification will not be so easy in practice. Serious obstacles concerning nanoengineering speed, nanocomponent size–shape–position reproducibility, and accessible chip size need to be overcome before practical atomic-scale logic and memory devices are realized using the sharp tips of scanning probe microscopes (2, 9–11).

Chemists, on the other hand, pride themselves in being able to synthesize small and perfect molecule-size (<10 Å) objects. However, to be able to routinely and reproducibly make 10–300-Å atomically perfect nanostructures, useful in future-generation electronic and optical devices, chemists have to develop new types of synthetic methods that have the ability to not only assemble but also position these tiny objects in appropriately organized and coupled arrays. The chemical alternatives for meeting this challenge and building such nanoscale devices involve a new way of thinking. Two possibly useful approaches involve what are called "patterning and templating" methods. In patterning, nanolithography is used to spatially define physically or chemically active foundation sites usually on planar substrates, upon which subsequent site-specific chemical synthesis allows the growth of nanoscale objects (12). Templating, on the other hand, exploits the preexisting, perfectly periodic, single-size and single-shape void spaces found in nanoporous materials for performing host–guest inclusion chemistry (13–17). Both approaches benefit from the principle of synthesis and self-organization in preexisting nanoscale regions on a planar substrate or within a porous solid (8).

Endosemiconductors and Exosemiconductors: New Types of Nanomaterials

Several new types of nanomaterials have recently emerged from my research group. In particular, we employed metal–organic chemical vapor deposition (MOCVD) self-assembly host–guest inclusion chemistry (18–20) and template-mediated hydrothermal chemistry (21, 22) for synthesizing new classes of QSE nanomaterials, which we refer to as endosemiconductors and exosemiconductors, respectively (Figure 2). Endosemiconductors are composed of the atomic constituents of traditional bulk semiconductors like AgBr, CdSe, Ge, TiO_2, SnS_2, and WO_3, but are instead reorganized into periodic assemblies of monodispersed nanoclus-

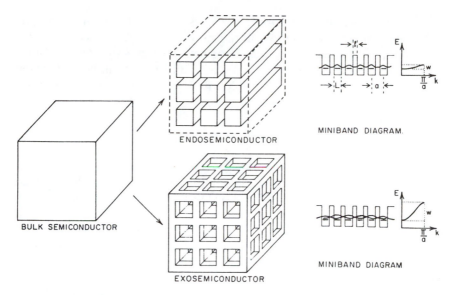

Figure 2. "Reconstitution" of a bulk semiconductor into an endosemiconductor and to an exosemiconductor. Qualitative miniband diagrams, constructed from electronic coupling between periodic arrays of electron-hole-in-a-box wave functions, for endosemiconductors (weak coupling) and exosemiconductors (strong coupling) are shown, together with appropriate energy E(k) versus wave vector k dispersion curves. (Reproduced with permission from reference 8. Copyright 1992.)

ters, housed within a crystalline nanoporous wide-based, gap-insulating host lattice, which usually comprises a zeolite or molecular sieve. Exosemiconductors can be most easily conceptualized as the geometric complement or "inside-out-version" of endosemiconductors, in which the semiconductor atomic constituents are instead juxtaposed into the form of a crystalline, open-framework nanostructure.

Interestingly, in the 1992 nanoengineering physics literature, structures that resemble these kinds of endo- and exosemiconductor materials are visualized as crystal lattices of quantum dots and quantum antidots (and wires) (5). The similarities between the properties of these nanostructures have yet to be demonstrated. Electronic coupling between nanoclusters in endosemiconductors is thought to operate either directly through the bonds of the host framework or indirectly through space (Figure 2). These interactions will likely decide the nature of the electronic band structure of the endosemiconductor and its potential value in, for example, quantum electronic, nonlinear optical, and photonic devices, like resonant tunneling transistors, transphasors, and multiple quantum dot lasers, respectively (discussed later).

In exosemiconductors, carriers are considered to be spatially and possibly quantum-confined to the wall thickness of the open-framework structure, the key difference being that the electronic band structure and optical and electronic properties derive from direct through-bond coupling (Figure 2). These materials probably will demonstrate molecule-discriminating transistor behavior and light-emitting diode—laser action for exosemiconductor based npn- and np-junctions, where n and p refer to electron and hole doping, respectively (discussed later). This behavior makes the development of chemoselective nanoscale devices an interesting and possible scenario (22).

In our work, various synthetic approaches for tuning the structure and composition of these QSE nanomaterials have been devised to engineer the degree of quantum, spatial, and dielectric confinement of charge carriers, their electronic band gaps, doping levels, and conductivities, so as to achieve desirable electronic and optical functions. The development of innovative methods for making and characterizing these kinds of nanomaterials in the form of large, perfect single crystals (0.4–5 mm) (23–25) and multicrystalline and oriented single-crystal thin films (25–28) are considered pivotal for any future production of nanoscale electronic, optical, and photonic devices and circuitry based on these materials.

Nanochemistry Synthesis and Nanophysics Fabrication Approaches to Endosemiconductors and Exosemiconductors

Within the context of endo- and exosemiconductor materials, it is useful to think about the nanophysics fabrication philosophy and the nanochemistry synthesis method for achieving a semiconductor quantum dot array and a nanoporous semiconductor, as illustrated in Figures 3 and 4, respectively. The scheme shown in Figure 3 brings forth the interrelationship between the commonly encountered planar deposition and lateral engineering methodologies of the semiconductor nanophysicist and our particular type of host—guest inclusion semiconductor nanochemistry for producing an organized assembly of, for example, GaAs quantum dots.

The nanophysicist arranges the atomic components of GaAs from $Me_3Ga—AsH_3$ gaseous precursors by using a one-step MOCVD procedure, followed by scanning-beam nanolithographic procedures, to form an array of GaAs quantum dots on a planar substrate. Our nanochemistry route involves a two-step MOCVD encapsulation—anchoring reaction sequence of events to create an assembly of GaAs quantum dots (nanoclusters) inside a crystalline nanoporous host lattice.

The ability to create lattice-matched quantum dot arrays of precise size and shape benefits enormously from a detailed knowledge of the

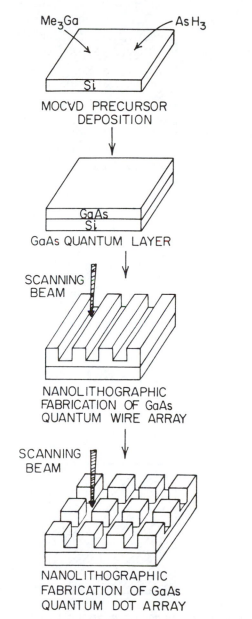

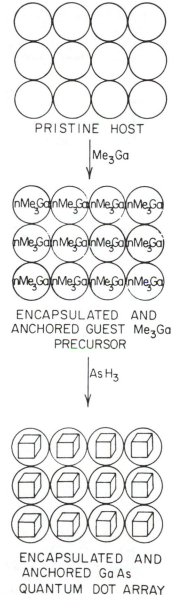

Figure 3. Nanophysics fabrication compared to nanochemistry synthesis of a GaAs quantum dot array. (Reproduced with permission from reference 8. Copyright 1992.)

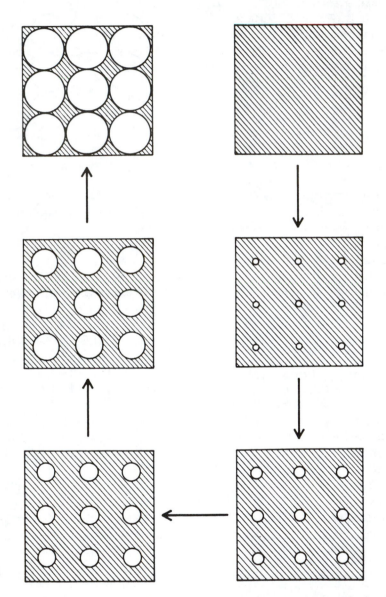

Figure 4. Nanophysics fabrication compared to nanochemistry synthesis of a nanoporous semiconductor. (Reproduced with permission from reference 8. Copyright 1992.)

structural, energetic, and dynamic aspects of the surface chemistry of the precursors used in MOCVD routes to such nanoengineered structures (*29*). It is equally important to apply reasoning of a similar type to the structural, anchoring (coordination and capping), and reactivity properties of the internal surfaces of nanoporous hosts toward MOCVD-type precursors and semiconductor nanocluster products (*30*).

By contrast, the scheme shown in Figure 4 attempts to compare the approach of the nanochemist and nanophysicist to the fabrication of a nanoporous semiconductor. Thus beginning at the top left of Figure 4, the nanophysicist can create an organized array of nanometer-dimension holes in a thin film or wafer of a semiconductor using, for example, chemical–electrochemical dissolution or scanning beam techniques (*11*). A Euclidean idealistic, rather than a more realistic fractal view, of these processes to create a nanoporous semiconductor is shown in Figure 4. Exceedingly small initially formed pores less than 200 Å wide can be grown. At low porosities, an array of noninteracting cylindrical pores runs perpendicular to the surface. In practice, some variation is to be expected in pore size, shape, and separation. Increasing the porosity further eventually causes merging of adjacent pores and the creation of arrays of isolated semiconductor columns. This approach clearly has the ability to transform a bulk semiconductor first into a nanoporous semiconductor and then into an array of nanodimension semiconductor objects.

A recent and exciting case in point concerns nanoscale silicon-based structures that have been shown to have the intriguing ability to emit visible light when optically or electrically excited (*31–33*). The mechanism of light emission is still somewhat controversial (*34, 35*). However, current thinking is that it originates from QSEs (rather than surface impurities, defects, or siloxenes) that have somehow altered the electron–hole indirect band-gap behavior of bulk Si that serves to produce only "useless" heat in the form of phonons, to that of a modified band-gap material that instead yields "useful" photons (*33*).

From the standpoint of a nanochemistry approach to crystalline nanoporous semiconductors, numerous inorganic solids can be fabricated with open structures of a one-dimensional (1-D) channel, two-dimensional (2-D) layer, and three-dimensional (3-D) framework type. Classic examples of these include the tungsten bronzes (WO_3) and Chevrel phases (Mo_6Se_8) (*36, 37*). Their electronic and optical properties are readily tuned through chemical or electrochemical doping methods to the semiconducting, metallic, and superconducting states. Their entrance window dimensions and internal void spaces are usually too small to allow entry of anything other than small and simple alkali metal ion guests.

However, the recent discovery of template-mediated hydrothermal synthetic routes to "nonoxide" zeotype framework materials, such as those based on tin(IV) and antimony(III) chalcogenides (*21, 22, 38*), makes possible the self-assembly synthesis of perfectly crystalline nanoporous solids, built up from the atomic components of bulk semiconductors having molecule-dimension window and internal void spaces, of the type illustrated in Figures 2 and 4. With the wide range of easily accessible source materials, "one-pot" syntheses could lead directly to crystalline nanoporous semiconductors (intrinsic and extrinsic), metals, and even superconductors.

Nanometer-Sized Semiconductor Clusters

As alluded to earlier, nanometer-dimension clusters of uniform size and shape, composed of the atomic components of bulk semiconductors have evoked great interest in the solid-state chemistry and condensed-matter physics communities because of their unique electronic and optical properties that are a consequence of quantum, spatial, and dielectric confinement effects involving exciton, electron, and hole carriers (1–11). One impediment to the practical utilization of these kinds of materials revolves around the synthetic challenge of preparing organized arrays and narrow size distributions of the clusters. In the ideal case of a perfect semiconductor "cluster" crystal, deleterious inhomogeneous size and shape contributions to their electronic and optical properties are reduced to a minimum, and thus optimum evaluation and exploitation of confinement effects is permitted.

Host–guest inclusion chemistry provides an attractive method for achieving the goal of monodispersion for these nanoscale semiconductor cluster materials. The selection of host has encompassed crystalline and amorphous organic, inorganic, polymeric, and biological materials (8, 13, 17). Here chemical protection and stabilization (capping), together with spatial restrictions imposed by the host, allow the nanochemist to synthesize, nucleate, grow, and arrest the "crystallization" of semiconductor clusters in the desirable nanometer QSE regime. The following sections describe some recent semiconductor nanochemistry in zeolite hosts, with examples taken mainly from our own research.

Assembly of Endosemiconductors

Zeolite and molecular-sieve hosts allow the ordered inclusion of nanoscale semiconductor clusters with dimensions in the range of 6 to 30 Å and having a very uniform size distribution (8, 39–43). Using liquid crystal templates, it has recently proven possible to synthesize ultralarge-pore, molecular-sieve type materials with channel structures having dimensions tunable over the range of about 15 to 100 Å (44). As a result of this seminal breakthrough, one can anticipate the discovery of a new generation of 15–100-Å mesoscopic materials.

Different architecture hosts permit fine control over the spatial arrangement of, and interactions between, these quantum size clusters. The zeolite framework provides a stabilizing medium, constraining the cluster growth to nanometer dimensions. Modification of framework charge density and intracavity electric-field gradients by manipulation of the host composition and extraframework cations enables one to fine-tune the electronic and optical properties of the encapsulated clusters (dielectric confinement effect).

Nanochemistry techniques for synthesizing zeolite-encaged semiconductor clusters are summarized as follows, with a selected example for each case. In these examples, SOD is sodalite, A is zeolite A, and Y is zeolite Y.

- aqueous ion exchange (*45*)

$$Na_{56}Y \xrightarrow{Cd^{2+}} Cd_{28}Y \xrightarrow{H_2S} (CdS)_{28}H_{56}Y$$

- metal ion exchange (*46*), where SOD is sodalite

$$8Na,2Cl\text{-SOD} \xrightarrow{Ag^+} 8Ag,2Cl\text{-SOD}$$

- vapor-phase impregnation (*47*), where A is zeolite A

$$Na_{12}A \xrightarrow{PbI_2} (PbI_2)_n Na_{12}A$$

- intrazeolite MOCVD (*18–20*), where Y is zeolite Y

$$H_{56}Y \xrightarrow{Me_2Cd} (MeCd)_{48}H_8Y \xrightarrow{H_2Se} (Cd_6Se_4)_8H_{24}Y$$

- intrazeolite CVD (*48*)

$$n\{Si_xH_{2x+2}\}-H_{56}Y \xrightarrow{\Delta} Si_{nx}-Y$$

- phototopotaxy (*49–56*)

$$n\{W(CO)_6\}-Na_{56}Y \xrightarrow{O_2} n\{WO_3\}-Na_{56}Y$$

We will illustrate the methodology with reference to the MOCVD type synthesis of Groups II–VI and IV–VI metal chalcogenide clusters inside zeolite Y supercages (18–20), the phototopotactic synthesis of tungsten oxide supralattices in zeolite Y (49–56), and sodalite supralattices (46, 57–65).

Stepwise Synthesis of II–VI Metal Chalcogenide Nanoclusters Inside the Supercages of Zeolite Y Using MOCVD-Type Precursors

In this section we describe some of our recent work on the self-organization of MOCVD-type precursors and their subsequent transformation to uniform arrays of single-size and single-shape semiconductor nanoclusters inside the diamond network of 13-Å diameter supercages found in zeolite Y (18–20). MOCVD is traditionally a "planar" technique for the deposition of very thin layers of lattice-matched materials onto a surface from the vapor phase (epitaxy) (9–11, 29). This approach has been modified in our research to "deposit" MOCVD precursor materials necessary for the in situ synthesis of semiconductor nanoclusters onto the "curved" internal surface of the zeolite Y supercage (topotaxy) (30, 40–44). Here the oxide framework is considered to act as a macrospheroidal, multisite, multidentate "zeolate" ligand toward both the MOCVD-type precursors and the resulting semiconductor nanoclusters products (30). This "crown ether zeolate ligand analogy" allows one to better understand and exploit the reactivity, complexing, coordinating, structure-directing, stabilizing, and capping properties of the zeolite internal surface for the anchoring and self-assembly of MOCVD-type precursors and semiconductor nanocluster products (18–20).

The MOCVD approach provides the nanochemist with a mild, controlled, and versatile method for the synthesis of semiconductor nanoclusters in zeolite Y. One can select precursors that are too large to enter the sodalite cages, thereby directing the precursor anchoring, cluster self-assembly, and deposition processes specifically to the supercages of the zeolite Y host. The loading of precursor is controlled through its steric requirements and reactivity with respect to the host lattice. This governs the chemistry leading up to the desired semiconductor nanocluster product (18–20).

The anchoring of $(CH_3)_n M$ (M is metal) MOCVD-type precursors to specific crystallographic sites in the supercage of zeolite Y proceeds according to the following reaction:

$$Z-OH + (CH_3)_n M \longrightarrow ZO-M(CH_3)_{n-1} + CH_4 \qquad (1)$$

The in situ mid-IR spectrum of the protons in zeolite Y allow close monitoring and quantification of the anchoring reactions. Protons can move from sodalite cages to the adjoining supercages, but there can be no net migration through the structure that would leave areas of unbalanced charge. Therefore, they report individually on the centers they occupy and the number of protons reacted. In conjunction with the number of CH_4 molecules simultaneously evolved and the elemental analysis of the materials, the Me_2Zn-, Me_2Cd-, and Me_4Sn-anchored precursors in acid zeolite Y are found to contain six methylzinc species, six methylcadmium species, or four dimethyl–trimethyltin species per supercage, respectively, homogeneously distributed throughout the lattice (*18–20*). The structures of these precursors were determined by a multiprong approach, including low-temperature extended X-ray absorption fine structure (EXAFS) analysis and low-temperature Rietveld refinement of synchrotron powder X-ray diffraction (PXRD) data (*18–20*). These results are summarized in Figure 5.

The semiconductor cluster self-assembly process is initiated upon addition of H_2S or H_2Se to the anchored MOCVD-type precursors. As in the precursor-anchoring step, the course of this reaction was also followed by in situ mid-IR spectroscopy. Here the remaining methylmetal groups release further CH_4 to form a labile metal hydrosulfide (or selenide) species according to the reaction:

$$Z-OM(CH_3)_{n-1} + (n-1)H_2X \longrightarrow Z-OM(XH)_{n-1} + (n-1)CH_4 \quad (2)$$

where X is S or Se. This reaction is followed by dehydrosulfurization–condensation cluster self-assembly reactions described by

$$Z-OM(XH) + (HX)MO-Z \longrightarrow Z-OMXMO-Z \quad (3)$$

which are driven to completion when excess H_2X is pumped away, leaving behind the charge-balancing, semiconductor nanocluster guests. These were determined to be $Cd_6S_4^{4+}$, $Cd_6Se_4^{4+}$, $Zn_6S_4^{4+}$, $Zn_6Se_4^{4+}$, and $Sn_4S_6^{4+}$ (*18–20*). The overall reaction pathways, and EXAFS–Rietveld determined structures appropriate for these nanomaterials are also summarized in Figure 5. The "zeolate ligand" (*19, 30*) offers the thermodynamic stability of a macrospheroidal multidentate anion and the kinetic control of a structure-directing template. Thus the zeolate ligand (*8*) controls the metal chalcogenide nanocluster growth process and stabilizes the nanocluster product in much the same way as phenylate and phenylthiolate capping ligands function in the solution-phase semiconductor cluster

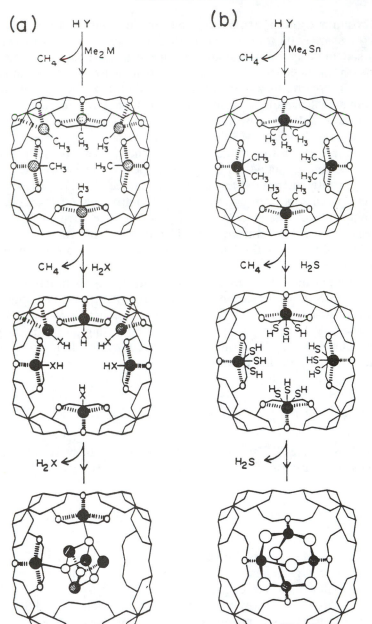

Figure 5. Intrazeolite MOCVD-type synthesis of precursors and semiconductor nanocluster products in the II–VI and IV–VI systems (a) $M_6X_4^{4+}$ (where M is Zn or Cd and X is S or Se) and (b) $Sn_4S_6^{4+}$ housed in the supercages of zeolite Y (18–20). (Reproduced with permission from reference 19. Copyright 1992.)

synthetic approaches employed by Steigerwald and Brus (*6*) and Herron and co-workers (*44, 66*).

Endosemiconductors of this type are envisioned to combine the properties of quantum, spatial, and dielectric confinement of bulk semiconductors. The physics literature teaches one to anticipate enhanced quantum electronic and excitonic optical nonlinearities for endosemiconductors whose guest clusters have dimensions comparable to electron, hole, and exciton length scales of the bulk semiconductors and dimensionalities of zero, known as quantum dots (*6–11*). Therefore, endosemiconductor single crystals and thin films could possibly be designed and fabricated to function as, for example, resonance tunneling multiple quantum dot transistors and nonlinear optical switches called transphasors. These speculative ideas are sketched in Figure 6.

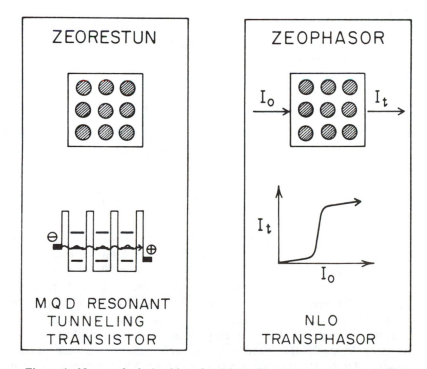

Figure 6. Nanoscale device ideas for Cd_6S_4–Y endosemiconductors. MQD is multiple quantum dot; NLO is nonlinear optical; zeorestun and zeophasor are zeotype-based device ideas as discussed in reference 8. (Reproduced with permission from reference 8. Copyright 1992.)

Synthesis of Redox-Interconvertible Tungsten Oxides Inside the Supercages of Zeolite Y

The volatile hexacarbonyl of tungsten has been used as a precursor in the synthesis of highly organized assemblies of molecular-dimension tungsten oxides, encapsulated exclusively within the supercages of zeolite Y (*49–56*). Following sublimation of the precursor into the host, it is next converted in an O_2 atmosphere to WO_3 by photo-oxidation, and may be subsequently thermally reduced in vacuum to yield WO_{3-x} clusters (where $0 \leq x \leq 1$) and then reversibly oxidized by heating in O_2 at 300–400 °C. The electronic properties of these molecular metal oxides can be easily manipulated as a result of their facile redox interconvertibility, and the further capability of fine-tuning their electronic environments by choosing which charge-balancing cation is present (*49–56*).

A maximum of two hexacarbonyl precursor molecules can be anchored in each supercage of the zeolite. Following photo-oxidation, half of the supercage void volume is freed so that subsequent precursor impregnations–photo-oxidations can be carried out. Thus, the stepwise loading proceeds according to

$$16W(CO)_6 - Na_{56}Y \longrightarrow 16WO_3 - Na_{56}Y \tag{4}$$

$$8W(CO)_6, 16WO_3 - Na_{56}Y \longrightarrow 24WO_3 - Na_{56}Y \tag{5}$$

$$4W(CO)_6, 24WO_3 - Na_{56}Y \longrightarrow 28WO_3 - Na_{56}Y \tag{6}$$

$$\cdots \qquad\qquad \cdots$$

$$1W(CO)_6, 31WO_3 - Na_{56}Y \longrightarrow 32WO_3 - Na_{56}Y \tag{7}$$

Altogether, the series of materials accessible by this method and those obtained by subsequent vacuum thermal treatment conform to the general unit cell formula $n(WO_{3-x})-M_{56}Y$, where $0 < n \leq 32$, $0 \leq x \leq 1$ and M is H, Li, Na, K, Rb, or Cs. Structural and electronic details of the various tungsten oxides were elucidated through the use of high-resolution transmission electron microscopy (HR-TEM), gravimetry, EXAFS, Fourier transform infrared (FTIR), ^{29}Si magic-angle spinning nuclear magnetic resonance (MAS NMR), ^{27}Al, ^{23}Na double-rotation nuclear magnetic resonance (DOR NMR), X-ray photoelectron (XPS), ultraviolet–visible (UV–vis), and electron spin resonance (EPR) spectroscopies (*44–56*). In all of these materials, the tungsten oxide moieties are strictly confined within the internal void space of the zeolite host, and their presence results in negligible perturbation of the host lattice crystallinity or the integrity of the framework, and only very slight changes in the unit cell size (*49–56*).

For the various tungsten oxide materials, structural characterization revealed that well-defined monomeric, dimeric, and tetrameric molecular tungsten oxides $n(WO_{3-x})-Na_{56}Y$ can be produced (Figure 7). When $x = 0$, $Na(II)^+$-cation-anchored W_2O_6 dimers were observed for values of n in the range 16 to 32. In the first-stage reduction products (wherein $x = \frac{1}{2}$), the structure was a $Na(II)^+$-cation-anchored W_2O_5 dimer for $n = 16$, but a $Na(II)^+$-cation-anchored W_4O_{10} tetramer when $n = 32$. The second-stage reduction products corresponding to $x = 1$ were all of a monomeric, framework, oxygen-anchored structure type with secondary anchoring interactions between the oxygen end of the oxotungsten bonds and a $Na(II)^+$ cation, over the loading range of $n = 16-32$. Measurements made by XPS spectroscopy clearly demonstrated that the oxidation states W^{6+}, W^{5+}, and W^{4+} (representing $x = 0$, $x = \frac{1}{2}$, and $x = 1$, respectively) can be assigned to the tungsten centers in $n(WO_{3-x})-Na_{56}Y$. In the special case of the W_2O_5 dimers, XPS, NMR, EPR, and UV–vis results indicated that both tungsten centers were in the spin-paired $W^{5+}-W^{5+}$ oxidation states rather than members of a mixed-valence $W^{4+}-W^{6+}$ moiety.

Great interest in these high loading materials ($16 \leq n \leq 32$) revolves around the concept of doping, that is, being able to control the carrier concentration, and band-gap engineering, that is, being able to fine-tune the electronic band properties of these zeolite-encapsulated metal oxide semiconductor materials. Methods of achieving this include adjustment of the number of oxygen vacancies (x) and alteration of the local electrostatic fields through the substitution of different extraframework cations (M) in the surrounding supercage (*49–56*). This capability is particularly attractive when one considers that oxide materials like bulk WO_3 and WO_3^{-x} are well known for having pervasive solid-state applications such as in rechargeable solid-state batteries; electrochromic devices; smart mirrors, windows, and displays; pH-sensitive microelectrochemical transistors; chemical sensors; electrochemical cells; and selective hydrocarbon oxidation catalysts (*36*).

Sodalite Supralattices: From Molecules to Clusters to Expanded Insulators, Semiconductors, and Metals

Sodalite is an ancient material with great potential for advanced applications. In the context of solid-state chemistry and condensed-matter physics, this Federovian framework material is considered to provide a unique opportunity for studying metal–nonmetal transitions; quantum, spatial, and dielectric confinement effects; and quantum electronic and nonlinear optic phenomena in expanded insulators, semiconductors, and metals (*46, 57–65*). One of the major goals of some of our recent research was to assemble novel nanostructures by encapsulating clusters consisting of the

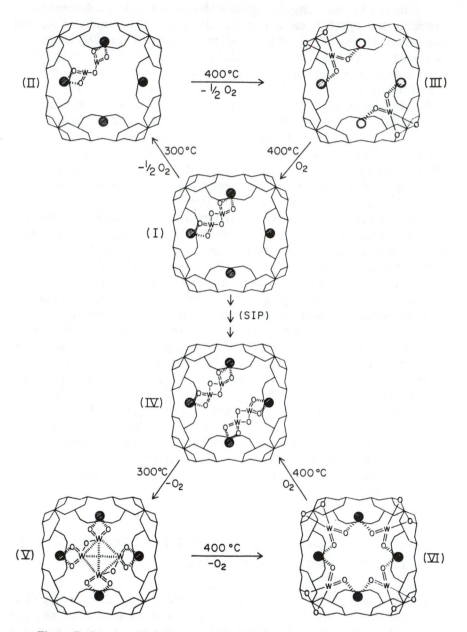

Figure 7. Structural information derived for redox-interconvertible $\mathrm{n}(WO_{3-x})-$ $Na_{56}Y$, *where* $0 < \mathrm{n} \leq 32$ *and* $\mathrm{x} = 0, 0.5,$ *or* 1.0 *(49–56). (Reproduced with permission from reference 8. Copyright 1992.)*

components of insulators, semiconductors, and metals inside the framework aluminosilicate sodalite, (*46, 57–65*). This nanoporous host acts as a stabilizing dielectric matrix, capable of organizing single-size and single-shape clusters in perfectly periodic arrays. Advanced nanomaterials of this type as well as a variety of potential applications for them were reviewed (*46, 57–65*).

In our research, the sodalite framework was used as a host material to confine clusters composed of the components of insulators, semiconductors, and metals. Sodalites are aluminosilicates of the type $M_8X_2(SiAlO_4)_6$, where M is a cation and X is an anion. The structure of sodalites is best described by considering the archetype, sodium chlorosodalite (NaCl-SOD) (Figure 8), which shows a single sodalite cage. It consists of 12 AlO_4^- and 12 SiO_4 tetrahedral units liked together by single-oxygen bridges in an alternating pattern to form a truncated octahedron with eight single six-ring and six single four-ring openings. Typically, the cage has a diameter of 6.6 Å, and the diameters of the hexagonal and tetragonal ring openings are quoted as 2.2–2.6 and 1.5–1.6 Å, respectively. Sodalite cages are fundamental building units of many zeolites.

Because of the valence difference between aluminum and silicon, the sodalite lattice possesses a negative charge equal to the number of aluminum atoms. This charge is balanced by exchangeable cations at C_3 sites near the six rings. Another cation and an anion at the center of the cage are often present as well. To maintain a charge balance, divalent anions require the presence of divalent cations or a complementary cage lacking an anion. Different kinds of both cations and anions may be mixed throughout the lattice. Thus a vast range of guests can be incorporated during the synthesis of the sodalites or produced by thermal, photochemical, and other reaction processes. These guests include materials that are insulators, semiconductors, photoconductors, and metals in their bulk form. The unit cell dimensions, charge-balance requirements, and cage-filling can be tuned by incorporating the appropriate ions during the sodalite synthesis and cation-exchange processes.

The sodalites must be synthesized in the sodium forms according to the recipes summarized in Figure 9. Depending on the choice and relative amounts of X and Y one can generate sodalite compositions described as Class A, B, and C, as depicted in Figure 9. Except for differences in the degree of hydration, only single crystallographic phases were found in as-synthesized Class A, B, and C sodium sodalites. Combined results from PXRD, Rietveld refinement, ^{23}Na DOR MAS NMR, FT–far-IR, and FT–mid-IR spectroscopies indicate that the anions (or small domains of anions) are distributed statistically throughout the sodalite lattice, forming a homogeneous solid solution (*46, 57–65*).

In Class A materials, all cages are filled with M_4X clusters. Class C sodalites are closely related, except that they can contain two different

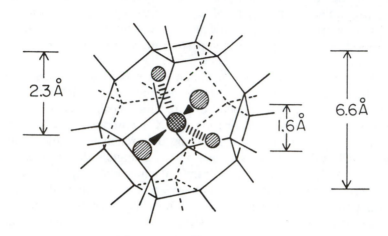

(A)   Sodalite Cage

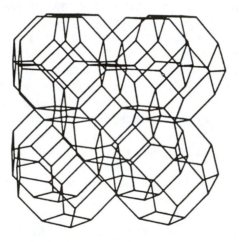

(B)   Sodalite

Figure 8. Part A: Sodalite cage showing a single cubooctahedron, a central anion, and four tetrahedrally disposed cations in the six-ring C_3 sites. Each corner represents an AlO_4^- or SiO_4 unit. Part B: Sodalite framework, emphasizing the close-packing of cages (46, 57–65). (Reproduced with permission from reference 8. Copyright 1992.)

SYNTHESIS

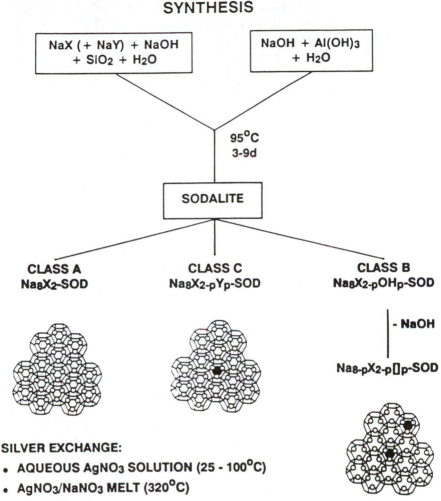

SILVER EXCHANGE:

- AQUEOUS AgNO₃ SOLUTION (25 - 100°C)
- AgNO₃/NaNO₃ MELT (320°C)
- STOICHIOMETRIC

Figure 9. Synthesis of Class A, B, and C sodalite supralattices (46, 57–65).
(Reproduced with permission from reference 8. Copyright 1992.)

anions, that is, mixed M_4X and M_4Y semiconductor component clusters. Class B sodalites contain isolated M_4X clusters "diluted" in cages with M_3 "spectator triangles" or M_4OH clusters. The sodalite framework itself is an insulator. Materials trapped inside the sodalite cages were isolated and exhibited molecular behavior at low loading levels, and they also communicated through the sodalite framework at higher concentrations and thereby formed an expanded cluster lattice or quantum supralattice after a threshold had been reached.

For a Class B sodalite containing sodium with a very low bromide concentration, the material contains isolated ionic Na_4Br^{3+} units with sodium bromide distances shorter than in the bulk salt. As the NaBr concentration is increased, greater coupling occurs between the Na_4Br^{3+} units until a certain threshold is reached and a miniband structure is formed. At this point the product obviously still does not have the rock-salt structure of bulk sodium bromide, but instead one can propose the formation of an expanded insulator within a sodalite framework.

A similar scheme is found for trapped semiconductor species such as the Group I–VII silver halides. By tuning the silver and halide loading in a series of sodium and silver chloro-, bromo-, and iodosodalites, it is possible to span the range found from an isolated silver halide AgX molecule, to an isolated silver halide Ag_4X^{3+} nanocluster, to an expanded supralattice of Ag_4X^{3+} nanoclusters of what are normally I–VII bulk semiconductors with different band gaps. In this context the key issue of intercavity electronic coupling between sodalite cage-encapsulated $Na_{4-n}Ag_nX^{3+}$ clusters was effectively probed through the compositional dependence of the unit cell dimensions (Rietveld PXRD), cation translatory mode frequencies (far-IR), quadrupole coupling constants (^{23}Na DOR NMR), optical excitation energies (UV–vis), halide chemical shifts (^{81}Br and ^{35}Cl MAS NMR), and orbital overlaps and density of states (extended Hückel molecular orbital (EHMO) calculations) (46, 57–65).

A final point worth briefly mentioning is that silver ions in sodalites intentionally containing an internal electron source (such as OH^- or $C_2O_4^{2-}$) can be reduced within the sodalite cages to form trapped silver clusters. Interaction between silver clusters results in the formation of an expanded metal, where intercage silver–silver distances are significantly longer than in the regular metal, but electronic interaction is still possible (46, 57–65, 67, 68).

With this information for sodalite supralattices concerning insulator, semiconductor, and metal nanocluster components, one can speculate that the mixed silver halide sodalite endosemiconductors in the form of single crystals or thin films could be arranged to function as a multiple quantum dot laser array, as envisioned in Figure 10. Laser action in this perceived sodalite-based device occurs through the generation of electrons and holes localized in the wide band-gap constituents of the nanocluster array, which subsequently undergo radiative recombination in the narrow band-gap nanocluster constituents.

The scheme of storing and processing information at extremely high densities in a crystalline nanoporous material was first demonstrated in the silver sodalites (46, 57–65, 67, 68). The idea is sketched in Figure 10. A system containing oxalate or hydroxide anions, which function as intrasodalite cage-reducing agents for Ag^+ cations, could be reversibly marked with a laser beam over many cycles and was suggested as a novel medium for reversible optical data storage. The proposed mechanism for reversible color changes involves laser-induced photothermal electron transfer between two types of tetrahedral silver cluster chromophoric centers, occluded in the sodalite framework (*46, 57–65, 67, 68*).

Clearly, all possible device applications for sodalites require much further study. However, the unique structural properties of the sodalite host and the variety of guests that can be occluded in controlled ways certainly make this ancient material an ideal "model system" for probing the structural, physical, and chemical properties of assemblies of single-size and single-shape encapsulated clusters built of the components of bulk insulators, semiconductors, and metals (expanded materials), as well as a promising candidate for advanced materials in the 21st century.

Assembly of Exosemiconductors

As mentioned earlier, the process of transforming a bulk semiconductor through to a nanoporous one, as conceptualized in Figure 4, turns out to

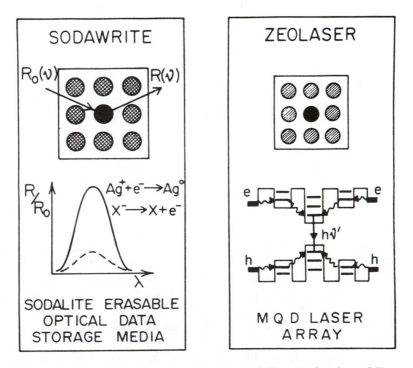

Figure 10. Nanoscale device ideas for silver sodalites. Sodawrite and Zeo-laser are zeotype-based device ideas as discussed in reference 8. (Reproduced with permission from reference 8. Copyright 1992.)

be a rather insightful way of considering the interrelationship between endosemiconductors and exosemiconductors composed of identical atomic constituents. In essence, one can be thought of as the geometric complement of the other. Before adjacent pores merge (lower porosity), one has a crystalline open-framework nanomaterial built of the atomic components of a bulk semiconductor. After they merge (higher porosity), one has a nanocluster crystal built of the atomic components of a bulk semiconductor. Therefore, the endosemiconductor and exosemiconductor can be visualized as quantum objects (wires or dots) with different electronic coupling strengths, as reflected in, for example, their electronic band gaps, bandwidths, and band curvature, as illustrated in Figure 2, resulting in distinct and novel electronic transport and optical properties.

A case in point concerns tin(IV) sulfide, a material that can exist in the molecular cluster $Me_4Sn_4S_6$ adamantane form containing tetrahedrally coordinated Sn(IV) centers, as the bulk semiconductor SnS_2 composed of parallel stacked van der Waals layers of octahedrally coordinated Sn(IV) centers, as the endosemiconductor form with a diamond lattice of $Sn_4S_6^{4+}$ adamantane clusters housed in the supercages of zeolite Y (19), and pertinent to the present discussion, as an exosemiconductor (21, 22, 69). These four forms of solid tin(IV) sulfide are illustrated in Figure 11.

The discovery of crystalline nanoporous semiconductors, a new class of nanomaterials, can be considered to originate with the template-mediated hydrothermal synthetic work of Bedard and co-workers (21). These represented the first examples of nonoxide open-framework materials. Other cases rapidly followed Bedard's work (21) from the groups of Parise (38) and Kanatzidis (70), involving antimony(III) and mercury(II) chalcogenide zeotypes.

The synthesis of the nanoporous tin(IV) sulfides, denoted SnS-n, where n represents distinct structure types resulting usually from the use of different quaternary ammonium templates, is shown in a general form in Figure 12. The intention of this general scheme is to illustrate the type of chemical and physical processes speculated to occur in the reconstitution of the bulk SnS_2 precursor (source material) to the crystalline exosemiconductor SnS-n products (22, 69, 71). Under hydrothermal reaction conditions the added OH^- or SH^- probably has the duel function of slowly mineralizing and complexing the SnS_2 solid source to produce a solution-phase $SnS_x(OH)_y(SH)_z^{q-}$ transporting agent. The labile hydrosulfide groups likely coordinated in these species appear to undergo a template-mediated (possibly space-filling, structure directing, charge-balancing roles) hydrosulfide condensation reaction to create an SnS-n seed (nucleation), which grows (crystallization) to the final well-formed crystalline SnS-n product (Figure 12).

The exosemiconductor denoted SnS-1 crystallizes with the distorted hexagonal platelike morphology seen in the scanning electron micrograph

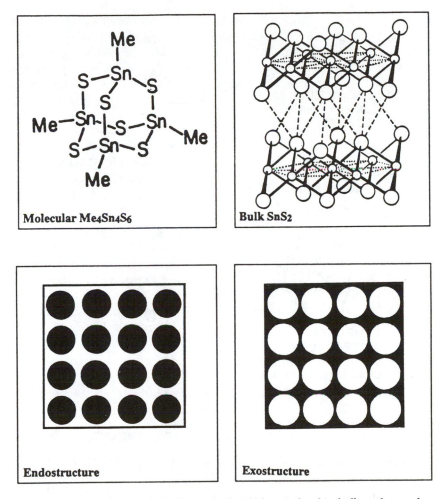

Figure 11. Four forms of solid tin(IV) sulfide: molecular, bulk, endo-, and exostructural (22, 69, 71). (Reproduced with permission from reference 69. Copyright 1992.)

(SEM) shown in Figure 13 (*22, 69, 71*). The elemental and scanning transmission electron microscopy with energy-dispersive X-ray (STEM–EDX) analysis of this material is 0.58 TMA · $SnS_{2.12}$ · 0.86 H_2O (TMA is tetramethylammonium cation). The 150 °C dehydrated TMA template-containing form has molecular-sieving properties (*21*), being able to discriminate between CO_2 (adsorbed 7–8 wt%) and Ar (excluded) having kinetic diameters of 3.3 and 3.4 Å, respectively, differing only by 0.1 Å!

A multiprong diffraction, spectroscopy, microscopy, and thermal analysis approach defines the structure of SnS-1 to be that illustrated in Figure 14. It is based on an open framework containing a hexagonal array of

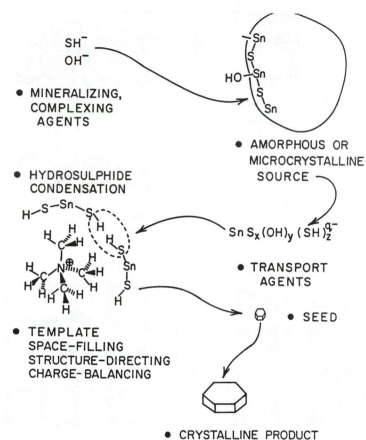

Figure 12. General scheme illustrating the synthesis of SnS-1 (22, 69, 71).

doubly sulfur-bridged [Sn(μ-S)$_2$Sn], broken-cube Sn$_3$S$_4$ cluster basic-building units with occluded TMA template. The space group of SnS-1 is determined to be monoclinic Cc, with unit cell dimensions $a = 22.941$ Å, $b = 13.207$ Å, $c = 16.053$ Å, $\beta = 108.36°$, cell volume 4616.06 Å^3, and unit cell formula 0.67TMA · SnS$_{2.33}$ (22, 69, 71). The free diameter of the hexagonal-shaped channels that run throughout the structure is about 8 Å. The optical spectrum of this material, when compared to Me$_4$Sn$_4$S$_6$, Sn$_4$S$_6$–Y, and SnS$_2$, is especially interesting with respect to quantum size effects and electronic coupling strengths in endotin(IV) sulfide and exo-tin(IV) sulfide (Figure 15). Inspection of these data reveal that the optical absorption onsets are essentially the same for Me$_4$Sn$_4$S$_6$ and Sn$_4$S$_6$–Y but progressively red-shift on passing to SnS-1 and then to SnS$_2$. Molecular orbital calculations define the electronic transitions responsible for these absorption onsets to be S(II) $\rightarrow$ Sn(IV) ligand-to-metal-charge

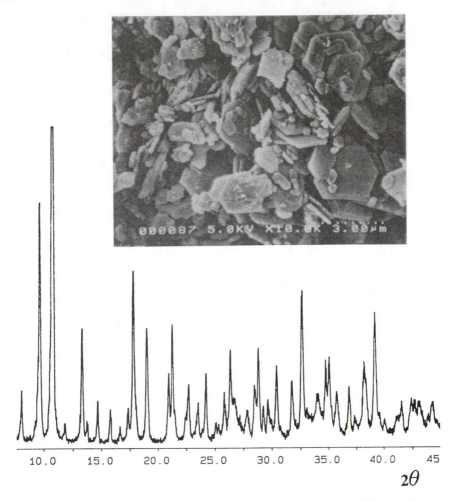

Figure 13. Typical SEM (top) and PXRD (bottom) pattern of SnS-1 (22, 69, 71).

transfer or valence-to-conduction-band-gap excitation for molecular and bulk situations, respectively (*22, 69, 72*).

The trends displayed in Figure 15 indicate that quantum and spatial confinement of electrons and holes are likely responsible for the observed blue spectral shift on passing from bulk SnS_2 to nanocluster $Me_4Sn_4S_6$. The close similarity of the optical absorption onsets of $Me_4Sn_4S_6$ and Sn_4S_6–Y indicate that only weak or negligible electronic coupling (direct through framework or indirect through-space) exists between $Sn_4S_6^{4+}$ adamantane nanoclusters housed in the diamond lattice of supercages of zeolite Y. The observed red-shift on passing from endotin(IV) sulfide Sn_4S_6–Y to exotin(IV) sulfide SnS-1 reveals the expected increase in elec-

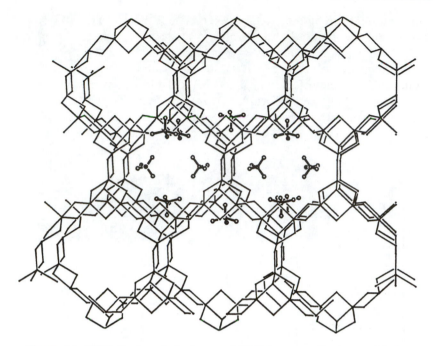

Figure 14. SEM (top) and single-crystal XRD structure (bottom) of SnS-1 (22, 69, 71).

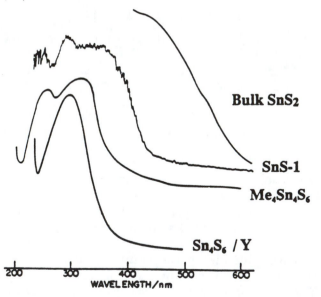

Figure 15. Optical reflectance spectra of Me₄Sn₄S₆, Sn₄S₆–Y, SnS-1, and SnS₂ (22, 69, 71). (Reproduced with permission from reference 69. Copyright 1992.)

tronic coupling strength between broken-cube Sn_3S_4 cluster, building-units (direct, through-bond) in the exotin(IV) sulfide compared to between $Sn_4S_6^{4+}$ adamantane nanocluster components in the endotin(IV) sulfide. Confinement of electrons and holes to the wall thickness of exotin(IV) sulfide SnS-1 is indicated from the QSE red-shift of the absorption edge on passing to bulk SnS_2 (Figure 15). Many other novel and fascinating open-framework structure types have been characterized in the SnS-n system by using different templates with pore sizes so far as large as 12 Å.

An especially exciting and significant recent development with exotin(IV) sulfide SnS-1 is the ability to synthesize mixed thioselenide polytypes denoted $SnS_{1-x}Se_x$-1, where the Sn to S,Se ratio is maintained over the entire composition range $0 \leq x \leq 1$ between SnS-1 and SnSe-1 isostructural end members (*74*). Because of the decreased band gap on passing from the lighter to the heavier metal chalcogenides as well as the corresponding progression of their properties from the semiconducting toward the metallic and even superconducting states (*36, 73*), this discovery of "exotin(IV) chalcogenide alloys" provides a unique opportunity to achieve for the first time compositional tuning of the electronic and optical properties of a crystalline nanoporous semiconductor and the possibility of nanoscale device applications (*74*).

In the classic case of bulk semiconductor alloys, exemplified by $Ga_{1-x}Al_xAs$ (*75*), the observed compositional dependence of both the band gap and unit cell dimension displays a linear Vegard law behavior expected for a random distribution (solid-solution) of Ga and Al components according to the atomic fraction weighted average of the two end-members:

$$E_g(Ga_{1-x}Al_xAs) = xE_g(AlAs) + (1-x)E_g(GaAs) \qquad (8)$$
$$a(Ga_{1-x}Al_xAs) = xa(AlAs) + (1-x)a(GaAs) \qquad (9)$$

where E_g is the band gap, a is the unit cell dimension of a cubic diamond lattice, and x is the atomic fraction of the AlAs component in $Ga_{1-x}Al_xAs$. This behavior implies average lattice constants, crystal potentials, and band structures for these kinds of bulk semiconductor alloys. The ability to engineer the band properties of an exosemiconductor alloy can be appreciated from the observed monotonic red-shifting of the absorption edges of $SnS_{1-x}Se_x$-1 materials on progressively increasing the selenium content in the compositional range $0 \leq x \leq 1$, seen in Figure 16. A solid-solution model for $SnS_{1-x}Se_x$-1 materials is favored by spatially resolved STEM–EDX analysis; Sn, S, and Se single-crystal compositional profiles; PXRD; and Raman and UV–vis analytical methods (*69, 74*). Especially relevant are preliminary results that demonstrate linear Vegard law behavior for the band gaps and d-spacings of $SnS_{1-x}Se_x$-1 materials (Fig-

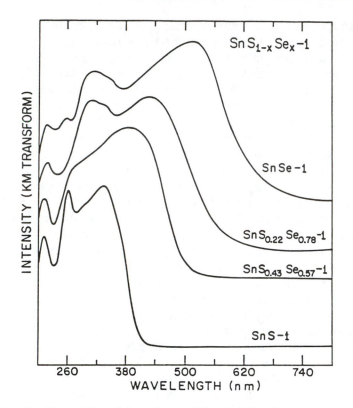

Figure 16. Compositional dependence of the optical reflectance spectra of the tin(IV) thioselenide $SnS_{1-x}Se_x$-1 exosemiconductor polytypes (22, 74).

ure 17), and thereby bring to practice for the first time the ability to engineer the electronic and optical properties of a nanoporous semiconductor, as illustrated in Figure 18 *(22, 69, 74).*

Now that we have gained a synthetic entry to exosemiconductors, let us proceed to fanciful speculation as to how they might be used in the future. For instance, molecule-discriminating transistor behavior and light-emitting diode–laser action can be imagined for pnp- and np-junction devices in which one of the components is an exosemiconductor (Figure 19). This possibility makes the development of molecule-discriminating electronic and optical sensing devices a possible scenario *(74).* Here one attempts to design the system so as to display chemoselective electronic switching or chemoselective electroluminescence, possibly originating from changes in the valence–conduction electronic band states, induced by the

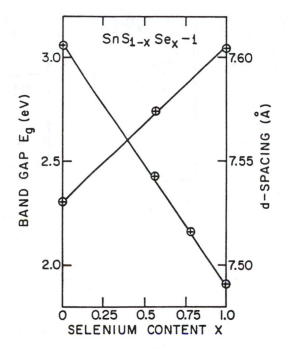

Figure 17. Band-gap engineering of the tin(IV) thioselenide $SnS_{1-x}Se_x$-1 exosemiconductor polytypes showing linear Vegard law behavior for the band gaps and d-spacings with changing selenium content (22, 74).

adsorption of specific molecules into the void spaces of the exosemiconductor junction material. These concepts are sketched in Figure 19.

A Nanoscale Future

We are entering an exciting and challenging era of solid-state chemistry and physics in which there will be increasing demands for structured nanophase materials with stringent requirements of size, shape, and dimensionality, as well as the type and concentration of dopants, defects, and impurities. In such a world of tiny objects, processes, and devices, the undisputed production workhorse of the nanophysicist over the next decade or so will be sophisticated forms of planar deposition and lateral engineering techniques. The practical limit of these methods appears to be around 200 Å with a dispersion of about 50 Å. Beyond this size, the nanotips of scanning probe microscopes will continue to be developed toward possibly achieving the ultimate in miniaturization, namely atomic- and molecular-

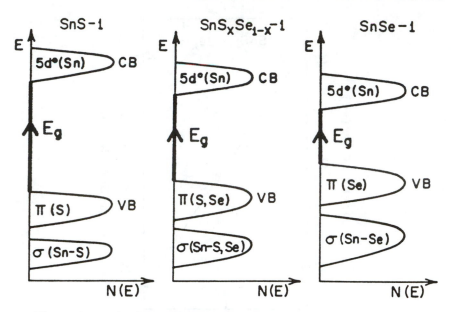

Figure 18. Qualitative energy E *versus density of states* N(E) *diagram illustrating the tunability of the electronic and optical properties of nanoporous tin(IV) thioselenides with changing selenium content (22, 74).*

scale devices. This promising technology could show some practical utility in the 21st century providing that the huge challenge of rapidly and reproducibly (probably in a massively parallel manner) moving and monitoring matter at the atomic level can be surmounted. Meanwhile the elegant patterning and templating methods of the chemist for producing spatially controlled nanophase materials with high degrees of perfection are likely to receive increasing attention in the exciting nanoscale world of the future (*8*).

With respect to zeotypes and their host–guest inclusion chemistry, one can anticipate an exciting nanochemistry future. Key issues and developments will likely include the growth of large single crystals of zeotypes, the growth of zeotype-oriented single-crystal thin films on planar substrates, the self-assembly of endosemiconductors in zeotype single crystals and thin films, property measurements on endosemiconductor and exosemiconductor single crystals and thin films and their doping to the metallic and maybe even superconducting states, and finally the fabrication and evaluation of endosemiconductor and exosemiconductor single-crystal and thin-film-based nanoscale devices of the type speculated about in this chapter and elsewhere (*8*). We all look forward to a nanoscale future and, we hope, the birth of the nanocomputer!

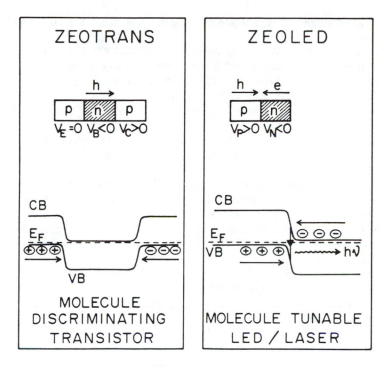

Figure 19. Nanoscale device ideas for exosemiconductors. Zeotrans and zeoled are zeotype-based device ideas as discussed in reference 8. (Reproduced with permission from reference 8. Copyright 1992.)

Acknowledgments

The generous financial assistance of the Natural Sciences and Engineering Research Council of Canada is gratefully appreciated. Informative and stimulating technical discussions with Robert Bedard and Edith Flanigen, as well as consistent and generous research support from their organizations, Union Carbide Corporation and Universal Oil Products, Tarrytown, New York, proved to be of great value. I am deeply indebted to all of my co-workers who have contributed to the work referenced to and or presented in this chapter.

References

1. Corcoran, E. "Diminishing Dimensions," *Sci. Am.* **1990,** *November,* 122.
2. "Engineering a Small World: From Atomic Manipulation to Microfabrication" *Science (Washington, D.C.)* **1991,** *254,* 1300–1335.

3. Wickramasinghe, H. K. "Scanned-Probe Microscopes," *Sci. Am.* **1989,** *October,* 98.
4. Avouris, P.; Lyo, I. W. *Science (Washington, D.C.)* **1990,** *253,* 173.
5. Reed, M. A. "Quantum Dots," *Sci. Am.* **1993,** *January,* p 118.
6. Steigerwald, M. C.; Brus, L. E. *Acc. Chem. Res.* **1990,** *23,* 183.
7. Wang, Y. *Acc. Chem. Res.* **1991,** *24,* 133.
8. Ozin, G. A. "Nanochemistry: Synthesis in Diminishing Dimensions," *Adv. Mater.* **1992,** *4,* 612.
9. "Nanostructure Physics and Fabrication." In *Proc. Int. Symp.* (College Station, TX, March 1989); Reed, M. A.; Kirk, W., Eds.; Academic: Orlando, FL, 1989.
10. *Science and Engineering of One and Zero-Dimensional Semiconductors;* In Nato ASI Series B: Physics; Beaumont, S. P.; Sotomayor Torres, C. M., Eds.; Plenum: New York, 1990, Vol. 214.
11. "Nanostructures and Mesoscopic Systems." In *Proc. Int. Symp.* (Sante Fe, New Mexico, May 1991); Kirk, W. P.; Reed, M. A., Eds.; Academic: Orlando, FL, 1992.
12. Nishizawa, M.; Shibuya, M.; Sawaguchi, T.; Matsue, T.; Uchida, I. *J. Phys. Chem.* **1991,** *95,* 9042, and references cited therein.
13. *Inclusion Compounds;* Atwood, J. L.; Davies, J. E. D.; MacNicol, D. D., Eds.; Academic: Orlando, FL, 1984, Vol. 1.
14. *Inclusion Compounds;* Atwood, J. L.; Davies, J. E. D.; MacNicol, D. D., Eds.; Oxford University Press: Oxford, England, 1991, Vol. 5.
15. *Inclusion Phenomena and Molecular Recognition;* Atwood, J. L., Ed.; Plenum: New York, 1990.
16. *Intercalation Chemistry;* Whittingham, M. S.; Jacobson, A. J., Eds.; Academic: Orlando, FL, 1982.
17. Mana, S.; Meldrum, F. C. *Adv. Mater.* **1991,** *3,* 316.
18. Ozin, G. A.; Steele, M. "Stepwise Synthesis of II-VI Semiconductor Nanoclusters Inside the Supercages of Zeolite Y Using MOCVD-Type Precursors." In *Proceedings of the 9th International Zeolite Conference;* Von Ballmoos, R; Higgens, J. B. H.; Treacy, M. M. J., Eds.; Butterworth–Heinemann: Boston, MA, 1993, Vol. II, p 185.
19. Ozin, G. A.; Bowes, C. L.; Steele, M. "Zeolates: A Coordination Chemistry View of Metal–Ligand Bonding in Intrazeolite MOCVD-Type Precursors and Semiconductor Nanocluster Products," *Mater. Res. Soc. Symp. Ser.* **1992,** *277,* 1050.
20. Ozin, G. A.; Steele, M.; Macdonald, P. M. *J. Am. Chem. Soc.* **1993,** *115,* 7285. Ozin, G. A.; Steele, M.; Holmes, A. J. *Chem. Mater.* **1994,** *6,* 999.
21. Bedard, R. L.; Wilson, S. T.; Vail, L. D.; Bennett, J. M.; Flanigen, E. M. "The Next Generation: Synthesis, Characterization, and Structure of Metal Sulphide-Based Microporous Solids." In *Zeolites: Facts, Figures, and Future;* Jacobs, P. A.; Van Santen; R. A., Eds.; Elsevier: Amsterdam, Netherlands, 1989.
22. Ahari, H.; Bedard, R. L.; Bowes, C. L.; Enzel, P.; Jiang, T.; Ozin, G. A.; Young, D. "Recent Advances in the Chemistry of Nanoporous Chalcogenides." In *Recent Advances in Molecular Sieve Materials;* ACS Symposium Series; American Chemical Society: Washington, DC, in press.

23. Ozin, G. A.; Kuperman, A.; Nadimi, S. "Non-Aqueous Synthesis of Zeolites," U.S. Patent 795,354.

24. Ozin, G. A.; Kuperman, A.; Nadimi, S.; Garces, J.; Olken, M. M. "Giant Synthetic Zeolite and Molecular Sieve Single Crystals," *Nature (London)* **1993**, *365*, 239.

25. Ozin, G. A.; Kuperman, A.; Nadimi, S.; Garces, J.; Olken, M. M.; Young, D. "Giant Synthetic Zeolite Single Crystals and Zeolite Thin Films for Advanced Zeolite Materials Applications." In *Recent Advances in Molecular Sieve Materials;* ACS Symposium Series; American Chemical Society: Washington, DC, in press.

26. Tsikoyiannis, J. G.; Haag, W. O. *Zeolites* **1992**, *12*, 126.

27. Sano, T.; Kiyozumi, Y.; Mizukami, F.; Takaya, H.; Mouri, T.; Watanabe, M. *Zeolites* **1992**, *12*, 131.

28. Sano, T.; Mizukami, F.; Takaya, H.; Mouri, T.; Watanabe, M. *Bull. Chem. Soc. Jpn.* **1992**, *65*, 146.

29. Stringfellow, G. B. *Organometallic Vapour Phase Epitaxy: Theory and Practice;* Academic: Boston, MA, 1989.

30. Ozin, G. A.; Özkar, S. *Chem. Mater.* **1992**, *4*, 511. Ozin, G. A. *Adv. Mater.* **1994**, *6*, 71.

31. Canham, L. T.; Houlton, M. R.; Leong, W. Y.; Pickering, C.; Keen, J. H. *J. Appl. Phys.* **1992**, *40*, 422.

32. Lehmann, V.; Gosele, U. *Adv. Mater.* **1992**, *4*, 114.

33. Iyer, S. S.; Xie, Y. H. *Science (Washington, D.C.)* **1993**, *260*, 40. Brus, L. E. *J. Phys. Chem.* **1994**, *98*, 3575. Brus, L. E. et al. *Phys. Rev. Lett.* **1994**, *72*, 2648.

34. Brandt, M. S.; Fuchs, H. D.; Stutzmann, M.; Weber, J.; Cardona, M. *Solid State Commun.* **1992**, *81*, 307.

35. McCord, P.; Yau, S. L.; Bard, A. J. *Science (Washington, D.C.)* **1992**, *257*, 68.

36. Rao, C. N. R.; Gopalakrishnan, J. *New Directions in Solid State Chemistry: Structure, Synthesis, Properties, Reactivity, and Materials Design;* Cambridge University Press: Cambridge, MA, 1986.

37. Schöllhorn, R. *Angew. Chem. Int. Ed. Engl.* **1988**, *27*, 1392.

38. Parise, J. B. *J. Chem. Soc. Chem. Commun.* **1991**, *22*, 1553; *Science (Washington, D.C.)* **1991**, *251*, 293.

39. Ozin, G. A.; Kuperman, A.; Stein, A. *Angew. Chem. Int. Ed.* **1989**, *28*, 359.

40. Stucky, G. D.; MacDougall, J. E. *Science (Washington, D.C.)* **1991**, *247*, 669.

41. Ozin, G. A.; Özkar, S. *Adv. Mater.* **1992**, *4*, 11.

42. Ozin, G. A.; Özkar, S.; Prokopowicz, R. A. *Acc. Chem. Res.* **1992**, *25*, 558.

43. Wang, Y.; Herron, N. *J. Phys. Chem.* **1991**, *95*, 525.

44. Beck, J. S.; Vartuli, J. C.; Roth, W. T.; Leonowicz, M. E.; Kresge, C. T.; Schmitt, K. D.; Chu, C. T. W.; Olson, D. H.; Sheppard, E. W.; McCullen, S. B.; Higgens, J. B.; Schlenker, J. L. *J. Am. Chem. Soc.* **1992**, *114*, 10834, and references cited therein.

45. Herron, N.; Wang, K.; Eddy, M.; Stucky, G. D.; Cox, D. E.; Moller, K.; Bein, T. *J. Am. Chem. Soc.* **1989**, *111*, 530.

46. Stein, G. A.; Ozin, G. A.; Stucky, G. D. *J. Am. Chem. Soc.* **1990**, *112*, 904.

47. Terasaki, O.; Tang, Z. K.; Nozue, Y.; Goto, T. *Mater. Res. Soc. Symp. Proc.* **1991**, *233*, 139.

48. Ozin, G. A.; Dag, O.; Kuperman, A.; Macdonald, P. M. In *Zeolites and Related Microporous Materials: State of the Art 1994;* Weitkamp, J.; Karge, H. G.; Pfeifer, H.; Höldrich, W., Eds.; Studies in Surface Science Catalysis, Vol. 84, 1994, p 1107; *Adv. Mater.* **1994,** in press; Ozin, G. A.; Kuperman, A.; Dag, O. *Adv. Mater.* **1994,** *6,* 147.

49. Ozin, G. A.; Özkar, S. *J. Phys. Chem.* **1990,** *94,* 7556.

50. Ozin, G. A.; Özkar, S.; Moller, K.; Bein, T. *J. Phys. Chem.* **1991,** *95,* 5376.

51. Özkar, S.; Ozin, G. A.; Moller, K.; Bein, T. *J. Am. Chem. Soc.* **1990,** *112,* 9575.

52. Ozin, G. A.; Özkar, S.; Macdonald, P. M. *J. Phys. Chem.* **1990,** *94,* 6939.

53. Ozin, G. A.; Özkar, S.; Prokopowicz, R. A. *Acc. Chem. Res.* **1992,** *25,* 558.

54. Ozin, G. A.; Özkar, S.; Jelinek, R. *J. Phys Chem.* **1992,** *96,* 5449.

55. Ozin, G. A.; Özkar, S.; Prokopowicz, R. A. *J. Am. Chem. Soc.* **1992,** *114,* 8953.

56. Ozin, G. A.; Malek, A.; Prokopowicz, R. A.; Macdonald, R. M.; Özkar, S.; Moller, K.; Bein, T. In *Synthesis, Characterization, and Novel Applications of Molecular Sieve Materials;* Bedard, R. L.; Bein, T.; Davis, M. E.; Garces, J.; Maroni, V. A.; Stucky, G. D., Eds.; Materials Research Society Symposium Proceedings **1991,** *233,* 109.

57. Ozin, G. A.; Kirkby, S.; Meszaros, M.; Özkar, S.; Stein, A.; Stucky, G. D. In *Materials for Nonlinear Optics: Chemical Perspectives;* Marder, S. R.; Sohn, J. E.; Stucky, G. D., Eds.; ACS Symposium Series 455; American Chemical Society: Washington, DC, 1991; p 554.

58. Ozin, G. A.; Stein, A. In *Proceedings of the 9th International Zeolite Conference;* Von Ballmoos, R; Higgens, J. B. H.; Treacy, M. M. J., Eds.; Butterworth–Heinemann: Boston, MA, 1993, Vol. I, p 93.

59. Stein, A.; Macdonald, P. M.; Ozin, G. A.; Stucky, G. D. *J. Phys Chem.* **1990,** *94,* 6943.

60. Stein A.; Meszaros, M.; Macdonald, P. M.; Ozin, G. A.; Stucky, G. D. *Adv. Mater.* **1991,** *3,* 306.

61. Stein, A.; Ozin, G. A.; Stucky, G. D.; Macdonald, P. M.; Jelinek, R. *J. Am. Chem. Soc.* **1992,** *114,* 5171.

62. Stein, A.; Ozin, G. A.; Stucky, G. D. *J. Am. Chem. Soc.* **1992,** *114,* 8119.

63. Stein, A.; Ozin, G. A.; Jelinek, R.; Chmelka, B. F. *J. Phys. Chem.* **1992,** *96,* 6744.

64. Stein, A.; Ozin, G. A.; Jelinek, R. *J. Am. Chem. Soc.* **1993,** *115,* 2390.

65. Stein, A.; Ozin, G. A. "Sodalites: Ancient Materials for Advanced Applications," In *Advances in the Synthesis and Reactivity of Solids;* Mallouk, T., Ed.; JAI Press: Greenwich, CT, 1994, Vol. 2, pp 93–154.

66. Herron, N.; Wang, Y.; Eckert, H. *J. Am. Chem. Soc.* **1990,** *112,* 1322.

67. "Photosensitive, Radiation Sensitive, Thermally Sensitive, and Pressure Sensitive Silver Sodalite Materials." Ozin, G. A.; Godber, J.; Stein, A. U.S. Patent 4,942,119, July 17, 1990.

68. Stein, A.; Ozin, G. A.; Stucky, G. D. *Soc. J. Photogr. Sci. Technol. Jpn.* **1990,** *53,* 322.

69. Bowes, C. L.; Ozin, G. A. "Nanomaterials: Tin(IV) Sulphide Endo- and Exosemiconductors." In *Nanophase and Nanocomposite Materials;* Komarneski, S; Parker, J. C.; Thomas, G. J., Eds.; Materials Research Society Symposium Proceedings, **1993,** *286,* 93.

70. Kanatzidis, M. G. *Chem. Mater.* **1990,** *2,* 353, and references cited therein.
71. Lough, A.; Ahari, H.; Bedard, R. L.; Bowes, C. L.; Jiang, T.; Ozin, G. A.; Young, D. Supramolec. Chem. and *Adv. Mater.* **1994,** in press.
72. Lifshitz, E.; Chen, Z.; Bykov, L. *J. Phys. Chem.* **1993,** *97,* 238.
73. Formtone, C. A.; FizGerald, E. T.; Cox, P. A.; O'Hare, D. *Inorg. Chem.* **1990,** *29,* 3860.
74. Ahari, H.; Bedard, R. L.; Bowes, C. L.; Jiang, T.; Ozin, G. A.; Young, D. in preparation.
75. Jaros, M. *Physics and Applications of Semiconductor Microstructures;* Oxford University Press: Oxford, England, 1989.

RECEIVED for review November 9, 1992. ACCEPTED revised manuscript May 12, 1993.

Molecule-Based Syntheses of Extended Inorganic Solids

Michael L. Steigerwald

AT&T Bell Laboratories, 600 Mountain Avenue, Murray Hill, NJ 07947

Herein we describe and analyze a number of reactions that lead from molecular reagents to solid-state products. The processes that take molecules to solids can be arrested before the final solid product is fully formed. Intermediates in these reactions are nanometer-sized fragments of the bulk solids in several instances and monodisperse molecular clusters in other cases. The cluster compounds that result from these arrested reactions are analyzed in the context of solid-state materials chemistry in order to reveal some of the similarities and differences between the molecular and solid-state regimes.

Molecules-to-Solids Reactions

As the size or dimensionality of a piece of a solid material shrinks, the molecular chemistry that is or can be associated with the fabrication of the solid becomes more important. For example, molecular chemistry is irrelevant to the preparation of three-dimensional ingots of gallium arsenide, but, as other chapters in this volume verify, the microscopic reactions of atomic and molecular reagents are critical to the growth of two-dimensional thin films of that same solid product. In comparison to the amount that is known about reactions of molecules that give strictly molecular products, little is known in detail about reactions that lead from molecular reagents to solid-state products. The more that is known about these reactions, the more effectively they can be used in the preparation of useful materials. This chapter is devoted to the discussion of molecules-to-solids reactions, focusing particularly on the role of molecular clusters as reaction intermediates.

0065−2393/95/0245−0373$13.00/0

Molecule-Based Syntheses of II−VI Compounds

Members of the family of so-called II−VI compounds are compound semi-conductors (*1*) of the general formula ME where M is a Group IIB metal (Zn, Cd, or Hg), and E is a Group VIa element (S, Se, or Te). The compounds are "tetrahedral" solids, meaning that the coordination geometry around each element is fourfold and tetrahedral. The compounds are three-dimensionally infinite solids. The two limiting structure types are zincblende and wurtzite; zincblende is based on cubic close-packing, and wurtzite is based on hexagonal close-packing (*2, 3*). The two structure-type names come from the names of two different naturally occurring forms of ZnS. The solids are all semiconducting except for HgTe, which is semimetallic. Members of this family of compounds have been used in several applications; for example, ZnSe is being studied as a photo-luminescent material, and $Hg_xCd_{1-x}Te$ (solid solutions of HgTe and CdTe) have been used as long-wavelength photodetectors.

For these applications the compounds are required in thin-film form. One of the important methods of II−VI thin-film preparation is or-ganometallic vapor-phase epitaxy (OMVPE) (*4*). In this method volatile molecular precursors are introduced to the gas phase in a reactor vessel that contains the substrate upon which the II−VI material is to be grown. Conditions within the vessel are maintained such that the molecular pre-cursors react to form the II−VI solid as a thin (nanometer to micrometer thick) film uniformly on the substrate surface. An example of a process of this type is shown in equation 1.

$$Cd(CH_3)_2 \ + \ Te(C_2H_5)_2 \ \longrightarrow \ CdTe \ + \ \cdots \tag{1}$$

Processes like that in equation 1 are certainly quite complex, and mecha-nisms are subjects of considerable study and speculation.

In the attempt to find processes that would give thin films of II−VI materials at significantly lower growth temperatures, replacements for $Te(C_2H_5)_2$ were sought (*5*). One alternative process uses dialkylditellu-rides in place of dialkylmonotellurides (equation 2) (*6*).

$$Cd(CH_3)_2 \ + \ H_3CTeTeCH_3 \ \longrightarrow \ CdTe \ + \ \cdots \tag{2}$$

The rationale in this case was simple: If the molecular chemistry that is relevant to the solid-forming process is the homolytic cleavage of covalent bonds to the central atom of the molecular precursor, then the solid-forming reactions will occur at lower temperatures if those covalent bonds are weaker. The Te–Te bond in a ditelluride is weaker than a Te–C bond in a monotelluride, so the process shown in equation 2 should proceed at a temperature lower than that in equation 1. This was found to be the case.

The mechanisms of the reactions involved in the process that is summarized by equation 2 are also not understood. In the attempt to outline some of the reaction pathways that might be involved, it was noticed (7) that simple diorganoditellurides react with elemental Hg to give organometallic compounds that contain both Hg and Te, the constituent elements of the semimetal HgTe (equation 3).

$$RTe–TeR + Hg \longrightarrow RTe–Hg–TeR \tag{3}$$

Furthermore, the organometallic intermediate, $Hg(TeR)_2$, converts to HgTe on heating (equation 4).

$$RTe–Hg–TeR \longrightarrow HgTe + TeR_2 \tag{4}$$

Similar reactions were reported (8–10) for the generation of several II–VI compounds, including ZnS, CdS, ZnSe, CdSe, CdTe, and HgSe. These studies do not prove that compounds such as $Cd(TeR)_2$ are involved in equation 2, but they do show that the di(organochalcogenato)metal compounds are possible intermediates in the molecules-to-solids process.

Compounds such as $Cd(TeR)_2$, which have all of the constituent elements of a product solid incorporated in a single molecule, are referred to as "single-source" compounds (11). Single-source compounds can offer advantages such as control over reaction stoichiometry and lowered reaction temperatures. Compounds of this variety have been used in the fabrication of a number of solid-state materials.

Even if it is granted that $Cd(TeCH_3)_2$ is an intermediate in equation 2, the question of how the solid-state compound, CdTe, forms therefrom remains to be answered.

Clusters of II–VI Semiconductor Materials

One of the many reasons that thin films of semiconductor materials are of scientific and technological interest is that the physical properties of the film vary in a controllable way when the thickness of the film is less than some value that is characteristic of the related bulk solid (*12*). (This characteristic value is the exciton diameter and is between 50 and 200 Å for most semiconductors of present interest.) This variation of properties with film dimension means that the behavior of the material or of the device of which the film is a part can be tuned by adjusting the thickness of the film. In the physics literature, this type of film is called a quantum well, and this type of behavior is referred to as quantum confinement because the band-gap (i.e., lowest energy) excitations that are characteristic of the semiconductor can no longer be considered to be occurring in effectively infinite solids.

Thin films of semiconducting solids are quantum-confined in just one dimension. In view of the importance of thin films in present electronics technology, it is natural to investigate the utility of materials that are confined in two or three dimensions (*13, 14*). Fragments of semiconductor solids in which each dimension is smaller than some characteristic value (the exciton diameter) are called quantum dots (or, equivalently, Q-state materials, semiconductor nanoclusters, or semiconductor crystallites.). As a fragment of a semiconductor material is made smaller than the exciton size, the separation between the highest occupied electronic level (known as the top of the valence band in the extended solid) and the lowest unoccupied electronic level (the bottom of the conduction band) is expected to increase (*15, 16*). The most direct physical result of this increased separation will be in the electronic spectra of the samples. As the size of the particle shrinks, the energy of the lowest electronic excitation in the particle is expected to increase.

To verify this and other predictions, methods for the synthesis of these essentially molecular materials must be devised. Some important requirements are as follows:

1. Internal crystallinity. The most fundamental requirement of the synthesis method is that the correct product be formed. When the desired product is a small fragment of a crystalline, extended solid-state compound, this requirement means that the atomic connectivity within each cluster must be identical to that in the solid. The reason for this requirement is that the properties of the clusters are to be rationalized in terms of those of the extended solid. The properties of amorphous (i.e., noncrystalline or poorly crystalline) materials differ from those of the associated crystalline materials (*17*). If the cluster compounds are internally noncrystalline, there is another level of uncertainty in the understanding of the cluster properties. For the same reason it is not sufficient that the cluster be simply internally crystalline. The internal crystal lattice must be the same as that of the extended solid.

2. Size control. Inasmuch as it is stipulated that the interest in these large cluster materials is based on the presumed variation in physical properties with size, it is important to be able to prepare samples of the cluster materials in which all of the constituent clusters are the same size. In molecular synthesis this is usually a trivial requirement because a molecule containing, for example, 20 atoms is physically and chemically so much different from a related molecule having 19 or 21 atoms. On the other hand, a semiconductor crystallite that is 100 Å in diameter can contain 10^4 to 10^6 atoms. The presence or absence of a few atoms more or less will be very difficult to either measure or control. For this reason a method of synthesis should involve some mechanism for the in situ control of the size of the product particle.

3. Shape control. Quantum confinement can occur in the three spatial dimensions independently, so it is important to be able to control the size of the cluster in all three dimensions, that is, to control the shape of the crystallite. The simple solution to this problem is to prepare spherical particles and thereby reduce the three-dimensional problem to a one-dimensional problem.

4. Passivation. The properties of the semiconductor clusters that are of greatest interest are those of the clusters when physically and chemically isolated from one another. For this reason the synthesis method should include a mechanism for keeping the particles distinct. An added benefit would be a passivation layer that would protect the clusters from degradation and allow the clusters to be dissolved or dispersed.

Formation of II–VI Clusters via Arrested Precipitation

The earliest preparations (*13, 14*) of semiconductor crystallites, per se, directly exploited the insolubility of metal sulfides and metal selenides. Solutions of, for example, Zn^{2+} or Cd^{2+} were treated with H_2S or H_2Se to generate the corresponding metal sulfides or selenides. When these reactions are run at high concentrations or at room temperature, the solid-state compounds precipitate immediately as polymicrocrystalline solids. When the same reactions are run at high dilution and at lower temperatures, the precipitation may be arrested at the point where nanometer-sized crystallites of the II–VI compounds have formed but before those crystallites can grow and fuse to give the polycrystalline solid material.

The results of these first studies verified the prediction that as the particle size shrinks, the onset of the optical absorption is shifted to higher energy.

Even though the arrested precipitation reactions are conducted at low temperatures, the semiconductor clusters that are formed are internally crystalline, as verified by transmission electron microscopy (TEM), and therefore the first of the synthesis requirements mentioned is satisfied (at least in part) by this method. The size and shape of the particles can be roughly controlled by the manipulation of the conditions of the precipitation, but there is significant polydispersity in samples prepared via this route. The passivation of the clusters in this synthesis is problematic. The colloidal particles are held apart by the intervening solvent, and the removal of the solvent gives a polycrystalline solid that cannot be redispersed to clusters.

The precipitation of metal sulfides and selenides is very rapid and gives nanocrystalline particles, so the method has been used widely. One large family of applications involved microinhomogeneous reaction media. Nanometer-sized particles of II–VI compounds have been prepared in polymer films (18, 19), zeolites (20), lipid bilayers (21), vesicles (21), and micelles (21–23), for example. In closely related work, inorganic polymers have been used as stabilizing agents (24, 25). In each case the structured reaction environment allowed the precipitation reaction to be conducted at more convenient temperatures and given more control over the distribution of sizes and shapes in the resulting clusters.

When the ionic precipitations are conducted in porous glasses or in zeolites, the mechanism of size control appears to be the filling of voids. For example when zeolite-Y is used as a host for CdS the $(CdS)_4$ clusters (20) that are produced just fill the voids in the zeolite structure. In other cases the mechanism of size control is less direct.

A micelle phase (26) results when a surfactant is used to mix otherwise immiscible solvents such as a saturated hydrocarbon and water. When the proper ratios of the three components are used, droplets of the minority liquid are encased by the surfactant and the conglomerates are dispersed in the majority liquid, apparently forming a solution. The droplet together with its surfactant coating is a "normal" micelle if the droplet is the hydrocarbon and a "reverse" micelle if it is water. Reverse micelles have been used as synthesis auxiliaries in the arrested precipitation of II–VI crystallites by dissolving the constituent ions (for example, Cd^{2+} and Se^{2-}) separately in reverse micelle microemulsions and subsequently combining the two. Nanocrystalline particles are formed in the ensuing reaction, and the average particle size is governed by the molar ratio of water to surfactant (22).

Because "void filling" has been used successfully as a mechanism of particle size control in other systems, the micelle reaction vessel might serve a similar function. The average micelle size in a microemulsion is determined by the molar ratio of water to surfactant; therefore, the II–VI particle might grow within a host micelle, simultaneously displacing the water pool, until the micelle is filled, at which point the surfactant coating adheres to the particle surface and terminates particle growth.

This mechanism apparently does not operate. If additional ionic feedstock (Cd^{2+} and Se^{2-}) is added sequentially to a previously prepared micelle-stabilized CdSe colloid (all else being equal), the average particle size increases (*27*). This increase can be observed directly because, as the particle size increases, the color of the colloidal sample is shifted to the red (Figure 1). If void filling were the size-control mechanism, the result of the feedstock addition would be the formation of more particles of the same average size, and therefore color, as the original particles. This experiment shows that the micelle-encapsulated particles are stable but selectively reactive: Cadmium ions on the surface of a particle react with added selenide, and surface selenide ions react with added cadmium. The mechanism of particle size control must be statistical–kinetic in nature (*28*).

A dividend from the particle-growth study is the ability to exploit the surface chemistry of the particles. The micelle-encapsulated, Cd^{2+}-rich particles react with Se^{2-} ions, so it is reasonable to expect similar reactions with organoselenide monoanions, and this type of reaction has been

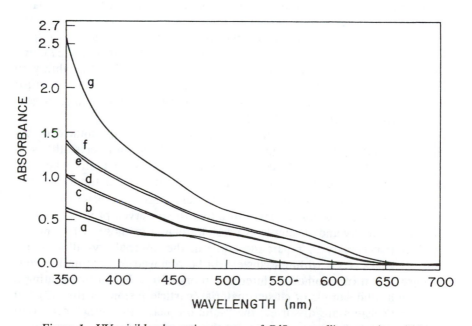

Figure 1. UV–visible absorption spectra of CdSe crystallites growing within inverse micelles. Spectrum a shows the absorption due to the original clusters, and spectra b through g show the effect of added chemical feedstock. (Reproduced from reference 27. Copyright 1988 American Chemical Society.)

observed (27). Micelle-encapsulated, Cd-rich nanocrystallites of CdSe react with phenyl selenide anion to give CdSe nanoparticles that are protectively coated with covalently attached phenyl groups (27).

The organic coating is important in several respects. First, it allows the surfactant to be removed. The surfactant in the arrested precipitation reaction is important not only in providing a micellar host for crystallite nucleation and growth, but also in providing a protective coating for the particles once they have formed. In the absence of a protective layer the particles, which are selectively quite reactive, would fuse together and lose their independent identities. The surfactant sheath is inconvenient because it is bound only weakly to the particle surface, and therefore a large amount of surfactant is required to ensure that the particle surfaces are effectively covered. The large excess of surfactant is often a nuisance when the physical properties of the particles themselves are being investigated. When the weakly held surfactant is replaced as the surface ligand by the covalently attached organic group, the surfactant can be washed away (27).

When the surfactant is removed the organically passivated semiconductor cluster compounds can be isolated as pure materials. (Pure in this case means that the solids that are isolated are made up of only "capped" semiconductor clusters. In each sample there is yet a distribution in particle size and shape.) With the appropriate choice of organic ligand, the capped clusters are soluble in organic solvents. Optical absorption spectroscopy (29), NMR spectroscopy (30), vibrational spectroscopy (31), and chromatographic investigations (27) are all facilitated by the ability to redisperse the isolated cluster material. Electron microscopy has verified that the soluble, capped clusters are single particles, there being, in general, no evidence for interparticle fusion.

Cluster solubility allows the particles to be internally annealed. As mentioned, internal crystallinity is a critical specification in crystallite synthesis. Crystallites of II–VI compounds contain growth faults when the particles are prepared at room temperature. In these cases the local tetrahedral coordination around each constituent atom is correct, but the longer range order is faulty. To ensure that the observed properties of the cluster are clearly and as simply as possible related to those of the infinite solid, it is necessary to repair the faults in the internal crystalline structure. Crystal defects within a cluster can be removed by heating the cluster such that it can isomerize (internally reorganize). When this heating is done on a solid sample of cluster material, particle fusion results (32), but when the clusters are dissolved, the "solution-phase annealing" can occur without particle fusion. Annealing of this type has afforded cluster materials of high quality, and comparison of the annealed material to the defective material shows the effect of the crystal faults directly.

The preparative approach described here, the combination of constituent ions followed by the attachment of an organic terminus, is closely related to molecular cluster chemistry of the type advanced by Dance and co-workers (*33, 34*) and You and Holm (*35*). In these cases the constituent ions are combined in the presence of capping agents in homogeneous reaction media, and molecular clusters are isolated by crystallization. Because crystallization is the method of isolation, the cluster compounds prepared via this route are perfectly monodisperse and have been structurally characterized by single-crystal crystallography. Clusters prepared and characterized by this approach are at the smaller limit of the "quantum size regime"; however, this general method has been extended recently to larger clusters (*36*).

Semiconductor Clusters via Arrested Thermolysis

Thus far we have established that bulk II–VI materials can be prepared by the thermolysis of molecular precursors such as $Cd(SeC_6H_5)_2$, and that isolated, passivated, nanometer-sized crystallites of II–VI materials can be prepared by arrested ionic precipitation. Questions naturally arise concerning possible relationships between the two processes. Are molecular reagents such as $Cd(SeC_6H_5)_2$ useful precursors to nanoclusters?

In fact, when a dilute solution of $Cd(SeC_6H_5)_2$ in a coordinating solvent is heated to moderately high temperatures, nanometer-sized clusters of CdSe are formed (*9*). Short heating times give small clusters, and longer heating times give larger clusters, the ultimate particle size being determined by solubility. The progress of this set of reactions can be followed by optical absorption spectroscopy (Figure 2). As the process evolves, the characteristic optical absorptions move steadily to lower energy. Electron microscope investigation of the solution at any point along the process path shows that the solution-phase thermolysis reaction generates nanometer-sized crystallites of CdSe and thereby supports the assertion that the shifts in optical absorption are due the growth of CdSe quantum crystallites.

Arrested thermolysis is a valuable alternative to arrested precipitation. Arrested ionic precipitation reactions most typically require aqueous solvent. Because many semiconductor materials of interest are sensitive to water, this requirement for aqueous solvent is a serious limitation. In the same vein many of the candidate ionic reagents react vigorously with water. Arrested thermolysis reactions allow a wider variety of solvents to be used. As distinct from the arrested precipitation reactions that yield II–VI products, arrested thermolysis reactions occur at substantially higher temperatures. This behavior can be an advantage by giving a more (internally) crystalline product because the cluster-formation and cluster-annealing steps can be combined into a single process. A drawback to the arrested thermolysis reaction type is the apparent lack of size-control mechanisms.

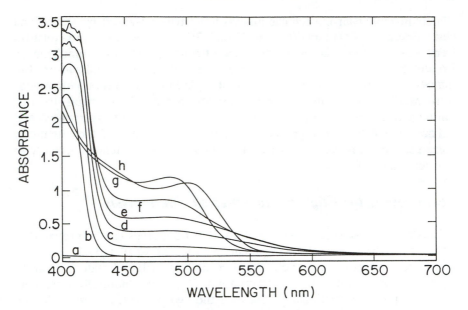

Figure 2. UV–visible absorption spectra taken during the solution-phase thermolysis of Cd(SeC$_6$H$_5$)$_2$. Spectrum a shows the absorption of the molecular starting material, and spectra b through h show the nucleation and growth of CdSe clusters as the solution of the molecular precursor is heated. (Reproduced from reference 9. Copyright 1989 American Chemical Society.)

Molecular compounds of the form M(ER)$_2$ (M = Zn, Cd, or Hg; E = S, Se, or Te; R = hydrocarbon) can be used to make II–VI semiconductor materials in either bulk or nanocluster form (*10*). In view of the previous discussion of the molecular synthesis of II–VI compounds, these studies form a direct link between molecular chemistry, nanophase chemistry, and semiconductor crystal growth.

Nanoscale Clusters of Other Semiconductor Materials

Much of the research in semiconductor nanoclusters has centered on II–VI materials. This fact is largely owing to the facility with which these materials form. Not only do the metal–nonmetal bonds form quickly from readily available reagents, but internal crystallization occurs under equally mild conditions. As mentioned, the internal crystallinity of semiconductor crystallites is crucial to the analysis and understanding of their properties; therefore, the internal crystallinity of the preparative sequences is of great importance.

Other families of semiconductor materials that have been studied include, but have in no way been restricted to, I–VII materials (e.g., silver halides) (*37*), III–V materials [e.g., GaAs (*38*) and InP (*39*)] and Group IV semiconductors [primarily Si (*40*)]. Silver halides form quickly at room temperature from aqueous systems to give internally crystalline large clusters. Much more forcing conditions have been required to produce crystalline particles of the III–V materials and Si. The feature most often invoked to explain the difference in crystallization behavior is the ionicity of the solid: The more ionic the solid (equivalently, the greater the electronegativity difference between the constituent elements), the greater are the ionic mobilities and the lower the energy that is required for the isomerization of the initially formed particle to the perfectly crystalline cluster.

The notion of directed valence supplements this explanation. In idealized terms (i.e., ignoring valence-bond resonance), each atom in a II–VI material forms two directed valence bonds and two donor–acceptor bonds to its nearest neighbors. In a III–V material each atom forms three directed valence bonds and one donor–acceptor bond. In a Group IV material all four bonds are of the directed valence type. Coordination chemistry and organometallic chemistry precepts indicate that donor–acceptor bonds reorganize more easily than do directed valence bonds; therefore, materials in which the bonding is closer to the donor–acceptor type should isomerize (crystallize in the present terms) more readily, all else being equal. Silicon and carbon, materials in which the bonding is entirely of the directed valence type, are difficult to recrystallize because any isomerization requires, at best, the homolytic cleavage of covalent bonds. Such is not the case for compounds such as CdSe.

Molecular Clusters as Stepping Stones to Solid-State Compounds

The discussion just presented indicates that simple solids such as II–VI semiconductors can be prepared from molecular reagents such as $M(ER)_2$ and suggests that nanometer-sized fragments of the bulk solid can occur as intermediates in the molecules-to-solids reactions. Thermal reactions of the sort summarized by equation 4 are difficult to characterize more fully because they are, in a sense, thermal runaway reactions and as such are difficult to control. To learn more in detail about molecules-to-solids transformations, simpler chemical reactions are more valuable.

One type of reaction that is conceptually much simpler than the molecular pyrolysis reactions typified by equation 4 is ionic precipitation. This set has already been described in detail. A complementary reaction type is "atomic" precipitation. In a reaction of this type the reagents that are combined are not ionic salts but molecular complexes of the zero-valent (or low-valent) complexes of the constituent elements. As the examples that follow show, reactions of this ilk have several advantages:

1. Using the appropriate reagents, molecules-to-solids reactions can occur at very low temperatures.

2. No ionic by-products are formed. Ionic by-products can be a particular problem when dealing with solid-state products.

3. Reactions that are irrelevant to the solid-growing process are avoided. An example of this type of irrelevant reaction is implied in equation 4. In that process the tellurium-containing by-product, TeR_2, is formed in essentially quantitative yield. This result implies that Te–C bonds are moved in reactions that must be rather complicated but that are nonetheless not directly related to the formation of the HgTe crystalline product. This complication is unnecessary in principle.

4. The neutral molecule reactions are adaptable to vapor-phase processing.

5. The zerovalent complexes used are coordination complexes, and therefore coordination chemistry can be used directly to control the reactions and their products.

In each of the examples to follow, a zerovalent complex of a transition metal is combined with a "zerovalent" complex of a chalcogen, namely, a phosphine chalcogenide (41). The ultimate product of each reaction is a solid-state, transition metal chalcogenide. Molecular cluster compounds have been isolated from each reaction by moderating the conditions under which the reaction is conducted. When they are subsequently subjected to the more vigorous reaction conditions, the isolated molecular clusters form the same extended solid-state transition metal chalcogenide. In this sense, the isolated clusters are intermediates in the molecules-to-solids reactions. Inasmuch as a goal of this work it to discover how solid-state compounds form from molecular reagents, we rationalize the structures of the isolated clusters' intermediates in terms of the structures of the extended solids, and note that as the clusters grow the structure of the ultimate solid emerges.

Palladium Telluride and Small Fragments of the NiAs Structure

Trialkylphosphine tellurides are remarkable in their ability to deposit elemental tellurium reversibly (42):

$$\text{TePR}_3 \; \rightleftarrows \; \text{Te} + \text{PR}_3 \tag{5}$$

The ease of this equilibrium prompted the study of the reactions of phosphine chalcogenides with zerovalent complexes of transition metals.

Phosphine complexes of zerovalent palladium react with triethylphosphine telluride in toluene at reflux to give polycrystalline PdTe as the only observed solid-state product (*43, 44*):

$$\text{Pd(PR}_3)_4 + \text{TePR}_3 \; \longrightarrow \; \text{PdTe} + \text{PR}_3 \tag{6}$$

When the two reagents are combined in toluene at room temperature, the molecular compound $\text{Pd}_2\text{Te}_2(\text{PEt}_3)_4$, **1**, forms in high yield:

$$\text{Pd(PEt}_3)_4 + \text{TePEt}_3 \; \longrightarrow \; \text{Pd}_2\text{Te}_2(\text{PEt}_3)_4 \tag{7}$$

The structure of **1** shows a planar Pd_2Te_2 parallelogram terminated at the Pd vertices by phosphine ligands such that the coordination around each Pd is square planar. When it is heated to reflux in toluene, compound **1** is converted to the solid-state compound PdTe, and therefore the sequence outlined in equation 8 occurs.

$$\text{Pd(PR}_3)_4 + \text{TePR}_3 \; \longrightarrow \; \text{Pd}_2\text{Te}_2(\text{PEt}_3)_4 \; \longrightarrow \; \text{PdTe} \tag{8}$$

In terms of studying molecules-to-solids pathways, it is important to see if **1** can be considered in any way to be a fragment of bulk PdTe. Stoichiometric PdTe crystallizes in the NiAs structure. This structure type is described as hexagonally close-packed anions with cations filling the octahedral holes in the anion lattice. A ball-and-stick representation of the PdTe structure is shown in Figure 3a. In Figure 3b a single valence-bond resonance structure of the same solid is shown. The particular resonance form was chosen to emphasize the divalent nature of each Pd and Te atom, and in this representation the PdTe structure can be seen as being formed from Pd_2Te_2 units that are assembled in a coherent array.

Given this construction, compound **1** can be appreciated as a very small building block of the PdTe solid. The Pd_2Te_2 core of **1** and of PdTe are compared in Figure 4. Comparison of the two shows many metrical differences. These are expected because the coordination about each atom is quite different in the two instances.

It is tempting to suggest that bulk PdTe grows by the stacking of Pd_2Te_2 units from a fundamental $\text{Pd}_2\text{Te}_2(\text{PEt}_3)_4$ precursor, but this mechanism does not appear to be the case, at least in solution. When compound **1** and an excess of supporting phosphine ligand are allowed to

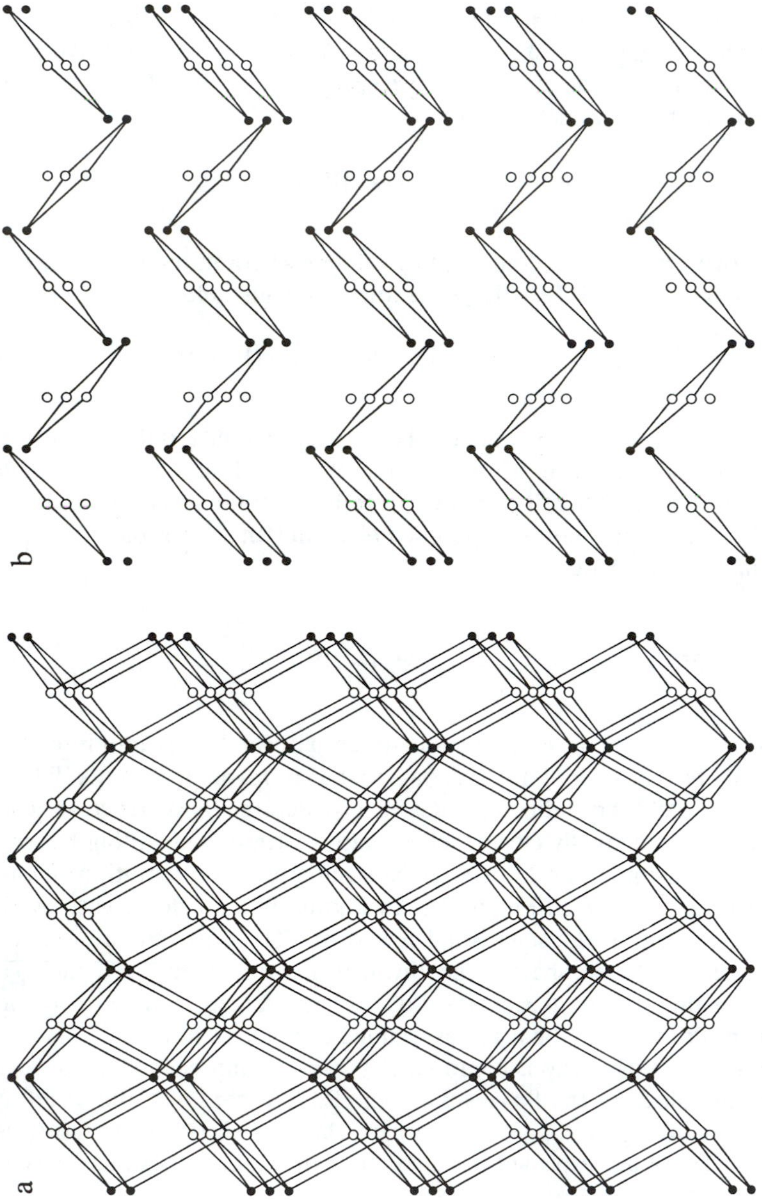

Figure 3. Ball-and-stick representation of bulk PdTe. The open circles represent Te atoms, and the filled circles represent Pd atoms. Part a includes all of the (equivalent) Pd–Te bonds. Part b is simplified by deleting all but two bonds to each atom. (Reproduced from reference 43. Copyright 1990 American Chemical Society.)

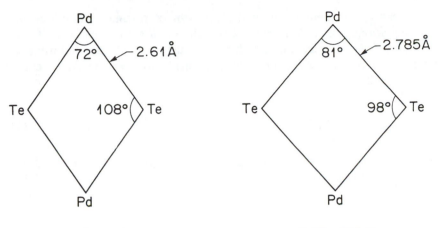

$Pd_2Te_2(PEt_3)_4$ PdTe BULK

Figure 4. Comparison of the Pd_2Te_2 units from $Pd_2Te_2(PEt_3)_4$ and solid PdTe. (Reproduced from reference 43. Copyright 1990 American Chemical Society.)

stand at room temperature in toluene for several days, a larger molecular cluster, $Pd_6Te_6(PEt_3)_8$, **2**, results (*44*). This more complicated cluster cannot be dissected directly into three Pd_2Te_2 parallelograms, but it is also related to a fundamental unit of the PdTe extended lattice, in this case to a Te-centered fragment of the bulk. (The structure of **2** and its relationship to that of PdTe are discussed in detail in reference 43.)

Cobalt Telluride and the Initial Steps in the Formation of a Solid

Dicobalt octacarbonyl is a readily available and convenient source of low-valent cobalt. It reacts with triethylphosphine telluride in the presence of additional triethylphosphine (*45*) to give the dicobalt ditelluride compound $[(CO)_2(Et_3P)_2CoTe]_2$, **3**:

$$Co_2(CO)_8 + 2TePEt_3 + 2PEt_3 \longrightarrow [(CO)_2(Et_3P)_2CoTe]_2 \quad (9)$$

The dicobalt ditelluride reacts with added dicobalt octacarbonyl to give a cluster compound, **4**, having the Co:Te stoichiometry ratio of 4:2:

$$[(CO)_2(Et_3P)_2CoTe]_2 + Co_2(CO)_8 \longrightarrow Co_4Te_2(PEt_3)_4(CO)_6 \quad (10)$$

This reaction is reminiscent of the reaction of organic ditellurides with Hg (equations 3 and 4), because **4** is a dimer of a Co–Te–Co unit. Furthermore, the tetracobalt ditellurium cluster, **4**, being tellurium-deficient, reacts with added phosphine telluride to give the larger cluster $Co_6Te_8(PEt_3)_6$, **5**:

$$3Co_4Te_2(PEt_3)_4(CO)_6 + 10TePEt_3 \longrightarrow 2Co_6Te_8(PEt_3)_6 \quad (11)$$

Finally, the Co_6Te_8 cluster can be converted to the solid-state compound Co_3Te_4 thermally (46):

$$Co_6Te_8(PEt_3)_6 \longrightarrow Co_3Te_4 + PEt_3 \quad (12)$$

Each of these Co–Te-containing molecular compounds is a stopping point on the path from the separate molecular precursors to the solid-state compound; consequently, structural relationships may exist between the molecules and the ultimate solid. (Similar to PdTe, Co_3Te_4 forms in the NiAs structure type (47). Because the stoichiometry is not 1:1, 25% of the octahedral holes in the hexagonal close-packed Te lattice are empty.)

The structures of **3**, **4**, and **5** were determined crystallographically. Compound **3** is a simple ditelluride having a Te_2 bridge linking two otherwise noninteracting Co_4 units. The only structural feature this compound could have in common with the Co_3Te_4 solid is the Co–Te bond length, and in the two cases the bond lengths are essentially identical. A complete structural analysis of **4** is rather complicated and is presented elsewhere (45). For the present discussion it is sufficient to say that the greater size and complexity of **4** afford more structural features to compare between the molecule and the solid. Several of these features, the general placements of the Co and Te atoms and the characteristic Co–Te–Co angle, compare well with corresponding sites in the solid; others, most notably the Co–Te distances, differ significantly from those in the solid.

In the structure of the Co_6Te_8 cluster, a severely distorted fragment of the Co_3Te_4 lattice can be discerned (46) (Figure 5). The molecular cluster is formed by an octahedron of six Co atoms concentric with a cube of eight Te atoms. The structure is terminated by a set of six phosphine ligands, one at each cobalt. A void site in the Co_3Te_4 is characterized by a Co_6Te_8 perimeter in which the six Co atoms are arranged in a trigonal prism. One may imagine a correspondence between the molecular Co_6Te_8 and the solid-state Co_6Te_8; the two are related by the process that distorts a trigonal prism to an octahedron. (The structural relationships between **5** and Co_3Te_4 are discussed in detail in references 45 and 46.)

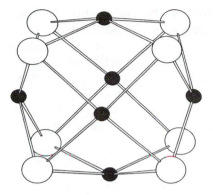

Figure 5. The molecular structure of $Co_6Te_8(PEt_3)_6$. The open circles represent the Te atoms. Taken together the eight form a cube. The filled circles represent Co atoms. The six together form an octahedron. The terminal phosphine ligands, one coordinated to each Co atom, are omitted. (Reproduced from reference 46. Copyright 1991 American Chemical Society.)

As many differences as similarities exist between the molecular compounds and the associated solid-state compounds, but as the first few steps from the original molecular ("atomic") precursors are made, aspects of the ultimate solid structure begin to develop.

Nickel Telluride and Large Molecular Clusters as Small Solids

For PdTe and Co_xTe_y just discussed, the clusters that have been isolated are fairly small. The isolation of larger clusters from "combination of atoms" reactions would make the hypothesis of the solid structure evolving with cluster size more clear.

Bis(cyclooctadiene)Ni [$Ni(cod)_2$] and triethylphosphine telluride react in refluxing toluene to form the extended solid δ-NiTe (48):

$$Ni(cod)_2 + TePEt_3 \longrightarrow NiTe + PEt_3 + cod \qquad (13)$$

By any one of several methods, this runaway reaction can be moderated, and molecular clusters can be isolated. When $Ni(cod)_2$ and triethylphosphine telluride are combined at room temperature in the presence of restrictive, chelating phosphines such as bis(dimethylphosphino)ethane [dmpe, $(CH_3)_2PC_2H_4P(CH_3)_2$], the NiTe-forming process is shut down almost entirely, the reaction yielding (49) the small-molecule $(dmpe)_2Ni_2Te_2$, **6**, (equation 14), a Ni-based relative of compound **1**:

$$Ni(cod)_2 + TePEt_3 + dmpe \longrightarrow (dmpe)_2Ni_2Te_2 \qquad (14)$$

(Like PdTe, the phase δ-NiTe crystallizes in the NiAs structure type; therefore, the identification of **6** as a fragment of NiTe follows the same reasoning as that for **1**.)

When $Ni(cod)_2$ is combined with a stoichiometric deficiency of phosphine telluride in the presence of an excess of supporting triethylphosphine, the cluster $Ni_9Te_6(PEt_3)_8$, **7**, forms (equation 15); and when equimolar amounts of $Ni(cod)_2$ and $TePEt_3$ are combined in toluene at room temperature there results the larger cluster $Ni_{20}Te_{18}(PEt_3)_{12}$, **8** (equation 16).

$$2Ni(cod)_2 + TePEt_3 + PEt_3 \longrightarrow Ni_9Te_6(PEt_3)_8 \qquad (15)$$

$$Ni(cod)_2 + TePEt_3 \longrightarrow Ni_{20}Te_{18}(PEt_3)_{12} \qquad (16)$$

These two clusters are larger than those already discussed, so it is valuable to determine whether or not they can be viewed as fragments of extended solids to which they are related.

A representation of the structure of $Ni_9Te_6(PEt_3)_8$ is shown in Figure 6. It shows a Ni atom in the center of a Ni_8 cube that is concentric with a Te_6 octahedron. The molecular structure is completed by the set of PEt_3 terminating ligands, one coordinated to each of the Ni atoms of the cube. Ignoring the Ni_8 cube, the central Ni atom in this cluster is surrounded by the octahedron of six Te atoms. This coordination geometry is precisely that seen around Ni in the NiTe (NiAs-type) solid, and in this sense **7** can be identified as a fragment of the NiTe solid (*48, 50, 51*).

The comparison between **7** and NiTe can be extended further: The stoichiometry of the fragment of NiTe generated by excising a Ni atom together with its nearest neighbors (six Te atoms) and next nearest neighbors (eight Ni atoms) is Ni_9Te_6—just that of **7**. The eight Ni atoms in the NiTe fragment are arranged in an hexagonal bipyramid; therefore, the rearrangement (reconstruction) needed to go from **7** to a fragment of NiTe is that from a Ni_8 cube to a Ni_8 hexagonal bipyramid. Attending this are significant changes in bond lengths. These bond-length changes can be rationalized by citing the absence of the external lattice in **7**.

Cluster **7** can also be appreciated as a fragment of the solid Ni_3Te_2 (*52*). A Ni-centered fragment of Ni_3Te_2 is shown in Figure 7. The reorganization needed to generate the nuclear structure of **7** is the symmetrization of the four Ni atoms labeled with asterisks in the figure. The relative placements of the other 11 atoms are common to the two structures.

Significant Ni–Ni bonding is apparent in **7**. The average distance between the central Ni atom and the eight external Ni atoms in **7** is 2.47 Å, a value that is quite close to the Ni–Ni distance in elemental Ni (2.49 Å). This fact is all the more important when it is noted that the distance

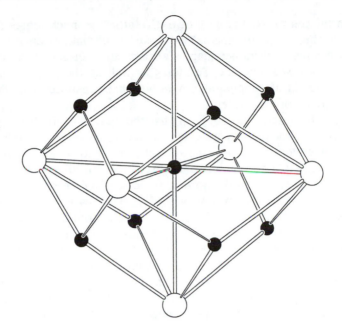

Figure 6. The molecular structure of $Ni_9Te_6(PEt_3)_8$. The open circles represent Te atoms, and the filled circles represent Ni atoms. The terminal phosphine ligands, one coordinated to each outer Ni atom, have been omitted. (Reproduced from reference 48. Copyright 1989 American Chemical Society.)

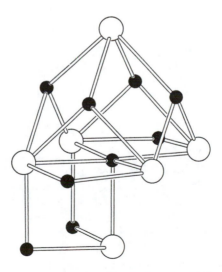

Figure 7. Ni-centered fragment of the bulk solid Ni_3Te_2. The six large open circles represent Te atoms, and the nine filled circles represent Ni atoms.

between the central Ni and each of the Te atoms is much longer (2.98 Å). In view of these bond distances, **7** be viewed as a small, reconstructed fragment of elemental Ni rather than as a fragment of either of the tellurides of nickel discussed, and the Te atoms as well as the phosphine ligands might be viewed as capping agents rather that constituents of the solid-mimicking core of the cluster.

From this discussion it is clear that the cluster $Ni_9Te_6(PEt_3)_8$ can be identified as a fragment of an extended inorganic solid, but it is equally clear that the identification of the subject solid is not unique. The lack of uniqueness is implicit in the previous discussions of PdTe and CoTe.

The larger cluster, **8**, is more complicated, and therefore the discussion of its structure is more involved (*51*). The key to the interpretation of the structure is the central Te atom that is coordinated by eight Ni atoms (labeled b and c in Figure 8). The six b Ni atoms form a distorted trigonal prism around the Te atom. This configuration is just the disposition of Ni atoms around each Te atom in the solid δ-NiTe. This knowledge motivates the identification of the $Ni_{20}Te_{18}$ core of **8** with a Te-centered fragment of δ-NiTe. In fact, if the two Ni atoms labeled c are ignored, the $Ni_{18}Te_{18}$ array can be analyzed (*51*) atom-for-atom as a piece of the NiAs structure type. Considering the earlier discussion of the Ni_9Te_6 fragment, this identification with the NiTe lattice is unique: The $Ni_{18}Te_{18}$ array cannot be a piece of either Ni_3Te_2 or of elemental Ni.

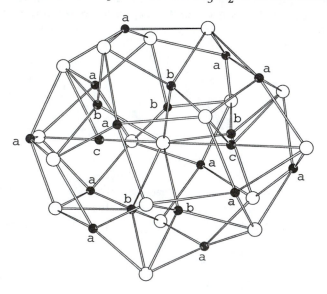

Figure 8. The molecular structure of $Ni_{20}Te_{18}(PEt_3)_{12}$. The open circles represent Te atoms, and the filled circles represent Ni atoms. The phosphine ligands are omitted. (Reproduced from reference 48. Copyright 1989 American Chemical Society.)

The two Ni c atoms may appear to be annoying impedimenta to this structural analysis, but their presence is rationalized easily by the identification of **8** as a fragment of δ-NiTe. This phase of NiTe is not a line phase but is stable over a wide range of Ni–Te stoichiometries near Ni_1Te_1 *(52)*. On the Ni-rich side the extra Ni atoms are accommodated by filling voids in the stoichiometric structure. [Complete occupation of these voids in the NiAs structure leads to the Cu_2Sb structure, the so-called "filled" NiAs structure *(53)*.] If the $Ni_{18}Te_{18}$ core of **8** is identified with the NiAs structure type, then the sites occupied by the two Ni c atoms are exactly those that would be occupied by Ni on the Ni-rich side of the δ-NiTe phase. In this sense the presence of the two "extra" Ni atoms in **8** helps to confirm the identification of this cluster as a small, Te-centered fragment of the associated extended solid.

Being deficient in Te, **7** may be expected to react with added $TePEt_3$. In fact, when **7** is combined with $TePEt_3$ in toluene at room temperature, the larger cluster, **8**, results *(44)*. This finding is important because it shows that although **7** is structurally related to several solid-state compounds, it reacts with added chemical feedstock to give a product whose structure is unambiguously related to only one of those solids.

For this restricted set of reactions, the formation of transition metal chalcogenides via the reaction of low-valent complexes of the constituent elements, several conclusions can be made:

1. Well-defined molecular clusters can occur as reaction intermediates.
2. Structural features of the ultimate solid-state products can be recognized in the cluster intermediates.
3. Significant differences exist between the structures of the intermediate clusters and the extended solids, and these differences (reconstructions) can be rationalized in local (i.e., molecular) terms.
4. The growth of the solid can occur in well-defined steps, with smaller isolated clusters growing to larger ones, and as the clusters become larger the similarities between the cluster and the solid become more refined.

The hope is that stepwise growth reactions of the type alluded to here will be generalized and become useful in the synthesis of, for example, monodisperse large clusters similar to those already described for the II–VI materials.

Summary

This review presented several different aspects of molecules-to-solids reactions. The first section described the use of molecular reagents in the preparation of thin films of II–VI compounds. The product of this type of reaction is a film of macroscopic extent, at least in two dimensions, and in order to gain insight into how the macroscopic product forms from microscopic reagents I stopped the process midstream.

The second section discussed the materials of intermediate size that result from interrupting the molecules-to-solids process: nanometer-sized fragments of the bulk II–VI materials. These compounds are intermediates not only in the chemical sense, but also in the physical sense, inasmuch as certain physical properties (e.g., crystalline structure and optical behavior) of these molecules–solids can be related directly to those of the bulk solid.

The study of these nanoscale materials has taught much about the formation and development of solid materials, but has been hampered by the distribution in cluster shapes and sizes that is intrinsic to the methods of cluster preparation. The third general section examined several molecules-to-solids reactions that were designed to be controllable. A variety of intermediate clusters can be isolated and identified, and each can be related to associated extended solids. In this way some of the pathways leading from molecules to solids can be elucidated in considerable detail.

References

1. An excellent introductory discussion of semiconductors is given in Kittel, C. *Introduction to Solid State Physics;* John Wiley & Sons: New York, 1986.
2. West, A. R. *Solid State Chemistry and Its Applications;* John Wiley & Sons: Chichester, England, 1984; Chapter 7.
3. Rao, C. N. R.; Gopalakrishnan, J. *New Directions in Solid State Chemistry;* Cambridge University Press: Cambridge, England, 1986; Section 1.8.
4. Stringfellow, G. B. *Organometallic Vapor-Phase Epitaxy: Theory and Practice;* Academic: San Diego, CA, 1989.
5. Steigerwald, M. L. In *Inorganometallic Chemistry;* Fehlner, T. P., Ed.; Plenum: New York, 1992, and references therein.
6. Kisker, D. W.; Steigerwald, M. L.; Kometani, T. Y.; Jeffers, K. S. *Appl. Phys. Lett.* **1987,** *50,* 1681.
7. Steigerwald, M. L.; Sprinkle, C. R. *J. Am. Chem. Soc.* **1987,** *109,* 7200.
8. Osakada, K.; Yamamoto, T. *J. Chem. Soc. Chem. Commun.* **1987,** 1117.
9. Brennan, J. G.; Siegrist, T.; Carroll, P. J.; Stuczynski, S. M.; Brus, L. E.; Steigerwald, M. L. *J. Am. Chem. Soc.* **1989,** *111,* 4141.
10. Brennan, J. G.; Siegrist, T.; Carroll, P. J.; Stuczynski, S. M.; Reynders, P.; Brus, L. E.; Steigerwald, M. L. *Chem. Mater.* **1990,** *2,* 403.
11. Cowley, A. H.; Jones, R. A. *Angew. Chem. Int. Ed. Engl.* **1989,** *28,* 1215.
12. See, for example, Ploog, K. *Angew. Chem. Int. Ed. Engl.* **1988,** *27,* 593.
13. Brus, L. E. *J. Phys. Chem.* **1986,** *90,* 2555.
14. Brus, L. E. *IEEE J. Quantum Electron.* **1986,** *22,* 1909.

15. Bawendi, M. G.; Steigerwald, M. L.; Brus, L. E. *Annu. Rev. Phys. Chem.* **1990,** *41,* 477.
16. Steigerwald, M. L.; Brus, L. E. *Acc. Chem. Res.* **1990,** *23,* 183.
17. See, for example, Kittel, C. *Introduction to Solid State Physics;* John Wiley & Sons: New York, 1986; Chapter 17.
18. Wozniak, M. E.; Sen, A.; Rheingold, A. L. *Chem. Mater.* **1992,** *4,* 753.
19. Chan, Y. N. C.; Craig, G. S. W.; Schrock, R. R.; Cohen, R. E. *Chem. Mater.* **1992,** *4,* 885.
20. Wang, Y.; Herron, N. *J. Phys. Chem.* **1987,** *91,* 257.
21. Fendler, J. H. *Chem. Rev.* **1987,** *87,* 877.
22. Meyer, M.; Wallberg, C.; Kurihara, K.; Fendler, J. H. *J. Chem. Soc. Chem. Commun.* **1984,** 90.
23. Petit, C.; Pileni, M. P. *J. Phys. Chem.* **1988,** *92,* 2282.
24. Fojitk, A.; Weller, H.; Koch, U.; Henglein, A. *Ber. Bunsenges. Phys. Chem.* **1984,** *88,* 969.
25. Henglein, A. *Chem. Rev.* **1989,** *89,* 1861.
26. Hoffmann, H.; Ebert, G. *Angew. Chem. Int. Ed. Engl.* **1988,** *21,* 902.
27. Steigerwald, M. L.; Alivisatos, A. P.; Gibson, J. M.; Harris, T. D.; Kortan, R.; Muller, A. J.; Thayer, A. M.; Duncan, T. M.; Douglass, D. C.; Brus, L. E. *J. Am. Chem. Soc.* **1988,** *110,* 3046.
28. Steigerwald, M. L.; Brus, L. E. *Annu. Rev. Mater. Sci.* **1989,** *19,* 471.
29. Alivisatos, A. P.; Harris, A. L.; Levinos, N. J.; Steigerwald, M. L.; Brus, L. E. *J. Chem. Phys.* **1988,** *89,* 4001.
30. Thayer, A. M.; Steigerwald, M. L.; Duncan, T. M.; Douglass, D. C. *Phys. Rev. Lett.* **1988,** *60,* 2673.
31. Alivisatos, A. P.; Harris, T. D.; Carroll, P. J.; Steigerwald, M. L.; Brus, L. E. *J. Chem. Phys.* **1989,** *90,* 3463.
32. Bawendi, M. G.; Kortan, A. R.; Steigerwald, M. L.; Brus, L. E. *J. Chem. Phys.* **1989,** *91,* 7282.
33. Dance, I. G. *Polyhedron* **1986,** *5,* 1037.
34. Lee, G. S. H.; Craig, D. C.; Ma, I.; Scudder, M. L.; Bailey, T. D.; Dance, I. G. *J. Am. Chem. Soc.* **1988,** *110,* 4863.
35. You, J.-F.; Holm, R. H. *Inorg. Chem.* **1991,** *30,* 1431, and references therein. See also: Lee, S. C.; Holm, R. H. *Angew. Chem. Int. Ed. Engl.* **1990,** *29,* 840.
36. Wang, Y.; Herron, N. *J. Phys. Chem.* **1991,** *95,* 525.
37. Johansson, K. P.; McLendon, G.; Marchetti, A. P. *Chem. Phys. Lett.* **1991,** *179,* 321.
38. Olshavsky, M. A.; Goldstein, A. N.; Alivisatos, A. P. *J. Am. Chem. Soc.* **1990,** *112,* 9438.
39. Douglas, T.; Theopold, K. H. *Inorg. Chem.* **1991,** *30,* 594.
40. Jasinski, J. M.; LeGoues, F. K. *Chem. Mater.* **1991,** *3,* 989.
41. Austad, T.; Rød, T.; Åse, K.; Sonstad, J.; Norbury, A. H. *Acta Chem. Scand.* **1973,** *27,* 1939.
42. Zingaro, R. A. *J. Organomet. Chem.* **1963,** *1,* 200.
43. Brennan, J. G.; Siegrist, T.; Stuczynski, S. M.; Steigerwald, M. L. *J. Am. Chem. Soc.* **1990,** *112,* 9233.
44. Steigerwald, M. L.; Stuczynski, S. M.; Kwon, Y.-U.; Vennos, D. A.; Brennan, J. G. *Inorg. Chim. Acta* **1993,** *212,* 219–224.

45. Steigerwald, M. L.; Siegrist, T.; Stuczynski, S. M. *Inorg. Chem.* **1991,** *30,* 4940.

46. Steigerwald, M. L.; Siegrist, T.; Stuczynski, S. M. *Inorg. Chem.* **1991,** *30,* 2256.

47. Wells, A. F. *Structural Inorganic Chemistry,* 5th ed.; Clarendon: Cambridge, United Kingdom, 1984; Chapter 17.

48. Brennan, J. G.; Siegrist, T.; Stuczynski, S. M.; Steigerwald, M. L. *J. Am. Chem. Soc.* **1989,** *111,* 9240.

49. Hessen, B.; Steigerwald, M. L., unpublished results.

50. Wheeler, R. A. *J. Am. Chem. Soc.* **1990,** *112,* 8737.

51. Nomikou, Z.; Schubert, B.; Hoffmann, R.; Steigerwald, M. L. *Inorg. Chem.* **1992,** *31,* 2201.

52. Stevels, A. L. N. *Philips Res. Rep. Suppl.* **1969(9),** 1.

53. Pearson, W. B. *The Crystal Chemistry and Physics of Metals and Alloys;* Wiley-Interscience: New York, 1972; pp 530–531.

RECEIVED for review November 9, 1992. ACCEPTED revised manuscript March 18, 1993.

Organometallic Chemical Vapor Deposition of Compound Semiconductors

A Chemical Perspective

Klavs F. Jensen

Departments of Chemical Engineering and Materials Science and Engineering, Massachusetts Institute of Technology, Cambridge, MA 02139

Application of organometallic chemical vapor deposition (OMCVD) for the growth of thin films of compound semiconductors with controlled structures and specific materials properties is reviewed. The influence of the molecular structure and reactivity of the organometallic precursor on properties of OMCVD-grown films is illustrated with examples of growth of III–V (e.g., GaAs) and II–VI compound semiconductors (e.g., ZnSe and ZnTe). These studies demonstrate the considerable control of film structure, stoichiometry, and impurity concentration levels that may be achieved through understanding of the underlying gas-phase and surface deposition chemistry. The use of controlled surface reactions to achieve unique quantum confinement structures is also illustrated. Opportunities for exploiting the flexibility and potential of OMCVD in future thin-film optical and electronic applications are suggested. Finally, research efforts needed to expand the current understanding of precursor selection, reaction pathways, kinetics, and surface nucleation phenomena are discussed.

0065–2393/95/0245–0397$13.75/0

The OMCVD Process

Organometallic chemical vapor deposition (OMCVD) of compound semiconductors represents an area of materials synthesis and processing in
which chemistry continues to play an increasingly important role (1, 2).
The process involves the transport of organometallic reagents in the gas
phase to a heated substrate where a thin solid film is subsequently formed
via reactions of adsorbed precursors (see Figure 1). Depending on processing conditions, organometallic species are transported intact or undergo gas-phase reactions. Gas-phase reactions can be beneficial by increasing the reactivity of species reaching the growth surface, but more often
they lead to the formation of impurities and particulates that prevent realization of desired film electronic and optical properties. Understanding
the manner in which structure and reactivity of the starting reagents affect
the final thin-film properties, therefore, is critical to the OMCVD process.

Physical transport processes, that is, fluid flow, heat transfer, and
mass transfer within the reactor enclosure, also play a key role in deter-

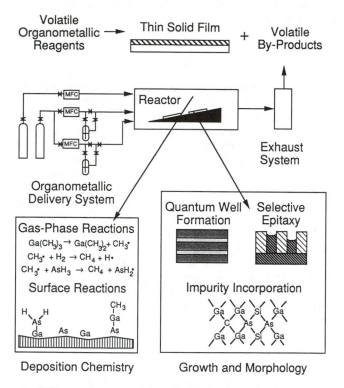

*Figure 1. Different elements of the OMCVD process, with growth of GaAs
from AsH$_3$ and Ga(CH$_3$)$_3$ as an example.*

mining the extent of gas-phase reactions and the access of precursors to the growth surface. Finally, the growth morphology must be controlled on the scale of nanometers to realize quantum confinement effects used in many compound semiconductor optical and electronic devices. The use of quantum effects and high-mobility electron structures in optical and high-speed devices requires the deposition of very thin (~5 nm) single-crystalline layers of high-purity materials surrounded by layers of different composition with a precisely controlled impurity (dopant) concentration. A typical example of a quantum-well laser structure for applications in telecommunication is shown in Figure 2 (*3*). The single-crystalline nature of the film is emphasized in the physics and electronics literature by the use of the other common name for the vapor-phase growth process, namely, organometallic vapor-phase epitaxy (OMVPE). If the process is operated under high-vacuum conditions, only surface reactions occur, and it is then called chemical beam epitaxy (CBE) or metal–organic molecular beam epitaxy (MOMBE) (*4*).

The wide range of organometallic reagents available as film precursors makes OMCVD a highly versatile process for realizing unique optical

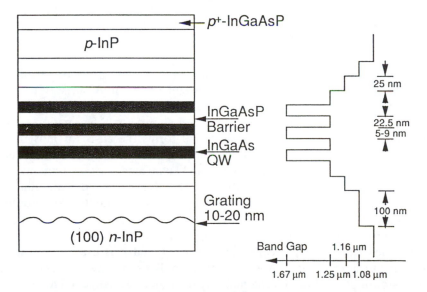

Figure 2. Example of a compound semiconductor device, a multiple quantum well, graded-index, separate confinement, distributed feedback (MQW–GRIN–SCH–DFB) laser. The active regions are made of $In_xGa_{1-x}As$, with $In_xGa_{1-x}As_yP_{1-y}$ as the barrier layers. The compositions (x and y) are adjusted to yield band gaps corresponding to the wavelengths shown in the figure, and they are further chosen to have a lattice mismatch of less than 3×10^{-4} to InP. (Reproduced with permission from reference 3. Copyright 1991 Elsevier.)

and electronic devices, including solid-state lasers, detectors, and transistors with high electron mobility. OMCVD is also, however, a complex materials synthesis scheme involving chemical and physical processes that must be controlled on the molecular level to achieve materials useful for device applications. A complete survey of the many facets of the process is beyond the scope of this volume, focused as it is on materials chemistry. General reviews of the process (1, 2 and references within) and a number of specific surveys of transport phenomena in OMCVD reactors (5, 6) and development of organometallic reagents (7–12, and additional references within refs. 1 and 2) have been published. The proceedings of the biennial International Conference on Metal–Organic Vapor-Phase Epitaxy (13–16) provide further insights into the different aspects of the process by combining advances in basic studies, new organometallic precursors, control of growth morphology, and device applications.

After a brief description of OMCVD applications, this review focuses on aspects of the process affected by chemistry, specifically, organometallic reagent development, gas-phase and surface kinetics, and finally, molecular control of growth process through surface chemistry.

OMCVD Applications

The flexibility of the OMCVD technique and the extensive opportunities for optical and electronic compound semiconductor devices is perhaps best illustrated by the energy-gap versus lattice-parameter diagram shown in Figure 3 (17). Composition curves are shown for a few important examples, such as the InGaAsP system that forms the basis for most fiber-optics communication systems. The device example, Figure 2, illustrates how the composition of this compound semiconductor system may be varied to produce layers with different band gaps, while maintaining a good lattice match to the InP substrate and thus avoiding the formation of dislocations.

Film growth on a single-crystalline substrate of the same material (e.g., GaAs on GaAs) is called homoepitaxy, and it is used to synthesize higher purity material than is available by bulk growth and to achieve thin layers of doped material. The majority of compound semiconductor devices, including the one exemplified in Figure 2, are based on heteroepitaxy, in which the growing film has a different composition from that of the substrate material (e.g., $Al_xGa_{1-x}As$ on GaAs and $In_xGa_{1-x}As_yP_{1-y}$ on InP). If the lattice parameters of the film and the underlying substrate are very close, as in $Al_xGa_{1-x}As$ on GaAs, the growth proceeds similarly to homoepitaxy. However, if there is a mismatch of lattice parameters, strain will develop with increasing film thickness and eventually be relaxed

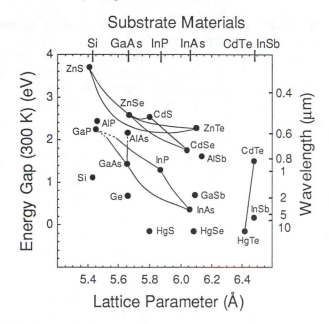

Figure 3. *Band-gap (in electron-volts on the left axis and corresponding wavelength in micrometers on right axis) vs. lattice parameters for elemental and compound semiconductors. Typical substrate materials are also displayed. Connecting band-gap and lattice-parameter curves are shown for selected systems. (Data are from reference 17.)*

through the formation of dislocations that severely limit device performance. Strained, very thin layers ($\sim$10 nm), without dislocations, so-called pseudomorphic layers, provide yet another option for device designs (*18–20*). In addition, unique quantum confinement device geometries can be realized by using the often large variation growth rate exhibited by different crystal planes of the substrate (*21–23*). This use of surface chemistry to tailor unique quantum confinement devices is discussed further later.

The energy-gap–lattice-parameter diagram (Figure 3) suggests it would be feasible, in principle, to realize light-emitting devices and detectors covering the spectrum from the far IR ($\lambda \sim$12 μm) to the near UV ($\sim$300 nm) by OMCVD of compound semiconductors. Narrow-band-gap Group II–VI materials, specifically, CdHgTe, are currently being used for far-IR detectors (*11, 12*), and efforts are underway to use strained-layer structures to extend the operation of InSb-based devices to the same wavelength region (*24*). Devices operating in the mid-IR region ($\lambda \sim$2–5 μm) have potential for both long-wavelength fluoride optical fiber communications (*25*) and environmental applications. For example, the availability of bright-light sources would enhance the use of IR spectroscopy in the detection of trace chemical pollutants.

By far, most OMCVD-grown devices have been developed for the near-IR regions ($\lambda \sim 1.3–1.5$ and ~ 0.8 μm corresponding to InGaAsP and AlGaAs materials systems, respectively). As already mentioned, InGaAsP layers are used in solid-state lasers and detectors for fiber-optics communications (26). AlGaAs has the processing advantage of a nearly complete lattice match between AlAs and GaAs (cf. Figure 3). OMCVD-grown heterojunction lasers were first demonstrated for this system (27). Current applications include lasers for printers and compact disc players, as well as high-power units for pumping of solid-state lasers, such as YAG lasers (28).

Recently visible diodes and lasers, emitting in the red to yellow region, have been developed on the basis of the materials system AlInGaP (28, 29). These lasers are replacing HeNe gas lasers in bar-code scanners, among other applications. High-power diodes are also finding increased use in the top center brake light on automobiles. Blue–green light emission is feasible with wide-band-gap II–VI (e.g., ZnSSe) (30 and references within) and III–V systems (e.g., GaInN) (31, 32), but the doping and growth morphologies of these systems have proven difficult to control. For example, ZnSe is easily doped n-type, whereas p-type conversion to a dopant level required for device purposes is an obstacle. p-Type doping has been achieved by nitrogen plasma doping in molecular beam epitaxy (MBE), and laser structures have been demonstrated (33, 34). However, sufficient doping levels have yet to be realized by OMCVD. Hydrogen incorporation has been identified as a possible cause for the problem (35, 36). The growth of high-quality GaN films is limited by nucleation effects, and materials typically have networks of extended defects (37). As for ZnSe, p-type doping and activation is difficult. Nevertheless, bright blue GaN-based diodes have been reported (38). The addition of Al further increases the band gap and gives promise of devices operating in the UV region of the spectrum.

OMCVD Reagents

Traditional reagents for the OMCVD process are alkyls (e.g., AlR_3, GaR_3, InR_3, and ZnR_2 with R = CH_3 or C_2H_5) and hydrides (e.g., AsH_3, PH_3, and H_2Se) (1, 2). These reagents have been purified and deposition conditions have been optimized to such an extent that high-purity compound semiconductors are routinely manufactured for device applications of the commercially significant systems, AlGaAs, GaInPAs, and AlInGaP. Nevertheless, considerable effort has been expended in the development of new sources motivated by the following issues:

- reduced carbon incorporation, in particular in AlGaAs
- elimination of toxic, gaseous hydrides, specifically AsH_3 and PH_3
- lower growth temperatures, particularly for II–VI compound semiconductors
- reagent purification

Source purity is a concern common to testing of all new organometallic precursors because of the extreme purity constraints (parts-per-billion or ppb range) of compound semiconductors. The problem is further exacerbated by impurities influencing electrical and optical properties at low concentrations (parts per billion), at which chemical identification may be extremely difficult even with modern analytical techniques such as secondary ion mass spectroscopy (SIMS). An important step, therefore, is to identify whether impurities in a deposited layer are intrinsic to the deposition chemistry or may be removed by further purification of the reagents. Carbon and oxygen are typical examples of impurities that may result from unintentional reactions of a precursor. Improvements in purification and detection schemes have alleviated some of the difficulties in evaluating new organometallic sources (*2, 9*).

Selected examples will be used to illustrate progress made in reducing carbon incorporation, replacing toxic gases, and lowering growth temperatures through synthesis of alternative reagents and understanding of growth kinetics. Additional assessments of the state of novel OMCVD sources are given in other surveys (*7–12*).

Carbon Incorporation. The unintentional incorporation of carbon into the deposited film is an inherent problem associated with the use of organometallic reagents. It has been a particular concern in the growth of GaAs and AlGaAs, where carbon tends to incorporate on the Group V site as an electrically active impurity. The problem is especially pronounced when trimethylgallium is used, and high ratios of arsine to trimethylgallium (~100:1) are then necessary to reduce the carbon content to acceptable levels for common device applications ($\leq 10^{14}$ C atoms/cm^3). The mechanism underlying carbon incorporation has not been established, but the proposed reaction sequence shown schematically in Figure 4 is consistent with the experimental observations of preferential incorporation on the As site, and reduced carbon content with increasing presence of arsenic hydrides (*39, 40*). Carbon incorporation is viewed as occurring through surface reactions of $GaCH_3$, leading to a gallium methylene ($GaCH_2$) that reacts with an exposed Ga atom to place carbon on an As lattice site.

Alternatively, the highly reactive methylene may become rehydrogenated to $GaCH_3$, or it may react with an adsorbed As–H to form GaAs and a desorbing methyl radical. Thus, an increase in available H through

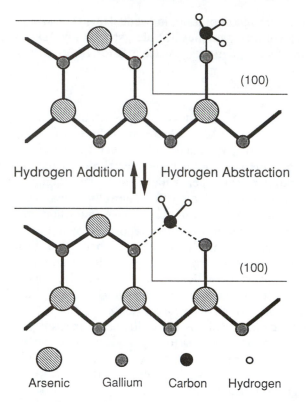

Figure 4. Proposed mechanism for carbon incorporation via a carbene intermediate.

introduction of more arsine would be expected to reduce carbon incorporation, as indeed is observed. The methylene species is essential to
predicting the site preference for carbon, as well as the relatively low carbon concentration in the solid film (≤ 1 ppm), compared to available
methyl groups. The process appears to be mainly controlled by surface reactions, because addition of methyl radicals in the gas phase has no significant effect (41).

When trimethylgallium is replaced by triethylgallium, carbon contamination is significantly reduced, and large ratios of arsine to trimethylgallium are no longer needed (42). This behavior has been attributed to the
β-hydride elimination of C_2H_4, offering an additional pathway for desorption of hydrocarbons from the growth surface besides the desorption of
radicals, which is the only option available to methyl species. The difference in carbon incorporation between trimethylgallium and triethylgallium
is especially apparent in MOMBE, in which arsine precracked to As_2 and
As_4 is typically used, and therefore, little surface H is available. Under
those conditions, trimethylgallium produces heavily carbon-contaminated

films, whereas impurity concentrations below 10^{15} atoms/cm^3 can be achieved with triethylgallium (*43*).

The replacement of trimethylgallium with triethylgallium may be viewed as a simple chemical solution to the carbon incorporation problem, obtained by the use of a reactant allowing for a more facile elimination of hydrocarbon ligands. An even more elegant alleviation of carbon problems in the ternary system, AlGaAs, has recently been achieved by replacing the Al–C bonds with Al–H bonds in the form of alane transported as a stable adduct with trimethylamine (*44–46*). Carbon incorporation in Al-GaAs is more severe than in GaAs, presumably because of the larger strength of the Al–C bond as compared to the Ga–C bond (*47*). Analogous to the use of triethylgallium, $Al(C_2H_5)_3$ has been explored as an alternative to $Al(CH_3)_3$ to obtain less carbon in AlGaAs. However, the low vapor pressure of $Al(C_2H_5)_3$ and premature decomposition of the reagent before the deposition have made its use difficult (*47*). Alane adducts have the advantage of having no carbon directly bonded to the Al, but transalkylation reactions in the gas phase with trimethylgallium may lead to the formation of $AlH_x(CH_3)_{3-x}$, which contributes to carbon incorporation (*48*). This interaction is reduced by lowering the growth pressure, and its effect is further diminished by using triethylgallium. The replacement of Ga alkyl sources with gallane would clearly be desirable, but gallane adducts have so far proven to polymerize too readily.

Carbon incorporation into AlGaSb is also an issue because there is no stable hydride source. On the other hand, carbon is much less of a concern with InP-based compounds, presumably because of the lower strength of the In–C bond as compared to Ga–C and Al–C. Carbon contamination has also been detected in II–VI systems such as ZnSe (*49*), but carbon is then no longer present as an acceptor-type defect, and carbon incorporation is generally less of a concern as compared to the AlGaAs system (*50*).

Replacement of Toxic, Gaseous Hydrides. Because of the potential environmental impact of an accidental release of highly toxic arsine and phosphine from their storage in high-pressure cylinders, extensive research has gone into the development of replacement organometallic As and P sources (*1, 2, 7, 8, 51*). A wide variety of alkylarsenic compounds have been evaluated, including trimethylarsine, triethylarsine, monoethylarsine, tetraethyldiarsine, and *tert*-butylarsine. Special structures such as phenylarsine, tris(trifluoromethyl)arsine, and dimethylaminoarsine have also been explored. Comparatively less effort has been devoted to organometallic P compounds. The trialkyls, trimethylphosphine and triethylphosphine, are too stable and form adducts with trimethylindium instead of yielding InP (*52*). Therefore, research has focused on *sec*-butylphosphine and *tert*-butylphosphine; *tert*-butylphosphine is preferred for the growth of InP (*53*).

The investigations of organometallic reagents as an alternative to the hydrides illustrate two key points related to OMCVD source development, namely, (1) the effect of molecular structure on film properties, specifically carbon incorporation, and (2) the need for high-purity organometallic precursors. Trialkyl sources, specifically trimethylarsine, led to extensive carbon incorporation (*54*) and the reagent was, in fact, a useful C-doping source (*55*). Hydrogen directly attached to arsenic, as in *tert*-butylarsine (H_2As-*t*-C_4H_9), appears to be necessary to remove hydrocarbon fragments and to avoid carbon incorporation, as discussed for arsine (*7, 8, 51*). Although the molecular structure and the decomposition mechanisms of *tert*-butylarsine appeared to be promising, the usefulness of the reagent for device applications was not resolved until the source could be synthesized in sufficient purity. Early reports (*56*) showed low electron mobilities (~5000 cm^2/V·s at 77 K) indicative of large impurity incorporation, but with improved purification, high electron mobilities (~100,000 cm^2/V·s at 77 K) were obtained a couple of years later (*57*). Both *tert*-butylarsine and *tert*-butylphosphine are now viable alternatives to arsine and phosphine, and device applications are being reported (*16*).

With MOMBE, the use of organometallic arsenic reagents has been limited not only by carbon incorporation, but also by the high-temperature precracking needed to break the C–As bond. Studies (*46*) with a new arsenic compound, tris(dimethylamino)arsine ($As[N(CH_3)_2]_3$), demonstrated that GaAs films, with no detectable carbon, may be grown at relatively low temperatures (~450 °C) without precracking when using trimethylgallium. This result is in contrast to the growth with arsine and trimethylgallium that leads to very high carbon incorporation (*43*). Considering that this new arsenic reagent has only As–N bonds, the absence of carbon seemingly contradicts the general observation that As–H bonds are needed to avoid carbon incorporation. Mass spectroscopy studies (*58*) revealed that tris(dimethylamino)arsine decomposes at relatively low temperatures (~450 °C), which rationalizes the growth without precracking of the arsenic source. The experiments also suggest a growth mechanism (*see* Figure 5) that involves transfer of hydrogen from the amine ligand to the surface. The resulting adsorbed hydrogen atoms may then act to remove carbon from the growth interface and prevent carbon incorporation, as already discussed (cf. Figure 4). Incidentally, no trimethylamine formation, which would have been a possible pathway for removal of methyl from trimethylgallium, was detected under growth conditions. This example further emphasizes the role of surface H in reducing carbon contamination and further demonstrates the need for caution in predicting the performance of organometallic precursors simply on the basis of their molecular structures.

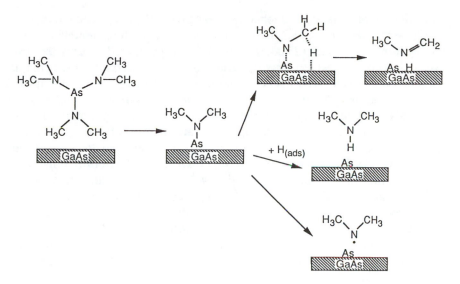

Figure 5. Proposed mechanism for the decomposition of tris(dimethylamino)arsine. The β-hydrogen transfer from the amino group to the surface is thought to be responsible for the low carbon incorporation observed for this precursor.

Organometallic Reagents for Low Growth Temperature.

The synthesis of new precursors, which would allow growth at the lowest possible temperatures, have been of particular interest for II–VI compound semiconductors because these tend to have relatively low-temperature stability and to be prone to temperature-activated defects (e.g., vacancies). Emphasis has been placed on tellurium reagents because of the commercial importance of $Cd_xHg_{1-x}Te$. By substituting increasingly branched hydrocarbon ligands in place of methyl groups, the Te–C bond strength has been reduced relative to dimethyltellurium, which has allowed a lowering of the growth temperature for CdTe and HgTe from ~500 °C for dimethyltellurium and ~410 °C for diethyltellurium to ~350 °C for diisopropyltellurium and ~300 °C for diisobutyltellurium (*59*). A further reduction in the growth temperature has been achieved using the allyl ligand, which upon homolysis produces the resonance-stabilized allyl radical and, therefore, has a relatively weak Te–C bond. For example, HgTe has been deposited at ~300 °C with methyl(allyl)tellurium (*60*) and at ~200 °C with diallyltellurium (*61*).

Unfortunately, organotellurium reagents, such as diallyltellurium and di-*tert*-butyltellurium, have very low vapor pressures (~1 torr (133.3 Pa) at room temperature), a property that makes them very difficult to use in OMCVD (*62*). Unsymmetrical perfluoroalkyltellurium species (*63*) have been proposed as precursors, and such use would combine the increased volatility characteristic of fluorinated reagents with a low decomposition temperature. For example, the vapor pressure of trifluoromethyl-*tert*-butyltellurium is 173 torr (23,061 Pa) at 17 °C, as compared to only 0.8 torr (106.6 Pa) for methyl-*tert*-butyltellurium at the same conditions, and the presence of the *tert*-butyl ligand is expected to result in a low decomposition temperature (*63*). Recent growth and mass spectroscopy studies (*64*) with this reagent nicely exemplify the often surprising complexity of OMCVD source decomposition chemistry and the possible danger of designing precursors by simple first-order arguments applied to each ligand separately. OMCVD of ZnTe with dimethylzinc and trifluoromethyl-*tert*-butyltellurium at low temperature (below 300 °C) results in formation of ZnF_2 (Figure 6a). Unique film morphologies of ZnTe with embedded micrometer-size crystallites of ZnF_2 (Figure 6b) arise at intermediate temperatures (~300–400 °C), and a sharp transition to pure ZnTe films (Figure 6c) appears above 400 °C. Thus, this example also illustrates the dramatic influence precursor chemistry may have on the deposited film.

a

*Figure 6. SEM micrographs of film morphologies resulting from OMCVD with dimethylzinc (20 μmol/h) and trifluoromethyl-*tert-butyltellurium (20 μmol/h) in hydrogen carrier gas at 10 torr (1333 Pa) on GaAs substrate (60). Part a: Deposition temperature = 250 °C, large ZnF_2 crystals and thin ZnTe film. Part b: Deposition temperature = 350 °C, ZnTe film with embedded ZF_2 crystals. Part c: Deposition temperature = 400 °C, ZnTe film with F content less than 0.1%.*

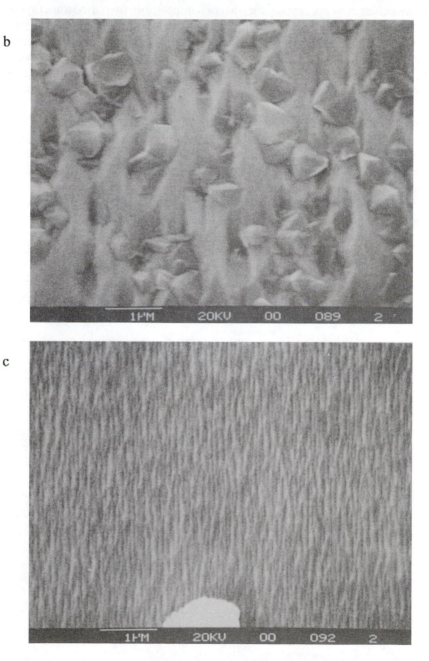

Figure 6. Continued.

Mass spectroscopy studies, including trapping experiments, suggest three competing pathways for the decomposition of trifluoromethyl-*tert*-butyltellurium: β-hydrogen elimination, homolytic fission, and difluoro-carbene elimination, as shown schematically in Figure 7. At low temperatures, β-hydrogen elimination leads to the formation of isobutene and the reactive intermediate, CF_3TeH, which either recombines with itself to bis-(trifluoromethyl)ditelluride, or it decomposes to hydrogen fluoride, Te, and difluorocarbene. The difluorocarbene, in turn, recombines to tetra-fluoroethene and adds to isobutene to yield 1,1-difluoro-2,2-dimethylcyclo-propane. The presence of hydrogen fluoride explains the formation of ZnF_2. The generated ditelluride readily eliminates Te that participates in the co-deposition of ZnTe. At higher temperatures, homolytic fission and radical reactions become an important mechanism of decomposition. This step favors the formation of the ditelluride and, consequently, the growth of ZnTe. At further elevated temperatures (>450 °C) the bis(trifluoromethyl)tellurium also contributes to the growth of ZnTe (*64*).

Difluorocarbene elimination could also be involved in the initial de-composition step at high temperatures. However, because trifluoromethyl-benzyltellurium, which has no β-hydrogen, displays greater thermal sta-bility than the *tert*-butyl source (*63*), β-hydrogen elimination appears to dominate for the present system.

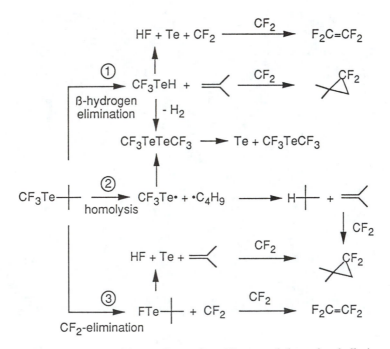

*Figure 7. Decomposition pathways for trifluoromethyl-*tert*-butyltellurium.*

Research into new organoselenium precursors for OMCVD has been driven primarily by issues in the growth of the wide-band-gap semiconductor, ZnSe, because of its potential applications in blue-light emitters. ZnSe films with good electronic and optical properties can be deposited at relatively low temperatures (300–350 °C) through the use of H_2Se and dimethylzinc (*65*). However, the films exhibit poor surface morphology and uniformity attributed to gas-phase adduct formation between H_2Se and dimethylzinc, followed by methane elimination reactions and oligomerization to large clusters (*10, 66*). This behavior, and the extreme toxicity of H_2Se, has motivated the investigation of less reactive organometallic selenium sources. Alkyl selenides, such as dimethyl selenide and diethyl selenide, have been used with dimethylzinc to deposit ZnSe layers at temperatures around 450–500 °C, with good surface morphology and optical properties (*67, 68*). However, as with tellurium containing compounds, lowering the deposition temperature is necessary to reduce native defect formation.

Following the low growth temperatures obtained through the use of allyl ligands in tellurium precursors, methyl(allyl) (*49, 69*) and diallyl (*70*) selenides were explored for growth of ZnSe. However, the species unexpectedly produced significantly higher levels of carbon incorporation than films grown from alkylselenium sources, which may be explained in terms of the decomposition mechanisms of the allyl precursors.

Gas-phase decomposition studies of allyl sulfides and ethers suggest that pyrolysis occurs by a rearrangement mechanism that involves a H-atom transfer in a six-centered ring-transition state (*71, 72*). The primary products are propene and the corresponding aldehydes for ethers, and thioaldehydes for sulfides. A study of the pyrolysis of organotellurium sources (*63*) showed that substituted hexadienes were the predominant decomposition product from allyltellurium reagents and indicated that decomposition occurred primarily by bond homolysis. Because the C–Se bond strength lies between that of C–Te and C–S, the pyrolysis of allyl selenides might be expected to occur by a combination of rearrangement and homolysis pathways; that is, for methyl(allyl) selenide:

rearrangement:

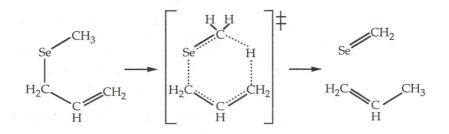

and homolysis:

$$H_3CSeCH_2CH=CH_2 \longrightarrow H_3CSe\cdot + \cdot CH_2CH=CH_2$$

$$2\cdot CH_2CH=CH_2 \longrightarrow CH_2=CHCH_2CH_2CH=CH_2$$

Mass spectroscopy studies revealed that a mixture of rearrangement and homolysis reactions did indeed take place for the allyl selenides, in contrast to allyltellurium compounds, and the excess carbon incorporation was explained in terms of the generation of selenoaldehyde in the rearrangement reaction (73). The allyl-Te· radicals formed in the homolysis of allyltellurium species lead to no significant levels of carbon incorporation, nor have alkyl-Se· radicals produced by pyrolysis of alkylselenium sources been reported to introduce carbon, except at very high ratios of selenide to zinc alkyls (69). On incorporation at the growth interface, the selenoaldehyde is less likely to lose its organic ligand and is then the most probable source of carbon in the ZnSe films grown from allyl selenides. This example again demonstrates the importance of understanding the decomposition chemistry and further illustrates that the presence of minor pathways leading to impurity incorporation can make a promising source useless for OMCVD applications.

Other Precursor Developments. In addition to the example just given, efforts have been extended to develop new organometallic Group II and Group III sources. Triisopropylgallium and tri-*tert*-butylgallium have been investigated as replacements for triethylgallium in MOMBE applications (74), but the tri-*tert*-butyl group yielded very low growth rates that were attributed to steric effects hindering the adsorption process. Research into new In precursors has considered both new reagents, such as triisopropylindium (75), and new ways to deliver a constant amount of the solid trimethylindium (76). This reagent is used in most applications, but its delivery rate is difficult to control because of changes in the surface area of the material as it evaporates. A new, chemical-based approach to this problem is to form a saturated solution of trimethylindium in a complexing solvent, for example, an amine, and then evaporate the indium precursor from the solution (76).

A number of coordination-saturated Ga and In species, for example, $(CH_2)_5Ga(CH_2)_3N(CH_3)_2$ and $(CH_3)_3InNH(iso\text{-}C_3H_7)_2$, have been investigated (77, 78) to alleviate the extreme sensitivity of Group III precursors to moisture and oxygen, as well as to reduce adduct formation with Group V precursors (e.g., arsine), which are Lewis bases. The Group III metal is saturated by coordination of sterically accessible nitrogen, which makes these so-called internal adducts nonpyrophoric and less sensitive toward alkoxide formation. Very promising materials properties have been obtained for GaAs, InP, and related alloys with this chemical approach to reducing problems with conventional precursors. Adducts of amines and Group II precursors, in particular, dimethylzinc, have also been used successfully to reduce prereactions in the growth of II–VI semiconductors, such as ZnSe, and to obtain improved electronic properties (*10, 79*).

Single-source precursors that directly yield the desired compound semiconductor, for example, $[(CH_3)_2Ga\text{-}\mu\text{-}As\text{-}(tert\text{-}C_4H_9)_2]_2$ → GaAs + hydrocarbons (*80*), have been developed with the aim of simplifying the OMCVD process and offering improved control of stoichiometry (*79, 80*). The concept has been demonstrated for both III–V and II–VI compound semiconductors including GaAs (*80*), InP (*81*), GaN (*82*), CdSe (*83*), and HgTe (*84*). Unfortunately, the vapor pressures of most of the single-source precursors are so low that the species are not practical for use in OMCVD. Nevertheless, these reagents may provide additional new insight into the growth processes underlying conventional OMCVD.

Current Trends in OMCVD

In addition to the development of new organometallic reagents for OMCVD, increasingly active efforts are directed toward understanding reaction mechanisms, developing in situ diagnostics, and controlling surface chemistry on the scale of a monolayer. Each of these areas could form the basis for lengthy reviews, so we will focus on a few, recent illustrative examples of these issues.

OMCVD Reactions: Spectroscopy and Kinetics. The preceding discussion of organometallic precursors demonstrated the strong link between precursor molecular structure of the precursor and resulting materials properties of the deposited film. Understanding this relationship is necessary to guide further source development and for the identification of optimal growth conditions. This need, in turn, provides the impetus for fundamental gas-phase and surface spectroscopy studies of reaction mechanisms and kinetics (*2, 5*). Issues and developments in CVD gas-phase measurements over the past decade recently were evaluated (*85*). Ultra-high-vacuum surface science techniques have begun to be applied to selected OMCVD systems, in particular, the surface reactions of trimethyl- and triethylgallium (*86–89*). These studies are directly relevant to MOMBE, but it is not clear to what extent the results will carry over to OMCVD processes running at near atmospheric pressure. Consequently, spectroscopies that could be used for in situ monitoring are also of interest (*90*).

Chemical reaction mechanisms and kinetics derived from these studies are essential to the development of quantitative models for simulating the combined effects of transport and reaction processes on OMCVD of thin films (5, 91) (see Figure 4). These models provide new insight into rate-limiting processes and enable the design of new OMCVD reactor configurations by delineating the factors controlling film thickness and composition uniformity, as well as impurity incorporation. Figure 8 exemplifies the complexity of the kinetic mechanisms involved in detailed simulations of deposition processes, even for simple systems such as growth of GaAs from trimethylgallium and arsine (40). The mechanism has more features than would be needed to correlate GaAs growth rate data, but the details are necessary to understand and predict the carbon incorporation behavior just described (40, 91).

Kinetic data are available only for a few OMCVD systems, so it is usually necessary to estimate a large number of the kinetic parameters in reactor models. Two approaches to this problem are possible. The easiest method is to select a small number of species participating in a simple mechanism and then fit a rate expression to experimental data. This empirical procedure has the advantage of being simple, but the model is not able to predict behavior outside the range of conditions used to fit the rate parameters.

The second and fundamental approach is to consider all plausible species and elementary reactions and then estimate rate parameters by transition-state theory (92) based on thermochemical computations. A sensitivity analysis is then usually needed to reduce the mechanism to a manageable size. Estimating surface reaction rate is considerably more difficult than estimating gas-phase reaction rates. The surface species are often poorly characterized, and the electronic structure of the solid makes it difficult to employ standard approaches for estimating reaction rates. Currently, it is necessary to resort to simple estimates based on bond strengths, steric effects, and dangling bonds, but continued advances in computation chemistry are expected to provide the tools necessary to quantify OMCVD reaction mechanisms.

In Situ Diagnostics. The complexity of the OMCVD process causes reactors to be operated in open loop, and composition variations due to changes in organometallic source delivery are then detected only in a "post mortem" analysis of the final layered structure. This lack of feedback control places the technique at a disadvantage relative to molecular beam epitaxy (MBE), in which reflection high-energy electron diffraction (RHEED) allows closed-loop control of device structures at the level of each atomic layer (93). Thus, a considerable incentive exists for the development of in situ, noninvasive, real-time monitoring methods for OMCVD (90).

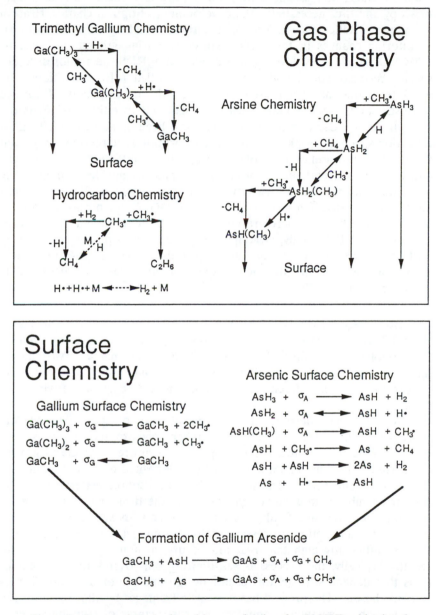

Figure 8. Proposed GaAs deposition mechanism for OMCVD with trimethylgallium and arsine.

The relatively high-pressure (0.01–1 atm), reactive OMCVD environment precludes the use of conventional ultra-high-vacuum surface spectroscopy and thus necessitates the development of optical characterization
techniques. Reflectance-difference spectroscopy (RDS) is a useful tool for
monitoring changes in the surface structure during growth of GaAs by
OMCVD (*90, 94*). By developing a catalogue of RDS spectra for different
surface reconstructions under precisely controlled MBE conditions, the
RDS technique has been used to explore the development of surface
reconstructions when a GaAs surface is exposed to either arsine or trimethylgallium. Useful growth data have also been obtained by differential
reflection spectroscopy, known as photoabsorption spectroscopy (*95*), even
though the interpretation of observed spectra is difficult. Spectroscopic
ellipsometry is a particularly promising technique capable of monitoring
both film thickness and alloy composition. Spectroscopic ellipsometry has
been demonstrated in a feedback control of AlGaAs quantum-well devices
by MOMBE (*96*), and there are no fundamental limitations to its application in OMCVD. Finally, although not practical as a process monitoring
tool, in situ synchrotron X-ray scattering studies of GaAs growth provided
significant new insight into the surface structure under actual OMCVD
growth conditions (*97*).

Surface-Controlled Growth. An exciting, recently applied area
of OMCVD is the synthesis of structures by growth processes tailored on
the monolayer scale through the use of surface chemistry. Two characteristic examples of the techniques used are atomic layer epitaxy and use
of growth rates on different crystal planes to define quantum wires and
dots.

Atomic layer epitaxy (ALE) is a special mode of operating OMCVD
processes such that the substrate is alternately exposed to the source
gases. Figure 9 illustrates the principle of the process for GaAs growth
from trimethylgallium and arsine. GaAs is deposited by first exposing the
substrate to trimethylgallium, purging the excess source material from the
reactor chamber with a carrier gas, exposing the trimethylgallium-covered
substrate to arsine, and finally, purging the excess arsine from the system.
If one of the adsorption processes is self-limiting and the growth temperature is sufficiently high that reaction occurs, a monolayer of GaAs will
result. Typically, the process is self-limiting only over a range of conditions that depend strongly on reactor design and the organometallic precursors chosen. The process has obvious advantages by allowing layer-by-
layer growth of completely uniform heterostructures of compound semiconductors.

The process was originally developed for the growth of ZnS (*98*) and
has since been expanded to OMCVD of III–V compounds (e.g., GaAs,
GaP, and InAs) (*99, 100*). The exact surface mechanism underlying the

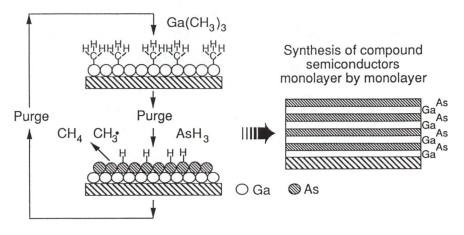

Figure 9.   Steps involved in atomic layer epitaxy of GaAs with trimethylgallium and arsine.

self-limiting growth mechanism has not been established. Competing theories based on selective adsorption, adsorbate inhibition, and a flux balance between etching and adsorption have been proposed (*86, 88*). ALE has not yet been produced at the same purity as OMVPE-grown material; specifically, carbon contamination is a problem. Nevertheless, the uniformity and atomic layer control inherent in the surface chemical control show promise for a wide variety of OMCVD applications.

The synthesis of unique quantum confinement structures, through use of the variation in surface reaction rate with crystallographic planes, is best illustrated in terms of the sample quantum wire and dot structures displayed in Figures 10a and 10b, respectively. The quantum wire is fabricated by using a selective etch to define grooves in a (100) GaAs substrate (*21*). AlGaAs grows faster on the (111)A side walls of the groove, whereas GaAs, under the condition used, grows faster on the (100) surface. This difference in growth habits results in the development of sharp corners between {111}A planes of the AlGaAs layers and the crescent-shaped GaAs quantum-well wire at the bottom of the form V groove (*see* Figure 10a).

The quantum dot example uses selective OMCVD in triangular openings defined in silicon oxide on (111)B GaAs to achieve three-dimensional carrier confinement (*22*). At high temperatures, AlGaAs and GaAs grow selectively on the exposed (111)B surface with (110) facets as sidewalls. AlGaAs buffer layers are grown to a thickness just below that necessary to completely form the tetrahedron. GaAs is then deposited on the tip, and the whole structure is therewith clad in AlGaAs (*see* Figure 10b). This strategy produces a three-dimensional confinement of carriers in the GaAs.

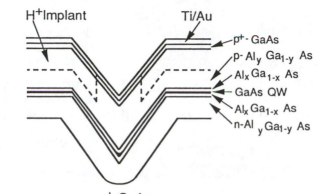

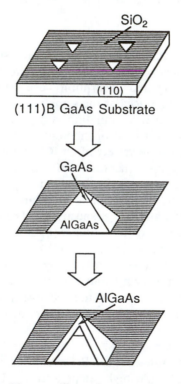

Figure 10. Part a: Diagram illustrating quantum wire structures formed on grooved substrates by OMCVD. (Reproduced with permission from reference 21. Copyright 1990 Elsevier.) Part b: Diagram illustrating the fabrication procedure for tetrahedral quantum dot structures. (Reproduced with permission from reference 22. Copyright 1992 Elsevier.)

Summary

OMCVD of compound semiconductors has been reviewed, with particular emphasis on the process chemistry. A considerable effort to synthesize new precursors is motivated by environmental concerns, reduced impurity incorporation, and the need for low-temperature processing, specifically for II–VI compound semiconductors. Significant advances in the development of new organometallic reagent precursors have been made for both III–V and II–VI materials. Moreover, an understanding is emerging for the relationship between precursor molecular structures and the resulting film materials properties. The use of surface kinetics controlled-growth processes on nonplanar and pattern surfaces is opening opportunities for new, unique structures based on electronic quantum confinement effects. Additional research into decomposition pathways of organometallic reagents, surface growth modes, and in situ spectroscopy techniques will be needed to understand and fully exploit the flexibility and power of OMCVD in the production of novel device structures for a wide range of optical and electronic applications.

Acknowledgments

The author is grateful to co-workers who have contributed to this chapter through discussions and research. Critical reading of the manuscript by M. Danek and S. Salim is also appreciated. The examples mentioned were supported by the National Science Foundation (DMR 9023162). Advanced Technology Materials, Air Products and Chemicals, and Epichem made donations of special organometallic precursors, and members of their research staffs are thanked for stimulating collaborations.

References

1. Stringfellow, G. B. *Organometallic Vapor Phase Epitaxy;* Academic: San Diego, CA, 1989.
2. Kuech, T. F.; Jensen, K. F. In *Thin Film Processes II;* Vossen, J. L.; Kern, W., Eds.; Academic: San Diego, CA, 1991; p 369.
3. Tanbun-Ek, T.; Logan, R. A.; Olsson, N. A.; Wu, M. C., Sergent, A. M.; Wecht, K. W. *J. Cryst. Growth* **1990,** *107,* 751–756.
4. Tsang, W. T. *J. Cryst. Growth* **1991,** *107,* 960.
5. Jensen, K. F. *J. Cryst. Growth* **1989,** *98,* 148.
6. Jensen, K. F.; Einset, E. O.; Fotiadis, D. I. *Annu. Rev. Fluid Mech.* **1991,** *23,* 199.
7. Brauers, A. *J. Cryst. Growth* **1990,** *107,* 281.

8. Lum, R. M.; Klingert, J. K. *J. Cryst. Growth* **1990**, *107*, 290.
9. Jones, A. C. *Chemtronics* **1989**, *4*, 5.
10. Jones, A. C.; Wright, P. J.; Cockayne, B. *J. Cryst. Growth* **1990**, *107*, 297.
11. Mullin, J. B.; Cole-Hamilton, D. J.; Irvine, S. J. C.; Hails, J. E.; Giess, J.; Gough, J. S. *J. Cryst. Growth* **1990**, *101*, 1.
12. Kisker, D. W. *J. Cryst. Growth* **1989**, *98*, 127.
13. *Proc. Third Int. Conf. Metal–Organic Vapor-Phase Epitaxy;* Stringfellow, G. B., Ed.; *J. Cryst. Growth* **1986,** 77.
14. *Proc. Fourth Int. Conf. Metal–Organic Vapor-Phase Epitaxy;* Watanabe, N.; Nakanisi, T.; Dapkus, P. D., Eds.; *J. Cryst. Growth* **1988,** 93.
15. *Proc. Fifth Int. Conf. Metal–Organic Vapor-Phase Epitaxy;* Richter, W.; Mullin, J. B., Eds.; *J. Cryst. Growth* **1991,** 107.
16. *Proc. Sixth Int. Conf. Metal–Organic Vapor-Phase Epitaxy;* Stringfellow, G. B.; Coleman, J. J., Eds.; *J. Cryst. Growth* **1992,** 124.
17. Landalt, H. H.; Börnstein, R. *Zahlenwerte and Functionenaris Physik, Chemie, Astronomie, Geophysik, Technik;* Springer: Berlin, Germany, 1977.
18. Matthews, J. W.; Blakeslee, A. E. *J. Cryst. Growth* **1975**, *29*, 273.
19. Matthews, J. W.; Blakeslee, A. E. *J. Cryst. Growth* **1976**, *32*, 265.
20. Osbourn, G. C. *J. Appl. Phys.* **1982**, *53*, 1586.
21. Bhat, R.; Kapon, E.; Simhony, S.; Colas, E.; Hwang, D. M.; Stoffel, N. G.; Koza, M. A. *J. Cryst. Growth* **1990**, *107*, 716–723.
22. Fukui, T.; Saito, H.; Kasu, M.; Ando, S. *J. Cryst. Growth* **1992**, *124*, 493–496.
23. Nishioka, M.; Tsukamoto, S.; Nagamune Y.; Tanaka, T.; Arakawa, Y. *J. Cryst. Growth* **1992**, *124*, 502.
24. Biefeld, R. M.; Hills, C. R.; Lee, S. R. *J. Cryst. Growth* **1988**, *91*, 515.
25. Bougnot, G.; Delannoy, F.; Pascal, F.; Grosse P.; Giani, A.; Kaoukab, J.; Bougnot, J.; Fourcade, R.; Walker, P. J.; Mason, N. J.; Lambert, B. *J. Cryst. Growth* **1991**, *107*, 502.
26. Pütz, N. *J. Cryst. Growth* **1991**, *107*, 806.
27. Dupuis, R. D.; Dapkus, P. D. *Appl. Phys. Lett.* **1977**, *31*, 466.
28. Crawford, G. *IEEE Circuits and Devices* **1992**, *September*, 24.
29. Katsuyama, T.; Yoshida, I.; Hashimoto, J.; Taniguchi, Y.; Hayashi, H. *J. Cryst. Growth* **1992**, *124*, 697.
30. Kukimoto, H. *J. Cryst. Growth* **1991**, *107*, 637.
31. Hiramatsu, K.; Amano, H.; Akasaki, I.; Kato, H.; Koide, N.; Manabe, K. *J. Cryst. Growth* **1991**, *107*, 509.
32. Matsuoka, T. *J. Cryst. Growth* **1991**, *124*, 433.
33. Park, R. M.; Troffer, M. B.; Rouleau, C. M.; Depuydt J. M.; Haase, M. A. *Appl. Phys. Lett.* **1990**, *57*, 2127.
34. Jeon, H.; Ding, J.; Patterson, W.; Nurmikko, D. V.; Xie, W.; Grillo, D. C.; Kobayashi, M.; Gunshor, R. L. *Appl. Phys. Lett.* **1991**, *59*, 259.
35. Kamata, A.; Mitsuhashi, H.; Fujita, H. *Appl. Phys. Lett.* **1993**, *63*, 3353.
36. Wolk, J. A.; Ager, J. W.; Duxstad, K. J.; Haller, E. E.; Taskar, N. R.; Dorman, D. R.; Olego, D. J. *Appl. Phys. Lett.* **1993**, *63*, 2756.
37. Amano, H.; Kito, M.; Hiramatsu, K.; Akasaki, I. *Jpn. J. Appl. Phys.* **1989**, *28*, L2112.
38. Skromme, B. J.; Liu, W.; Jensen, K. F.; Giapis, K. P. *J. Cryst. Growth* **1994**, *134*, 338.

39.   Mountziaris, T. J.; Jensen, K. F. *J. Electrochem. Soc.* **1991**, *138*, 2426.
40.   Masi, M.; Simka, H.; Jensen, K. F.; Kuech, T. F.; Potemski, R. *J. Cryst. Growth* **1992**, *124*, 483.
41.   Buchan, N. I.; Kuech, T. F.; Beach, D.; Scilla, G.; Cardone, F. *J. Appl. Phys.* **1991**, *69*, 2156.
42.   Kuech, T. F.; Potemski, R. *Appl. Phys. Lett.* **1985**, *47*, 821.
43.   Pütz, N.; Heinecke, H.; Weyers, M.; Heyen, M.; Lütz, H.; Balk, P. *J. Cryst. Growth* **1986**, *74*, 292.
44.   Jones, A. C.; Rushworth, S. A.; Bohling, D. A.; Muhr. G. T. *J. Cryst. Growth* **1990**, *106*, 246.
45.   Hobson, W. S.; Ren, F.; Lamont Schnones, M.; Sutz, S. K.; Harris, T. D.; Pearton, S. J.; Abernathy, C. R.; Jones, K. S. *Appl. Phys. Lett.* **1991**, *59*, 1975.
46.   Abernathy, C. A.; Jordan, A. S.; Pearton, S. J.; Hobson, W. S.; Bohling, D. A.; Muir, G. T. *Appl. Phys. Lett.* **1990**, *56*, 2654.
47.   Kuech, T. F.; Veuhoff, E.; Kuan, T. S.; Deline, V.; Potemski, R. J. *J. Cryst. Growth* **1986**, *77*, 257.
48.   Grady, A. S.; Markweel, R. D.; Russell, D. K.; Jones, A. C. *J. Cryst. Growth* **1990**, *106*, 239.
49.   Giapis, K.; Jensen, K. F.; Potts, J. E.; Pachuta, S. J. *J. Electron. Mater.* **1990**, *19*, 453.
50.   Nakamura, S.; Mukai, T.; Senoh, M. *Appl Phys. Lett.* **1994**, *64*, 1687.
51.   Stringfellow, G. B. *J. Electron. Mater.* **1988**, *17*, 327.
52.   Moss, R. H.; Evans, J. S. *J. Cryst. Growth* **1981**, *55*, 129.
53.   Larsen, C. A.; Chen, C. H.; Kitamura, M.; Stringfellow, G. B.; Brown, D. W.; Robertson, A. J. *Appl. Phys. Lett.* **1986**, *48*, 1531.
54.   Lum, K. M.; Klingert, J. K.; Kisker, D. W.; Abys, S. M.; Stevie, F. A. *J. Cryst. Growth* **1988**, *93*, 120.
55.   Kuech, T. F.; Tischler, M. A.; Wang, P. J.; Scilla, G.; Potemski, R.; Cardone, F. *Appl. Phys. Lett* **1988**, *53*, 1317.
56.   Chen, C. H.; Larsen, C. A.; Stringfellow, G. B. *Appl. Phys. Lett.* **1987**, *50*, 218.
57.   Haacke, G.; Watkins, S.; Burkhard, H. *Appl. Phys. Lett.* **1989**, *54*, 2029.
58.   Salim, S.; Lu, J. P., Jensen, K. F.; Bohling, D. A. *J. Cryst. Growth* **1992**, *124*, 16.
59.   Hoke, W. E.; Lemonias, P. J.; Korenstein, R. *J. Mater. Res.* **1988**, *3*, 329.
60.   Parsons; J. D.; Lichtman, L. S. *J. Cryst. Growth* **1988**, *86*, 222.
61.   Korenstein, R.; Hoke, W. H.; Lemonias, P. J.; Higa, K. T.; Harris, D. C. *J. Appl. Phys.* **1988**, *68*, 4929.
62.   Kirss, R. U.; Brown, D. W.; Higa, K. T.; Gedridge, R. W. *Organometallics* **1991**, *10*, 3589.
63.   Gordon, D. G.; Kirss, R. U.; Brown, D. W. *Organometallics* **1992**, *11*, 2947.
64.   Danek, M.; Patnaik, S.; Jensen, K. F.; Gordon, D. G.; Kirss, R. U.; Brown, D. W. *Chem. Mater.* **1993**, *5*, 1321.
65.   Giapis, K.; Lu, D. C.; Jensen, K. F. *Appl. Phys. Lett.* **1989**, *54*, 353.
66.   Stutius, W. *Appl. Phys. Lett.* **1978**, *33*, 656.
67.   Mitsuhashi, H.; Mitsuishi, I.; Kukimoto, H. *J. Cryst. Growth* **1986**, *77*, 219.
68.   Giapis, K. P.; Lu, D. C.; Fotiadis, D. I.; Jensen, K. F. *J. Cryst. Growth* **1990**, *104*, 629.

69. Giapis, K. P.; Jensen, K. F.; Potts, J. E.; Pachuta, S. J. *Appl. Phys. Lett.* **1989,** *55,* 463.

70. Patnaik, S.; Jensen, K. F.; Giapis, K. *J. Cryst. Growth* **1991,** *107,* 390.

71. Martin, G.; Ropero, M.; Avila, R. *Phosphorus and Sulfur* **1982,** *13,* 213.

72. Kwart, H.; Sarner, S. F.; Slutsky, J. *J. Am. Chem. Soc.* **1973,** *95,* 5234.

73. Patnaik, S.; Ho, K.-L.; Jensen, K. F.; Gordon, D. G; Kirss, R. U.; Brown, D. W. *Chem. Mater.* **1993,** *5,* 305.

74. Jones, A. C.; Lane, P. A.; Martin, T.; Freer, R. W.; Calcott, P. D. J.; Houlton, M. R.; Whitehouse, C. R. *J. Cryst. Growth* **1992,** *124,* 81.

75. Chen, C. H.; Chiu, C. T.; Stringfellow, G. B.; Gedridge, R. W. *J Cryst. Growth* **1992,** *124,* 88.

76. Frigo, D. M.; van Berkel, W. W.; Maassen, W. A. H.; van Mier, G. P. M.; Wilkie, J. H.; Gal, A. W. *J Cryst. Growth* **1992,** *124,* 99.

77. Hövel, R.; Brysch, W., Neuman, N.; Heime, K.; Pohl, L. *J Cryst. Growth* **1992,** *124,* 106.

78. Pohl, L.; Hostalek, M.; Schuman, H.; Hartmann, U.; Wasserman, W.; Brauers, A.; Regel, G. K.; Hövel, R.; Balk, P.; Scholz, F. *J Cryst. Growth* **1991,** *107,* 309.

79. O'Brien, P. *Chemtronics* **1991,** *5,* 61.

80. Ekerdt, J. G.; Sun, Y. M.; Jackson, M. S.; Lakhotia, V.; Pacheco, K. A.; Koschmieder, S. U.; Cowley, A. H.; Jones, R. A. *J Cryst. Growth* **1992,** *124,* 158.

81. Andrews, D. A.; Davies, G. A.; Bradley, D. C.; Faktor, M. M.; Frigo, D. M.; White, E. A. D. *Semicond. Sci. Technol.* **1988,** *3,* 1053.

82. Ho, K. L.; Jensen, K. F.; Hwang, J. W.; Gladfelter, W. L.; Evans, J. F. *J Cryst. Growth* **1991,** *107,* 376.

83. Brennan, J. G., Segrist, T.; Caroll, P. J.; Stuczynski, S. M.; Reynders, P.; Brus, L. E.; Steigerwald, M. L. *J. Am. Chem. Soc.* **1989,** *111,* 4141.

84. Arnold, J.; Walker, J. M.; Yu, K. M.; Bonasia, P. J.; Seligson, A. L.; Bourret, E. D. *J Cryst. Growth* **1992,** *124,* 647.

85. Breiland, W. G.; Ho, P. In *Chemical Vapour Deposition: Principles and Applications;* Hitchman, L. M.; Jensen K. F., Eds.; Academic: New York, 1993.

86. Yu, M. L., Buchan, N. I.; Souda, R.; Keuch, T. F. *Mater. Res. Soc. Symp. Proc.* **1991,** *222,* 3.

87. Donnelly, V. M.; McCaulley, J. A. *Surf. Sci.* **1990,** *238,* 34.

88. Greighton, J. R.; Banse, B. A. *Mater. Res. Soc. Symp. Proc.* **1991,** *222,* 15.

89. Greighton, J. R. *Surf. Sci.* **1990,** *234,* 287.

90. Aspnes, D. *Mater. Res. Soc. Symp. Proc.* **1990,** *198,* 341.

91. Jensen, K. F.; Fotiadis, D. I.; Mountziaris, T. J. *J. Cryst. Growth* **1990,** *107,* 1.

92. Benson, S. W. *Thermochemical Kinetics;* Wiley: New York, 1976.

93. Neare, J. H.; Joyce, B. A., Dobson, P. J.; Norton, N. *Appl. Phys.* **1983,** *A31,* 1.

94. Kamiya, I.; Aspnes, D. E.; Tanaka, H.; Florez, L. T.; Harbison, J.; Bhat, R. *Phys. Rev. Lett.* **1992,** *60,* 627.

95. Kobayashi, N.; Makimoto, T.; Yamauchi, Y.; Horikoshi, Y. *J. Cryst. Growth* **1990,** *107,* 62.

96.   Aspnes, D. E.; Quinn, W. E.; Tamargo, M. C.; Pudensi, M. A. A.; Schwartz, S. A.; Brasil, M. J. S. P.; Nahory, R. E. *Appl. Phys. Lett.* **1992,** *60,* 1244.
97.   Kisker, D. W.; Stephenson, G. B.; Fuoss, P. H.; Lamelas, F. J.; Brennan, S.; Imperatori, P. *J. Cryst. Growth* **1992,** *124,* 1.
98.   Suntola, T. In *Atomic Layer Epitaxy;* Suntola, T.; Simpson, M., Eds.; Chapman and Hall: New York, 1990.
99.   Tischler M. A.; Bedair, S. M. In *Atomic Layer Epitaxy;* Suntola, T.; Simpson, M., Eds.; Chapman and Hall: New York, 1990.
100. Dapkus, P. D.; Maa, B. Y.; Chen, Q.; Jeong, W. G.; DenBaars, S. P. *J. Cryst. Growth* **1991,** *64,* 1687.

RECEIVED for review January 15, 1993. ACCEPTED revised manuscript June 15, 1993.

Interfaces, Interfacial Reactions, and Superlattice Reactants

Thomas Novet, David C. Johnson*, and Loreli Fister

Materials Science Institute and Department of Chemistry, University of Oregon, Eugene, OR 97403

The structure and properties of interfaces are crucial in such key technological areas as electronic devices, structural composites, and nanocrystalline materials. Nevertheless, surprisingly little is known about controlling interfacial reactions. This chapter briefly overviews what is known about interfacial reactions between bulk reactants and describes the use of superlattice reactants in the study and control of interfacial reactions. Unlike previous methods, superlattice reactants allow for the control of superlattice structure via layer-by-layer deposition, the ability to follow the thermodynamics of the initial stages of interfacial reactions by using calorimetry, and the ability to follow the evolution of interfacial structure by using the low-angle superlattice diffraction pattern. Several examples are presented that demonstrate these advantages.

The Importance of Interfaces in Technology

Interfacial research focuses on the unique properties of the boundary regions and seeks to understand how they differ from bulk properties and how they influence the behavior of the composite (*1*). An interfacial boundary, or zone, is the transition region containing the composition gradient resulting from any interfacial reaction between two different materi-

*Corresponding author

als. This interfacial zone might be the boundary between two solid-state reactants, between silicon carbide and a metal alloy in a reinforced metal–ceramic composite, or between a metal interconnect and silicon in an integrated circuit. The width of this transition region depends on the conditions used when forming the interface, impurities present at the interface, and subsequent processing such as thermal annealing.

The study of interfaces and interfacial reactions is of increasing importance because of the development of ever-smaller microelectronic structures, complex heterogeneous catalysts, biologically compatible implants, and exotic multicomponent composites. In these systems, the composition and structure of the interfacial region are often crucial to performance and device–composite properties. The pervasiveness and importance of interfaces is illustrated in the following two examples: a single transistor from an integrated circuit board and a structural metal–matrix composite. Also, the fundamental importance of interfaces in solid-state reactions is discussed.

Microelectronic Devices. Figure 1 contains a schematic of the structure of a gallium arsenide based, modulation-doped, field-effect transistor. The performance of this device depends on the creation of a variety of interfaces (2). A Schottky barrier is created between the titanium–silicon composite and the undoped AlGaAs in the gate. Ohmic contacts are found at the source and the drain. Tungsten layers within the transistor act as diffusion barriers that prevent the solid-state reaction between the ohmic contacts and the gold-conducting layers. The properties of all of these interfaces depend on interfacial structure and are affected by the various processing steps used in the creation of the device.

The importance of interfaces in electronic devices prompted considerable research effort toward understanding device properties that are in-

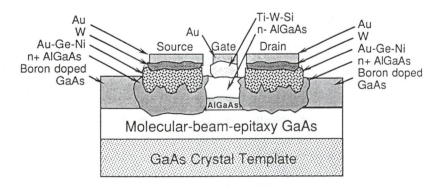

Figure 1. Structure of a gallium arsenide based, modulation-doped, field effect transistor.

fluenced by interfaces (*1*). Much of this effort was directed at optimizing such a multilayer device by determining the elements, compounds, or alloys most suitable for solving the immediate problem. This problem-by-problem approach contributed to the complexity of the final device, which contains many different materials. As the sizes of devices continue to shrink, interfaces will represent a larger fraction of the total volume, and their properties will become increasingly important in determining the characteristics of the device. Thus, an understanding of interfacial structure and reactivity needs to be developed to optimize performance and reduce the complexity of the final device.

Metal-Matrix Structural Composite Materials.

Interfacial properties are also crucial in determining the properties of advanced structural composite materials (*3*). The idea of creating a composite material combining the best properties of the individual components is not new. Steel-reinforced concrete and straw-reinforced adobe brick are two examples of this old concept. This ability to enhance the properties of existing materials via composites is used in many technological applications. In the aerospace industry, lightweight alloys or composites with higher strength and stiffness compared with conventional materials are needed to enhance performance. For example, the maximum operating temperature of a gas turbine is limited by the high-temperature distortion and oxidation of the nickel-based alloys from which they are made. Current research efforts are directed at duplicating the success of bulk composites on a much shorter scale to obtain new materials that can meet increasingly stringent technological demands (*4*).

Figure 2 illustrates one such promising composite—NiAl reinforced with silicon carbide fibers in which the diameter of the fibers is on the order of micrometers. The goal of this composite material is to use the strength of the ceramic stiffener and overcome the brittleness of the ceramic with the ductility of the host metal matrix. To achieve this goal, the interface between the metal matrix and the composite host must be strong enough to adsorb the energy of a crack propagating through the metal matrix and be weak enough to prevent the fracturing of the ceramic reinforcement. If the interface is too weak, a crack will propagate around the ceramic fiber. If the interface is too strong, the ceramic fiber will be severed as the crack propagates through the ceramic and continues in the metal matrix. With a suitably tailored interface, the composite will fail through a mode known as *fiber pullout*. This failure mode should be familiar to anyone who has broken a fiberglass- or graphite-reinforced bulk composite material such as a tennis racket. The overall strength is optimized because the crack energy is dissipated in the interfacial region surrounding the fiber in the composite.

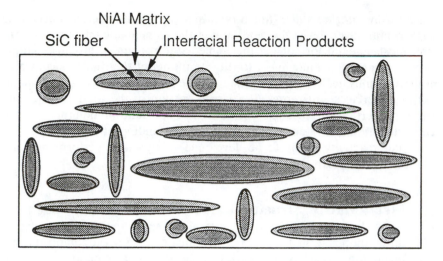

Figure 2. Structure of an NiAl-based, SiC-reinforced, metal–ceramic composite.

The length scale of the particles in a composite is even smaller in nanocrystalline materials. In these materials, the volume of material present at interfacial grain boundaries is approximately equal to that of the bulk grains. Nanocrystalline materials have enhanced properties relative to those of the bulk grains because of the extremely small grain size (*4*). These enhanced properties must result, in part, from the interfacial regions where metastable compositions or compounds may be stabilized. Success of this composite research is even more intimately dependent on controlling the properties of the interfaces between the components.

Solid-State Reaction at Interfaces. As seen in the two examples previously mentioned, performance of a composite material or device often depends on controlling the extent of the interfacial reaction. The control of interfacial reactions is also very important in the preparation of compounds via solid-state reactions. In a typical solid-state reaction, a product layer forms between the reactants and grows in a planar fashion (*5*). Because the reactants must diffuse longer distances through this product layer as it grows, the reaction rate decreases as a function of time. Obviously, the identity of the product layer, the diffusion rates of the reactants through this layer, and the distance the reactants need to move are all important to the overall rate of the reaction.

The reaction conditions used in traditional solid-state synthesis—high temperatures and long times—result from the desire and need to increase the diffusion rate of the reactants through the forming product layers at the interface (*6*). These conditions, however, also result in a loss of control of reaction intermediates (Figure 3). The high reaction temperatures permit any and all phases to nucleate and grow during the interdiffusion process. Although significant progress has been made in understanding model diffusion couples (*7*), no detailed understanding exists of either interdiffusion or nucleation in "real" synthetic systems containing multiple diffusion pathways through complicated, low-angle, grain boundary contacts. Solid-state chemists, consequently, have had to settle for finding a set of experimental conditions under which the desired product is thermodynamically stable. As recently summarized by DiSalvo (*8*), "The synthesis of novel extended solids is as much an art as a science."

The state of the art in solid-state synthesis is in significant contrast to synthetic molecular chemistry, where researchers have focused on kinetic approaches to desired products. In synthetic molecular chemistry, the details of the reaction mechanism are regarded as being as important as the product itself, in that they provide an important conceptual framework. Reaction mechanisms promote an understanding of cause-and-effect relationships between changes in synthetic parameters and product distributions. This kinetic approach to synthesis led to the great richness of small-molecule chemistry, highlighted by the isolation of structural and geometric isomers for materials with the same composition.

A similar, kinetic-based approach to the synthesis of extended solids that permits the preparation of metastable compounds is needed. In such a solid-state reaction pathway, it will be crucial to understand and control interfacial reactions. In this way, synthetic solid-state chemistry is linked to microelectronic devices, structural composites, and other interface-dependent technologies. In all of these areas, the development of techniques that monitor and control interfacial reactions, and therefore result in the ability to control reaction intermediates, will be a significant breakthrough.

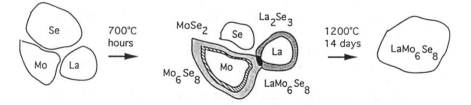

Figure 3. Reaction pathway found in a typical, diffusion-limited, solid-state reaction in which all stable compounds nucleate and grow at the interfaces between the reacting elements.

Interfacial Reactions

Interfaces are inherently difficult to study, because the amount of material in the interfacial region is very small relative to the bulk. In addition, the interfaces of interest are often buried within the sample, so common surface analysis probes cannot be used. The complexity of typical reacting mixtures also have thwarted efforts of investigators to examine interfacial reactions and determine reaction mechanisms in situ.

Diffusion Couples. One method of simplifying the aforementioned problem is to restrict the interface to a known location and geometry by creating a diffusion couple. A diffusion couple is made by placing two blocks of dissimilar materials in intimate contact with one another. A well-defined and planer interface results between the two materials. This geometry allows the study of the interdiffusion reaction as well as phase formation as a function of temperature and time. The length scale of the product layers usually investigated via bulk diffusion couples is on the order of micrometers, and detailed reaction mechanisms were proposed to explain the growth of the product layers (7, 9–13).

As in traditional solid-state preparative reactions, the interfacial reaction is diffusion-limited. High reaction temperatures and long reaction times are usually necessary as a result of the high activation energies of diffusion in the solid state. An undesirable consequence of these reaction conditions is that every thermodynamically stable binary phase in the phase diagram will nucleate and grow. The relative amounts of the various compounds will be determined by the relative diffusion rates of the elements through each of the compounds formed. Consequently, bulk diffusion couples have long been used to probe phase diagrams and to determine the existence of stable binary phases (14). A comparison of an iron–silicon diffusion couple both before and after prolonged heating is illustrated in Figure 4. A comparison of the phases formed in Figure 4 with the equilibrium phase diagram shown in Figure 5 illustrates the utility of this technique for exploring equilibrium phase diagrams.

Although the rate of growth of product layers in bulk diffusion couples as a function of time and temperature has been addressed, a continuing question has been how the initial products form at the interfaces. The two key steps in this process are interdiffusion of the initial reactants and nucleation of the product.

The kinetic aspects of interdiffusion reactions were dramatically illustrated in the 1970s and early 1980s when several research groups began to explore the reaction of thin-film diffusion couples of metal–silicon and metal–metal systems (9, 11, 15–20). In a typical investigation, crystalline films of ∼500 Å of each element were deposited to produce a diffusion

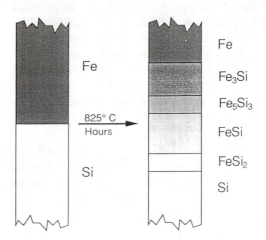

Figure 4. Bulk iron–silicon diffusion couple after prolonged annealing at 750 °C.

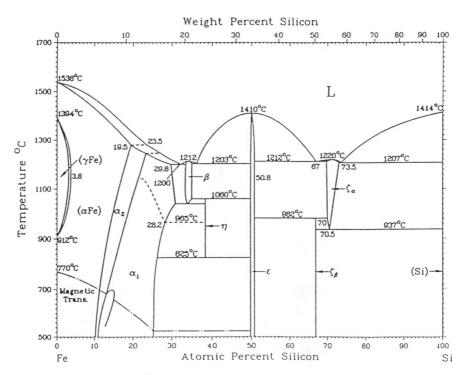

Figure 5. The iron–silicon phase diagram. (Reproduced with permission from reference 65. Copyright 1990.)

couple. This diffusion couple was subsequently annealed at temperatures of several hundred degrees Celsius to interdiffuse the elements. If the thickness of the individual layers was less than some critical distance, a compound nucleated at the interface and grew until it exhausted one of the reactants. Only then did another crystalline phase nucleate at the compound interface where some element remained. This second compound grew until it exhausted the supply of compound or element. In this sequential evolution, not all of the compounds in the phase diagram were necessarily formed (*21, 22*).

This sequential evolution results from diffusion and nucleation alternately determining the progress of the reaction. The rate of growth of a particular phase, and hence the time scale of the reaction, is governed by diffusion rates. However, at these length scales, nucleation of a new phase does not occur until one of the reacting layers is consumed. The next phase to form is then determined solely by nucleation energetics. Figure 6 illustrates the behavior observed in a prototypical thin-film diffusion couple, iron—silicon, as a function of temperature and time.

Solid-State Amorphization Reactions. The next obvious length-scale regime is for the entire film to become amorphous before the nucleation of any crystalline compound can occur at an interface. In 1983, Schwartz and Johnson (*23*) observed this behavior, reporting that crystal-

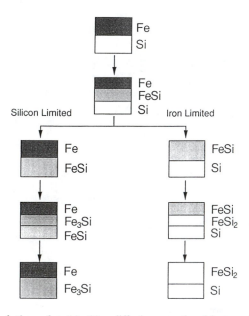

Figure 6. Evolution of a thin-film diffusion couple of iron and silicon as a function of temperature and time.

line Au–La multilayer composites with elemental layers of 300 Å interdiffused at low temperatures to form an homogeneous amorphous alloy. In a later publication (*24*), striking calorimetry and diffraction data were presented for the Ni–Zr system that clearly demonstrated the formation of an amorphous intermediate from the interdiffusion reaction of crystalline nickel and crystalline zirconium layers. Subsequent research (*25*) found that several metal–metal, metal–silicon, and metal–carbon systems also react at low temperatures via a solid-state amorphization reaction pathway.

A surprising aspect of these solid-state amorphization reactions is the stability of the amorphous alloy with respect to the unreacted elemental components. This stability was attributed to a large negative heat of mixing of the elements, which is supported by the many solid-state amorphization reactions in which the formation of the amorphous alloy produces a majority of the heat of formation of the final crystalline compound (*24*).

A second surprising aspect of solid-state amorphization reactions is the inability of the system to nucleate a crystalline compound from the amorphous intermediate. Although the crystalline product is more thermodynamically stable than the amorphous alloy intermediate, it does not form because of kinetic limitations resulting from a nucleation barrier (*26*). This barrier to the nucleation of a crystalline compound is affected by many factors within a solid matrix.

Nucleation. The transformation of a metastable amorphous intermediate into a thermodynamically more stable crystalline product initially begins on a very small scale that arises from entropy considerations. This nucleation step involves the assemblage of the proper kinds of atoms via diffusion, structural rearrangement into intermediates, and the formation of stable nuclei.

Nucleation from an homogeneous fluid system provides a simple model for understanding factors that will affect nucleation within a solid matrix. Suppose a small region of a stable crystalline compound, referred to as an embryo, appears in the middle of a melt with stoichiometry identical to that of the crystalline compound. A free energy decrease, ΔF_v per unit volume, would be expected as a result of the conversion of the metastable liquid to the crystalline compound. This embryo is bounded by a surface that has a positive free energy, ΔF_s per unit area, associated with it. If the embryo grows above a "critical" radial size, ΔF_v dominates and the embryo survives to nucleate; otherwise, ΔF_s dominates and the embryo disappears back into the melt, as shown pictorially in Figure 7 (*14*).

Atomic mobility, stresses and strains within a solid, impurities, surfaces, and the composition difference between the background and the embryo will all affect the magnitude of the barrier hindering nucleation of a

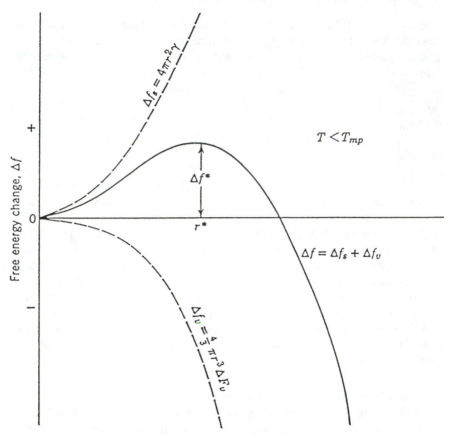

Figure 7. The free energy of a spherical crystalline embryo relative to a melt of identical stoichiometry as a function of its radius. The total free energy is the sum of the surface and volume free energies, whose dependence on radius is also shown. Symbols: Δf_s, surface free energy; r, radius of nucleus; γ, surface free energy per unit area; Δf^*, activation energy for nucleation; r^*, critical radius of nucleus; T, temperature; Δf_v, volume free energy; and ΔF_v, free energy change.

crystalline solid from an amorphous intermediate. Indeed, most of these factors have been used to explain the formation of amorphous alloys from solid-state amorphization reactions as well as search for new amorphous alloy systems that are kinetically stable with respect to nucleation (25). Many of these energies are inherently offset at an interface, lessening the activation barrier for nucleation and resulting in a smaller critical radius than that found in the amorphous bulk. (The common practice of scratching the sides of a flask to induce crystallization used by synthetic molecular chemists reflects the lessening of nucleation barriers at surfaces and interfaces.)

The thickness of the amorphous phase formed at an interface between two reactants has an upper limit (27). This upper limit is thought to result from competition between diffusion and nucleation. Initially, as a result of the short diffusion path and diffusion rate, the interface moves at a quicker rate than the rate for nucleation of a crystalline compound at the interface. Thus, the interface moves before a nucleus can reach a critical size. As the thickness of the amorphous alloy increases, the rate of movement of the interface decreases. At some critical thickness of the amorphous phase, the rate of interfacial movement will become comparable to the nucleation rate, and the amorphous phase will crystallize.

Summary. The large amount of research done on reactions of bulk and thin-film diffusion couples provided an important background for studies of interfacial reactions on shorter length scales. The research on thin-film diffusion couples demonstrated that the progression of a solid-state reaction was affected by diffusion lengths in the reacting systems. An amorphous alloy was suggested as an important reaction intermediate. Nucleation from this intermediate was suggested to be the determining factor in the sequence of products formed by the reaction. Also important was the idea that the composition of the amorphous alloy strongly influenced the crystalline product formed. The research conducted on solid-state amorphization reactions showed that if diffusion lengths were reduced further, the system could interdiffuse completely before a crystalline compound had time to nucleate at an interface.

Superlattice Reactants

Our approach to probing interfacial reactions has been to take advantage of several unique properties of modulated composites, an example of which is shown in Figure 8. The four most important advantages of these superlattices for studying interfacial reactions are listed.

1. They are prepared in a layer-by-layer manner allowing control of individual layer thicknesses on an angstrom-length scale.

2. If the repeat distance of the elemental layers is held constant, the composite behaves like a one-dimensional crystal. Analysis of the resulting low-angle diffraction pattern yields the structure of the interfacial region.

3. The very high density of interfaces in these modulated composites permits the thermodynamics of the interfacial reactions to be explored using differential scanning calorimetry.

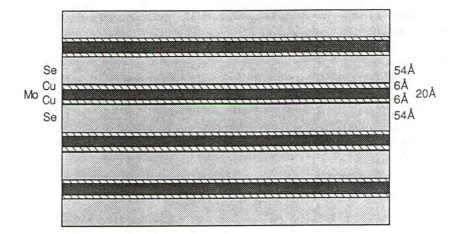

Figure 8. Typical ternary superlattice reactant prepared at the University of Oregon.

4. The time scale of the interfacial reactions in a multilayer permits in situ studies of the evolution of interfacial structure by using both low- and high-angle X-ray diffraction.

Preparation of Superlattice Reactants. Superlattices are most commonly produced by the sequential deposition of elemental layers in a high-vacuum environment (*28, 29*). The design of a deposition system to make superlattice reactants must permit any sequence of individual elemental layers to be deposited, allow for the independent control of the individual layer thicknesses, and also produce a usable quantity (1–10 mg) of superlattice materials for subsequent study (*30*). This custom-built, ultra-high vacuum chamber has independently controlled deposition sources. High melting point elements are deposited by using electron-beam evaporation sources independently controlled by quartz crystal thickness monitors. Knudsen sources controlled by a temperature controller are used to evaporate low melting point elements and are monitored by separate quartz crystal monitors.

The superlattice structure shown in Figure 8 greatly simplifies the actual structure of the final composite. Insight as to the real structure of the composites can be gained by examining the growth process involved in preparing a multilayer via sequential evaporation in high vacuum. The deposition substrate is never atomically flat but has hills and valleys on a larger length scale (hundreds of angstroms) in addition to sharper features on a shorter length scale (angstroms). The height of these features varies with the substrate used. The deposition begins by depositing the first element (A) on the substrate at ~1 Å/s. At this deposition rate, 10^{15} atoms per second hit each square centimeter of substrate surface, where they can react with the substrate, coat the substrate evenly, fill in valleys preferentially, and cluster into islands during deposition. The deposition of the second element (B) involves the same topological questions. In general, the A–B and B–A interfaces are different because of the different reactivities of the depositing A atoms onto B and B atoms onto A. Because the methods used to control the layer thicknesses are only accurate to ~1 Å, variations in the average layer thicknesses for each of the elements will exist. In addition to this variation, a larger local variation of layer thicknesses throughout the multilayer will result from island formation during growth. These local variations can be arranged coherently between the layers, or these local inconsistencies can be spatially incoherent. The effects of interdiffusion and both coherent and incoherent roughness on multilayer structure are shown in Figure 9.

Interfacial Structure via Low-Angle X-ray Diffraction. X-ray diffraction is well-suited to the study of superlattice structures, because it can provide structural information on the atomic scale without damaging the sample. Both low- and high-angle diffraction effects can result from the periodicity introduced via the superlattice structure. The high-angle diffraction patterns of superlattices containing crystalline layers have satellites around the diffraction peaks of the crystalline components. These satellites result from the interaction of diffracted waves from each crystalline layer within the superlattice. Although significant progress has been made in modeling the high-angle diffraction patterns (*31–45*), the low-angle diffraction pattern resulting from the chemical modulation of the superlattice is the primary tool used to determine how layers interdiffuse and react (*46–53*). The rest of this section focuses on the extraction of the chemical modulation of the interfacial regions within superlattices from the low-angle diffraction pattern.

A multilayer composite diffracts X-rays at low angles because of the difference in electron density of the individual elemental layers. The repeating unit of elemental layers can be thought of as a unit cell. Therefore, the superlattice will diffract X-rays even if the individual elemental layers are themselves amorphous. An example of such a low-angle diffraction pattern for an iron–silicon composite is shown in Figure 10. This low-angle diffraction pattern contains much information concerning the structure of the repeat unit. As in conventional single-crystal diffraction studies, the position of the Bragg diffraction peaks permits the size of the repeating unit to be determined. The intensities of the Bragg peaks give information about the width and shape of the concentration profiles at the interface of the elemental layers. The diffraction pattern can, in principle,

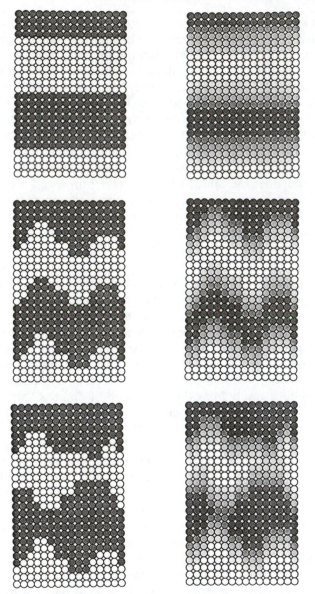

Figure 9. Structural parameters influencing the intensity of X-ray diffraction patterns: (a) perfect superlattice with atomically smooth and abrupt interfaces between element A (dark shading) and element B (light shading), (b) coherent superlattice with smooth and diffused interfaces (shading is used to indicate percent site occupancy of A and B), (c) superlattice with coherently rough interfaces, (d) superlattice with coherently rough and diffused interfaces, (e) superlattice with incoherently rough interfaces, and (f) superlattice with incoherently rough and diffused interfaces.

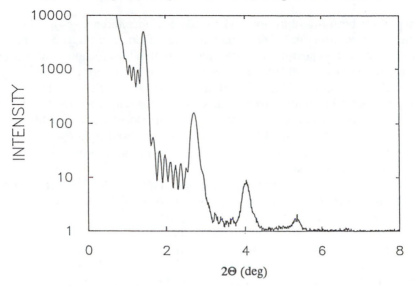

Figure 10. Diffraction pattern of an as-deposited iron–silicon superlattice consisting of 10 repeating units, each containing 22 ± 1 Å of silicon and 44 ± 1 Å of iron.

be deconvoluted to yield detailed structural information about the interfaces.

The fundamental difference between multilayers and naturally occurring crystals creates difficulties in analyzing intensity data and understanding the desired structural information of both the interfacial region and the composite as a whole (54). In a natural crystal, a unit cell varies by about 0.001 Å, and the atoms are located in specific and localized atomic positions within the unit cell. Chemical criteria, such as reasonable coordination numbers and bond distances, are used to help solve the inherent phase problem in crystallography. In a multilayer, the unit-cell size varies on an angstrom scale, the topography of the substrate and subsequently deposited layers can have significant incoherence, the interdiffusion at the interfaces potentially can vary from layer to layer, and the individual layers can themselves be amorphous. Thus, the traditional crystallographic models used for natural crystals need to be modified.

The diffraction pattern in Figure 10 contains several intense Bragg reflections as well as weak subsidiary maxima between these Bragg reflections resulting from the modulated structure of the multilayer. The analysis of this spectra must explain the positions and intensities of the Bragg and subsidiary maxima, the rapid decline of the subsidiary maxima relative to the Bragg maxima with increasing angle, and the increasing line width of the Bragg maxima with increasing angle.

The structural model we used to simulate the experimental data incorporates both interdiffusion and topological incoherence (54). The variables in this structural model include the thicknesses of the elemental layers within a repeating unit, the amount of interdiffusion at the interfaces, and the extent of incoherent roughness at the interfaces.

The intensities of the diffraction maxima are sensitive to the ratio of the individual elemental layer thicknesses, as shown in Figure 11. Fortunately, the total thickness is determined by the positions of the diffraction maxima, and the individual layer thicknesses are fixed from the deposition process used to make the sample. These constraints severely limit the range of possible solutions.

Interdiffusion preferentially attenuates higher order Bragg reflections as the interfaces become less abrupt or rounded. This attenuation is illustrated in Figure 12. The interdiffusion process can be modeled by using two adjustable parameters that approximate the amount of interdiffusion (for example, silicon into iron and iron into silicon) to duplicate the observed intensities of the Bragg maxima. The elemental profiles at the interfaces are calculated by assuming that the diffusion process is in the form of each atom taking a random walk at the interface.

Topological incoherence is responsible for attenuation of the subsidiary maxima relative to the Bragg maxima and the increasing width of the Bragg profiles with increasing angle (55). This condition is illustrated in Figure 13. Topological incoherence caused by domains larger than the X-ray coherence length can be approximated by averaging the calculated diffraction patterns of a collection of multilayers with various errors in their layer spacings. This calculation is done for a collection of layer thicknesses in which the deviation of each layer is chosen randomly within a given distribution width. The intensities of a number of these patterns calculated as a function of diffraction angle are then added together and divided by the number of patterns to determine an average diffraction profile. This approach is similar in philosophy to the method of Warren and Averback (56) for determining the domain size and degree of imperfection in ductile materials.

The concentration profile obtained for each of the n multilayers is put into a recursive calculation by dividing each unit cell into a large number of thin slabs whose refractive index $n(z)$ is taken to be constant and to be that of the center of the slab. The Parratt computational recursion method (57), further developed by Underwood and Barbee (44), is used to calculate the reflectivity of the multilayer as a function of diffraction angle.

This model, containing a limited number of adjustable parameters, was successful in simulating experimental diffraction patterns. As an example, Figure 14 contains both an experimental and calculated diffraction pattern for an iron–silicon superlattice. The parameters obtained from the calculated pattern shown in Figure 14 are summarized in Table I. The

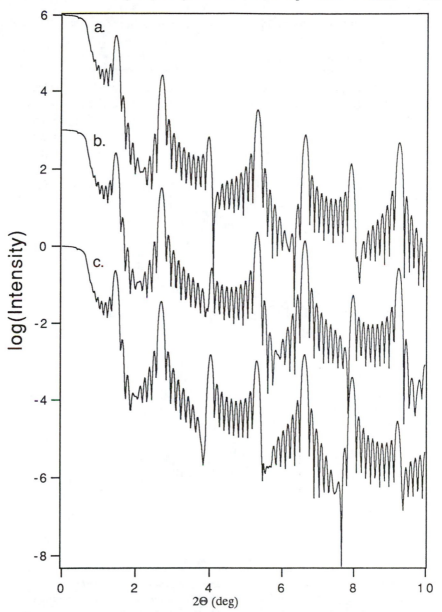

Figure 11. Calculated diffraction patterns for iron–silicon superlattices illustrating the sensitivity of the diffraction pattern to relative layer thicknesses. Diffraction pattern a was from a multilayer containing 10 repeating units consisting of 21.5 Å of silicon and 45.7 Å of iron. Diffraction pattern b was from a multilayer containing 10 repeating units consisting of 22.5 Å of silicon and 44.7 Å of iron. Diffraction pattern c was from a multilayer containing 10 repeating units consisting of 23.5 Å of silicon and 43.7 Å of iron.

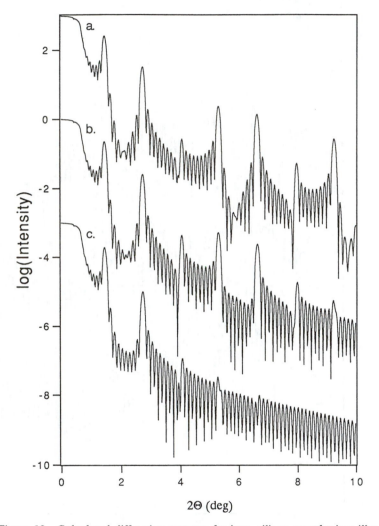

Figure 12. Calculated diffraction patterns for iron–silicon superlattices illustrating the effect of interdiffusion. Each calculation was done on a multilayer containing 10 repeating units consisting of 22.5 Å of silicon and 44.7 Å of iron. In pattern a, the interfaces were abrupt (silicon and iron diffusion lengths were both zero); in pattern b, iron and silicon were interdiffused 3 Å; and in pattern c, iron and silicon were interdiffused 6 Å.

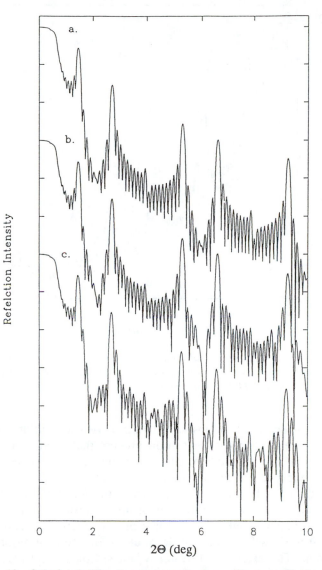

Figure 13. Calculated diffraction patterns for iron–silicon superlattices illustrating the sensitivity of the diffraction pattern to topological incoherence. Diffraction pattern a was from a perfect multilayer containing 10 repeating units consisting of 21.5 Å of silicon and 45.7 Å of iron with abrupt interfaces. Diffraction pattern b was from an identical multilayer, except one of the elemental layers in one of the repeating units was made 1 Å larger. Diffraction pattern c was from a multilayer containing 10 repeating units in which a random error in layer thickness of less than 1 Å was used for each elemental layer.

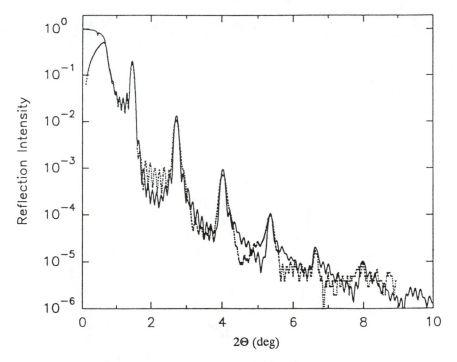

Figure 14. The experimental and calculated diffraction patterns for an iron–silicon superlattice containing 10 repeating units each consisting of 22 ± 1 Å of silicon and 44 ± 1 Å of iron.

layer thicknesses used in this calculation were fixed within experimental error to those used in the deposition process that created the multilayer. The structural incoherence obtained from the data analysis is on the order of two atomic layers. This incoherence is most likely caused by a variation in the local layer thickness resulting from the formation of elemental islands during deposition. The extent of interdiffusion of silicon at the interfaces is independent of whether iron or silicon is being deposited, whereas iron shows a significant dependence on the identity of the depositing atom. The extent of diffusion when silicon is deposited on iron is twice that calculated for iron deposited on silicon. This marked asymmetry was observed by other investigators (*58*). Further studies in which deposition conditions are varied are required to understand the underlying cause of this asymmetry.

The accuracy of the incoherent roughness obtained via the analysis of low-angle diffraction patterns was qualitatively assessed in a diffraction study of Pt–Co multilayers. These multilayers were prepared with identical thicknesses of Pt and Co by using a variety of sputtering gases. Cross-sectional, high-resolution transmission electron microscopy studies demon-

Table I. Summary of Parameters Used in Fitting the Diffraction Pattern of the Iron–Silicon Multilayer Shown in Figure 14

Parameter[a]	As Deposited	*4 h at 100 °C* *Then 2 h at 110 °C*	*5 h at 177 °C*
dSi	22.5	20.5	19
dFe	44.7	44.7	44
σSi1	3	6.3	7
σSi2	3	6.3	7
σFe1	3.3	5.5	7.5
σFe2	6	7	8
Δ	6	5	4

[a]dSi and dFe are the average thicknesses (Å) of silicon and iron layers within the composite. σSi1 and σSi2 are the silicon diffusion lengths (Å) for interfaces of silicon grown on iron and on silicon, respectively. σFe1 and σFe2 are the iron diffusion lengths (Å) for interfaces of iron grown on silicon and silicon grown on iron, respectively. Δ is the distribution width (Å) of the random errors in each elemental layer used to approximate the roughness in the direction perpendicular to the plane of the layers.

strated that the interfacial structure of these samples depended on the sputtering gas. The magnitude of the roughness parameters obtained from fitting the diffraction patterns were in excellent agreement with those obtained directly from high-resolution transmission electron microscopy (*54*).

Interfacial Reactions Probed with Differential Scanning Calorimetry and X-ray Diffraction. The initial step in studying interfacial reactions by using multilayer composites is obtaining an overview of the reactions that occur on heating. We used differential scanning calorimetry (DSC) and X-ray diffraction to determine the thermochemical and structural changes, respectively. DSC is a fast, convenient method of characterizing the thermal evolution of ultrathin-film multilayer composites. It provides information about when (at what temperature) phase transformations and various reaction events occur. High-angle X-ray diffraction, meanwhile, may be used to determine what has occurred at these temperatures. Together, DSC and high-angle X-ray diffraction provide a powerful method for following solid-state reactions.

These experiments are complemented by variable-temperature, low-angle diffraction studies of the transition from the initial layered reactant to the homogeneous amorphous intermediate. The data obtained permit the interdiffusion reaction to be confirmed and quantified. The combined results of DSC and low-angle X-ray diffraction measurements performed on superlattices of different structures permit details of the general reaction mechanism of interfacial reactions to be uncovered.

In a DSC experiment, the temperatures of a sample and a reference are increased as a function of time. The temperature difference between the two is constantly monitored. If the sample has an exothermic transition, its temperature will be higher than that of the reference. Heat is then applied to the reference to raise the temperature to that of the sample. Conversely, if the sample has an endothermic transition, heat is supplied to the sample to eliminate the temperature difference. A thermogram thus consists of the difference between the heat supplied to the reference and to the sample as a function of temperature.

The key experimental factor in obtaining DSC data on superlattices is eliminating the signal damping resulting from the heat capacity of the substrate. We eliminate the substrate completely from our DSC samples by depositing our superlattices on a polymer-coated substrate and subsequently dissolving away the polymer, leaving a free standing superlattice.

Figure 15 shows the calorimetry data for a representative multilayer composite. These data, for a multilayer sample with an iron-to-silicon molar composition of 1:2, reveal a broad exotherm with an onset temperature of 80 °C and a sharp exotherm centered at 460 °C. Diffraction data collected on this sample as a function of temperature to determine the structural changes associated with these exotherms are shown in Figure 16. The diffraction pattern of the as-deposited sample contains low-angle Bragg peaks associated with the layered nature of the composite as dis-

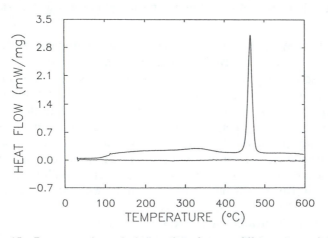

Figure 15. Representative calorimetry data for a multilayer composite. The data shown here were from a composite in which the layer thicknesses were chosen to arrive at an iron-to-silicon relative molar composition of 1:2.

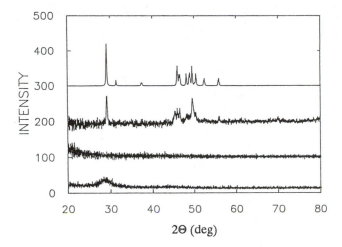

Figure 16. Diffraction data collected as a function of temperature on the iron–silicon (1:2) sample used in Figure 15. The bottom curve was obtained on an as-deposited sample, the next higher curve after annealing at 350 ° C, the third curve after heating to 600 ° C, and the top curve is from a sample of FeSi$_2$ from conventional synthesis.

cussed previously, whereas the high-angle diffraction data show only a broad maxima indicative of the radial electron density distribution within the sample. Diffraction data collected on a sample annealed at 300 °C and then cooled indicate that the sample was no longer layered and that no crystalline compound formed. Transmission electron microscopy experiments performed on similarly annealed iron–silicon composites confirmed the amorphous nature of the reaction intermediate. Diffraction data collected after heating a sample past the sharp exotherm at 460 °C showed that the sample crystallized and formed FeSi$_2$.

Temperature-dependent low-angle diffraction studies confirmed that the broad, low-temperature exotherm in the DSC scans is due to interdiffusion of the layers. The intensity of the low-angle diffraction peaks in iron–silicon superlattices remained constant as temperature was raised to 80 °C. Above this temperature, the diffraction peaks decreased in intensity with time, as shown in Figure 17. Analysis of the initial-change Bragg intensities as a function of annealing time suggested that at temperatures above 100 °C the interdiffusion was consistent with Fick's laws of diffusion and yielded an initial interdiffusion coefficient of 1×10^{-19} cm^2/s at 150 °C (52).

The rate of decay of the low-angle diffraction peaks began to decrease at long annealing times, however, and this decrease suggests that structural changes within the composite complicated the interdiffusion process. We believe that the decrease in the diffusion rate was caused by

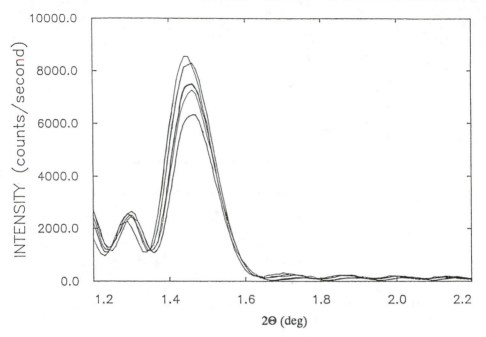

Figure 17. Change in intensity of the low-angle diffraction pattern of an iron–silicon superlattice as a function of time at 150 ° C. The curves from top to bottom represent data that was collected every hour sequentially.

the selective elimination of the most mobile and energetically unfavorable configurations within the multilayer, leading to a corresponding increase in the activation energy required for diffusion. The elimination of high-energy conformations and voids with time and temperature was given strong support by the decrease in the repeat unit length with time, as shown in Figure 18. As temperature was raised, the repeat unit length decreased at a greater rate, because more high-energy conformations and voids became accessible as diffusion pathways. This finding agrees with previous studies involving metal–metal superlattices, in which a consistent picture for the interdiffusion process was obtained by correcting the diffusion coefficient for the change in volume of the film during the anneal (*59, 60*).

More insight into the structural changes associated with the diffusion process was gained through a detailed analysis of the diffraction patterns taken during annealing. Table I contains the simulation results obtained at two points during the annealing of an iron–silicon superlattice. Significant changes occur in the interfacial structure after 6 h of annealing at or below 110 °C. The analysis of the diffraction data suggested that the change in repeat unit size was caused by a contraction of the silicon layer

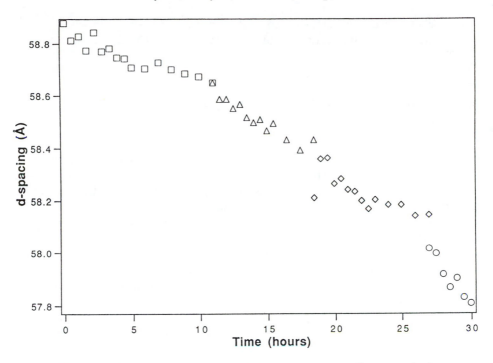

Figure 18. Decrease in the repeat-unit length of an iron–silicon superlattice as a function of time and annealing temperature. Key (°C): □, 100; △, 110; ◇, 135; and ○, 159.

as silicon diffused into iron, and this finding was consistent with the large increase in the silicon diffusion length. The iron diffusion length was also found to increase, but to a smaller degree. The extent of iron diffusion at the iron deposited on the silicon interface increased significantly more than it did at the silicon deposited on the iron interface. This result was expected from Fick's laws of diffusion, because the sharper concentration gradient found at the iron deposited on silicon provided a larger driving force for the interdiffusion process. The incoherence of the interfaces also decreased on annealing, a result implying that the interface became smoother. This finding was consistent with a planar-diffusion front and Fick's laws of diffusion.

Analysis of the diffraction pattern of the multilayer after annealing for 5 h at 177 °C showed that the parameters continued to change in a manner consistent with Fick's laws of diffusion. The repeat unit of the multilayer continued to decrease in size, and the silicon region contributed a larger proportion of the decrease. All the diffusion lengths continued to increase. The asymmetry found in the iron concentration gradients at the interfaces rapidly disappeared because of interdiffusion. The incoherence of the interfaces was again found to decrease on annealing, implying that the interface became smoother, as expected from Fick's laws of diffusion (52).

Controlling Solid-State Reactions by Limiting Interfacial Widths

Superlattice reactants combine the ability to follow interfacial reactions in detail, as demonstrated in the preceding sections, with the ability to control the individual layer thicknesses in the initial composite on an angstrom-length scale. This ability to control the length scale of the layering provides a new synthetic variable for controlling solid-state reactions. Reduction of the length scale of the individual layers below a critical thickness gives general access to a new reaction pathway involving a bulk amorphous intermediate. Control of the composition of this amorphous reaction intermediate permits the crystallization of metastable and stable compounds without the formation of other crystalline compounds either as intermediates or as products. This control has opened up horizons in the synthesis of new solid-state materials, as a new range of free-energy space is now kinetically accessible. As an example of this control, the last subsection describes the preparation of a ternary compound without the formation of crystalline binary compounds as reaction intermediates.

Effect of Layer Thicknesses on the Mechanism of Solid-State Reactions. Several events occur simultaneously in the early stages of a solid-state reaction between two elements: the elements interdiffuse, and crystalline compounds nucleate and grow in the interdiffused region. Diffusion and nucleation are essentially competing. Although the individual layer thicknesses in a superlattice reactant determine the time and the temperature required to interdiffuse the layers, the energetics of nucleation should be independent of the layer thickness to a first approximation. Therefore, a reduction in layer thicknesses below a critical thickness should be possible, so that diffusion is completed before nucleation of a product can occur. Because nucleation typically occurs at interfaces as a result of a lower nucleation barrier, this scenario eliminates the potential for interfacial nucleation and traps the amorphous alloy as a reaction intermediate.

To demonstrate this concept, a series of molybdenum–selenium samples with the same composition (a 1:2 molar ratio of molybdenum to selenium) but varying layer thicknesses were prepared. DSC was used to investigate the reaction pathway of the multilayer composites. Figure 19 shows representative thermograms for composites with individual layers

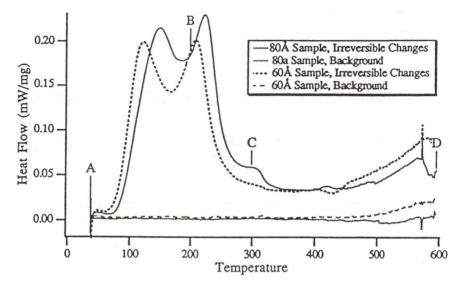

Figure 19. Differential scanning calorimetry data for the reaction of superlattices of molybdenum and selenium with repeat units of 60 and 80 Å, respectively.

thicker than the critical thicknesses ($\sim$10 Å for molybdenum and 20 Å for selenium) for the molybdenum–selenium system. Each of these data sets contains two overlapping exotherms below 250 °C. Diffraction data collected on the 80-Å sample heated to the points A, B, C, and D indicated in Figure 19 are shown in Figure 20. These data demonstrate that by the beginning of the second exotherm $MoSe_2$ has nucleated.

Low- and high-angle diffraction data collected on the 80-Å sample as a function of temperature from 100 to 184 °C support the conclusion that the first exotherm is the result of the initial interdiffusion of the layers. The first-order peak decreases as expected, but the second-order diffraction peak grows with time, as seen in Figure 21. This result suggests that a plateau of composition develops in the interface region as the amorphous interface expands. Once this plateau reaches a critical size, $MoSe_2$ is observed to nucleate and grow, resulting in the second exotherm. This behavior agrees with the generally accepted picture for the solid-state reaction between two elements. Thus, a layered Mo–Se composite with a repeat unit of $\sim$30 Å or greater behaves as if each interface were a thin-film diffusion couple.

The behaviors observed for composites with individual layers thinner than the critical thicknesses are distinctly different than those described previously for composites with thicker modulation distances. This difference is illustrated by the DSC data for a sample with a 26-Å repeat unit, shown in Figure 22. The samples with a total-layer spacing less than 30 Å

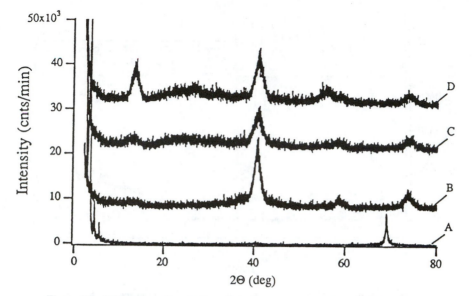

Figure 20. Diffraction data collected during reaction of the molybdenum–selenium composite with a repeat distance of 80 Å at (A) room temperature, (B) 200 °C, (C) 300 °C, and (D) 600 °C. Data sets are offset vertically for clarity. The diffraction peaks observed in scans B–D are the 00l lines of $MoSe_2$, a result indicating a preferred orientation of the product. The peak at ~72° on diffraction pattern A is from the silicon wafer substrate.

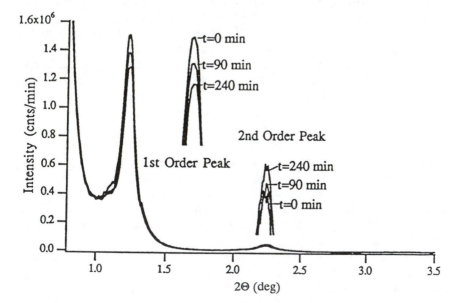

Figure 21. Low-angle diffraction patterns of an 80-Å sample at 0, 90, and 240 min while being annealed at 184 °C.

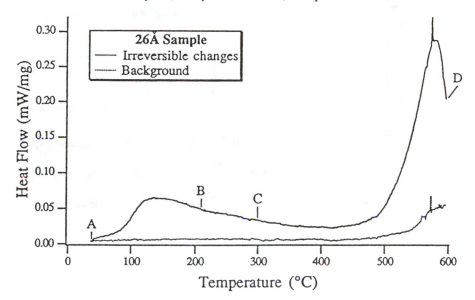

Figure 22. Differential scanning calorimetry data for sample with 26-Å modulation length. Key: A, room temperature, B, 200 °C, C, 300 °C, and D, 600 °C.

show a broad maximum beginning at 100 °C, followed by a large exotherm with a maximum at 575 °C. Figure 23 shows diffraction data collected on samples heated to the points A, B, C, and D in Figure 22. The broad maxima at ~41 in patterns B and C are the result of the distribution of molybdenum–selenium bond distances within the amorphous alloy. Extended heating of the sample at 350 °C for 26 h completely eliminated the low-angle diffraction peaks, and the intensity and line width of the broad maxima at 41 were not affected by this heating. Crystalline $MoSe_2$ was observed only after the large exotherm at 575 °C.

Low-angle diffraction data collected as a function of temperature from 109 to 222 °C confirm that the first exotherm results from the initial interdiffusion of the layers. The intensities of the low-angle diffraction peaks are plotted in Figure 24. Compared with the 80 °C composite discussed earlier, the plateau growth is severely depressed, as indicated by the decay of the second-order diffraction peak with time. High-angle diffraction scans indicate that the sample remains amorphous as the sample completely interdiffuses (*61*).

These results clearly demonstrate that the reaction path of a solid-state reaction can be controlled by adjusting the layer thickness of the initial superlattice reactant. In the ultrathin-layer regime, the interfaces disappear quickly because of the short length scale for diffusion. The composite interdiffuses completely and forms a homogeneous, amorphous

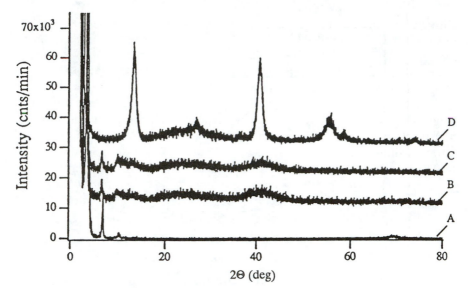

Figure 23. Diffraction data collected during the reaction of the 26-Å sample at (A) room temperature, (B) 200 °C, (C) 300 °C, and (D) 600 °C. Data sets are offset vertically for clarity. The broad maxima at ~24° in B–D are from the glass substrate. The diffraction peaks in D are the 00l diffraction lines of MoSe₂. The phase grows with a preferred orientation.

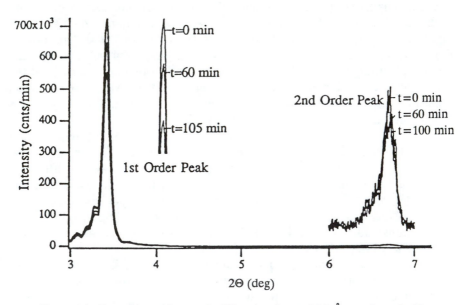

Figure 24. Intensities of low-angle diffraction data of 26-Å sample at 0, 60, and 105 min while being annealed at 222 °C.

alloy before any binary-phase nucleates, as shown in the reaction mechanism depicted in Figure 25. This pathway is distinctly different than what is observed in composites layered with a larger repeat distance. Above a critical layer thickness, the composites behave as thin-film diffusion couples with nucleation occurring at the interfaces.

The large difference in nucleation temperatures in the two different length-scale regimes increases the synthetic importance of the ultrathin-film reaction mechanism. The amorphous Mo–Se alloy formed from the ultrathin-film composites is surprisingly stable. The crystalline compound that forms from this homogeneous alloy depends only on the relative barriers to nucleation for the possible compounds, not on their final free-energy states. The key to obtaining the desired compound is controlling the nucleation of this amorphous alloy.

Composition Control of Nucleation. As described in the previous section, we were able to separate the mixing of the elements from the crystallization of a binary phase by using superlattice reactants in which the individual elemental layers are below a critical thickness. In this thickness regime, the rate-limiting step in the formation of a crystalline phase from the amorphous alloy is nucleation, not diffusion. The presence of a nucleation barrier introduces a kinetic factor into synthesis by using superlattice reactants not found in conventional solid-state synthetic approaches. This nucleation barrier depends on the free-energy gain per unit volume, surface energy of the growing embryo, internal stress in the film, and the energy necessary to rearrange the amorphous alloy. This last term is minimized for the crystalline phase closest to the composition of the amorphous alloy. If the rearrangement energy is large relative to the other terms, composition could control the crystalline phase that nucleates. To test this possibility, we prepared a series of iron–silicon superlattice reactants of varying stoichiometry, each with individual layer thicknesses below the critical thicknesses found for the iron–silicon system. DSC and X-ray diffraction data were collected on each these samples.

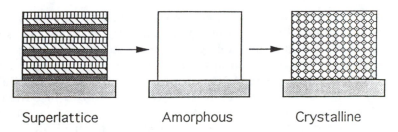

Superlattice   Amorphous   Crystalline

Figure 25. Reaction mechanism for ultrathin-film composites showing the formation of a homogeneous, amorphous alloy as the key reaction step.

The calorimetry data shown in Figure 13, which contain a broad, low-temperature exotherm and the sharp exotherm, are representative of what was observed for all of the samples studied. Onset temperatures of the broad exotherm were consistent for all of the samples studied, and these results suggest that this exotherm is connected with interdiffusion of the layers. The diffusion onset temperature is a measure of the activation energy for diffusion and is dependent on the structure of the iron–silicon interface. The structure of this interface depends only on the deposition conditions, which were held constant from sample to sample, rather than on the amount of iron or silicon away from the interface.

In addition to the broad, low-temperature diffusion exotherm, all of the samples showed a sharp exotherm. The temperature at which the sharp exotherm occurred, however, varied considerably as a function of composition. Before this exotherm, the samples were amorphous, as determined by X-ray diffraction and transmission electron microscopy. Diffraction data collected after these exotherms clearly indicate that the phase that crystallized was the phase closest in composition to the amorphous alloy. Excellent agreement was found between the observed and previously reported diffraction data for each of the synthesized iron silicides (62).

In addition to control of nucleation via stoichiometry of the amorphous intermediate, a second concept demonstrated in this iron–silicon study was that the rate-limiting step in the formation of a crystalline phase in this low-temperature synthesis method was nucleation. The kinetic importance of this fact is highlighted by the reaction pathway observed for the superlattice reactant with an iron-to-silicon molar ratio of 5:3. We observed the crystallization of Fe_5Si_3 at a temperature of 500 °C. From the bulk-phase diagram (Figure 6), Fe_5Si_3 is thermodynamically unstable with respect to a mixture of Fe_3Si and $FeSi$ at temperatures below 825 °C. The crystallization of Fe_5Si_3 demonstrated that metastable phases will form if they are easier to nucleate than the more thermodynamically stable phases. Superlattice reactants represent a new, controlled approach to the synthesis of metastable compounds via amorphous reaction intermediates.

Direct Formation of a Ternary Compound from an Amorphous Intermediate.

Traditional synthetic approaches to solid-state compounds involve high reaction temperatures and stable binary compounds that are reaction intermediates. Possible ternary products are therefore limited to compounds that are more thermodynamically stable than any intermediates under the reaction conditions. This limitation of these high-temperature techniques is severe. With the use of superlattice reactants, one can avoid thermodynamically stable intermediates such as binary compounds. The ability of superlattice reactants to access an amorphous intermediate is especially important when attempting to prepare ternary and other higher order compounds. A new range of free-energy space is now kinetically accessible and can be explored for new ternary compounds.

The results presented in the previous sections suggest a rational approach to preparing a ternary compound. The first step is to prepare a ternary superlattice designed to evolve into an amorphous ternary intermediate. The second step is to control the crystallization of this amorphous intermediate into the desired ternary compound. As described in the following paragraphs, we prepared $Cu_xMo_6Se_8$ from ternary superlattice reactants to demonstrate this capability.

First, we needed to prepare a ternary superlattice that would evolve into an amorphous reaction intermediate. Because an amorphous reaction intermediate was thought to form by avoiding interfacial nucleation, we hoped that information on critical layer thicknesses obtained on binary superlattices would be transferable to ternary superlattices. If the initial layer thicknesses in the ternary composite were greater than the critical values found in the binary-phase diagrams, we expected to form the binary compounds before the composite completely interdiffused. If the initial layer thicknesses were less than the critical thicknesses found in the binary phase systems, the composite should have evolved to an amorphous alloy. In the ternary Cu–Mo–Se system, the molybdenum–selenium system was found to have the thinnest critical distance—molybdenum layers must be 9 Å or less to avoid the nucleation of $MoSe_2$ at the Mo–Se interfaces in the ternary composite.

Second, the crystallization of this amorphous intermediate into the desired ternary compound needs to be controlled. The results on iron–silicon presented in the previous section suggest that the formation of a binary compound might be suppressed by significant additions of a ternary element. This suppression results from the difficulty in disproportionating the amorphous ternary intermediate to form the initial nuclei of the binary compound.

Two ternary superlattices of identical stoichiometry (a 2:6:8 molar ratio of copper to molybdenum to selenium) but varying layer thickness were prepared to test these hypotheses. The first superlattice had elemental layers above the critical layer thicknesses found for the binaries, and the second composite had layers thinner than these critical layer thicknesses.

Differential thermal analysis data and the corresponding diffraction patterns obtained on the "thick" sample are presented in Figures 26 and 27, respectively. The sample is clearly layered and X-ray amorphous as deposited. Heating past the two low-temperature exotherms crystallizes molybdenum diselenide. The temperatures of these exotherms correspond

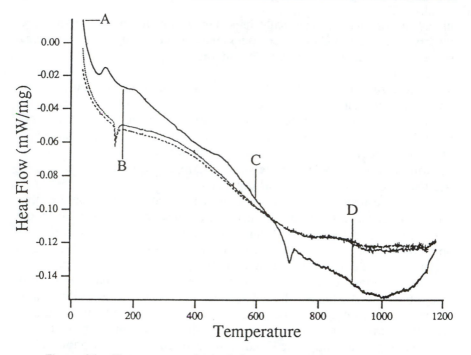

Figure 26. Thermograms obtained for a ternary copper–molybdenum–selenium superlattice consisting of a repeating unit containing 10 Å of copper, 82 Å of selenium, and 30 Å of molybdenum. The curve with points A–D indicated is the first thermogram and contains the irreversible changes in the sample as it is initially heated. The curve without indicated points is the second thermogram and is an effective baseline. At the points labeled A–D, diffraction data were collected, and the diffraction patterns are presented in Figure 27.

to those observed in the thick, binary molybdenum–selenium samples studied, and this observation suggests that the first exotherm results from the beginning of the interdiffusion of the layers, whereas the second corresponds to the crystallization of $MoSe_2$. From 300 to 600 °C, the binary compound $MoSe_2$ crystallizes, and the diffraction pattern contains the 001 lines of this binary reaction intermediate. Following the endotherm found at 700 °C, the binary compound converts into the expected, thermodynamically stable, ternary compound.

DSC data obtained for the "thin" sample are presented in Figure 28. A distinct difference exists between the calorimetry data obtained for the thick and thin ternary samples. Three broad exotherms are observed; the lowest temperature one occurs between 150 and 300 °C, the second between 600 and 800 °C, and the third between 850 and 1100 °C. The endothermic transition observed at 725 °C in the thicker samples is not found for the thin samples.

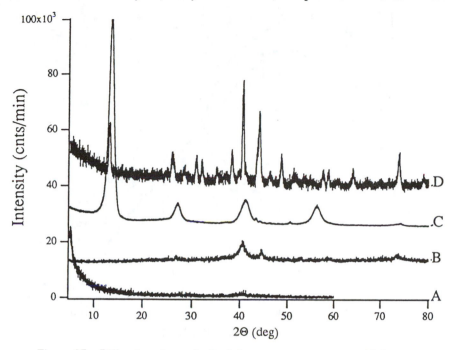

Figure 27. Diffraction data obtained for a ternary copper–molybdenum–selenium superlattice consisting of a repeating unit containing 10 Å of copper, 82 Å of selenium, and 30 Å of molybdenum at the temperatures and points indicated on Figure 26. The curves are offset vertically for clarity.

Diffraction data collected as a function of temperature, shown in Figure 29, confirm that thickness alters the intermediates in the reaction. As deposited, the superlattice is layered and X-ray amorphous. After heating to 170 °C, the sample begins to interdiffuse, and two small diffraction peaks are found in the high-angle diffraction data that suggest that something has begun to crystallize. The only change in diffraction patterns during extended annealing of the sample below 500 °C is the disappearance of the low-angle Bragg diffraction peaks as the elemental layers interdiffuse. The two small diffraction maxima evident in the high-angle diffraction scan at 170 °C become less evident as a result of this annealing. The diffraction pattern of the sample obtained after annealing at 600 °C indicates the ternary compound has begun to crystallize and suggests that the exotherm between 600 and 800 °C results from this crystallization. The diffraction patterns collected between 600 and 1200 °C show the gradual development of the full diffraction pattern of the ternary compound with no evidence for the formation of crystalline binary intermediates. The gradual improvement in the crystallinity of the ternary product is also evident by the smooth increase in the particle sizes obtained from the X-ray line widths (*61, 63*).

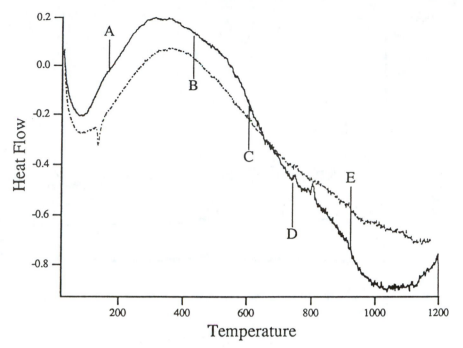

Figure 28. Thermograms obtained for a ternary copper–molybdenum–selenium superlattice consisting of a repeating unit containing 3 Å of copper, 25 Å of selenium, and 9 Å of molybdenum. The curve with points A–E indicated is the first thermogram and contains the irreversible changes in the sample as it is initially heated. The curve without indicated points is the second thermogram and is an effective baseline. At the points labeled A–E, diffraction data were collected and the diffraction patterns are presented in Figure 29.

Although the same product is formed from superlattices in both thickness regimes, the reaction sequences are completely different. The reaction sequence observed for the thick ternary superlattice is similar to that expected for a "bulk" reaction. In a bulk reaction, one reactant typically becomes mobile before the others and reacts with the stationary elements to form binary compounds. In the bulk reaction between copper, molybdenum, and selenium, selenium is the first element to become mobile and react with the metals to form binary compounds. The thick superlattice initially forms $MoSe_2$ at the molybdenum–selenium interfaces, as expected from our earlier investigation of the effect of layer thickness on the reaction sequence of binary molybdenum–selenium superlattices. The thin-superlattice reactant forms an amorphous reaction intermediate before crystallizing the ternary compound directly, as depicted in Figure 25. This ability to prepare a ternary compound via the direct nucleation

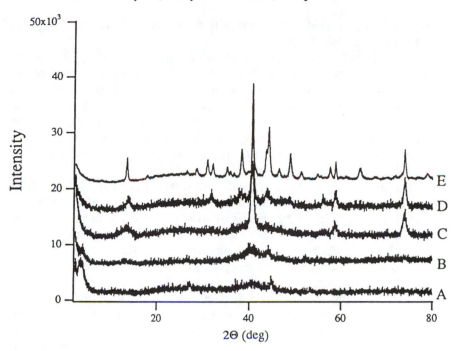

Figure 29. Diffraction data obtained for a ternary copper–molybdenum–selenium superlattice consisting of a repeating unit containing 3 Å of copper, 25 Å of selenium, and 9 Å of molybdenum at the temperatures and points indicated on Figure 28. The curves are offset vertically for clarity.

from an intimately mixed, metastable reaction intermediate and form a desired product represents a significant advance toward the development of a rational synthetic approach to new ternary and higher order compounds.

Future Research Directions

The research described in this chapter focused on the development of a new approach to interface study and to solid-state synthesis. This new approach uses superlattices as a unique starting point that is capable of being tailored to specific needs. We addressed the preparation of known compounds in established phase diagrams via desired reaction intermediates to determine the capabilities and limitations of superlattices as probes of interfacial reactions and as reactants themselves. The ability to tailor the structure of superlattices on an angstrom level has opened new reaction pathways, provided a means to study interfacial reactions on an angstrom level, and provided the capability of controlling the reaction pathway taken by the superlattice.

Several fundamental research directions are suggested by the unique capabilities of superlattice reactants. In principle, changes in the low-angle diffraction pattern with annealing temperature and time allow one to follow solid-state reactions at buried interfaces on an angstrom level. Also, these changes should permit a detailed description of how elements or compounds react in the solid state. Progress in this area is tied to an improvement in the ability to interpret the low-angle diffraction patterns and extract elemental composition profiles from this data.

The ability to follow interfacial reactions has considerable practical importance in addition to the scientific significance. Many physical phenomenon are affected by interfacial structure and interfacial reactions. An example of this importance is found in the physical properties of composite materials, which are often dominated by interfacial phases formed by the reaction of the individual components of the composite. A detailed knowledge of the structural and chemical development of the interfacial regions is crucial to improving performance in a rational manner.

The ability to prepare amorphous intermediates makes superlattice reactants an ideal way to explore low-temperature phase diagrams. Traditionally, phase diagrams were explored by slowly cooling high-temperature melts. Compounds that are only stable at low temperatures are extremely difficult to form via this procedure, as compounds formed at high temperatures must react. This reaction involves large diffusion distances, and, because solid-state diffusion is very limited at low temperatures, compounds stable only at low temperatures are not accessible kinetically. Superlattice reactants permit phase diagrams to be explored from the bottom up—from low temperatures to high temperature via the formation of a metastable, amorphous reaction intermediate. The composition of this amorphous intermediate is then used as a probe to favor the nucleation of compounds only stable at low temperatures.

The ability to form amorphous reaction intermediates also presents the opportunity to study nucleation phenomenon with the solid state. Solid-state nucleation is a complex process that has been investigated extensively only in glass-forming systems such as silicates. The ability to form amorphous intermediates within systems with widely varying structures and bonding tendencies can be used to expand on these initial studies. Maximum separation of the variables affecting the nucleation process will be crucial. The studies described in this chapter were all performed on free-standing samples to eliminate the effect of the substrate. Temperature and composition were the only variables that were systematically explored. Because the nucleation process often involves a volume change, an obvious additional parameter to explore is hydrostatic pressure. The changes in nucleation temperature with pressure can the be used to determine a "volume of activation" for the nucleation event.

An obvious initial synthetic area in which to apply the methodology discussed in this chapter is the synthesis of new ternary and higher order compounds that are metastable with respect to binary constituents. The first step is to determine the best values for the adjustable parameters in the initial superlattice reactant so that the superlattice will evolve into an bulk amorphous intermediate. The parameters that can be easily varied include the stoichiometry, the initial layer thicknesses, and the layering sequence in the superlattice.

Research summarized in this chapter demonstrates that the stoichiometry and initial layer thicknesses are crucial parameters in determining whether the superlattice will interdiffuse to form an amorphous reaction intermediate. Of particular importance, the ability to transfer knowledge about the critical thickness obtained from studies on binary systems provides a rational starting point.

The ability to deposit any sequence of layers provides the ability to sequentially react components. Creating complex layer sequences is necessary in some systems to avoid crystallization of a binary compound. An example of such a system is iron and aluminum, which react upon deposition to form FeAl (*64*). Whereas the reaction of a composite based on the repetition of the layer sequence ABC involves the simultaneous reaction at the AB, BC, and AC interfaces, the reaction of a composite based on the repetition of the layer sequence ABCB involves initial reaction of only AB and BC interfaces. Thus, the repeat unit of a superlattice composite can be used as an additional experimental variable to potentially avoid the formation of an unwanted reaction intermediate.

The second key step is controlling the nucleation of the amorphous reaction intermediate. In addition to stoichiometry of the amorphous intermediate, several other techniques can be used. The substrate can affect the nucleation process in at least two ways: by providing nucleation sites or by producing large internal stresses or strains in the superlattice film upon annealing. The importance of internal stress can be experimentally probed by depositing on substrates that are prestressed. The idea is to stress the substrate and cause it to bend. By depositing on the concave surface and then unstressing the substrate, a longitudinal stress is imposed on the film. By depositing on the convex surface and then unstressing the substrate, a compressive stress is imposed on the film. By varying the initial stress imposed on the substrate, the effect of internal stress on nucleation can be systematically explored. Changes in nucleation temperature with internal stress could then be used as an additional variable when using superlattice reactants to prepare new materials.

Substrates provide nucleation sites that result in epitaxial growth of the nucleated phase off the substrate. This process could be exploited to direct the nucleation of the amorphous intermediate to a desired structure, although one of the major experimental limitations of epitaxial growth historically has been the lack of high-quality substrate material of the desired structure.

Nucleation can also be controlled via the addition of impurities that act as nucleation centers. This phenomenon, referred to as seeding, is potentially a very powerful way to control the nucleation phenomenon when using superlattices as reactants. The basic idea is to take advantage of the limited diffusion rates within the low-temperature reactions of superlattices, as shown in Figure 30. A superlattice is prepared containing layers of the elements designed to produce an amorphous intermediate for the desired compound. On top of this superlattice, a limited amount of a second set of elements is then deposited, which is designed to produce an amorphous intermediate that is isostructural to the desired compound. Finally, on top of these isostructural layers, more superlattice is prepared containing layers of the elements designed to produce an amorphous intermediate for the desired compound. This superlattice composite is then heated to interdiffuse layers. The resulting trilayer of amorphous material is still segregated between the first, second, and third depositions because of limitations in diffusion rates at the low-annealing temperatures. Ideally, the middle layer then crystallizes and thereby causes the top and bottom layers to crystallize into the desired isostructural compound. Preliminary investigations in our laboratories using this approach were extremely encouraging.

Control of the nucleation event in the amorphous intermediate should also permit the formation of single-crystal films of desired compounds. One route to single-crystal films is adapting a "trick" used to

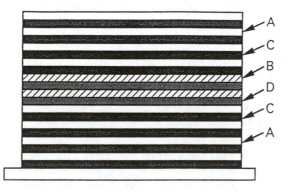

Figure 30. The thermal evolution of an ultrathin-film superlattice designed so that the middle layers of the resulting amorphous composite "seed" the crystallization of the desired compound.

grow single crystals from melts, in which a polycrystalline seed is initially used to begin the growth process. The growing polycrystalline boule is then thinned down to a very narrow neck, so that only one of the growing crystallites has the correct growth orientation. After this "necking" procedure, the remaining growth of the boule is a single crystal from the one crystalline that grew through the neck. The adaptation of this trick to our thin-film systems is shown in Figure 31 and involves a lithographically patterned substrate that divides the amorphous film into two regions separated by a narrow neck. Nucleation via a heat pulse in one region results in multiple nucleation events, but it is hoped that only one of the growing crystals will survive through the neck region. This growing crystalline then acts as the lone nucleation site for the second region. Volume changes during crystallization will probably set an upper size limit on the single-crystal area.

Superlattice reactants will also permit the preparation of "designer compounds" consisting of a chemically modulated crystalline structure, as shown in Figure 32. The synthesis of this new type of compound again takes advantage of the limited long-range diffusion that occurs during the low-temperature annealing of the initial superlattice. The chemically

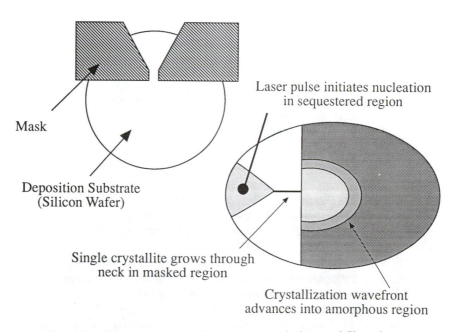

Laser pulse initiates nucleation in sequestered region

Mask

Deposition Substrate (Silicon Wafer)

Single crystallite grows through neck in masked region

Crystallization wavefront advances into amorphous region

Figure 31. The proposed procedure to prepare single-crystal films via patterning of the deposited region followed by localized heating in one of the sequestered regions. Only one crystallite will have the correct orientation to grow through the neck of the masked region of the amorphous film.

Figure 32. A proposed "designer compound" consisting of a superstructure of MoSe₂ and NbSe₂

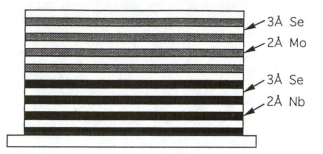

Figure 33. The proposed design of a "superlattice of superlattices" as an initial reactant to form the compound shown in Figure 32.

modulated structure shown in Figure 32 is prepared from an intergrown superlattice of superlattices, as shown in Figure 33. Preliminary investigations in our laboratories using this approach were encouraging.

Acknowledgments

This work was supported by a Young Investigator Award from the Office of Naval Research (N00014–87–K–0543) and the continuation grant (N0014-91-J-1288). Support by the National Science Foundation (Grants DMR–8704652 and DMR–8553291); the Department of Energy (Grant FG06–86ER45275); the donors of the Petroleum Research Fund administered by the American Chemical Society; the University of Oregon; and the U.S. Department of Education, Graduate Assistance in Areas of Need Program for Thomas Novet is also gratefully acknowledged. Finally, none of the results would have been possible without the efforts and persistence of the graduate students (Loreli Fister, John McConnell, Xiao-Mei Li, Chris Grant, and Thomas Novet), who actually performed the experiments.

References

1. Weaver, J. H. *Physics Today* **1986**, January, 24–30.
2. del Alamo, J. A. *MRS Bull.* **1992**, 27, 42–44.
3. *Materials Research Society Symposium Proceedings;* Clark, W. A. T.; Dahmen, U.; Briant, C. L., Eds.; Materials Research Society: Pittsburgh, PA, 1991, Vol. 238; pp 763–884.
4. Cohen, M. In *Advancing Materials Research;* Psaras, P. A.; Langford, H. D., Eds.; National Academy Press: Washington, DC, 1987; pp 51–110.
5. Novet, T.; McConnell, J. M.; Johnson, D. C. *Chem. Mater.* **1992**, 4, 473–478.
6. West, A. R. *Solid-State Chemistry and its Applications;* John Wiley and Sons: New York, 1989.
7. Schmalzried, H. *Solid-State Reactions;* Verlag Chemie: Deerfield Beach, FL, 1981, Vol. 12; p 254.
8. DiSalvo, F. J. *Science* **1990**, 247, 649–655.
9. Gas, P.; d'Heurle, F. M.; Goues, F. K.; La Placa, S. J. *J. Appl. Phys.* **1986**, 59, 3458–3466.
10. Majni, G.; Nobili, C.; Ottaviani, G.; Costato, M.; Galli, E. *Appl. Phys.* **1981**, 52, 4047–4054.
11. Nava, F.; Psaras, P. A.; Takai, H.; Tu, K. N. *J. Appl. Phys.* **1986**, 59, 2429–2438.
12. Wittmer, M. *J. Appl. Phys.* **1983**, 54, 5081–5086.
13. Ziegler, J. F.; Mayer, J. W.; Kircher, C. J.; Tu, K. N. *J. Appl. Phys.* **1973**, 44, 3851–3857.
14. Brophy, J. H.; Rose, R. M.; Wulff, J. *Thermodynamics of Structure;* John Wiley and Sons: New York, 1964, Vol. 2; pp 91–94.
15. Herd, S.; Tu, K. N.; Ahn, K. Y. *Appl. Phys. Lett.* **1983**, 42, 597.
16. Coulman, B.; Chen, H. *J. Appl. Phys.* **1986**, 59, 3467–3474.
17. Canali, C.; Catellani, F.; Ottaviani, G.; Prudenziati, M. *Appl. Phys. Lett.* **1978**, 33, 187–190.
18. Ziegler, E.; Lepetre, Y.; Schuller, I. K.; Spiller, E. In *Grazing Incidence Optics;* Osantowski, J. F.; Van Speybroeck, L. P., Eds.; International Society for Optical Engineering: Bellingham, WA, 1986, Vol. SPIE 640; pp 145–148.
19. Walser, R. M.; Bené, R. W. *Appl. Phys. Lett.* **1976**, 28, 624–625.
20. Bené, R. W. *Appl. Phys. Lett.* **1982**, 41, 529–531.
21. Pretorius, R. *Mater. Res. Soc. Symp. Proc.* **1984**, 25, 15–20.
22. Ottaviani, G. *Mater. Res. Soc. Symp. Proc.* **1984**, 25, 21–31.
23. Schwarz, R. B.; Johnson, W. L. *Phys. Rev. Lett.* **1983**, 51, 415–418.
24. Cotts, E. J.; Meng, W. J.; Johnson, W. L. *Phys. Rev. Lett.* **1986**, 57, 2295–2298.
25. Clemens, B. M.; Sinclair, R. *MRS Bull.* **1990**, XV(No. 2), 19–28.
26. Cahn, J. W.; Hilliard, J. E. *J. Chem. Phys.* **1959**, 31, 688–699.
27. Gösele, U.; Tu, K. N. *J. Appl. Phys.* **1989**, 66, 2619–2626.
28. Webb, R. J.; Goldman, A. M. *J. Vac. Sci. Technol.* A **1985**, 3, 1907–1912.
29. Nastasi, M.; Hung, L. S.; Johnson, H. H.; Mayer, J. W.; Williams, J. M. *J. Appl. Phys.* **1985**, 57, 1050–1054.

30. Fister, L.; Li, X.-M.; McConnell, J.; Novet, T.; Johnson, D. C. *J. Vac. Sci. Technol.* **1993,** *11,* 3014–3019.
31. Clemens, B. M.; Gay, J. G. *Phys. Rev. B* **1987,** *35,* 9337–9340.
32. Bartels, W. J.; Hornstra, J.; Lobeek, D. J. W. *Acta Cryst.* **1986,** *A42,* 539–545.
33. Fullerton, E. E.; Schuller, I. K.; Vanderstraeten, H.; Bruynseraede, Y. *Phys. Rev. B* **1992,** *45,* 9292–9310.
34. Gilfrich, J. V.; Brown, D. B.; Rosen, D. In *Multilayer Structures and Laboratory X-Ray Laser Research;* Ceglio, N. M.; Ohez, P., Eds.; International Society for Optical Engineering: Bellingham, WA, 1986, Vol. SPIE 688; pp 115–121.
35. McWhan, D. B. In *Synthetic Modulated Structures;* Chang, L. C.; Giessen, B. C., Eds.; Academic: New York, 1985; pp 43–74.
36. Névot, L.; Pardo, B.; Corno, J. *Rev. Phys. Appl.* **1988,** *23,* 1675–1686.
37. Pardo, B.; Megademini, T.; André, J. M. *Rev. Phys. Appl.* **1988,** *23,* 1579–1597.
38. Rosenbluth, A. E.; Lee, P. *Appl. Phys. Lett.* **1982,** *40,* 466–468.
39. Savage, D. E.; Kleiner, J.; Schimke, N.; Phang, Y.-H.; Jankowski, T.; Jacobs, J.; Kariotis, R.; Lagally, M. G. *J. Appl. Phys.* **1991,** *69,* 1411–1424.
40. Shaw, K. D.; Krieger, A. S. *Appl. Opt.* **1989,** *28,* 1052–1054.
41. Spiller, E.; Rosenbluth, A. E. In *S.P.I.E. Proceedings;* Marshall, G. F., Ed.; International Society for Optical Engineering: Bellingham, WA, 1985, Vol. 563; pp 221–236.
42. Spiller, E. *Rev. Phys. Appl.* **1988,** *23,* 1687–1700.
43. Spiller, E. In *Physics, Fabrication, and Applications of Multilayered Structures;* Dhez, P.; Weisbuch, C., Eds.; Plenum: New York, 1988; pp 271–309.
44. Underwood, J. H.; Barbee, T. W., Jr. *Appl. Opt.* **1981,** *20,* 3027–3034.
45. Underwood, J. H.; Barbee, T. W., Jr. *Appl. Opt.* **1988,** *20,* 302.
46. Cook, H. E.; Hilliard, J. E. *J. Appl. Phys.* **1969,** *40,* 2191.
47. Balluffi, R. W.; Blakely, J. M. *Thin Solid Films* **1975,** *25,* 363–392.
48. DuMond, J.; Youtz, J. P. *J. Appl. Phys.* **1940,** *11,* 357–365.
49. Fleming, R. M.; McWhan, D. B.; Gossard, A. C.; Wiegmann, W.; Logan, R. A. *J. Appl. Phys.* **1980,** *51,* 357–363.
50. Greer, A. L.; Spaepen, F. In *Synthetic Modulated Structures;* Chang, L. C.; Giessen, B. C., Eds.; Academic: New York, 1988; pp 419–486.
51. Murakami, M.; Segmuller, A.; Tu, K. N. In *Analytical Techniques for Thin Films;* Tu, K. N.; Rosenberg, R., Eds.; Academic: New York, 1988, Vol. 27; pp 201–248.
52. Novet, T.; Xu, Z.; Kevan, S. D.; Johnson, D. C. *Materials Science and Engineering A: Structural Materials: Properties, Microstructure, and Processing;* 1992, *A162,* 115–122.
53. Piecuch, M. *Rev. Phys. Appl.* **1988,** *23,* 1727–1732.
54. Xu, Z.; Tang, Z.; Kevan, S. D.; Novet, T.; Johnson, D. C. *J. Appl. Phys.* **1993,** *74,* 905–912.
55. Xu, Z.; Tang, Z.; Kevan, S. D.; Novet. T.; Johnson, D. C. *Materials Research Society Symposium Proceedings;* Atwater, H. A.; Chasen, E.; Grabon, M. H.; Legally, M. G., Eds.; Materials Research Society: Pittsburgh, PA, 1993, Vol. 280, pp 241–244.

56.   Warren, B. E.; Averback, B. L. *J. Appl. Phys.* **1950,** *21,* 595.

57.   Parratt, L. G. *Phys. Rev.* **1954,** *95,* 359.

58.   Falco, C. M. In *Physics, Fabrication, and Applications of Multilayered Structures;* Dhez, P.; Weisbuch, C., Eds.; Plenum: New York, 1988, Vol. 182; pp 3–15.

59.   Chasen, E.; Mizoguchi, T. *Z. Phys. Chem. (Munich)* **1988,** *156,* 397–401.

60.   Philophsky, E. M.; Hilliard, J. E. *J. Appl. Phys.* **1969,** *40,* 2198–2205.

61.   Fister, L.; Johnson, D. C. *J. Am. Chem. Soc.* **1992,** *114,* 4639–4644.

62.   Novet, T.; Johnson, D. C. *J. Am. Chem. Soc.* **1991,** *113,* 3398–3403.

63.   Fister, L.; Johnson, D. *J. Am. Chem. Soc.* **1993,** *116,* 629–633.

64.   Grant, C.; Johnson, D. C. *Chem. Mater.* **1994,** *6,* 1067–1071.

65.   *Binary Alloy Phase Diagrams,* 2nd ed.; Massalski, T. B.; Okamoto, H.; Subramanian, P. R.; Kacprzak, L., Eds.; ASM International: Materials Park, OH, 1989; Vol. 2, pp 2664–2665.

RECEIVED for review January 11, 1993. ACCEPTED revised manuscript August 3, 1993.

Oxide Superconductors

Arthur W. Sleight

Department of Chemistry, Oregon State University, Corvallis, OR 97331–4003

The discovery of superconductivity at temperatures higher than the boiling point of nitrogen led to a resurgence of interest in superconductors. All of these higher temperature superconductors are complex oxides of copper. About 100 different oxide superconductors of this type have been prepared, but only a few are superconducting at temperatures above the boiling point of nitrogen. Applications for the oxide superconductors appear promising in several areas.

WITH THE AVAILABILITY OF LIQUID HELIUM early in this century, elemental mercury could be cooled below 4 K; Hg was discovered to be a perfect conductor of electricity. This superconducting property of Hg had not been predicted and was, therefore, very surprising. Subsequently, superconductors were found to have unique magnetic and junction properties. The temperature at which superconductors become superconducting, T_c gradually increased with the discovery of this property in other elements and compounds. Most of the advances resulted from alloys or intermetallic compounds containing niobium. By 1973, T_c was raised to 23 K in Nb_3Ge (1).

During much of the development of superconducting materials, an adequate theory did not exist. However, in 1957, the Bardeen–Cooper–Schrieffer (BCS) theory appeared (2), and we then seemed to have a good fundamental understanding of the superconducting state. Unfortunately, on the basis of BCS theory, several theorists concluded that a T_c higher than ~30 K could not be achieved. From 1973 to 1986, T_c did not increase above 23 K, and much of the interest in the synthesis of new superconduc-

0065–2393/95/0245–0471$12.00/0

tors faded. The discovery in 1986 by Bednorz and Müller (3) of superconductivity above 30 K in complex oxides of copper led to a resurgence of
interest in superconducting materials.

Thousands of papers and dozens of reviews have appeared on oxide
superconductors in the past few years. Some of these are of particular interest to chemists (4–7). Details of work mentioned here can be found in
those reviews. This chapter is a mere synopsis.

Oxide Superconductors

Even before the discovery of cuprate superconductors, oxide superconductors attracted considerable attention. Superconductivity at ~13 K was
discovered in both the Li–Ti–O and Ba–Pb–Bi–O systems. Although
such T_cs were considered surprising for oxides, these materials contained
elements near the "magic" areas of the periodic table. An empirical rule
was that higher temperature superconductors existed only in materials
containing niobium, lead, or their nearest neighbors in the periodic table.
Copper itself was not superconducting at any temperature, and only a few
compounds of copper became superconducting at low temperatures.

All of the cuprate superconductors are based on CuO_2 sheets stacked
one on top of the other. Within these sheets, copper atoms are coordinated to four oxygens in a square-planar arrangement. These sheets can be
flat and have all the copper and oxygen atoms in a single plane, but usually these sheets are wrinkled in one of several ways. Superconductivity
does not seem to be influenced much by whether these sheets are flat.

The oxidation state of copper in the sheets is always close to 2.
However, if the oxidation state is exactly 2, these sheets support neither
superconductivity nor normal metallic conductivity. Sheets of CuO_2 containing only divalent copper have antiferromagnetic insulating properties.
Introduction of mixed valency of either the type Cu(II, III) or Cu(II, I)
seems to lead invariably to superconductivity for CuO_2 sheets. The optimum content of Cu(III) or Cu(I) is always close to 15%.

If copper is divalent, the CuO_2 sheets have a charge of 2–. Thus,
these $(CuO_2)^{2-}$ sheets must be charge-balanced in other parts of the
structure. The simplest such compound is $SrCuO_2$. In this structure, Sr^{2+}
cations are between the $(CuO_2)^{2-}$ sheets. For several years, we were unable to introduce the appropriately mixed valency to produce superconductivity in this structure. Now, this valency has been accomplished for
both the Cu(II, III) and Cu(II, I) situations. Interestingly, neither $SrCuO_2$
itself nor superconductors based on this structure are thermodynamically
stable at normal pressure. All of these materials are prepared either at
high pressure or by thin-film synthesis techniques. This structure is

known as the "infinite-layer structure", because the CuO_2 sheets stack directly one on the other without interruption.

Several one-layer cuprate structures exist. All initial studies of cuprate superconductors were based on La_2CuO_4, which is termed the "one-layer structure" because single $(CuO_2)^{2-}$ sheets are separated by $(La_2O_2)^{2+}$ sheets. Antiferromagnetic insulating properties are observed for La_2CuO_4 because all the copper is divalent. Mixed Cu(II, III) valency can be obtained by adding interstitial oxygen, by partial replacement of oxygen by fluorine, and, most commonly, by the partial substitution of a lower valent cation for lanthanum. Despite numerous attempts, mixed valency of the Cu(II, I) type was never produced in the La_2CuO_4 structure. However, mixed valency of this Cu(II, I) type led to superconductivity in another one-layer structure, which occurs with Nd_2CuO_4, for example.

The shorter Cu–O distances in La_2CuO_4 are widely believed to readily accommodate Cu(II, III) mixed valency but not Cu(II, I) mixed valency. On the other hand, the longer Cu–O distances in Nd_2CuO_4 accommodate Cu(II, I) mixed valency but not Cu(II, III) mixed valency. Another significant difference exists between the one-layer La_2CuO_4 and Nd_2CuO_4 structures. In the Nd_2CuO_4 structure, copper atoms are coordinated to only four oxygen atoms, as in the infinite-layer structure. However, in the La_2CuO_4 structure, two more oxygen atoms are coordinated to copper at considerably longer distances.

In principle, all structures between the one-layer and infinite-layer structures are possible. In fact, the only intermediate structures easily prepared are the two- and three-layer cases. The other cases generally require a layer-by-layer synthesis approach. The first multilayered cuprate superconductor studied was $YBa_2Cu_3O_7$, which has CuO_2 double layers that may be represented as YCu_2O_4. Separating these CuO_2 double layers are layers of Ba_2CuO_3, which contain Cu–O chains. Mixed valency of the type Cu(II, III) exists in both the CuO_2 sheets and the Cu–O chains. The $RBa_2Cu_3O_7$ two-layer cuprate, where R may be nearly any rare earth element, is superconducting and has a T_c of ~92 K.

The first three-layer structured cuprate was discovered in the Bi–Sr–Ca–Cu–O system, and its ideal composition is $Bi_2Sr_2Ca_2Cu_3O_{10}$. Soon after this discovery, another three-layer structure, $Tl_2Ba_2Ca_2Cu_3O_{10}$, was found. A trend in T_c was now apparent. As the number of layers increased, T_c increased. A T_c of 110 K was obtained for $Bi_2Sr_2Ca_2Cu_3O_{10}$, and a T_c of 125 K was found for $Tl_2Ba_2Ca_2Cu_3O_{10}$. The possibility that the route to an even higher T_c was to increase the number of CuO_2 layers stacked directly one on the other created some excitement. However, as the four-, five-, and six-layered materials were prepared, the discovery was made that the maximum T_c was already obtained at the three-layer structure.

Metastability

One of the characteristics of the high-temperature superconductors is that they are not thermodynamically stable at ambient conditions (5). This characteristic was already established before the discovery of cuprate superconductors. Achieving a T_c of 23 K in the Nb—Ge system clearly required the synthesis of a metastable material. Furthermore, BCS theory predicts that attempts to push T_c higher would ultimately lead to a material with either an electronic or a structural instability.

Numerous ways to make metastable materials exist. One is to quench from high pressure, as is usually done for the synthesis of diamond or the infinite-layer superconductors. Another approach is to quench from high temperature, as is done in the synthesis of $Bi_2Sr_2Ca_2Cu_3O_{10}$ or window glass. Still another way is through intercalation or deintercalation. We know that $YBa_2Cu_3O_7$ is not thermodynamically stable at any condition of pressure or temperature. On the other hand, $YBa_2Cu_3O_6$, which has the same basic structure as $YBa_2Cu_3O_7$, has a conveniently wide range of stability. Thus, we first prepare $YBa_2Cu_3O_{6+x}$ under equilibrium conditions, where x is small. Subsequently, x is increased through nonequilibrium oxidation to obtain a high-temperature superconductor. The opposite approach is taken to produce superconductors based on the Nd_2CuO_4 structure. These compounds are first prepared in air; then, oxygen is deintercalated to produce a superconductor. In these four cases, the starting point is thermodynamically stable. Another approach to producing metastable materials is through epitaxial growth, such as is being done for diamond films and many superconducting films.

The metastability of high-temperature superconductors should not normally be a concern for application of superconductors. Many technologically important materials such as diamond and various glasses are metastable. All living materials are metastable. The importance of recognizing the metastable characteristic of high-temperature superconductors is twofold. First, the synthesis approach is completely different depending on whether one is attempting to prepare a stable or a metastable material. Many material scientists feel that good equilibrium phase-diagram information is very important for the systems on which they work. Such information is always important, especially if composition, temperature, and pressure are included as variables. However, phase-diagram studies for six-component systems such as Bi—Pb—Sr—Ca—Cu—O become exceedingly difficult to comprehend, even if one does the several hundred individual experiments necessary. This six-component system is important, because it is currently the system of choice for wire made of high-temperature superconductors. Phase diagrams are important, but many of the materials with the most interesting properties do not exist in any equilibrium phase diagram. The second reason for recognizing the metastable nature of high-temperature superconductors is related to theory. Presumably, the correct theory for high-temperature superconductors will explain why such materials cannot exist at equilibrium.

Theory

Despite the intensive effort by many theorists, no consensus has been reached on a theory for the superconductivity found in the high-T_c oxide superconductors. This does not mean that the appropriate theory was not already proposed; it may well have been proposed during the first few weeks after the discovery of this class of materials. A substantial problem in attempting to verify any particular theory is the poor state of definition of these superconductors. Despite enormous efforts, many questions remain about the composition and structure of these materials. The structures are complex and have extraordinarily high concentrations of various defects. In many cases, the real structures remain elusive. The structures we generally show and discuss are highly idealized. A successful theory will probably not be based on such idealized structures.

However, progress has been made on a theory for high-temperature superconductors. Some experimental results are clearly inconsistent with some of the proposed theories. For example, we now know that theories based on the uniqueness of Cu(II, III) compared with Cu (II, I) cannot be correct. Thus, we are gradually narrowing the list of possible theories.

Applications

Applications for low-temperature superconductors exist today, and the applications for niobium-based materials are expanding. Magnets using superconducting wire are used in NMR spectroscopy. Magnetic resonance imaging, a medical imaging system, is also based on a magnet made from superconducting wire. Motors have been made for several applications. A 280-ton, superconductor-powered ship was recently launched. Storage rings built with superconductors have begun to replace batteries for protection against power disruption. Levitated trains using superconducting magnetics were built for demonstration purposes, and more are planned. Electrical power generation and transmission were demonstrated and may well be implemented in some cases. Superconducting quantum interference devices (SQUIDs) are in use as very sensitive detectors of magnetic fields, including brain waves.

Oxide superconductors bring about the promise of increased use of superconductors because of their higher T_c and because of the high magnetic fields they are capable of producing. Cooling to liquid nitrogen temperatures can be accomplished much more economically than cooling to liquid helium temperatures. For some space probes and satellites, the required operating temperature can be achieved without cooling. Currently, a very active program is underway to produce efficient refrigerators that will cool to the temperatures required for the oxide superconductors.

Several groups are now preparing wire from oxide superconductors in lengths up to ∼100 m. The material of choice is Bi–Pb–Sr–Ca–Cu–O, and it is usually sheathed in silver metal. Such wire could potentially be used for transmission lines operating at ∼77 K or for powerful magnets operating at ∼20 K.

High-quality thin films of oxide superconductors are now being prepared by many groups using a variety of different techniques. Numerous electronic applications exist for such films. Much of this activity has been in small start-up companies. One company announced the fabrication of the first superconducting multi-chip module, and another reported the fabrication of an integrated flux magnetometer based on a SQUID using an oxide superconductor.

Role of Chemists

Chemists have played a critical role in the area of oxide superconductors. Creative synthesis routes were developed in several cases. Organometallic compounds were prepared for use as precursors to ceramic bodies or for use in metal organic chemical vapor deposition. A continuing role for chemists is to provide improved synthesis of the known oxide superconductors. An even more important role lies in the search for new superconductors, which may not be oxides and may not contain copper. Given the status of the theory of superconductivity, exploratory synthesis holds the best promise for producing the next good surprise in superconductivity.

References

1. Gavaler, J. R. *Proc. LT13,* **1974,** *3,* 558–562.
2. Bardeen, J.; Cooper, L. N.; Schrieffer, J. R. *Phys. Rev.* **1957,** *505,* 1175–204.
3. Bednorz, J. G.; Muller, K. A. *Z. Phys.* **1986,** *64,* 189–196.
4. Sleight, A. W. *Science (Washington, D.C.)* **1988,** *545,* 1519–1527.
5. Sleight, A. W. *Phys. Today* **1991,** *44,* 24–30.

6. *Chemistry of Superconductor Materials, Preparation, Chemistry, Characteriza-
     tion, and Theory;* Vanderah, T., Ed.; Noyes Publications: Park Ridge, NJ,
     1992; pp 714–734.
7. *Concise Encyclopedia of Magnetic and Superconducting Materials;* Pergamon:
     New York, 1992; pp 423–431.

RECEIVED for review November 9, 1992. ACCEPTED revised manuscript May 11,
1993.

Characterization of Complex Materials by Scanning Tunneling Microscopy

A Look at Superconductors with High Critical Temperatures

Zhe Zhang and Charles M. Lieber*

Department of Chemistry and Division of Applied Sciences, Harvard University, Cambridge, MA 02138

Scanning tunneling microscopy (STM) is becoming an increasingly important tool for materials research because STM can provide direct, real-space information about the local structure and electronic properties of complex solids. In this chapter, we illustrate new and important information obtained from STM studies of copper oxide superconductors with high critical temperatures. As a prelude to the experimental results, the basic theoretical concepts needed to understand and interpret the STM experiment are discussed. Then, STM investigations of the normal and superconducting states of $Bi_2Sr_2CaCu_2O_8$ solids are reviewed. STM measurements were used to elucidate the nature of structural disorder in the BiO layer and the corresponding changes in the electronic states as a function of oxygen doping. In addition, studies of the superconducting state were used to illuminate the superconducting energy gap of $Bi_2Sr_2CaCu_2O_8$ and other low-temperature phenomena.

*Corresponding author

0065–2393/95/0245–0479$14.25/0

UNDERSTANDING THE ELECTRONIC AND STRUCTURAL PROPER-
TIES of materials from the knowledge of atomic- and molecular-level properties is a focus of current research efforts in solid-state chemistry. This
understanding is essential to the materials sciences because it will lead the
way to the rational design and preparation of new solids with predictable
properties. Two new approaches that have begun to provide key atomic-
level structural and electronic information are scanning tunneling microscopy (STM) and scanning tunneling spectroscopy (STS) (1–7). STM and
STS can be used to directly image the atomic structures and electronic
properties of the interfaces of conductors and semiconductors.

As a surface-sensitive technique, STM provides experimental information that may be of questionable value when used to provide insight
into the bulk properties of materials. The surfaces of low-dimensional
solids such as the transition metal dichalcogenides and the copper oxide
superconductors often terminate with bulk structure and bonding. In
these cases, STM was shown to provide key insight into the structure and
electronic properties of the bulk (7–17).

The focus of this chapter is the applications of STM and STS to the
elucidation of the complex structural and electronic properties of the
copper oxide superconductor $Bi_2Sr_2CaCu_2O_8$ (18, 19). First, the basic experimental and theoretical concepts of STM will be reviewed. The
remainder of the chapter will then concentrate on several experimental
studies of the $Bi_2Sr_2CaCu_2O_8$ system, including (1) the atomic-level nature of structural disorder in the BiO layer of $Bi_2Sr_2CaCu_2O_8$ and the
low-energy electronic states associated with this structure, (2) the structural and electronic consequences of oxygen doping of $Bi_2Sr_2CaCu_2O_8$, and
(3) low-temperature STS studies of the $Bi_2Sr_2CaCu_2O_8$ superconducting
energy gap.

An Introduction to STM and Tunneling

In this section, we briefly review the instrumentation and theoretical concepts that are essential for understanding the STM studies that will be discussed. More detailed reviews can be found elsewhere (4, 20–22). A typical tunneling microscope is illustrated schematically in Figure 1. The
underlying basis for the operation of the microscope is electron tunneling
between a sharp metal tip and a conducting sample. When the tip and
sample are sufficiently close, their wave functions can overlap. If a bias
voltage, V, is then applied to the sample, a tunneling current, I, will flow
between the sample and tip.

Electrons will tunnel from filled electronic states in the tip to empty
states in the sample when V is positive; conversely, electrons will tunnel

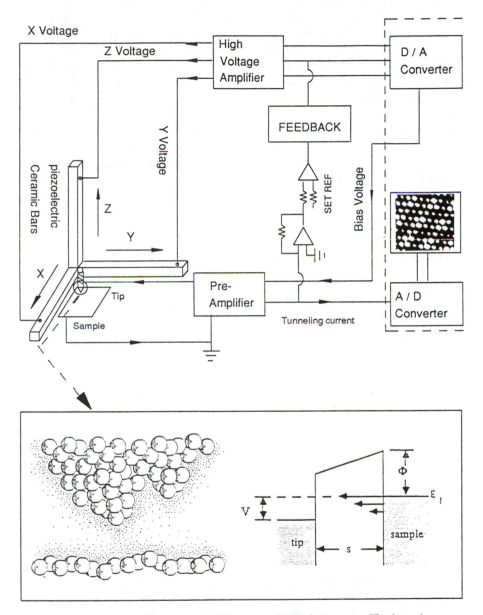

Figure 1. Schematic illustration of a tunneling microscope. The inset is a one-dimensional representation of tunneling between a metallic sample and tip, where s is the tip–sample separation, V is the applied voltage, E$_f$ is the Fermi energy, and ϕ is the average work function.

from filled sample states to empty tip states when V is negative (Figure 1). The tunneling current that flows when V is applied varies exponentially with the tip–sample separation. For a typical work function of 4 eV, I decreases 10-fold for a 1-Å increase in separation. The actual decay rate can, however, vary significantly with the barrier properties. The strong exponential dependence of the tunneling current on distance enables STM to achieve high vertical resolution. An atomic resolution map of the surface can then be generated by rastering the tip over the sample with angstrom-level control by using piezoceramic positioners. Experimental images are typically acquired in the constant-current mode in which a feedback loop controls the vertical position of the tip above the sample so that I is equal to a reference current (I_{ref}) at all coordinates on the surface. Therefore, features in constant-current-mode images correspond to vertical displacements of the piezoceramic positioner needed to maintain a constant tunneling current.

An understanding of the response of the tunneling current to the barrier properties, applied voltage, and so on, is essential to the interpretation of such STM data; insight into these problems can be obtained from theoretical analyses of the tunneling problem (23–28). As first discussed by Tersoff and Hamann (23, 24), an expression for I can be derived readily by assuming unperturbed sample and tip wave functions and then by using perturbation theory. In the limit of small bias voltage and low temperature, this treatment yields

$$I = (2\pi/\hbar)e^2 V \sum |M_{st}|^2 \delta(E_s - E_f)\delta(E_t - E_f) \qquad (1)$$

where M_{st} is the tunneling matrix element between wave functions on the tip, ψ_t, and sample, ψ_s; V is the bias voltage; $\hbar$ is Planck's constant; δ is the delta function; and E_s, E_t, and E_f are the sample, tip, and Fermi energies, respectively. As shown by Bardeen (29), the tunneling matrix can be written

$$M_{st} = (\hbar^2/2m) \int (\psi_t^* \nabla \psi_s - \psi_s \nabla \psi_t^*) dS \qquad (2)$$

where m is the mass of the electron; ∇ is the differential operator; and the integral corresponds to a surface, S, within the barrier region between the sample and tip. To evaluate M_{st} in a way that the resulting expression for I can be compared quantitatively to STM images in general (i.e., not for one specific choice of sample and tip) requires several approximations. Tersoff and Hamann (23, 24) showed that by assuming the tip forms a locally spherical potential well with only s-wave functions, I could be expressed as

$$I \propto \sum_s |\psi_s(r_o)|^2 \delta(E_s - E_f) \qquad (3)$$

where r_o is the center of curvature of the tip. By definition, the summation is the local density of sample electronic states, $\rho(r_o, E)$, at the center of curvature of the tip

$$\rho(r_o, E) \equiv \sum_s |\psi_s(r_o)|^2 \delta(E_s - E_f) \qquad (4)$$

and thus, constant-current images correspond to contours of constant density of sample electronic states.

In view of the simplicity of this result, the effect of the approximations made in deriving equation 4 are considered. Treatment of the tip as a spherical potential is reasonable because in almost all cases the experimental tip shape is unknown, although it probably terminates in a cluster of atoms; that is, the tip is approximately spherical. The s-wave function approximation for the tip is more significant because it leads to a cancellation in equation 2 such that I depends only on the square of the sample wave function. Recently, Tersoff (27) examined more general descriptions of the tip and found that for metals, constant-current images still correspond to contours of constant density of sample electronic states. STM images of semiconductor surfaces at low V could, however, deviate significantly from this simple picture because only a small pocket of the surface Brillouin zone contributes to the tunneling. This deviation is limited, however, to the lowest Fourier component of the image, and the model of the s-wave function tip may still be used in many cases to interpret images.

In contrast, Chen (28) suggested that the s-wave function approximation is unable to explain the resolution obtained in many experiments and that the tip function should be taken as a p_z or d_{z^2} dangling-bond state. Using this tip model, Chen was able to consistently explain atomic resolution images of close-packed metal surfaces. Although these results indicate that significant progress has been made in understanding the role of the tip, future development of these theoretical models is important so that experimental data can be quantitatively interpreted.

In the foregoing discussion, we showed that for reasonable approximations, constant-current STM images correspond to contours of constant local density of sample electronic states. We now examine the tunneling current expression further to determine the explicit dependence of I on tip–sample separation and V. First, if we account for the exponential decay of the sample and tip wave functions into the tunneling gap, we can rewrite equation 3 as

$$I \propto \sum_s |\psi_s|^2 \exp[-2\kappa(R + s)]\delta(E_s - E_f) \tag{5}$$

where R is the radius of the tip, and s is the tip–sample separation. The decay parameter, κ, can be written as

$$\kappa = [(2m\phi/\hbar^2) + k^2]^{1/2} \tag{6}$$

where ϕ is the average work function, and k is the parallel wave vector component of ψ_s (30). For tunneling between planar, free-electron, metal electrodes at small bias (i.e., $V << \phi$), equation 5 can be written as

$$I \propto (V/s) \exp(-2\kappa s) \tag{7}$$

with $\kappa = 1.025\sqrt{\phi}$ (30). If we consider only the dominant exponential part of equation 7, we can express the work function as

$$\phi = 0.952(d \ln I / ds)^2 \tag{8}$$

Equation 8 indicates that the work function can be determined by measuring the distance dependence of the tunneling current; however, the derivation of equation 8 does not consider modifications of the barrier due to the close proximity of the two electrodes (i.e., the STM tip and sample). Lang (31) demonstrated that the slow decay of the exchange correlation potential causes the apparent work function to be less than the true sample work function for electrode separations appropriate to the STM experiment. Hence, the barrier height measured by STM is generally smaller than the sample work function.

In addition, we consider tunneling simultaneously from two distinct layers in a solid, such as the BiO and CuO layers of $Bi_2Sr_2CaCu_2O_8$. If we assume that the work functions of layers 1 and 2 are the same, then the ratio of the tunneling currents, I_1/I_2, from these layers is

$$I_1/I_2 = \exp\left[-1.025\sqrt{\phi}(s_2 - s_1)\right] \tag{9}$$

This result shows that if the densities of states (DOS) of the two layers are similar, but layer 2 is 3 Å below a surface composed of layer 1, then the tunneling contribution from layer 2 can be neglected. However, when the DOS differ significantly for the two layers, this simplification may be inappropriate.

Lastly, we analyze the response of I to variations in V at fixed tip–sample separations. Because I is proportional to the local density of sample electronic states (LDOS), the suggestion was made that I–V data could provide a direct measure of the LDOS versus energy. Because the wave-function decay depends on V, equation 3 should be rewritten as

$$I \propto \int_\rho (r,E)T(E,eV)dE \qquad (10)$$

where $T(E,eV)$ is the transmission probability that takes into account the energy (eV)-dependent wave-function decay. In this expression, the assumption is made that the densities of tip states do not vary significantly with energy. In the limit of small bias, $dI/dV \propto$ LDOS (i.e., the typical assumption for conventional tunneling spectroscopy); however, for finite bias, the exponential dependence of $T(E,eV)$ on V and s becomes important. As shown by Feenstra et al. (32), the exponential dependence can be effectively removed by normalizing dI/dV by V/I; that is

$$(V/I)dI/dV \propto \text{LDOS} \qquad (11)$$

Comparisons of the LDOS data obtained by STS with photoemission and inverse photoemission results show that the approximations made in arriving at equation 11 are reasonable. STS data typically provide only a qualitative measure of the LDOS, although we hope that further theoretical work will overcome this shortcoming. With this background in hand, we now turn to our STM studies of the high-temperature superconductors.

Copper Oxide Superconductors

The copper oxide superconductors are highly anisotropic materials that all contain two-dimensional copper oxide planes as a key structural element (18, 19). The focus of the studies discussed herein will be on the $Bi_2Sr_2Ca_{n-1}Cu_nO_{2n+4}$ family of materials and, in particular, the $n = 2$ or double-layer compound (33–40). This solid consists of quasi-two-dimensional repeat units containing two BiO, two SrO, one Ca, and two CuO_2 layers (*see* Figure 2).

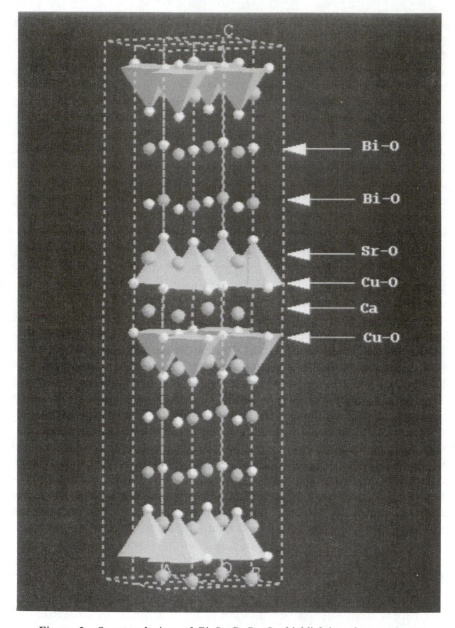

Figure 2. Structural view of $Bi_2Sr_2CaCu_2O_8$ highlighting the quasi-two-dimensional nature of this solid. Crystals cleave preferentially between the Bi—O double layers in the structure (a—b plane); the Cu—O planes correspond to the bases of the square-pyramid polyhedra and are separated by Ca^{2+} ions. The metals are gray and the oxygen ions are white The c-axis is vertical with respect to the page in this model.

The average structure of $Bi_2Sr_2CaCu_2O_8$ obtained from diffraction studies is quite straightforward, although in reality these materials exhibit a wide range of structural complexities that are not yet well-characterized. For example, $Bi_2Sr_2CaCu_2O_8$ shows considerable disorder in the BiO layer's atomic structure, oxygen nonstoichiometry, and substitution of metals between the idealized layers shown in Figure 2 (*33–40*). The detailed nature of such local disorder is intrinsically difficult to address by conventional diffraction techniques, as are the electronic consequences of these structural variations. A complete picture of the microscopic properties of these materials is, however, almost certainly necessary to understand high-temperature superconductivity, because the structural disorder has a dimensionality similar to the superconducting-pair coherence length. To characterize the structure and low-energy electronic states of $Bi_2Sr_2CaCu_2O_8$ in detail, we have been using STM, and in the next section we summarize the status of our work on this system.

The BiO Layer of $Bi_2Sr_2CaCu_2O_8$. The crystallographic separation between adjacent BiO layers in $Bi_2Sr_2CaCu_2O_8$ is quite large, >3 Å, and as such provides a natural cleavage plane in crystals of this solid (*36, 37*). Because covalent bonds are not broken when crystals are cleaved between the weakly interacting BiO layers, the BiO surface layer does not reconstruct; therefore, the STM experiment probes properties similar to those found in the bulk (*41, 42*).

A large-area, gray-scale image of this freshly cleaved BiO layer of $Bi_2Sr_2CaCu_2O_8$ (critical temperature, T_c, = 85 K) crystal is shown in Figure 3. This image exhibits a one-dimensional superstructure with an average modulation of 27 Å along the *a*-axis. Images acquired simultaneously with positive (empty sample electronic states) and negative (filled sample states) bias voltages show the same features. These bias, voltage-dependent images demonstrate that the observed superstructure approximates the variation in the total DOS; hence, we can conclude that this modulation is a structural feature. Bulk X-ray and electron diffraction studies (*36–40*) show a similar modulation and support this conclusion. The following models were proposed to explain the origin of this unique structural modulation in $Bi_2Sr_2CaCu_2O_8$: (1) lattice mismatch between the BiO and CuO layers, (2) extra oxygen substituted into the BiO layer that causes it to periodically buckle, and (3) periodic Cu or Sr substitution for Bi in the BiO layer (*36–40*). At least three problems intrinsic to these materials have hampered diffraction-based efforts to elucidate the origin of this structure. First, the superstructure period is incommensurate with respect to the lattice; second, the BiO layer has considerable disorder; and third, the X-ray and electron-scattering cross sections for oxygen are significantly smaller than those for Bi.

Figure 3. Image measuring 1000 × 1000 Å² of Bi₂Sr₂CaCu₂O₈ recorded with a bias voltage of 690 mV and a tunneling current of 0.8 nA.

STM is not limited by these problems and can, therefore, provide unique insight into the origin and effects of this modulation *(41–45)*. For example, analyses of real-space STM images such as Figure 3 show that the superstructure period is not a sinusoidal modulation. The period varies from 22 to 27 Å, and the distribution of periods about the average is broad and non-Gaussian *(42)*. These results strongly indicate that the superstructure is not due to simply BiO–CuO lattice mismatch (which would yield a sinusoidal modulation), but must have some substitutional component that causes local fluctuations in the superstructure period. Further evidence for the importance of substitution in determining the properties of the superstructure will be discussed in the context of our studies of oxygen doping in this system.

The large area images clearly indicate that substitution must play a role in the one-dimensional superstructure. In principle, atomically resolved images of the BiO lattice sites should resolve, in detail, the substitutional contribution to the modulation. A typical high-resolution image is shown in Figure 4. The atomic structure in this image has tetragonal

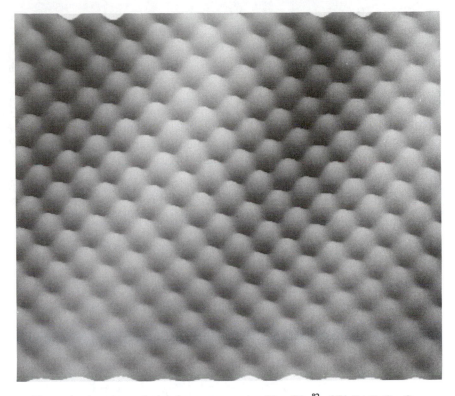

Figure 4. Atomic resolution image measuring 50 × 50 $Å^2$ of $Bi_2Sr_2CaCu_2O_8$ recorded with a bias voltage of 300 mV and a tunneling current of 1.4 nA.

symmetry with a period of 3.8 Å. This distance is consistent with both the average Bi–Bi and O–O distances determined by crystallography; that is, alternate atomic sites, either Bi or O, are imaged by STM. Because only one site is imaged by STM, these data cannot be used to unambiguously define the complete atomic structure associated with the superstructure modulation. The atomic resolution images do show, however, that significant positional disorder can exist in the BiO layer, and we believe this disorder reflects inhomogeneous substitution in the BiO layer. The atomic-scale disorder can be reduced by careful crystal-growth procedures.

The observation of alternate atomic sites indicates that the DOS are peaked over only one of the lattice sites, and hence, that the BiO layer may be semiconducting. STS is used to directly characterize the electronic character of the BiO layer. For finite bias-voltage spectroscopic measurements, our interpretation of the STS data is based on eqs 10 and 11. Applying this analysis [i.e., that $(V/I)dI/dV \propto LDOS$)] requires that the tunneling barrier, ϕ, be greater than the bias voltage.

Tunneling Barrier Height. To determine the magnitude of ϕ, we characterized the dependence of I on tip–sample separation and on V. From the section "An Introduction to STM and Tunneling", we know that in the limit of free electron metals and small bias voltage, $I \propto \exp(-2\kappa s)$, where $\kappa = 1.025\sqrt{\phi}$. Hence, ϕ can be determined by measuring the distance dependence of the tunneling current. Figure 5 shows a typical I–s curve measured on $Bi_2Sr_2CaCu_2O_8$ taken at room temperature. The inset is ln (I)–s relationship. Analysis of these data yields a tunneling barrier of 2.6 eV; thus, the STS measurement should be confined between ± 1.5 V for a meaningful interpretation.

Distance-Dependent Electronic States in $Bi_2Sr_2CaCu_2O_8$.

To directly characterize the electronic character of the BiO layer, tunnel-

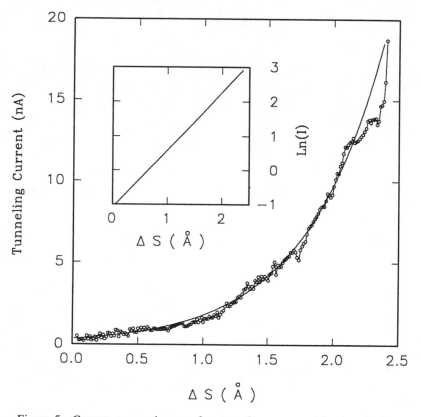

Figure 5. Current versus tip–sample separation measured for a typical as-grown sample. The tip was moved toward the surface after opening the feedback loop; the stabilized tunneling resistance was 10^9 Ω. The inset shows a plot of ln(I) versus distance. The apparent work function determined from these data is 2.6 eV.

ing spectroscopic measurements were performed *(42–45)*. In contrast to other electronic spectroscopic methods, STM can provide a direct measure of the BiO layer electronic states and not a convolution of both BiO and CuO$_2$ layers. Such data are essential to the development of models of the electronic structure for this system. Current and normalized conductivity versus sample bias voltage curves obtained with a feedback-stabilized tunneling resistance of 10^9 Ω are shown in Figure 6. An important feature exhibited by these spectroscopic data is the apparent gap in the DOS at the Fermi level (E_f). Such data are reproducible on as-grown crystals (T_c = 85 K) and indicate that the BiO layer is semiconducting in these materials. In contrast, band-structure calculations using the full-potential, linearized, augmented, plane wave method suggested that the BiO layer is metallic *(46)*, although these calculations do not consider the superstructure and other structural disorder present in the BiO layer. Interestingly, recent tight-binding calculations indicate that distortions in the BiO layer raise the energy of the Bi-derived band above E_f and drive the layer to a semiconducting state *(47)*.

One striking phenomenon in this study is that the measured DOS change with the tip–sample separation (after correction for distance-dependent factors). In preliminary work, we found that STS measurements made at tip–sample separations closer than those used to record Figure 6 showed a finite DOS at E_f *(44)*. More systematic measurements made in our group *(44)* and elsewhere *(45)* confirm this report (Figure 7). Specifically, when the tip is moved ~2 Å closer to the sample, the data show that the conductivity equals zero only at the origin. These results

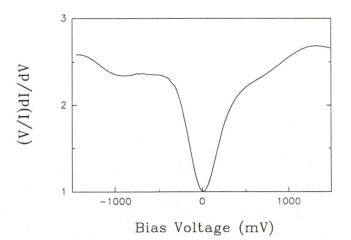

Figure 6. Plot of the normalized conductance versus sample bias voltage for Bi$_2$Sr$_2$CaCu$_2$O$_8$ sample. The voltage corresponds to the energy relative to E$_f$ (V = 0).

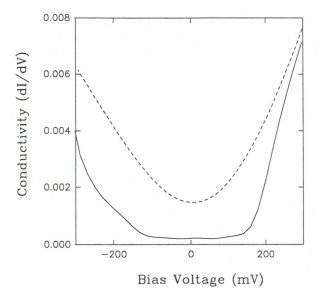

Figure 7. Conductance versus voltage curves recorded at two different tip–sample separations. The solid line was recorded with a tunneling resistance of 10^9 Ω, whereas the dashed line was measured with a tunneling resistance of 10^8 Ω.

demonstrate that for small tip–sample separations, no apparent gap in the DOS exists; that is, the BiO layer appears weakly metallic. One explanation that we (*44*) suggested for these observations is that the metallic states correspond to a tunneling contribution from the CuO layer that is 4.5 Å below the surface. Alternatively, the surface wave function may have an unusually short decay length (because k is >0). From the value of the apparent work function (2.6 eV) obtained from the section "An Introduction to STM and Tunneling", the decay length is not unusually short; therefore, we believe that this alternative hypothesis is unlikely.

Electronic and Structural Effects of Oxygen Doping. Oxygen doping plays a widely recognized, key role in determining the properties of copper oxide superconductors (*48–50*). For example, results from many studies established that variations in oxygen concentration (i.e., in $Bi_2Sr_2CaCu_2O_{8+\delta}$ where δ corresponds to a small change in oxygen stoichiometry) change the average carrier concentration and, thereby, T_c. Recent observations (*50*) also suggest that subtle oxygen rearrangements can lead to superconductivity and enhanced fractions of superconducting material even when the average oxygen stoichiometry remains constant. The physical properties depend not only on the absolute amount of oxygen, but also on the detailed arrangement of oxygen in the lattice. To understand these and other important effects, variations in the microstructure and local electronic structure associated with oxygen doping must be characterized. Determination of the local structure, including oxygen positions by diffraction techniques, is inherently difficult because of crystal disorder and the small, scattering cross section of oxygen. In addition, spatial variations in the electronic properties due to oxygen doping will be especially important in determining superconductivity in short-coherence-length materials such as $Bi_2Sr_2CaCu_2O_{8+\delta}$, where δ is the change in oxygen concentration caused by doping.

To address this important problem, we used STM to characterize the local structural and electronic changes that occur in vacuum-annealed $Bi_2Sr_2CaCu_2O_{8+\delta}$ crystals (*51*). Oxygen was removed from as-grown superconducting crystals ($T_c \approx 85$ K) by vacuum annealing (pressure, P, $\approx 10^{-3}$ torr or 133 mPa) at 400 °C. We found that these conditions yield systematic decreases in T_c as a function of annealing time. The suppression of superconductivity through vacuum annealing is also reversible; that is, annealing nonsuperconducting crystals in air at 400 °C for more than 8 h yields $T_c \approx 85$ K superconducting material. Thermogravimetric analyses of our crystals show that these changes in T_c correspond to a reversible, 2.5–3.0% change in the oxygen concentration.

STM images recorded on nonsuperconducting samples from which ~3% oxygen was removed are shown in Figure 8. These images were recorded at 900 and 150 mV by switching the bias voltage on alternate scan lines; therefore, the two images are recorded in an identical spatial location on the sample surface. The image recorded at high V exhibits the one-dimensional superstructure characteristic of $Bi_2Sr_2CaCu_2O_8$. Notably, the superstructure period is unchanged in oxygen-deficient, nonsuperconducting samples compared with $Bi_2Sr_2CaCu_2O_8$. These results indicate that oxygen removed from the samples either does not come from the BiO layer or that oxygen loss from this layer has little effect on the superstructure. To address the structural location of oxygen loss more directly, we compared atomic resolution images of the BiO layer of nonsuperconducting and superconducting crystals (Figure 9) (*51*). Notably, images of the BiO layer of the oxygen-deficient crystals do not exhibit vacancies; therefore, oxygen might be lost from the SrO or CuO_2 layers. More studies are needed to resolve this important issue.

Images recorded at small V provide important additional information about the electronic effects of oxygen loss. Specifically, these data exhibit nonperiodic features in addition to a contribution from the one-dimensional superlattice. Analysis of images acquired over a range of V show that the superstructure has the same spatial location irrespective of V and represents a true structural feature. The fact that the irregular features are observed only for small V indicates that these features are due to variations in the electronic states near E_f. Because the low bias-voltage elec-

Figure 8. Image measuring 450 × 450 $\mathring{A}^2$ of nonsuperconducting $Bi_2Sr_2CaCu_2O_8$ recorded with (A) V = 900 mV and (B) V = 150 mV. Adjacent maxima of the superstructure are separated by about 27 $\mathring{A}$ in A and B. Irregular electronic features are also detected at low bias in B.

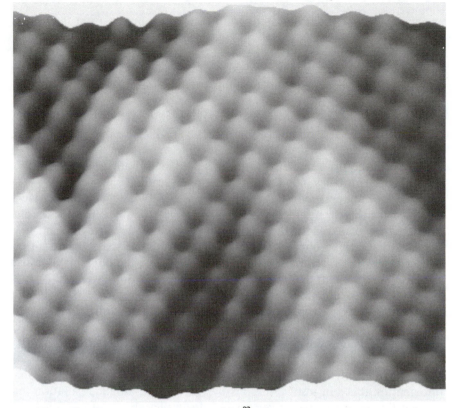

Figure 9. Image measuring 50 × 50 Å² of a nonsuperconducting sample recorded with a bias voltage of 360 mV and a tunneling current of 2 nA.

tronic states are not observed in images of superconducting samples, we conclude that they reflect spatial variations in the electronic properties due to oxygen loss. These electronic variations occur on the same scale as the coherence length (~20 Å) and may affect pairing in intermediate T_c crystals.

We also used STM and STS to characterize how the structure and electronic states evolve when additional oxygen is introduced into the lattice through high-pressure annealing (52). Recent X-ray photoemission spectroscopy studies of $Bi_2Sr_2CaCu_2O_{8+\delta}$ crystals doped in high-oxygen pressure show that the DOS of E_f increase with increasing δ (53). To explain this increase in the DOS, the suggestion was made that oxygen doping transformed the BiO structural element into a metallic layer (53). However, X-ray photoemission spectroscopy probes several unit cells in depth; therefore, these observed changes may or may not arise from variations in the BiO- or CuO_2-layer DOS. STM is uniquely suited to elucidate the origin of high-pressure doping because it can directly probe the surface BiO-layer electronic states with little contribution from the underlying CuO_2 planes.

Spectroscopic data obtained on as-grown and oxygen-annealed $Bi_2Sr_2CaCu_2O_8$ crystals that were cleaved are shown in Figure 10. A key result that is immediately evident upon comparison of the data is the distinctly different $I–V$ behavior near E_f for the as-grown and oxygen-annealed samples. The as-grown crystals exhibit a low current within ±200 mV of E_f and relatively sharp increases in I beyond these points, whereas the oxygen-annealed samples show a smooth increase in I for all V. Because the tip–sample separation is similar in both experiments, this difference is probably not due to a distance-scaling effect. Indeed, the normalized conductivity, $(V/I)dI/dV$, shows a 330-mV gap in BiO-layer DOS for the as-grown sample but no obvious gap for the annealed sample. The absence of a gap in the oxygen-annealed samples is significant and suggests that oxygen doping introduces impurity states that obscure an intrinsic gap in the BiO-layer DOS. Alternatively, oxygen doping may cause the BiO band to shift and cross E_f; in this case, the BiO layer will be metallic. STS measurements provide only a relative measure of the DOS; therefore, these two hypotheses are difficult to distinguish without additional experiments. Qualitatively, we (54) noted that the conductance, dI/dV, of the BiO layer appears to be lower than that of the metallic TlO layer of $Tl_2Ba_2CaCu_2O_8$ (Tl-2212) crystals.

STM imaging of the BiO layer's atomic structure provides additional information that can resolve these issues. However, atomic resolution images of the as-grown and oxygen-annealed samples show virtually identical surface structures (Figure 11). The lattices for both samples have periods of 3.8 ± 0.2 Å that correspond to either the Bi–Bi or O–O lattice sites. The observation of alternate lattice sites for a range of bias conditions indicates that the Bi–O layer is semiconducting in both as-grown and oxygen-annealed crystals. A recent STM study (54) of the related system Tl-2212 showed that simultaneous imaging of both Tl and O sites is possible when the surface is metallic; therefore, the present data probably did not result from an instrumental limitation. Most likely, the BiO-layer DOS near E_f increases with oxygen annealing but the layer remains semiconducting.

The mechanism of oxygen doping in the $Bi_2Sr_2CaCu_2O_8$ system is also important to consider. If the increase in the DOS were due to oxygen incorporated into the BiO layer, then we would expect either extra lattice sites or a locally distorted structure. Comparisons of the images of as-grown and oxygen-annealed samples do not exhibit evidence for such structural defects (Figure 11). These results suggest (but do not prove) that oxygen is not incorporated directly into the BiO layer of the annealed samples. Another explanation, which is consistent with our structural and electronic data, is that oxygen is incorporated into Cu–Bi vacancy or inter-

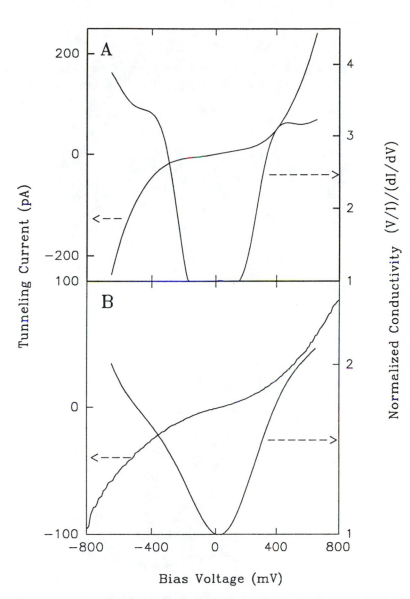

Figure 10. Typical I–V *data obtained on as-grown (A) and oxygen-annealed (B) samples. The feedback stabilized tunneling resistance, 10⁹ Ω, was the same for both experiments. Normalized conductance curves, (V/I)dI/dV, corresponding to the as-grown and oxygen-annealed samples are also shown.*

Figure 11. An STM image measuring 40 × 40 $\mathring{A}^2$ recorded on (A) as-grown and (B) 12-atm, oxygen-annealed $Bi_2Sr_2CaCu_2O_8$ cleaved single crystals. The tetragonal lattice spacing in the two images is 3.8 $\mathring{A}$. The images were recorded by using bias voltage and tunneling current, respectively, of (A) 300 mV and 0.9 nA, and (B) 250 mV and 1.8 nA.

stitial sites in the SrO layers. This oxygen would cause only subtle changes in the BiO layer structure and DOS, as observed in our experiments.

Studies of $Bi_2Sr_2CaCu_2O_8$ in the Superconducting State

A detailed picture of the energy gap (2Δ) and low-energy excitations is essential to understand the mechanism of superconductivity in the copper oxide materials. The Bardeen–Cooper–Schrieffer (BCS) theory of superconductivity predicts that a complete gap in the density of electronic states exists as T approaches zero, and that the magnitude of 2Δ has the universal value of $3.53kT_c$ (*55*). For conventional superconductors, these predictions have agreed well with measurements of the DOS by tunneling spectroscopy (*56*). For the copper oxide materials, however, tunneling spectroscopy measurements have led to suggested values of 2Δ between 0 and 12 kT_c, although recent studies (*57–67*) seem to converge toward an energy scale of 5–8 kT_c. Furthermore, large conductances within $\pm\Delta$, which suggest that a true energy gap may not exist, were reported by several groups (*58, 61–63*).

Material inhomogeneities such as oxygen nonstoichiometry probably have been responsible for many of the apparently conflicting measurements of 2Δ by tunneling spectroscopy (*63, 64, 67*). The short coherence lengths of the copper oxide materials make tunneling measurements especially susceptible to local variations in the superconducting properties. Hence, we (*67*) placed considerable effort on STS measurements of $Bi_2Sr_2CaCu_2O_{8+\delta}$ crystals that were carefully annealed to produce homogeneous samples. Our tunneling measurements were made in two distinct ways with low-temperature STM. First, we studied normal metal–insulator–superconductor (NIS) junctions formed between a Pt-tip and $Bi_2Sr_2CaCu_2O_{8+\delta}$ superconductor. These junctions provide a direct measure of 2Δ. Second, we investigated superconductor–insulator–superconductor (SIS) junctions formed between two pieces of $Bi_2Sr_2CaCu_2O_{8+\delta}$ material. These latter junctions provide a direct measure of 4Δ.

Typical $I-V$ and dI/dV versus V curves obtained by using NIS junctions at 4.2 K are shown in Figure 12. These results are representative of several hundred data sets obtained on the annealed $Bi_2Sr_2CaCu_2O_{8+\delta}$ crystals. The tunnel–junction resistances in our experiments varied from 10^7 to 10^9 Ω; within this range, similar results were observed. However, the current exhibits a weak, nonexponential dependence on tip–sample separation. This weak dependence on tip–sample separation indicates that the tip touches the sample surface during these experiments. In this contact geometry, the tunneling measurements average both $a–b$ and c directions (*68*).

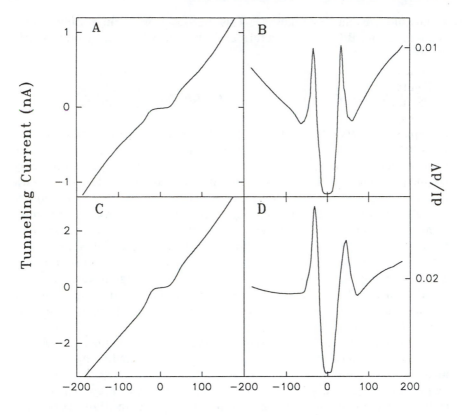

Figure 12. A and C: Typical I–V curves obtained on oxygen-annealed Bi₂Sr₂CaCu₂O₈ single crystals at 4.2 K. These NIS junctions were formed between a Pt–Ir tip and cleaved Bi₂Sr₂CaCu₂O₈ crystal surfaces. B and D: Conductance (dI/dV) versus voltage curves corresponding to the I–V data in A and C, respectively. The curves exhibit well-developed gap structure and linear background conductance for |V| > ±Δ. Conductances at V = 0 in B and D are 8 and 5%, respectively, of G(100).

Qualitatively, the *I–V* curves exhibit a flat, low-current region about the Fermi level (E_f, V = 0) and relatively sharp conductance onsets at ±25 mV. These features are characteristic of a conventional superconducting energy gap. Interestingly, this well-developed gap structure is observed reproducibly over the surfaces of oxygen-annealed $Bi_2Sr_2CaCu_2O_{8+\delta}$ samples. In contrast, we observe a wide range of *I–V* behavior on as-grown $Bi_2Sr_2CaCu_2O_{8+\delta}$ samples; these variations are similar to those found in other reports (*63*). Therefore, much of the uncertainty observed in previous work may be due to oxygen nonstoichiometry. In the following section, we confine our analysis solely to reproducible measurements obtained on oxygen-annealed $Bi_2Sr_2CaCu_2O_{8+\delta}$ samples.

The conductance, $G(V) = dI/dV$, versus voltage curves provide essential insight into the nature of the superconducting gap in these materials (Figures 12C and 12D). The conductance within the gap is low. For the NIS junctions, $G(0)/G(150)$ values, where $G(150)$ is representative of the normal-state conductance, are between 2 and 8%. These values can be compared with conductances of 30–50% reported previously (*59, 62, 63*).

We do find, however, large conductances in tunneling measurements made on the as-grown $Bi_2Sr_2CaCu_2O_{8+\delta}$ crystals. Because extrinsic effects such as sample inhomogeneity were not accounted for in the previous tunneling studies, these previous data are not necessarily indicative of d-wave pairing or gapless superconductivity. In our carefully annealed samples, $G(V)$ is very low at $V = 0$; however, the behavior of $G(V)$ for $V > 0$ is also important to consider.

First, the increase in $G(V)$ within the gap is not proportional to $|V|$. A linear increase in $G(V)$ would be a clear signature for gapless superconductivity (*60*); therefore, our results may argue against this possibility. Comparing the conductance at $V = 0$ and $V = \Delta/2$ predicted for s-wave BCS gap with our data, we find that the increase in $G(V)$ at $V = \Delta/2$ is close to or slightly larger than the increase predicted by a thermally broadened (4.2 K) BCS-gap expression (*see* Figures 12C and 12D, respectively). These results indicate close analogy to a conventional BCS-like gap; however, the divergence at the gap edge differs from conventional behavior.

Similar results were also obtained from SIS junctions; representative $I–V$ and $G(V)$ curves are shown in Figure 13. The $I–V$ curves exhibit a flat, low-current region about $V = 0$ and pronounced conductance onsets at approximately ± 50 mV. The near-zero current region about E_f detected in the SIS measurements is approximately 2 times larger than in the NIS data. This observation is consistent with the measurement of 2Δ and 4Δ in the NIS and SIS junctions, respectively, and indicates that other effects such as the Coulomb blockade do not contribute significantly to our data.

The $G(V)$ curves from the SIS junctions also exhibit low conductance at E_f (~4%). The background conductance within the gap increases more rapidly than does the background conductance found for the NIS junctions. The increase in conductance could be due to either poor junction quality or gapless superconductivity. Poor SIS junction quality is probably the major factor leading to the increase in $G(V)$ within the gap; however, we do not have sufficient experimental evidence at this time to resolve the observed differences in the gap excitations for the NIS and SIS junctions. Regardless of the origin of the excitations observed in the gap of these SIS junctions, these data confirm the magnitude of 2Δ for the oxygen-

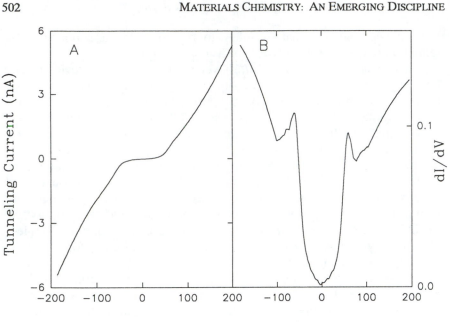

Figure 13. A: Representative I–V *curve obtained from an SIS junction formed between two oxygen-annealed* $Bi_2Sr_2CaCu_2O_8$ *crystals at 4.2 K. The junction geometry averages the* c *and* a–b *directions. B: Conductance versus voltage curve corresponding to the data in A. The conductance at* V *= 0 is 4% of G(100). The gap structure observed in the SIS junctions exhibits greater broadening than the data obtained from the NIS junction.*

annealed $Bi_2Sr_2CaCu_2O_{8+\delta}$ crystals and provide a consistent and reproducible energy scale for superconductivity.

In addition, we further analyzed the reproducible gap structure observed in our NIS measurements to quantitatively assign a value to 2Δ and to probe the energy dependence of the DOS. The experimental data were fit to the following modified BCS model for DOS:

$$N_S = R_e\{(eV - i\Gamma)/[(eV - i\Gamma)^2 - \Delta^2]^{1/2}\} \tag{12}$$

where Γ is a phenomenological parameter to account for broadening, N_S is the density of states in the superconducting state, R_e corresponds to the real part of the complex number in braces {}, and i is $\sqrt{-1}$.

A typical fit using this model is shown in Figure 14. The experimental data are well-fit for $|V| < \Delta/2$, but the experimental conductance peaks are broadened significantly compared with the model DOS. Similar

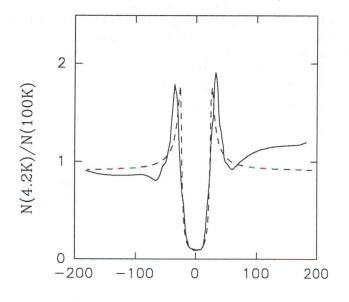

Figure 14. Conductance determined at 4.2 K normalized by the conductance at 100 K, G(4.2 K)/G(100 K), versus voltage for a typical NIS junction (solid line). The dashed line corresponds to a fit of this experimental data to the BCS model for DOS (see equation 12). This simple model for the superconducting state fits the gap structure well, although the experimental conductance peaks are significantly broader than the model fit. The values of Δ and Γ extracted from the model fit are 25 and 3 meV, respectively.

fits were also obtained for other $G(V)$ data with $\Gamma = 1$–3 meV; this broadening energy is greater than the thermal energy (0.36 meV). One should not, however, place too much significance on Γ because sample quality may still be limiting. That is, the finite transition widths indicate that the samples are not perfectly crystalline.

Nevertheless, this analysis has several different points. First, the well-defined gap in the NIS data can be fit at low energies by using a conventional model, although the broadening needed for a best-fit is greater than expected for only thermal effects. Second, the magnitudes of 2Δ and 4Δ extracted from our fits to the NIS and SIS data, 50 ± 5 and 98 ± 5 mV, respectively, show that the energy scale for superconductivity is $\sim 6.8kT_c$. The magnitude of 2Δ extracted from this analysis is, therefore, consistent with high-resolution, electron energy loss spectroscopic and photoemission spectroscopic measurements of $Bi_2Sr_2CaCu_2O_{8+\delta}$ crystals

(66, 69, 70). Lastly, the divergence in the experimental data at the gap edge is weak in comparison with the behavior expected for an s-wave BCS superconductor. This deviation from conventional behavior may be a signature of d-wave pairing or gapless superconductivity, although additional studies will be needed to confirm the intrinsic nature of the divergence.

These low-temperature STS studies of homogeneous, oxygen-annealed $Bi_2Sr_2CaCu_2O_{8+\delta}$ single crystals have shown that a well-developed gap structure can be observed reproducibly in high-quality samples. The low conductance and change of $G(V)$ observed within the gap region differ from the behavior expected for either a gapless or BCS-like superconductor and may indicate d-wave pairing. Analyses of data from both NIS and SIS junctions provide a consistent scale for superconductivity with $2\Delta \approx 6.8kT_c$.

Summary and Conclusions

STM and STS were used to characterize the electronic and structural properties of the high-temperature copper oxide superconductor $Bi_2Sr_2CaCu_2O_{8+\delta}$. In particular, these studies have done the following:

1. elucidated the local structural order in the BiO layer of $Bi_2Sr_2CaCuO_8$ and the low-energy electronic states associated with this disorder

2. characterized the structural and electronic effects caused by reducing and increasing the oxygen concentration in this material

3. determined a consistent value for the magnitude of the superconducting energy gap in $Bi_2Sr_2CaCuO_{8+\delta}$ single crystals

These data represent a firm beginning of a detailed microscopic picture of the structural and electronic properties for these complex materials. Continued STM and STS studies will undoubtedly lead to a much clearer understanding of superconductivity in the copper oxides. We believe that the use of STM and STS will also provide essential insight into many other problems in materials chemistry.

Acknowledgments

Charles M. Lieber thanks David and Lucile Packard of the National Science Foundation, Camille and Henry Dreyfus Foundation, and the A. P. Sloan Foundation for generous support of this work.

References

1. Binnig, G.; Rohrer, H.; Gerber, C.; Weibel, E. *Phys. Rev. Lett.* **1982**, *49*, 57.
2. Golovchenko, J. A. *Science (Washington, D.C.)* **1986**, *232*, 48.
3. Tromp, R. M.; Hamers, R. J.; Demuth, J. E. *Science (Washington, D.C.)* **1986**, *234*, 304.
4. Hamers, R. J. *Annu. Rev. Phys. Chem.* **1989**, *40*, 531.
5. Griffith, J. E.; Kochanski, G. P. *Annu. Rev. Mater. Sci.* **1990**, *20*, 194.
6. Avouris, P. *J. Phys. Chem.* **1990**, *94*, 2246.
7. Lieber, C. M.; Wu, X. L. *Acc. Chem. Res.* **1991**, *24*, 170.
8. Coleman, R. V.; Giambattista, B.; Hansma, P. K.; Johnson, A.; McNairy, W. W.; Slough, C. G. *Adv. Phys.* **1988**, *37*, 559.
9. Wu, X. L.; Lieber, C. M. *Science (Washington, D.C.)* **1989**, *243*, 1703.
10. Wu, X. L.; Lieber, C. M. *Phys. Rev. Lett.* **1990**, *64*, 1150.
11. Wu, X. L.; Lieber, C. M. *J. Am. Chem. Soc.* **1989**, *111*, 2731.
12. Chen, H.; Wu, X. L.; Lieber, C. M. *J. Am. Chem. Soc.* **1990**, *112*, 3326.
13. Wu, X. L.; Zhou, P.; Lieber, C. M. *Phys. Rev. Lett.* **1988**, *61*, 2604.
14. Wu, X. L.; Zhou, P.; Lieber, C. M. *Nature (London)* **1988**, *335*, 55.
15. Wu, X. L.; Lieber, C. M. *Phys. Rev. B.* **1990**, *41*, 1239.
16. Dai, H.; Chen, H.; Lieber, C. M. *Phys. Rev. Lett.* **1991**, *66*, 3183.
17. Wang, C.; Giambattista, B.; Slough, C. G.; Coleman, R. V. *Phys. Rev. B* **1990**, *42*, 8890.
18. Rao, C. N. R.; Raveau, B. *Acc. Chem. Res.* **1989**, *22*, 106.
19. Torardi, C. C.; Subramanian, M. A.; Calabrese, J. C.; Gopalakrishnan, J.; Morrissey, K. J.; Askew, T. R.; Flippen, R. B.; Chowdhry, U.; Sleight, A. W. *Science (Washington, D.C.)* **1988**, *240*, 631.
20. Hansma, P. K.; Tersoff, J. *J. Appl. Phys.* **1987**, *61*, R1. Hansma, P. K.; Elings, V. B.; Marti, O.; Bracker, C. E. *Science (Washington, D.C.)* **1988**, *242*, 209.
21. Kuk, Y.; Silverman, P. J. *Rev. Sci. Instrum.* **1989**, *60*, 165.
22. Sakurai, T.; Hashizume, T.; Kamiya, I.; Hasegawa, Y.; Sano, N.; Pickering, H. W.; Sakai, A. *Prog. Surf. Sci.* **1990**, *33*, 3.
23. Tersoff, J.; Hamann, D. R. *Phys. Rev. Lett.* **1983**, *50*, 1998.
24. Tersoff, J.; Hamann, D. R. *Phys. Rev. B* **1985**, *31*, 805.
25. Lang, N. D. *Phys. Rev. Lett.* **1986**, *56*, 1164.
26. Selloni, A.; Carnevalli, P.; Tosatti, P. E.; Chen, C. D. *Phys. Rev. B* **1986**, *33*, 5770.
27. Tersoff, J. *Phys. Rev. B* **1990**, *41*, 1235.
28. Chen, C. J. *Phys. Rev Lett.* **1990**, *65*, 448.
29. Bardeen, J. *Phys. Rev. Lett.* **1963**, *6*, 57.
30. Simmons, J. *J. Appl. Phys.* **1963**, *34*, 1793.
31. Lang, N. D. *Phys. Rev. B* **1988**, *37*, 10395.
32. Feenstra, R. M.; Stroscio, J. A.; Fein, A. P. *Surf. Sci.* **1987**, *181*, 295.
33. Whangbo, M.-H.; Torardi, C. C. *Acc. Chem. Res.* **1991**, *24*, 127.
34. Sleight, A. W. *Science (Washington, D.C.)* **1988**, *242*, 1539.
35. Ramakrishnan, T. V.; Rao, C. N. R. *J. Phys. Chem.* **1989**, *93*, 4414.
36. Subramanian, M. A.; Torardi, C. C.; Calabrese, J. C.; Gopalakrishnan, J.; Morrissey, K. J.; Askew, T. R.; Flippen, R. B.; Chowdhry, U.; Sleight, A. W. *Science (Washington, D.C.)* **1988**, *239*, 1015.

37. Gao, Y.; Lee, P.; Coppens, P.; Subramanian, M. A.; Sleight, A. W. *Science (Washington, D.C.)* **1988**, *241*, 954.

38. Zandbergen, H. W.; Groen, W. A.; Mijlhoff, F. C.; van Tendeloo, G.; Amelinckx, S. *Physica C (Amsterdam)* **1988**, *156*, 325.

39. Bordet, P.; Capponi, J. J.; Chaillout, C.; Chenavas, J.; Hewat, A. W.; Hewat, E. A.; Hodeau, J. L.; Marezio, M. *Stud. High Temp. Supercond.* **1989**, *2*, 171.

40. Eibl, O. *Physica C (Amsterdam)* **1991**, *175*, 419.

41. Kirk, M. D.; Nogami, J.; Baski, A. A.; Mitzi, D. B.; Kapitulnik, A.; Geballe, T. H.; Quate, C. F. *Science (Washington, D.C.)* **1988**, *242*, 1673.

42. Wu, X. L.; Zhang, Z.; Wang, Y. L.; Lieber, C. M. *Science (Washington, D.C.)* **1990**, *248*, 1211.

43. Tanaka, M.; Takahashi, T.; Katayama-Yoshida, H.; Yamazaki, S.; Fujinami, M.; Okabe, Y.; Mizutani, W.; Ono, M.; Kajimura, K. *Nature (London)* **1989**, *339*, 691.

44. Zhang, Z.; Wang, Y. L.; Wu, X. L.; Huang, J. L.; Lieber, C. M. *Phys. Rev. B* **1990**, *42*, 1082.

45. Shih, C. K.; Feenstra, R. M.; Chandrashekhar, G. V. *Phys. Rev. B* **1991**, *43*, 7913. Shih, C. K.; Feenstra, R. M.; Kirtley, J. R.; Chandrashekhar, G. V. *Phys. Rev. B* **1989**, *40*, 2682.

46. Massidda, S.; Yu, J.; Freeman, A. J.; Krakauer, H.; Pickett, W. E. *Phys. Rev. Lett.* **1988**, *60*, 1665.

47. Torardi, C. C.; Jung, D.; Kang, D. B.; Ren, J.; Whangbo, M.-H. *Mater. Res. Soc. Symp. Proc.* **1989**, *156*, 295.

48. Cava, R. J. *Science (Washington, D.C.)* **1990**, *247*, 656.

49. Daeumling, M.; Seuntjens, J. M.; Larbalestier, D. C. *Nature (London)* **1990**, *346*, 332.

50. Jorgensen, J. D.; Pei, S.; Lightfoot, P.; Shi, H.; Paulikas, A. P.; Veal, B. W. *Physica C (Amsterdam)* **1990**, *167*, 571.

51. Wu, X. L.; Wang, Y. L.; Zhang, Z; Lieber, C. M. *Phys. Rev. B* **1991**, *43*, 8729.

52. Zhang, Z.; Lieber, C. M. *Phys. Rev. B* **1992**, *46*, 5845.

53. Shen, Z.-X.; Dessau, D. S.; Wells, B. O.; Olson, C. G.; Mitzi, D. B.; Lombardo, L.; List, R. S.; Arko, A. J. *Phys. Rev. B* **1991**, *44*, 12098.

54. Zhang, Z.; Chen, C. C.; Lieber, C. M.; Morosin, B.; Venturini, E. L.; Ginley, D. S. *Phys. Rev. B* **1992**, *45*, 987.

55. Bardeen, J.; Cooper, L. N.; Schrieffer, J. R. *Phys. Rev.* **1957**, *108*, 1175.

56. Giaever, I. *Phys. Rev. Lett.* **1960**, *5*, 147.

57. Kirtley, J. R. *Int. J. Mod. Phys. B* **1990**, *4*, 201.

58. Hasegawa, T.; Ikuta, H.; Kitazawa, K. In *Physical Properties of High Temperature Superconductors III;* Ginsberg, D. M., Ed.; World Scientific Publishing: River Edge, NJ, 1992; p 525.

59. Valles, J. M.; Dynes, R. C.; Cucolo, A. M.; Gurvitch, M.; Schneemeyer, L. F.; Garno, J. P.; Waszczak, J. V. *Phys. Rev. B* **1991**, *44*, 11986.

60. Mandrus, D.; Forro, L.; Koller, D.; Mihaly, L. *Nature (London)* **1991**, *351*, 460.

61. Boekholt, M.; Hoffmann, M.; Guntherodt, G. *Physica C (Amsterdam)* **1991**, *175*, 127.

62. Huang, Q.; Zasadzinski, J. F.; Gray, K. E.; Liu, J. Z.; Claus, H. *Phys. Rev. B* **1989,** *40,* 9366.

63. Liu, J.-X.; Wan, J.-C.; Goldman, A. M.; Chang, Y. C.; Jiang, P. Z. *Phys. Rev. Lett.* **1991,** *67,* 2195.

64. Tao, H. J.; Chang, A.; Lu, F.; Wolf, E. L. *Phys. Rev. B* **1992,** *45,* 10622.

65. Chen Q.; Ng, K.-W. *Phys. Rev. B* **1992,** *45,* 2569.

66. Li, Y.; Huang, J. L.; Lieber, C. M. *Phys. Rev. Lett.* **1992,** *68,* 3240.

67. Zhang, Z.; Lieber, C. *Phys. Rev. B* **1993,** *47,* 3423.

68. Wolf, E. L. *Principles of Tunneling Spectroscopy;* Oxford University Press: New York, 1989.

69. Olson, C. G.; Liu, R.; Yang, A.-B.; Lynch, D. W.; Arko, A. J.; List, R. S.; Veal, B. W.; Chang, Y. C.; Jiang, P. Z.; Paulikas, A. P. *Science (Washington, D.C.)* **1989,** *245,* 731.

70. Dessau, D. S.; Wells, B. O.; Shen, Z.-X.; Spicer, W. E.; Arko, A. J.; List, R. S.; Mitzi, D. B.; Kapitulnik, A. *Phys. Rev. Lett.* **1991,** *66,* 2160.

RECEIVED for review November 9, 1992. ACCEPTED revised manuscript April 8, 1993.

Biomimetic Mineralization

Patricia A. Bianconi

Department of Chemistry, Pennsylvania State University,
University Park, PA 16801

*Researchers in biomimetic mineralization attempt to improve syn-
thetic materials design by adapting biological mechanisms of molecular control that are used to produce biological inorganic–organic composites. Crucial features of biomineralization that regulate and control the growth of organized crystalline arrays include
matrix mediation of nucleation, control of ion concentration at
the inorganic–organic interface, and control of crystal nucleation
and growth by soluble molecules via molecular recognition at the
inorganic–organic interface. Research in materials chemistry on
the synthesis of ordered synthetic composites by biomimetic routes
includes study of the in situ precipitations of inorganic materials
in polymeric matrices. These precipitations are synthetic analogues of matrix mediation. Other approaches seek to determine
the nature of the molecular-recognition processes at inorganic–organic interfaces by studying (1) the orientation of crystals
grown under Langmuir–Blodgett layers or in constrained environments, (2) the addition of additives to selectively control crystal
morphology, and (3) the growth of oriented crystals on surfaces by
cooperativeness of chemical groups.*

RESEARCH ON THE FABRICATION OF COMPOSITES, which are physical mixtures of several components, has been stimulated by the desire to
combine various physical properties into a single material. This approach
has been very successful in producing materials with improved physical
properties, such as reinforced elastomers (*1, 2*); dense, defect-free ceramics
(*3–6*); processable, nonlinear, optical materials (*7, 8*); and composite devices (*9–12*). The composites were fabricated by physical mixing of two
components or by initiating polymerization of an organic monomer dis-

0065–2393/95/0245–0509$12.00/0

persed in a preformed inorganic matrix [such as sol–gel-derived silica (3, 4) or layered FeOCl (13)].

However, composites in which two disparate materials (usually an inorganic phase and an organic polymer) are blended in an organized way on the molecular level improve physical properties and can behave as "alloys" of the inorganic and organic components. Extraordinarily high synergy of the two components' material properties as well as unique interactions at the phase interfaces have been seen (1, 14–20). Usually, such organized molecular-level composites are found only in nature as the results of biomineralization processes that combine inorganic crystals and organic biopolymers to produce, for example, bones, teeth, and shells (21–31). In such organized biological materials, inorganic crystals of various minerals are formed in ordered arrays in association with a matrix of organic macromolecules (22). These materials are distinguished from synthetic composites by the high degree of organization and regularity displayed by the inorganic phase; inorganic crystals of uniform size, morphology, and crystallographic orientation have been found (21). However, biomimetic mineralization attempts to chemically control the growth of inorganic crystals in organic matrices by adapting or imitating biological mineralization strategies.

The barrier to fabrication of synthetic composite alloys has been the high temperatures usually required to modify inorganic matrices and blend them molecularly with more fragile organic polymers. Thus, current techniques for composite fabrication usually modify only the organic component on a molecular level, and a high degree of blending with the preformed inorganic component is harder to achieve (an exception is when inorganic networks are fabricated at low temperatures by using the sol–gel technique). Also, the inorganic component cannot be ordered to a high degree within the synthetic matrix. Although composites fabricated by many methods of physical blending have wide areas of application (1, 2, 9–12), they are less intimately mixed on the molecular level and display less synergy of material properties than do the natural materials produced by biomineralization. Fabricated composites also do not display the desirable effects of having the inorganic phase arranged with a very high degree of order; the size, shape, and crystallographic orientation of the inorganic crystals is uniform only in biological materials (21–31). This structural regularity, especially in crystallographic orientation, would be highly desirable in synthetic thin films because many properties (magnetic, ferroelectric, and superconducting) are highly anisotropic (32).

At present, a primary goal of synthetic materials chemistry is achieving not only better control of the composition and morphology of new materials but also the controlled assembly of molecular species into organized arrays (9–12). Such control should allow the construction of well-defined surface or aggregate morphologies and of multicomponent films with anisotropic electronic, magnetic, or optical properties (*33*). A synthetic process that gives electronic, optical, and structural composite materials with the same degree of molecular alloying as biomaterials would be very valuable in controlling composition, morphology, and assembly. The degree of organization and hierarchical structure seen in biomineralized composites has never been achieved in synthetic materials and would add greatly to the properties synthetic materials could display. Orientation of the dipole or magnetic moments of inorganic crystals within the synthetic polymer matrix, for example, would allow new types of nonlinear optical, electronic, or magnetic behavior to be explored. The alignment of correctly shaped crystals within a supporting matrix and the resulting dependence of the mechanical properties of the material on the matrix–mineral composition and the type and level of organization (as is seen in shells) (*34*) could produce stronger yet less brittle structural materials and artificial biominerals. Also, the study of the fundamental interactions of the organic–inorganic phase interface on the molecular level could provide valuable information on the surface behavior of supramolecular composite arrays (*35*).

The underlying principles used by organisms in controlling mineralization have been studied and described in detail. Three key features that biology adopts in the controlled deposition of inorganic particles within polymer matrices to give organized crystalline arrays are ion-flux regulation at the matrix interface, growth and habit modification by soluble molecules present within the polymer matrix, and crystallochemical mediation of nucleation and growth by molecular-specific interactions at the polymer surface (*21–31*). These features result in nucleation at specific matrix sites and control of crystal growth, habit, and orientation by molecular interactions at the interface between the mineral and the organic matrix.

Research in materials chemistry on the synthesis of such ordered synthetic composites by biomimetic routes has taken many directions and produced many advances. The literature reporting inorganic–organic in situ synthesis, which is the synthetic method most analogous to natural biomineralization, is extensive (*21, 29, 36–63*). More specifically, biomimetic routes to in situ synthesized composites have produced metallic particle- (*64–66*) and metallic oxide-containing polymer films (*67, 68*); aligned crystal–polymer composites containing optically nonlinear compounds; polymer–ceramic composites in which the polymer catalyzes and controls the ceramic's formation; thin-film ceramics of superior properties; and oriented, size-controlled semiconductor and magnetic clusters within membranes and polymers.

Synthetic control of inorganic-phase properties by the organic matrix in which the material is synthesized is attainable and is analogous to that seen in natural minerals. Other approaches seek to determine the nature of the molecular recognition processes at inorganic–organic interfaces by studying the following: the orientation of crystals grown under Lang-muir–Blodgett layers or in constrained environments, the addition of addi-tives to control crystal morphology, and the growth of oriented crystals on surfaces by cooperativeness of chemical groups. Synthetic manipulation and control of inorganic crystals by specific molecular-bonding interac-tions has been demonstrated and shown to be analogous to biological sys-tems. Because these two approaches are physically possible in synthetic systems and some chemical systems combine aspects of both approaches, synthetic analogues of the biomineralization process, or biomimetic miner-alizations, may be possible and may provide a route to the fabrication of new materials that display greater degrees of desirable biomimetic prop-erties.

In Situ Synthesis: Matrix Mediation of Crystal Nucleation and Growth

In biological materials, the inorganic particles are grown in situ within a polymeric matrix and under the control of the matrix (21–31). The fabri-cation of synthetic composites by a biomimetic, controlled, in situ precipi-tation route has been used to prepare more highly organized materials than can be obtained by physical mixing of inorganic and organic com-ponents (21, 29, 48). Control of size, morphology, and orientation of crys-tals has been achieved.

The confinement and physical shaping of the mineralization zone by membrane and polymer assemblies is a general feature of most systems in-volving controlled biomineralization (21–31, 69). Similar confinement in synthetic systems has produced materials with controlled particles sizes. Extensive studies of in situ sol–gel syntheses were reported by Weiner (17) and Mark and co-workers (49–63). Alkoxysilanes were hydrolyzed in the solid state within polymer films such as polysiloxanes and polyoxides, which could previously have been cross-linked. These in situ reactions produced very small, well-dispersed particles of amorphous SiO_2. These particles provided reinforcement that was superior to the blending of pre-formed filler particles into the polymer.

Many variations of the process that gave the products more tailored properties were reported, such as use of a polymer matrix with hydroxide- or alkoxide-terminated chains that participated in the reaction and simul-taneously cured and reinforced the polymer on a molecular level (49–63). Monomers bearing one hydrolysis-inert group that could later be cross-linked were also used. Similar processes can be used with a variety of or-ganometallic monomers to produce polymer-blended ceramics such as TiO_2 (49–63, 70, 71), Al_2O_3 (49–63), ZrO_2 (49–63), and $BaTiO_3$ (49–63,

72). Moreover, these works (*49–63, 70–72*) also demonstrated that the inorganic phase generated by sol–gel synthesis can be incorporated into the composite in a relatively high concentration that is analogous to the very high mineral concentrations seen in biocomposites (*21*). Synthesis of ultrafine, multicomponent ceramic particles of mixed-metal oxides was also achieved using phospholipid vesicles as membrane-mimetic synthesis hosts (*73*). The vesicle membrane controlled precipitation kinetics, induced size control, and prevented agglomeration of the nanophase particles. The membrane also acted as a lubricant–dispersant that facilitated particle consolidation into a ceramic body.

In addition, another biomimetic property, orientation of inorganic particles synthesized in situ within polymer matrices, can be achieved in these systems by alignment of the films. Deformation at elevated temperatures of polymers containing ceramic particles or curing of polymers containing magnetic particles in the presence of a magnetic field produces alignment of the in situ generated particles, both spatially within the polymer and crystallographically, as shown by the anisotropic magnetic moments of some of the resulting films (*49–63*). Imposing anisotropy on a poly(vinyl chloride) matrix by drawing the matrix successfully imposed this same anisotropy on titania particles synthesized within the matrix. The resulting particles were highly elongated along the aligned polymer-chain axis (*48*).

Highly active, second-order, nonlinear optical materials were also fabricated by similar alignment of incorporated, optically active crystals by physical alignment of the polymer matrix in which they were grown or dispersed (*7, 8*). For example, films of poly(methyl methacrylate) containing the optically nonlinear compound 3-nitroaniline were aligned using temperature-gradient zone melting, in which a heated polymer sample is slowly drawn across a sharp thermal gradient. The guest compound crystallizes in a line within the polymer film as it traverses the temperature gradient, and the crystals are aligned parallel to the drawing axis. The optical-quality films displayed second-harmonic generation more that 400 times that of potassium dihydrogen phosphate, and thus they represent very efficient frequency doubling media (*74*).

Similar control of the size of semiconductor clusters and magnetic particles was achieved by using the biomimetic approach of confinement and physical shaping of the mineralization zone by membrane and polymer assemblies. Surfactant assemblies such as aqueous and reversed micelles, vesicles and polymerized vesicles, and bilayer membranes of amphiphilic lipids or cyclams were used as models for biological membranes. With these models, size-quantized microcrystals and ultrathin, particulate films of cadmium, induim, copper, lead, mercury, and zinc sulfides were synthesized by introducing solutions of metal salts on one side of the membrane and hydrogen sulfide on the other (*42–47, 75*). Diffusion and reaction of the inorganic reagents within the membrane was controlled by the membrane's particular properties, and the metal sulfide crystalline products were produced with far greater control and order than when solution reactions were used. The semiconductor films could be transferred to substrates; these films displayed the altered electrical, photochemical, and optical properties that are associated with quantum-size confinement in semiconductors.

The membranes combine the advantages of vesicles (thinness) with those of polymeric films (accessibility) to achieve molecular organization and compartmentalization analogous to those seen in biological mineralization. Steigerwald, Brus, and co-workers (76, 77) used inverse micelles as size-controlled host media to synthesize a variety of II—VI semiconductor clusters of uniform size. The clusters were stabilized and released as molecular semiconductor crystallites by capping their surfaces with a covalently bonded organic coating. Inorganic nanophase materials such as Ag_2O (78), $CoSiO_3$ (79), iron oxides (80), and alumina (81) were prepared in unilamellar vesicles composed of phospholipids. The vesicle provides a reaction environment of restricted size and ion concentration for inorganic precipitation, confers size control, and prevents agglomeration (69).

Recently, protein cavities were used as compartmentalized, host synthetic media for inorganic crystals, and these cavities had greater robustness and less sensitivity to changes in phase behavior as a function of ambient conditions than did unilamellar vesicles (69). Apoferritin molecules, native ferritin proteins from which the iron oxide core is removed and the central protein cavity is left empty, were found to be efficient hosts for the deposition of a variety of mineral phases, including iron and manganese oxides and sulfides (82). Because this protein shell can induce mineralization specifically within the preformed cavity as a result of the particular distribution of amino acid residues on and near the cavity surface, site-directed mutageneses that produced ferritins with modified cavity surfaces were found to alter the cavity's nucleation and precipitation abilities (83). Therefore, techniques of molecular biology might be effective in tailoring the synthesis of inorganic materials within these cavities.

In an analogous approach to biomimetic mineralization, size-restricted or spatially organized reaction environments within polymer matrices were used to fabricate composites of organic polymers with magnetic particles and quantum-sized semiconductor and metal clusters. A sulfonated, polystyrene, ion-exchange resin exchanged with Fe(II) or Fe(III) was used as a synthetic matrix, and nanometer-sized crystals of the magnetic oxide γ-Fe_2O_3 were fabricated within the resin to form a magnetic composite material (84). The matrix provided spatially localized sulfonate binding sites for crystal nucleation and minimized the degree of aggregation of the iron oxide crystals; therefore, an upper limit was placed on the crystals' sizes. The resulting crystals were uniform in size, morphology, and crystalline phase and exhibited unusual optical transparency in magnetic materials due to their nanometer-sized dimensions. Identical syntheses of the iron oxides carried out in the absence of a polymer matrix resulted in nonmagnetic, amorphous, or poorly crystalline aggregates of various oxide phases with dimensions in the micrometer range, a finding that underscores the high degree of control over crystal nucleation and growth that biomimetic, matrix-mediated syntheses can confer.

In an analogous approach, copolymers of monomers that contain both coordinating and noncoordinating groups for metal cations, such as ethylene–15% methacrylic acid copolymer, were used as synthetic matrices (*40, 85–90*). These phase-separated ionomers can be viewed as microemulsions consisting of polar microdomains of the methacrylic acid residues with ~5-nm diameters that are dispersed in a polyethylene matrix. Also, block copolymers that are organized as they are cast into lamellar, polar, and nonpolar microdomains were synthesized by ring-opening, metathesis copolymerization of cyclic olefins and norbornenes functionalized with metal-binding groups such as dimethoxyethane (*91*). Metal cations incorporated into these polymers by solution casting or by pressing bind selectively in the polar cluster or lamellar domains. When exposed to hydrogen sulfide, in situ reaction produces a variety of sulfur(II, VI) or sulfur(IV, VI) semiconductor clusters whose sizes are controlled by the size of the polar polymer domain or whose spatial distributions are controlled by the lamellar structure of the organized polymer. Self-organized block copolymers in which one monomer contains a metal-chelating ligand such as a bidentate phosphine can create composite films of polymers containing spatially organized arrays of metal clusters by an analogous, in situ synthetic route (*92*). Preorganization of the polymer matrix that imposes size or spatial organization on the inorganic phase grown in situ is a common feature of biological mineralization systems (*21–31*).

Molecular Recognition at the Organic–Inorganic Interface

Synthetic mimicking of the biological mechanisms that facilitate uniform morphology and crystallographic orientation in biological composites has also produced synthetic composites of greater order and organization in the crystalline phase than are normal. In addition to use of the organic matrix to constrain the inorganic reaction environment and regulate the crystal's size and morphology, an important biological strategy for obtaining a high degree of control over crystallization is mediation of a crystal's nucleation and growth by molecular-specific interactions at the inorganic–organic interface by both polymers and low-molecular-weight molecules (*22*). Synthetic mimicking of this strategy has included the study of organic surfaces as molecular templates for oriented nucleation, organic additives for control of crystal morphology, and chemical complementarity between the surface chemistry of organic matrices and nucleated crystal faces.

Polymer matrices with surfaces derivatized with specific functional groups that act to chemically control nucleation and orientation were used to deposit single crystals and polycrystalline films of $CaCO_3$ (93) and iron oxides (32), which can show preferred orientations. Highly oriented, needle-like crystals of goethite [α-FeO(OH)], all with the long axis perpendicular to the substrate, were formed from aqueous solutions on surface-sulfonated polystyrene films, although no deposition of iron oxides was seen on unsulfonated films. Increase in crystal deposition rate with an increase in degree of polymer sulfonation suggests that the sulfonate sites initiate nucleation and play a role in controlling crystallite size and orientation.

The modification of crystal habit by interaction with adsorbed molecules that poison the growth of specific faces is a well-known phenomenon (94–100). Biology controls the morphology of crystals by similar means and produces inhibitors, often acidic macromolecules, that slow the relative rates of growth of the crystal in different directions and selectively shape the crystal. These adsorbed molecules, along with specific nucleation-inducing sites on the polymer matrix, can also induce crystallographic orientation. A synthetic system with such chemical cooperativeness was reported by Addadi, Weiner, and co-workers (101, 102). They found that polystyrene films functionalized with sulfonate groups concentrated calcium ions from solution and created the supersaturation necessary for nucleation.

When β-sheet poly(aspartate), which presents a structured surface of carboxylate anion sites, is adsorbed onto the functionalized polystyrene, oriented $CaCO_3$ (calcite) crystal formation from solution occurs. Up to 60% of the calcite crystals deposited were oriented on the (0 0 1) face, and their c-axes were perpendicular to the surface of nucleation. Unsulfonated polystyrene films, sulfonated films, or adsorbed poly(aspartate) alone does not significantly induce this orientation; up to a limit the number of oriented crystals obtained was proportional, to the extent of film sulfonation. These results demonstrate that sulfonate groups on the film surface and adsorbed organized arrays of carboxylate groups act cooperatively to induce control of crystalline habit and orientation. An assembly of proteins rich in aspartic acid, which were extracted from mollusk shells and adsorbed on solid substrates, were also able to induce calcite nucleation with a (0 0 1) orientation (101, 102).

Langmuir mono- or bilayers of surfactants on the air–water interfaces of crystallizing solutions have been studied more widely than have polymers as biomimetic means of chemically controlling crystal orientation and growth by specific molecular interactions with organic molecules. The

charged, polar head groups of the surfactant molecules interact with, concentrate, and nucleate inorganic ions from solution; whereas the organized
nature of the structured monolayer film provides the means of controlling
orientation, phase, and habit (*103–105*).

Because the properties of surfactant assemblies can be understood at
the chemical level (far more so than the complex interfaces within polymeric matrices), more specific information can often be obtained about
geometric, stereochemical, and electrostatic factors important in the
recognition processes between the organic assembly and specific crystal
faces (*69*). For example, oriented growth of crystals of α-glycine and sodium chloride was achieved under chiral Langmuir monolayers of amphiphilic amino acids by a structural match between the monolayer and the
specific surface layer of the attached, growing crystals (*106, 107*). Mann
and co-workers (*108–111*) reported that the crystallization of calcium carbonate can be controlled by a monolayer of stearic acid on the surface of
the crystallizing solution. In the absence of this monolayer, crystals of the
most stable polymorph of calcium carbonate, calcite, form on the walls
and surface of the crystallization vessel. In the monolayer's presence, crystals of a less stable polymorph, vaterite, form exclusively on the organic
surface and are nucleated with their (0 0 0 1) faces parallel to the surface.

Changes in parameters such as the monolayer's head-group charge,
inter-head-group spacing, and solution conditions induced modifications in
the structure and orientation of the nucleated crystals. The experiments
suggest a stereochemical mechanism for the preferential formation of vaterite over calcite induced by the monolayer: vaterite should be formed if
the first layer of carbonate ions that is nucleated retains the orientation of
the carboxylate head groups of the monolayer (*108*).

In similar crystallizations that are directed by organic templates, Langmuir monolayers induced oriented nucleation of barium sulfate crystals of
unusual morphologies from aqueous solutions (*112, 113*). For these crystals, nucleation on a particular crystal face and development of specific
morphologies could be explained in terms of specific electrostatic, geometric, and stereochemical interactions at the organic–inorganic interface. In
addition, epitaxial growth of oriented crystals of lead sulfide under organized monolayers of arachidic acid was reported (*114*); the specific structure and orientation of the crystals arose from preferred ionic arrangement and nucleation of a specific crystal face because of its good lattice
match with the inter-head-group spacing of the monolayer.

The specificity and control on the molecular level of these biomimetic crystallizations suggest that organic assemblies of precise molecular design could be fabricated for use in the controlled growth of inorganic materials. Composite materials in which ordered orientation of inorganic
crystals was induced by surfactant–inorganic interactions were reported.
Examples of these materials are the iron oxide and multilayer surfactant

films in which magnetic anisotropy of the magnetite particles is achieved by in situ synthesis with preferred orientation in the ordered multilayers (*115*). Other examples are the cadmium sulfide–poly(ethylene oxide) films in which regular size, shape, crystalline phase, and crystallographic orientation are induced in the cadmium sulfide crystals by matrix mediation of nucleation and growth and interaction with a sulfonate head-group surfactant such as sodium bis(2-ethylhexyl)sulfosuccinate (*116*).

Summary

Biomimetic mineralization uses synthetic materials chemistry, which combines aspects of chemistry, biology, and materials science, to provide greater control over material properties. It seeks to develop synthetic control of the organized intermingling of the inorganic and organic components of composite materials to achieve the chemical synthesis of artificial materials that are analogous to biological materials in their structural complexity and organization. In addition to providing information concerning the basic chemistry and properties of these new, more organized composite materials, this approach should yield fundamental new insights into the nature of the interface between inorganic–organic phases and information on these interactions in ultrafine dimensions.

References

1. Manson, J. L.; Sperling, L. H. *Polymer Blends and Composites;* Plenum: New York, 1976; Chapters 10–12.
2. Sheldon, R. P. *Composite Polymeric Materials;* Applied Science Publishers: New York, 1982; Chapter 2.
3. Pope, E. J. A.; Mackenzie, J. D. In *Tailoring Multiphase and Composite Ceramics;* Tressler, R.; Messing, G.; Pantano, C.; Newnham, R., Eds.; Plenum: New York, 1986; pp 187–194.
4. Pope, E. J. A.; Mackenzie, J. D. *Mater. Res. Soc. Symp. Proc.* **1986**, *73*, 809.
5. Schmidt, H.; Seiferling, B. *Mater. Res. Soc. Symp. Proc.* **1986**, *73*, 739.
6. Moffott, W. C.; Bowen, K. H. *J. Mater. Sci. Lett.* **1987**, *6*, 383.
7. Calvert, P. D.; Moyle, B. D. *Mater. Res. Soc. Symp. Proc.* **1988**, *109*, 357.
8. Moyle, B. D.; Ellul, R. E.; Calvert, P. D. *J. Mater. Sci. Lett.* **1987**, *6*, 167.
9. Newnham, R. In *Design of New Materials;* Cocke, D. L.; Clearfield, A., Eds.; Plenum: New York, 1987; pp 70, 583.
10. Hu, K. A.; Moffatt, D.; Runt, J.; Safari, A.; Newnham, R. *J. Am. Ceram. Soc.* **1987**, *70*, 583.
11. Newnham, R. In *Tailoring Multiphase and Composite Ceramics;* Tressler, R.; Messing, G.; Pantano, C.; Newnham, R., Eds.; Plenum: New York, 1986; pp 385–394, and references therein.

12. Safari, A.; Sa-Gong, G.; Giniewice, J.; Newnham, R. In *Tailoring Multiphase and Composite Ceramics;* Tressler, R.; Messing, G.; Pantano, C.; Newnham, R., Eds.; Plenum: New York, 1986; pp 445–454.

13. Kanatzidis, M. G.; Tonge, L. M.; Marks, T. J.; Marcy, H. O.; Kannewurf, C. R. *J. Am. Chem. Soc.* **1987,** *109,* 3797.

14. Berman, A.; Addadi, L.; Weiner, S. *Nature (London)* **1988,** *331,* 546.

15. Klempner, D.; Berkowski, L. In *Encyclopedia of Polymer Science and Engineering,* 2nd ed.; Kroschwitz, J., Ed.; Wiley: New York, Vol. 8; pp 279–341.

16. Berman, A.; Addadi, L.; Kuick, A.; Leiserowitz, L.; Nelson, M. *Science (Washington, D.C.)* **1990,** *250,* 664.

17. Weiner, S. *Biochemistry* **1983,** *22,* 4139.

18. Traub, W.; Arad, T.; Weiner, S. *Proc. Natl. Acad. Sci. USA* **1989,** *86,* 9822.

19. Mann, S; Sparks, N. H. C. *Proc. R. Soc. London B* **1988,** *234,* 441.

20. Fink, D.; Caplan, A.; Heuer, A. H. *MRS Bull.* **1992,** *October,* 27.

21. Calvert, P.; Mann, S. *J. Mater. Sci.* **1988,** *23,* 3801.

22. Mann, S. *Nature (London)* **1988,** *330,* 119.

23. *Biomineralization: Chemical and Biological Perspectives;* Mann, S.; Webb, J.; Williams, R. J. P., Eds.; VCH Publishers: New York, 1989.

24. Krampitz, G.; Graser, G. *Angew. Chem. Int. Ed. Engl.* **1988,** *27,* 1145.

25. Mann, S. *Struct. Bond.* **1983,** *54,* 125.

26. Weiner, S. *CRC Crit. Rev. Biochem.* **1986,** *20,* 365.

27. Addadi, L; Weiner, S. *Angew. Chem. Int. Ed. Engl.* **1992,** *31,* 153.

28. Degens, E. T. *Topics Curr. Chem.* **1976,** *64,* 1.

29. Heuer, A. H.; Fink, D. J.; Laraia, V. J.; Arias, J. L.; Calvert, P. D.; Kendall, K.; Messing, G. L.; Blackwell, J.; Rieke, P. C.; Thompson, D. H.; Wheller, A. P.; Veis, A.; Caplan, A. I. *Science (Washington, D.C.)* **1992,** *255,* 1098.

30. "Materials Synthesis Based On Biological Processes." *Mater. Res. Soc. Symp. Proc.* Alper, M.; Calvert, P.; Frankel, R.; Rieke, P.; Tirrell, D., Eds.; **1991,** *218.*

31. "Materials Synthesis Utilizing Biological Processes." *Mater. Res. Soc. Symp. Proc.* Alper, M.; Calvert, P. D.; Rieke, P. C., Eds.; **1989,** *174.*

32. Tarasevich, B. J.; Rieke, P. C. In "Materials Synthesis Utilizing Biological Processes"; *Mater. Res. Soc. Symp. Proc.* **1989,** *174,* 51.

33. Lehn, J. M. *Angew. Chem. Int. Ed. Engl.* **1988,** *27,* 89.

34. Sarikaya, M.; Gunnison, K. E.; Yasrebi, M.; Aksay, I. A. In "Materials Synthesis Utilizing Biological Processes"; *Mater. Res. Soc. Symp. Proc.* **1989,** *174,* 109.

35. Andres, R. P.; Averback, R. S.; Brown, W. L.; Brus, L. E.; Goddard, W. A.; Kaldor, A.; Louie, S. G.; Moscovitz, M.; Peercy, P. S.; Riley, S. J.; Siegel, R. W.; Spaepen, F.; Wang, Y. *J. Mater. Res.* **1989,** *4,* 704.

36. Yamaoto, T.; Kubota, E.; Tanigucchi, A.; Tominaga, Y. *J. Mater. Sci. Lett.* **1986,** *5,* 132.

37. Kubota, E.; Yamamoto, K. *Jpn. J. Appl. Phys.* **1987,** *26,* L1601.

38. Ohtani, B.; Adzuma, S.; Nishimoto, S.; Kagiya, T. *J. Polym. Sci. Polym. Lett.* **1987,** *25,* 383.

39. Rancourt, S.; Porta, G. M.; Moyer, E. S; Madeleine, D. G.; Taylor, L. T. *J. Mater. Res.* **1988,** *3,* 996.

40. Wang, Y.; Herron, N. *J. Phys. Chem.* **1991,** *95,* 525.
41. Herron, N.; Wang, Y.; Eddy, M. M.; Stucky, G. D.; Cox. D. E.; Moller, K.; Bein, T. *J. Am. Chem. Soc.* **1989,** *111,* 530–540.
42. Fendler, J. H. *Chem. Rev.* **1987,** *87,* 877.
43. Zhao, X. K.; Baral, S.; Rolandi, R.; Fendler, J. H. *J. Am. Chem. Soc.* **1988,** *110,* 1012.
44. Yuan, Y.; Cabasso, I.; Fendler, J. H. *Chem. Mater.* **1990,** *2,* 226.
45. Xu, S.; Zhao, X.; Fendler, J. H. *Adv. Mater.* **1990,** *2,* 1983.
46. Zhao, X.; Herve, P.; Fendler, J. *J. Phys. Chem.* **1989,** *93,* 908.
47. Baral, S.; Zhao, X.; Rolandi, R.; Fendler, J. H. *J. Phys. Chem.* **1987,** *91,* 2701.
48. Calvert, P. *MRS Bull.* **1992,** *October,* 37.
49. Mark, J. *Chemtech* **1989,** 230, and references therein.
50. Mark, J. E. In *Ultrastructure Processing of Advanced Ceramics;* Mackenzie, J. D.; Ulrich, D. P., Eds.; Wiley: New York, 1988; pp 623–633.
51. Wang, S.; Mark, J. *Macromolecules* **1990,** *23,* 4288.
52. Mark, J.; Schaefer, D. *Mater. Res. Symp. Soc. Proc.* **1990,** *171,* 51.
53. Esplard, P.; Mark, J. E.; Guyot, A. *Polym. Bull.* **1990,** *24,* 173.
54. Clarson, S. J.; Mark, J. E.; Dodgson, K. *Polym. Commun.* **1988,** *29,* 208.
55. Sur, G. S.; Mark, J. E. *Polym. Bull.* **1988,** *20,* 131.
56. Mark, J. E.; Wang, S. B. *Polym. Bul.* **1988,** *20,* 443.
57. Sohoni, G. B.; Mark, J. E. *J. Appl. Polym. Sci.* **1987,** *34,* 2853.
58. Clarson, S. J.; Mark, J. E. *Polym. Commun.* **1987,** *28,* 249.
59. Sun, C. C.; Mark, J. E. *J. Polym. Sci. Polym. Phys. Ed.* **1987,** *17,* 197.
60. Wang, S. B.; Mark, J. E. *Polym. Bull.* **1987,** *17,* 271.
61. Liu, S.; Mark, J. E. *Polym. Bull.* **1987,** *18,* 33.
62. Sun, C. C.; Mark, J. E. *Polym. Bull.* **1987,** *18,* 259.
63. Sur, G. S.; Mark, J. E. *Polym. Bull.* **1987,** *18,* 369.
64. Calvert, P.; Broad, A.; Cloke, G. *Polym. Prepr. (Am. Chem. Soc. Div. Polym. Chem.)* **1988,** *27*(2), 246.
65. Wohlford, T. L.; Schaaf, J.; Taylor, L. T.; Furtsch, T. A.; Khor, E.; St. Clair, A. K. In *Conductive Polymers;* Seymour, R. B., Ed.; Plenum: New York, 1981; p 7.
66. Taylor, L. T.; Porta, G. M. *J. Mater. Res.* **1988,** *3,* 211.
67. Sobon, C. A.; Bowen, H. K.; Broad, A.; Calvert, P. D. *J. Mater. Sci. Lett.* **1987,** *25,* 383.
68. Okada, H.; Sakata, K.; Kunitake, T. *Chem. Mater.* **1990,** *2,* 89.
69. Mann, S. *MRS Bull.* **1992,** *October,* 32.
70. Burdon, J. W.; Calvert, P. D. In "Materials Synthesis Based on Biological Processes." *Mater. Res. Soc. Symp. Proc.* **1991,** *218,* 203.
71. Broad, A.; Calvert, P. D. In *Atomic and Molecular Processing of Electronic and Ceramic Materials;* Aksay, I.; McVay, G.; Stoebe, T.; Wager, J., Eds.; Materials Research Society: Pittsburgh, PA, 1987; pp 89–98.
72. Broad, A.; Calvert, P. D. In "Materials Synthesis Utilizing Biological Processes." *Mater. Res. Soc. Symp. Proc.* **1989,** *174,* 61.
73. Liu, H.; Graff, G. L.; Hyde, M.; Sarikaya, M.; Aksay, I. A. In "Materials Synthesis Based on Biological Processes." *Mater. Res. Soc. Symp. Proc.***1991,** *218,* 115.

74. Azoz, A.; Calvert, P.; Kadim, M.; McCaffery, A.; Seddon, K. *Nature (London)* **1990**, *334,* 49.
75. Kimizuka, N.; Miyoshi, T.; Ichinose, I.; Kunitake, T. *Chem. Lett.* **1991**, 2039.
76. Steigerwald, M. L.; Brus, L. E. *Acc. Chem. Res.* **1990**, *23,* 183.
77. Steigerwald, M. L.; Alivisatos, A. P.; Gibson, J. M.; Harris, T. D.; Kortan, R.; Muller, A. J.; Thayer, A. M.; Duncan, T. M.; Douglass, D. C.; Brus, L. E. *J. Am. Chem. Soc.* **1988**, *110,* 3046.
78. Mann, S.; Williams, R. J. P. *J. Chem. Soc. Dalton Trans.* **1983**, 311.
79. Mann, S.; Skarnulis, A. J.; Williams, R. J. P. *Isr. J. Chem.* **1981**, *21,* 3.
80. Mann, S.: Hannington, J. P.; Williams, R. J. P. *Nature (London)* **1986**, *324,* 565.
81. Bhandarkar, S.; Bose, A. *J. Colloid. Interface Sci.* **1990**, *135,* 531.
82. Meldrum, F. C.; Wade, V. J.; Nimmo, D. L.; Heywood, B. R.; Mann, S. *Nature (London)* **1991**, *349,* 684.
83. Wade, V. J.; Levi, S.; Arosio, P.; Treffry, A.; Harrison, P. M.; Mann, S. *J. Mol. Biol.* **1991**, 541.
84. Ziolo, R. F.; Giannelis, E. P.; Weinstein, B. A.; O'Horo, M. P.; Ganguly, B. N.; Mehrota, V.; Russell, M. W.; Huffman, D. R. *Science (Washington, D.C.)* **1992**, *257,* 219.
85. Mahler, W. *Inorg. Chem.* **1988**, *23,* 435.
86. Wang, Y.; Suna, A.; Mahler, W.; *Mater. Res. Soc. Symp. Proc.* **1988**, *109,* 187.
87. Wang, Y.; Suna, A.; Mahler, W.; Kasowski, R. *J. Chem. Phys.* **1987**, *87,* 7315.
88. Wang, Y.; Mahler, W. *Opt. Commun.* **1987**, *61,* 233.
89. Hilinski, E. F.; Lucas, P. A.; Wang, Y. *J. Chem. Phys.* **1988**, *89,* 3435.
90. Wang, Y.; Suna, A.; McHugh, J.; Hilinski, E. F.; Lucas, P. A.; Johnson, R. D. *J. Chem. Phys.* **1990**, *92,* 6927.
91. Cummins, C. C.; Schrock, R. R.; Cohen, R. E. *Chem. Mater.* **1992**, *4,* 27.
92. Chan, Y. N. C.; Schrock, R. R.; Cohen, R. E. *Chem. Mater.* **1992**, *4,* 24.
93. Rieke, P. In *Atomic and Molecular Processing of Electronic and Ceramic Materials;* Aksay, I.; McVay. G.; Stoebe, T.; Wager, J., Eds.; Materials Research Society: Pittsburgh, PA, 1987; pp 109–114.
94. Addadi, L.; Berkovitch-Yellin, Z.; Weissbuch, I.; van Mil, J.; Shimon, L. J. W.; Lahav, M.; Leiserowitz, L. *Angew. Chem. Int. Ed. Engl.* **1985**, *24,* 466.
95. Berkovitch-Yellin, Z.; van Mil, J.; Addadi, L.; Idelson, M.; Lahav, M.; Leiserowitz, L. *J. Am. Chem. Soc.* **1985**, *107,* 3111.
96. Mann, S.; Didymus, J. M.; Sanderson, N. P.; Heywood, B. R.; Samper, E. J. A. *J. Chem. Soc. Faraday Trans.* **1990**, 1873.
97. Mann, S.; Heywood, B. R.; Rajam, S.; Walker, J. B. A. *J. Appl. Phys.* **1991**, *24,* 154.
98. Weissbuch, I.; Addadi, A.; Lahav, M.; Leiserowitz, L. *Science (Washington, D.C.)* **1991**, *253,* 637.
99. Davey, R. J.; Black, S. N.; Bromley, L. A.; Cottier, D.; Dobbs, B.; Rout, J. E. *Nature (London)* **1991**, *353,* 549.
100. Cody, A. M.; Cody, R. D. *J. Cryst. Growth* **1991**, *113,* 508.
101. Addadi, L.; Moradian, J.; Shay, E.; Maroudas, N. G.; Weiner, S. *Proc. Natl. Acad. Sci. U.S.A.* **1987**, *84,* 2732.
102. Addadi, L.; Weiner, S. *Proc. Natl. Acad. Sci. U.S.A.* **1985**, *82,* 4110.
103. Jacquemain, D.; Wolf, S. G.; Leveiller, F.; Deutsch, M.; Kjaer, K.; Als-

     Nielsen, J.; Lahav, M.; Leiserowıtz, L. *Angew. Chem. Int. Ed. Engl.* **1992,** *31,*
     130.
104. Leveiller, F.; Jacquemain, D.; Lahav, M.; Leiserowitz, L.; Deutsch, M.;
     Kjaer, K.; Als-Nielsen, J. *Science (Washington, D.C.)* **1991,** *252,* 1532.
105. Gavish, M.; Wang, J. L.; Eisenstein, M.; Lahav, M.; Leiserowitz, L. *Science
     (Washington, D.C.)* **1992,** *256,* 815.
106. Landau, E. M.; Levanon, M.; Leiserowitz, L.; Lahav, M.; Sagiv, J. *Nature
     (London)* **1985,** *318,* 353.
107. Landau, E. M.; Popovitz-Biro, R.; Levanon, M.; Leiserowitz, L.; Lahav, M.;
     Sagiv, J. *Mol. Cryst. Liq. Cryst.* **1986,** *134,* 323.
108. Mann, S.; Heywood, B. R.; Rajam, S.; Birchall, J. D. *Nature (London)* **1988,**
     334, 692.
109. Mann, S.; Heywood, B. R.; Rajam, S.; Birchall, J. D. *Proc. R. Soc. London
     A* **1989,** *423,* 457.
110. Mann, S.; Heywood, B.; Rajam, S.; Walker, J.; Davey, R.; Birchall, J. D.
     Adv. Mater. **1990,** *2,* 257.
111. Walker, J. B.; Heywood, B. R.; Mann, S. *Mater. Chem. Commun.* **1991,** *1,*
     889.
112. Heywood, B. R.; Mann, S. *J. Am. Chem. Soc.* **1992,** *114,* 4681.
113. Heywood, B. R.; Mann, S. *Langmuir* **1992,** *8,* 1492.
114. Zhao, X. K.; McCormick, L. D. *Appl. Phys. Lett.* **1992** *61,* 849.
115. Okada, H.; Sakata, K.; Kunitake, T. *Chem. Mater.* **1990,** *2,* 89.
116. Bianconi, P. A.; Lin, J.; Strzelecki, A. *Nature (London),* **1991** *349,* 315.

RECEIVED for review November 9, 1992. ACCEPTED revised manuscript March 30,
1993.

Inorganic Biomaterials

Larry L. Hench

Materials Science and Engineering, University of Florida,
Gainesville, FL 32611

A new generation of inorganic implant materials has been developed and used to repair the human body. These bioceramic materials offer improvements in corrosion and wear resistance over metallic and polymeric implants that are used in total joint replacements. Bioactive glasses, ceramics, glass–ceramics, and composites bond to living bone and soft connective tissues and solve problems of interfacial stability and stress shielding of bone. Molecular-orbital modeling of inorganic–biological interfacial reactions combined with sol–gel processing of new bioactive gel glasses and composites offer the promise for the molecular design of biomaterials used for specific disease states and geriatric tissues. However, proof of long-term clinical reliability of the new biomaterials is still needed.

ONE OF THE GREAT CHALLENGES facing materials science today is the development and improvement of a new generation of biomaterials to repair the body. The population is aging, and millions of people receive implants to help maintain their quality of life as they age (*1*). A greater understanding of the chemistry of materials is needed to produce materials that will last as long as the patient does. This time period is often 15–30 years and is double or triple the expected lifetime of many spare parts in use today (*2*). In this chapter I review the current understanding and development of inorganic biomaterials. I also focus on the chemical requirements of materials with longer lifetimes and materials designed to be

[1]Current address: Advanced Materials Research Center, University of Florida, Alachua, FL 32615

0065–2393/95/0245–0523$13.25/0

used in patients with debilitating skeletal diseases such as osteoporosis and arthritis.

The primary use of inorganic biomaterials is the repair, replacement, or augmentation of diseased or damaged parts of the musculoskeletal system, such as bones, joints, and teeth (3, 4). In many of these applications, which are summarized in Figure 1, the materials are used in the form of devices called implants or prostheses. The form of the biomaterial depends on its intended function in the body, as indicated in Table I. Load-bearing implants are usually made from bulk, nonporous materials; but coatings or composite structures may also be used to achieve improved mechanical and interfacial chemical properties. Implants that serve only to fill space or augment existing bone tissue are used in the form of powders, particulates, or porous materials.

Inorganic biomaterials are also widely used in repair of the cardiovascular system as pyrolytic carbon coatings on prosthetic heart valves (5). Therapeutic applications of inorganic glasses in the form of radioactive glass beads for localized treatment of tumors are also in clinical use (6).

For many years, the guiding principle of inorganic biomaterial development was that the materials should be as chemically inert as possible (3). Body fluids are highly corrosive saline solutions. The first materials used in skeletal repair were metals optimized for strength and corrosion resistance (Table II). Metallic implants for orthopedic applications have been very successful, and hundreds of thousands are implanted annually (2). The original applications were as removable devices, such as those for stabilization of fractures. Use as permanent joint replacements began in the 1960s with the development of self-curing poly(methyl methacrylate) "bone cement", which provided a stable mechanical anchor for a metallic prosthesis in its bony bed (7). High levels of clinical success (>85% of implants are still operating after 5 years) of cemented metallic orthopedic implants led to rapid growth in the use of implants.

The increase in number of implants has been accompanied by an increase in the life expectancy of patients and a decrease in the average age of patients receiving an implant. This progression means that a growing proportion of patients will outlive the expected lifetime of their prostheses (8, 9).

Interfacial Instability

Many factors can contribute to the failure and shortened lifetime of an implant; however, instability of the interface between the implant and its host tissue is the dominant problem (8, 9). The primary causes of interfacial instability are chemical and mechanical mismatch between the implant

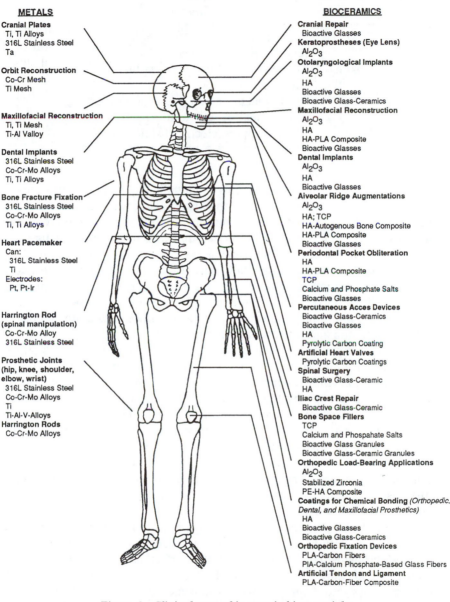

METALS

Cranial Plates
Ti, Ti Alloys
316L Stainless Steel
Ta

Orbit Reconstruction
Co-Cr Mesh
Ti Mesh

Maxillofacial Reconstruction
Ti, Ti Mesh
Ti-Al Valloy

Dental Implants
316L Stainless Steel
Co-Cr-Mo Alloys
Ti, Ti Alloys

Bone Fracture Fixation
316L Stainless Steel
Co-Cr-Mo Alloys
Ti, Ti Alloys

Heart Pacemaker
Can:
316L Stainless Steel
Ti
Electrodes:
Pt, Pt-Ir

Harrington Rod
(spinal manipulation)
Co-Cr-Mo Alloy
316L Stainless Steel

Prosthetic Joints
(hip, knee, shoulder,
elbow, wrist)
316L Stainless Steel
Co-Cr-Mo Alloys
Ti
Ti-Al-V-Alloys
Harrington Rods
Co-Cr-Mo Alloys

BIOCERAMICS

Cranial Repair
Bioactive Glasses
Keratoprostheses (Eye Lens)
Al_2O_3
Otolaryngological Implants
Al_2O_3
HA
Bioactive Glasses
Bioactive Glass-Ceramics
Maxillofacial Reconstruction
Al_2O_3
HA
HA-PLA Composite
Bioactive Glasses
Dental Implants
Al_2O_3
HA
Bioactive Glasses
Aiveolar Ridge Augmentations
Al_2O_3
HA; TCP
HA-Autogenous Bone Composite
HA-PLA Composite
Bioactive Glasses
Periodontal Pocket Obliteration
HA
HA-PLA Composite
TCP
Calcium and Phosphate Salts
Bioactive Glasses
Percutaneous Acces Devices
Bioactive Glass-Ceramics
Bioactive Glasses
HA
Pyrolytic Carbon Coating
Artificial Heart Valves
Pyrolytic Carbon Coatings
Spinal Surgery
Bioactive Glass-Ceramic
HA
Iliac Crest Repair
Bioactive Glass-Ceramic
Bone Space Fillers
TCP
Calcium and Phospahate Salts
Bioactive Glass Granules
Bioactive Glass-Ceramic Granules
Orthopedic Load-Bearing Applications
Al_2O_3
Stabilized Zirconia
PE-HA Composite
Coatings for Chemical Bonding *(Orthopedic,*
Dental, and Maxillofacial Prosthetics)
HA
Bioactive Glasses
Bioactive Glass-Ceramics
Orthopedic Fixation Devices
PLA-Carbon Fibers
PIA-Calcium Phosphate-Based Glass Fibers
Artificial Tendon and Ligament
PLA-Carbon-Fiber Composite

Figure 1. Clinical uses of inorganic biomaterials.

**Table I. Relation Between Form and Function
of a Biomaterial**

Form	Function
Powder	Space-filling, therapeutic treatment, regeneration of tissues
Coating	Tissue bonding, thromboresistance, corrosion protection
Bulk	Replacement and augmentation of tissue, replacement of functioning parts
Wire	Suture, staples, electrical stimulation

and living tissues. The body responds to metallic implants by developing a thin, nonadherent, fibrous capsule that isolates the device from its host tissue (Table III). The implant can move within the capsule and create local stress concentrations at the interface that lead to loosening, pain, and even fracture of the bone or implant.

All metals corrode within their fibrous capsule. Corrosion products have varying degrees of toxicity depending on the type of metal and the patient. Migration of metallic ions can occur and cause systemic effects (10). Motion and wear at the tissue–metal interface accelerate corrosion and the formation and migration of debris (2, 10).

Metallic implants have moduli of elasticity that are many times larger than those of either cortical or cancellous bone (compare Tables II and IV). The mismatch of moduli of elasticity means that most of the load of body weight is carried by the implant. However, bone must be continuously loaded in tension to remain healthy. Under compressive loads or no load, bone resorbs through a complex cellular process (11). Bone resorption decreases the cross-sectional area in contact with the prosthesis. This decrease weakens the support structure for the device and increases the probability of failure. This problem is called "stress shielding", and it gets worse as we age because of a diminished ability to grow new bone in response to mechanical stress (12).

The exothermic reaction during polymerization of bone cement kills bone cells and results in a brittle layer of dead bone adjacent to an implant, which increases the probability of interface failure. Invasion of the methyl methacrylate monomer into the bloodstream or forcible injection of fat particles into capillaries when the cement is injected into bone can also cause serious complications in surgery, including death. Obviously, alternatives to metals or bone cement are needed.

Table II. Metal Implant Materials

Element	Stainless Steel	Co–Cr Alloy	Ti–Al$_6$–V$_4$ Alloy
Al	—	—	5.5–6.5
C	0.3	0.05–0.15	0.08
Co	—	Bal[a]	—
Cr	17.00–20.00	19.00–21.00	—
Fe	Bal	3.00	0.25
H$_2$	—	—	0.013
Mg	—	2.0	—
Mn	2.00	—	—
Mo	2.00–4.00	—	—
Ni	10.00–14.00	9.00–11.00	—
N$_2$	—	—	0.05
O$_2$	—	—	0.13
P	0.03	—	—
S	0.03	—	—
Si	0.75	1.0	—
Ti	—	—	Bal
V	—	—	3.5–4.5
W	—	14.0–16.0	—
E (GPa)[b]	200	230	105
UTS (MPa)[c]	207–1,160	430–1,028	700–1,050
K$_{1c}$ (MPa·m$^{1/2}$)[d]	~100	~100	~80

NOTE: All values are composition weight percents unless otherwise indicated. Single values are the maximum weight percent. — means the material contains none of the specified element.

[a] Bal means balance of element in the material.

[b] E is Young's modulus.

[c] UTS is ultimate tensile strength.

[d] K_{1c} is fracture toughness.

Bioceramics

Ceramics, especially oxides, eliminate the problem of metallic corrosion of implants. Medical-grade aluminum oxide (alumina), with high purity and small grain size (Table IV), was developed for use in load-bearing orthopedic prostheses, especially as the head or ball of a total hip joint. The

Table III. Tissue Attachments of Different Biomaterials

Biomaterial	Attachment	Example	Typical Use
Nearly inert	Dense, nonporous nearly inert materials Al_2O_3 (single crystal and polycrystalline) attach by bone growth into surface irregularities by cementing the device into the tissues, or by press-fitting or screwing into a defect (called "morphological fixation").	Tooth implants; ball of hip 316L stainless steel Co–Cr–Mo alloy Ti and Ti alloy High-density polyethylene	Tooth and joint replacement Tooth and joint replacement Tooth and joint replacement Socket of hip replacement
Porous	For porous, inert implants, bone ingrowth occurs, which mechanically attaches the bone to the material (called "biological fixation").	Porous metal coatings Hydroxyapatite-coated porous metals Hydroxyapatite	Fixation of total joint replacement Fixation of total joints Bone replacement
Bioactive	Dense, nonporous surface-reactive ceramics, glasses, and glass–ceramics attach directly by chemical bonding with the bone (called "bioactive fixation").	Bioactive glasses Bioactive glass–ceramics Hydroxyapatite	Periodontal repair, ossicle replacement Vertebral replacement Bone repair
Resorbable	Dense, nonporous (or porous) resorbable ceramics are designed to be slowly replace by bone.	Tricalcium phosphate Calcium–phosphate salts Bioactive glasses	Bone augmentation Bone augmentation Bone augmentation

Table IV. Physical Characteristics of Alumina and PSZ Bioceramics Compared with Bone

Characteristic	Sapphire (Al_2O_3) Single Crystal	High-Alumina Ceramics	ISO Alumina Standard 6474	PSZ	Cortical Bone	Cancellous Bone
Content (wt %)	Al_2O_3 > 99.9	Al_2O_3 > 99.8	$Al_2O_3 \geq$ 99.50	ZrO_2 > 97	—	—
Density (g/cm^3)	3.95	>3.93	$\geq$3.90	5.6–6.12	1.6–2.1	0.9–1.2
AGS (μm)	—	3–6	<7	1	—	—
Ra (μm)	—	0.02	—	0.008	—	—
Hardness[a]	2300	2300	>2000	1300	—	—
CS (MPa)	5000[b]	4500	—	—	100–230	2–12
BS (MPa)[c]	1300[d]	550	400	1200	50–150	—
E (GPa)	380	380	—	200	7–30	—
K_{1c} ($MPa \cdot m^{1/2}$)	—	5–6	—	15	2–12	0.05–0.5
SCG (n)[e]	—	30–52	—	65	—	—

NOTE: — means not determined.

ABBREVIATIONS: ISO is International Standards Organization. PSZ is partially stabilized zirconia. AGS is average grain size. Ra is surface roughness. CS is compressive strength. BS is bending strength. E is Young's modulus. K_{1c} is fracture toughness. SCG is slow crack growth.

[a] Hardness was determined by the Vickers hardness test.
[b] Measured parallel to crystal.
[c] Measured after testing in Ringer's solution.
[d] Measured perpendicular to crystal.
[e] n is slope.

very high resistance to wear and low friction of alumina make it superior to metals in this application (*4, 12, 13*). Alumina is also used as a dental implant and in a variety of other applications (Figure 1 and Table I). Hulbert et al. (*4*) discussed the historical development of alumina implants.

The strength of alumina is sufficient for use under compressive loads, but even high-purity alumina exhibits slow crack growth, static and cyclic fatigue, and low toughness under tensile loads in physiological environments. Alumina offers no advantage over metals with regard to stress shielding of bone because of its high modulus of elasticity (Table IV). Alumina also offers no solution to the problem of interface stability because a thin, fibrous capsule forms at the interface with bone.

The higher toughness and lower modulus of elasticity of zirconia, in stabilized and transformation-toughened forms, provide some advantages over alumina and metals for use in orthopedics (*14*). Zirconia is also isolated from tissues by a thin, fibrous capsule, and therefore, it does not lead to improved interfacial stability over metal implants.

Porous Implants

Porous implants improve interfacial stability. Bone will grow into pores of >100-μm diameter and maintain a blood supply necessary for cell maintenance (*15, 16*). This method of fixation is often called "biological fixation" (Table III). The pores can exist throughout the implant; however, large pores severely degrade strength and toughness. Porous ceramics cannot be used for load-bearing devices but are suitable for space-filling applications.

Porous coatings applied to metals lead to porous ingrowth of bone at the interface, and the mechanical loads are still carried by the bulk-metal substrate (*17, 18*). The difficulties of stress shielding remain. The high surface area of the porous metal coatings also greatly increases the potential for corrosion. Coatings such as beads can come off or degrade. Sintering of the coatings can alter the microstructure of the substrate and the strength and toughness of the device. Removal of porous, coated prostheses, such as is required in cases of infection, is often more difficult than removal of cemented prostheses. Clinical results indicate that the prognoses for cemented prostheses are about the same as porous, coated prostheses (*5, 8, 9*).

Resorbable Bioceramics

The skeletal system has the capacity to repair itself, although the capability diminishes with age and disease (*11*). The ideal solution to the problem of interfacial stability is the use of biomaterials to augment the body's own reparative processes (*19*). Resorbable implants, such as tricalcium phosphate and other calcium salts and some bioactive glasses (Table III), are based on this concept.

Two chemical and mechanical problems exist in the development and use of resorbable biomaterials. First, the products of resorption must be compatible with cellular metabolic processes. This compatibility is a highly restrictive requirement. The rate of resorption must also be matched by the capacities of the cellular environment and cardiovascular and lymphatic systems to process and transport the resorption products. Few data are available to provide research guidance in these areas (*20*). Consequently, resorbable implants are presently used on a trial-and-error basis. The second problem is that the strength of a resorbable implant decreases as chemical breakdown occurs. Eventually, strength is restored by the growth of the replacement tissues, which is the object of the implantation. However, during the interim period, the mechanical properties required for function are difficult to maintain. These problems, as yet, have not been overcome.

Bioactive Ceramics

The fourth alternative to the achievement of a stable implant–tissue interface is the use of bioactive fixation (Table III). A bioactive material is one that elicits a specific biological response at the interface of the material, which results in the formation of a bond between the tissues and the material (*21*). This concept is based on control of the surface chemistry of the material. A bioactive implant reacts chemically with body fluids in a manner that is compatible with repair processes of the tissues. A fibrous capsule is prevented from forming by the adhesion of repairing tissues. Because the chemical reactions are restricted to the surface, the material does not degrade in strength the way resorbable or porous implants do.

The first bioactive material reported was a simple four-component glass composed of SiO_2, Na_2O, CaO, and P_2O_5 (composition 45S5 in Table V) (*22*). The low silica content and presence of calcium and phosphate ions in the glass result in very rapid ion exchange in physiological solutions and in rapid nucleation and crystallization of hydroxycarbonate apatite (HCA) bone mineral on the surface. The growing bone-mineral layer bonds to collagen, which is grown by the bone cells, and a strong interfacial bond is formed between the inorganic implant and the living tissues. During the past 20 years, many other materials were shown to develop bioactive bonding to bone (*23*). Selected properties of bioactive ceramics in current clinical use are summarized in Table V.

The interfacial strengths developed by a wide range of bioactive materials are equivalent to or greater than that of bone (*22–27*). After im-

**Table V. Composition and Mechanical Properties
of Bioactive Ceramics Used Clinically**

Composition/ Property	Bioglass 45S5	S53P4 Glass 9	Glass–Ceramic Ceravital	Glass–Ceramic Cerabone A–W
Composition				
Na_2O	24.5	23	5–10	0
K_2O	0	—	0.5–3.0	0
MgO	0	—	2.5–5.0	4.6
CaO	24.5	20	30–35	44.7
Al_2O_3	0	—	0	0
SiO_2	45.0	53	40–50	34.0
P_2O_5	6.0	4	10–50	16.2
CaF_2	0	—	—	0.5
B_2O_3	0	0	—	—
Property				
Phase	glass	glass	apatite	apatite [$Ca_{10}(PO_4)_6(O,F_2)$]
			glass	β-wollastonite (CaO·SiO_2)
				glass
Density (g/cm^3)	2.6572	—	—	3.07
Hardness[b]	458 ± 9.4	—	—	680
CS (MPa)[c]	—	—	500	1080
BS (MPa)[d]	42[e]	—	—	215
E (GPa)[f]	35	—	100–150	218
K_{1c} (MPa·m$^{1/2}$)[g]	—	—	—	2.0
SCG (n)[h]	—	—	—	33
Reference	3	51	52	53

NOTE: All composition values are weight percents. — means not determined or not present.

[a]Value is weight percent for F.

[b]Hardness measured by Vickers hardness test.

[c]CS means compressive strength.

[d]BS means bending strength.

Table V. Continued.

Glass–Ceramic Ilmaplant L1	Glass–Ceramic Bioverit	Sintered Hydroxyapatite $Ca_{10}(PO_4)_6(OH)_2$ >99.2%	Sintered β-$Ca_3(PO_4)_2$ >99.7%
4.6	3–8	—	—
0.2	3–8	—	—
2.8	2–21	—	—
31.9	10–34	—	—
0	8–15	—	—
44.3	19–54	—	—
11.2	2–10	—	—
5.0	3–23[a]	—	—
—	—	—	—
apatite	apatite $[Ca_{10}(PO_4)_6(OH)_2]$	apatite	whitlockite
β-wollastonite	phlogopite $[(Na,K)Mg_3(AlSiO_{10})F_2]$	β-$Ca_3(PO_4)_2$	
glass	glass		
—	2.8	3.16	3.07
	500	600	—
500	500–1000	460–687	
160	100–160	115–200	140–154
—	70–88	280–110	33–90
2.5	0.5–1.0	1.0	—
—	—	12–27	—
54	55	56–58	59, 60

[e]Value measured as tensile strength.
[f]E is Young's modulus.
[g]K_{1c} is fracture toughness.
[h]SCG means slow crack growth.

plantation in bone and testing to failure, fracture occurs in bone or the implant material depending upon their relative strength (Table VI). The important finding is that failure does not occur through the interface. A few compositions of bioactive glasses (Bioglass 45S5 in Table V) also bond to soft tissues, as well as bone, with an adherence strength greater than the cohesive strength of the collagen fiber bundles of the soft connective tissues (28, 29).

Bioactive implants have differing rates of bonding depending on their bulk composition (Figure 2) (21). The most rapid rates of bonding occur for bioactive glasses with SiO_2 contents of 45–52% by weight. Soft tissue and bone bonding take place within 5–10 days for these compositions. Bioactive glasses or glass–ceramics containing 55–60% SiO_2 by weight require more time to form a bond with bone (Figure 2) and *do not* bond to soft connective tissues.

The reason for this important difference of in vivo behavior between various bioactive implants is related to their surface reaction kinetics in physiological solutions. Figure 3 summarizes the sequence of 11 reactions that apparently occur on the surface of a bioactive glass as a bond with bone is formed. Details of these reactions were previously investigated by using a variety of analytical techniques (21, 30–33). Stages 1–5 are well-understood, but the overall understanding for stages 6–11 is sparse. For several stages, such as stages 8 and 9, we lack the basic knowledge of the biological processes that control the genetic expression of highly differentiated cells like osteoblasts. Bioactive substrates with known differences in surface chemistry and surface kinetics are unique model systems for exploring these biological phenomena.

The time scale for the surface reactions for a glass with high bioactivity is also shown in Figure 3. Glasses with the highest levels of bioac-

Table VI. Failure Loads of Some Bioceramics 8 Weeks After Implantation

Material	Ref.	Failure Load (kg)	Location of Fracture
Dense sintered alumina	61, 62	0.13 ± 0.02	interface
Bioglass 45S5-type glass	61, 62	2.75 ± 1.80	within material
Ceravital KGS-type glass–ceramic	63	3.52 ± 1.48	within material
Cerabone A–W glass–ceramic	61, 62	7.43 ± 1.19	within bone
Dense sintered hydroxyapatite	61, 62	6.28 ± 1.58	within material
Dense sintered β-$Ca_3(PO_4)_2$	64	7.58 ± 1.97	not specified
Natural polycrystalline calcite	65	4.11 ± 0.98	within material

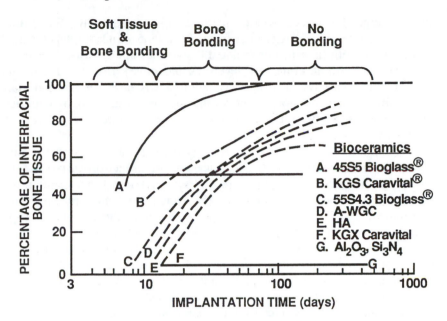

Figure 2. Bioactivity spectrum for various bioceramic implants as time dependence of bone and soft-tissue bonding at an implant interface. (Reproduced with permission from reference 19. Copyright 1991 American Ceramic Society.)

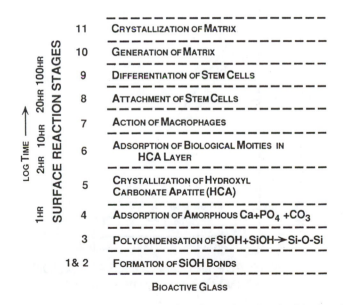

Figure 3. Proposed sequence of interfacial reactions involved in forming a bond between tissue and bioactive ceramics. (Reproduced with permission from reference 19. Copyright 1991 American Ceramic Society.)

tivity undergo the surface reaction in stages 1–5 very rapidly. For example, a polycrystalline HCA layer is formed on 45S5 Bioglass (stages 4 and 5) within 3 h both in vitro in simulated body fluids and in vivo. However, compositions with intermediate levels of bioactivity that bond only to bone require 2–3 days to form a crystalline HCA layer on the material. Compositions that are not bioactive and do not form a bond to either bone or soft tissues do not form a crystalline apatite layer even after 3–4 weeks in solution.

The differences in rates of change of the inorganic phases on the surface of the material alter which biological species are adsorbed (stage 6 in Figure 3) and subsequently the rate and types of interfacial cellular responses (stages 7–11). For example, the attachment and spreading time of certain types of fibroblasts, which can form a nonadherent fibrous capsule at an implant interface, are slowed down considerably on a bioactive glass surface compared with inert surfaces (Figure 4), whereas the cellular activity of bone-growing cells (osteoblasts) is enhanced (Figure 5) (34, 35).

As used in a new generation of implants with enhanced lifetimes, bioactive materials offer the opportunity for molecular tailoring of compositions to match the biochemical requirements of diseased and damaged tissues. Bioactive materials can also provide the flexibility to tailor biome-

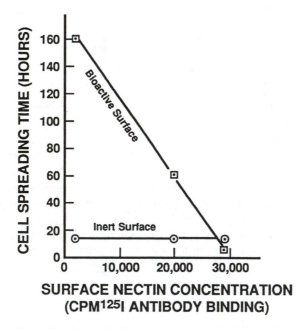

Figure 4. Spreading time of cells as a function of relative cell–nectin concentration. (Based on data in reference 3, p 285.)

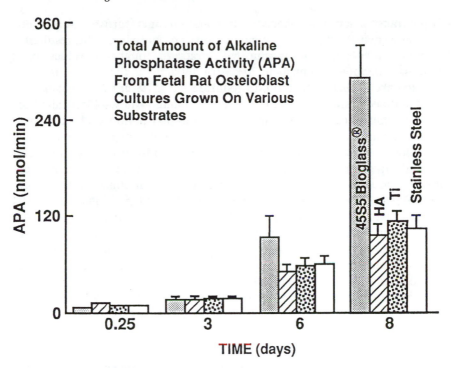

Figure 5. Total amount of alkaline phosphatase activity (APA) from fetal rat osteoblast cultures grown on various substrates. (Data courtesy of Vrouwenvelder et al., reference 35.)

chanical characteristics to match those of the natural host tissues by optimizing the microstructure of the material or by being used as a second phase in a composite.

Table V, which summarizes the physical properties and compositions of various bioactive ceramics, shows that the potential versatility of bioactive materials is beginning to be realized. Bioactivity is retained even in multiphase glass–ceramics. Strength and toughness have been increased greatly since co-workers and I (*22*) reported the first bioactive glasses 20 years ago. Thus, the composition of a bioactive matrix phase can be manipulated to optimize chemical reaction rates and match physiological requirements with an inactive reinforcing phase. If the inactive reinforcing phase is optimized with respect to size and distribution, then strength and toughness will be maximized. Apatite (A)–wollastonite (W), glass–ceramic, which is composed of apatite and wollastonite crystals in a silicate glass matrix, was developed by Kokubo, Yamamuro, and co-workers (*26*) and has more enhanced mechanical properties than 45S5 Bioglass (*23, 25, 26, 36*). Additional optimization of microstructure was done by Kasuga et al.

(*37*): transformation-toughened, tetragonal zirconia particles were added to a bioactive A–W, glass–ceramic before hot pressing. The result is a very tough bioactive composite with fracture toughness values of 4 MPa·m$^{1/2}$ and bend strengths of 703 MPa.

Excellent clinical success was reported (*1, 4, 19, 36*) for the bioactive glasses in nonloaded applications such as middle-ear implants, alveolar-ridge maintenance implants, and particulates for periodontal and maxillofacial repair. Bioactive A–W, glass–ceramic implants are used very successfully in vertebral repair and replacement of portions of the iliac crest in total hip surgery. As yet, very few data exist on the environmental sensitivity and fatigue life of such multiphase materials under the complex physiological loads of total-joint replacements, and multiphase materials are not used in these situations. The moduli of elasticity of all of the present generation of bioactive glass ceramics are too high (Table V) to avoid the problem of stress shielding. Specially designed bioactive composites appear to be the best way to optimize biomechanical properties with a biochemically optimized surface.

Bioactive Composites

Many research teams are attempting to develop composites and coatings that improve interfacial stability and mechanical properties. Yamamuro, Wilson, and I (*18, 36*) collected two volumes of papers and reviews of these materials. The interest in this approach to the interface instability problem is largely responsible for the large growth in world-wide research centers studying bioactive ceramics (Figure 6). Wilson (*38*) recently analyzed the growth in the field.

Table VII summarizes the properties of some of the more promising composite biomaterials. Few of the systems studied eliminate the critical problem of stress shielding. Bonfield (*39*) has made progress in this area by using a higher-modulus bioactive phase of hydroxyapatite (HA) powders dispersed in a low-modulus, high-density polyethylene (PE) matrix (*39*). The resulting composite has a Young's modulus of 1–8 GPa and a strain-to-failure value of >90 to 3% as the volume fraction of HA increases to 50%. The transition from ductile to brittle behavior occurs between 40 to 45% HA by volume (Figure 7). The ultimate tensile strength of the composite remained within 22–26 MPa. At 45% HA, the fracture toughness was 2.9 MPa·m$^{1/2}$. At <40% HA, the fracture toughness was considerably greater because of the ductile deformation associated with crack propagation in the two-phase system.

The mechanical properties of the PE–HA composite are close to or superior to those of bone. The bioactivity of the composite is less than

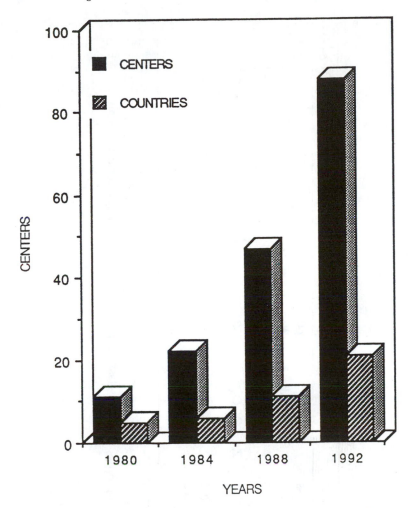

Figure 6. Number of research centers and countries presenting studies on bioactive ceramics in the last four World Biomaterials Congresses (compiled from the abstracts of papers presented).

optimal because the rate of bone bonding of HA is slow (*see* Figure 2) and the surface area of HA available for bonding to bone, exposed by machining the composites, is also relatively low. The next step in development of an optimized, two-phase composite is to use a high-modulus inorganic phase with higher rates of bioactivity. PE–bioactive glass composites are currently being developed and may yield the required combination of chemical and mechanical characteristics for load-bearing devices with enhanced lifetimes, interfacial bonding, and minimal stress shielding.

Table VII. Mechanical Properties of Bioactive Composites

Composite	Ref.	Matrix	DP	DP (vol %)	BS (MPa)	E (GPa)	K_{Ic} (MPa·m$^{1/2}$)
G–Fe	66	Bioglass	Stainless[a]	60	340	65	—
HA–Fe	67	HA	Fe–Cr–Al[a]	30	224	142	7.4
HA–ZrO$_2$	68	HA	ZrO$_2$(Y$_2$O$_3$)[b]	10–50	450	—	3.0
HA–TiO$_2$	69	HA	TiO$_2$[b]	15	252	238	3.0
GC–ZrO$_2$	70	A–W-type GC	ZrO$_2$(Y$_2$O$_3$)[b]	30	600	—	3.0
GC–Ti	71	Ceravital-type GC	Ti[b]	30	60[c]	100	—
PE–HA	72	PE	HA[b]	45	22–26[c]	6	3.0

NOTE: — means not determined.

ABBREVIATIONS: DP means dispersed phase. BS means bending strength. E is Young's modulus. K_{Ic} is fracture toughness. For composites, G means glass, HA means hydroxyapatite, C means ceramic, and PE means polyethylene.

[a]Dispersed phase is a fiber.
[b]Dispersed phase is a particle.
[c]Measurement is of tensile strength.

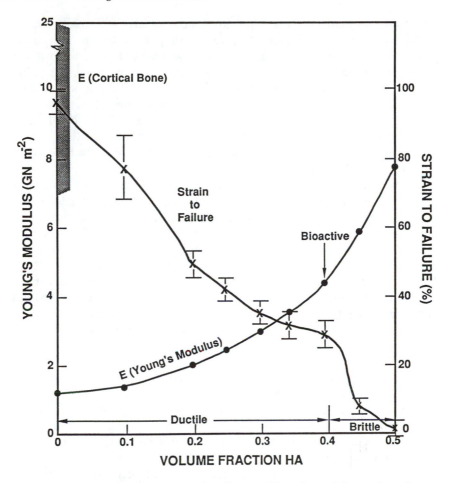

Figure 7. Effect of hydroxyapatite (HA) on Young's modulus and strain-to-failure value of a polyethylene–HA composite. (Reproduced with permission from reference 19. Copyright 1991 American Ceramic Society.)

Molecular Tailoring of Surface Chemistry

The greatest challenge for inorganic biomaterials is the design of the surface chemistry of materials to meet the requirements of aged, diseased, or damaged tissues. Most of the materials in use today were developed by trial and error. Almost no data exist on the in vivo response of the materials as a function of the age of tissues or the effects of disease states, such as osteoporosis or arthritis, on interfacial reactions or biomechanical behavior of interfaces. Only in the last few years have some principles been established to guide the development of new materials. Surface reaction kinetics of bioactive ceramics of various compositions were determined (stages 1–5 in Figure 3). The compositional effects responsible for controlling the reaction rates are now known reasonably well (40, 41). However, details of stages 6–11 in Figure 3 are only beginning to be investigated. The material variables responsible for controlling the in vivo bioactivity of a material (Figure 2) are poorly understood. This lack of understanding of the bone–material interface is a major barrier to the design of new materials with improved performance and has recently become the focus of many research programs (42).

Two new directions of research hold promise for improving the scientific basis for tailoring surface reactions of inorganic biomaterials. One is the discovery that sol–gel-derived glasses have an expanded compositional range of bioactivity relative to glasses made by traditional melting and casting processes (43). Sol–gel processing is one of the most important new methods for production of new, chemically derived materials (44, 45). The low temperatures of sol–gel processing (Figure 8) make it possible to control surface chemistry of the resulting materials with greater flexibility than high-temperature melting and casting of glasses or sintering or hot pressing of ceramics. Details of the seven processing steps in making bioactive gel glasses (Figure 8) were previously discussed (45). Advantages of sol–gel processing of inorganic biomaterials include the following: new compositions, greater homogeneity, higher levels of purity, net-shape casting of monoliths, low-temperature coating of substrates, control of powder size distribution, control of surface chemistry of the gel glasses, expanded ranges of glass formation, control of pore networks at a nanometer scale, and commercial advantages such as lower energy consumption and negligible environmental impact.

Figure 9 shows the extended range of compositions in the SiO_2–CaO–P_2O_5 system that are bioactive when made by alkoxide-based, sol–gel processing (43) compared with bioactive compositions made by melting and casting (46). Gel-derived glasses with as much as 88% SiO_2 develop HCA layers, whereas the SiO_2 limit for melt–glasses is 60%. This shift in compositional limit is huge. Melt–glasses with >55% SiO_2 require several days to form a polycrystalline HCA layer (31–33), whereas gel glasses do so in only a few minutes (43). The chemical origin of these important differences appears to be the large concentration of silanols on the surface of the gels after processing temperatures of 500–800 °C.

Semiempirical molecular-orbital calculations, using AM-1 and extended Hückel methods, show that metastable silica clusters formed from a condensation reaction of neighboring silanols (stage 3 in Figure 3) can act as heterogeneous nucleation sites for HCA crystals (stages 4 and 5) (46). The metastable silica clusters can also act as preferential adsorption sites for amino acids such as alanine (47). These calculational results indicate

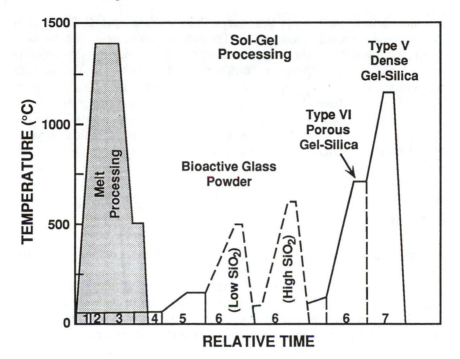

Figure 8. Sol–gel processing sequence: (1) mixing, (2) casting, (3) gelation, (4) aging, (5) drying, (6) stabilization, and (7) densification.

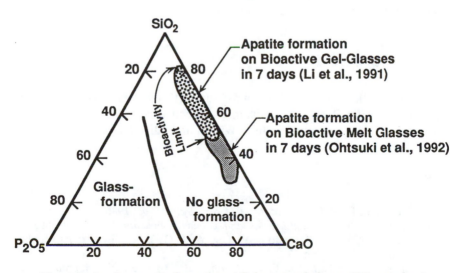

Figure 9. In vitro apatite formation at 7 days on gel glasses (50) compared with melt glasses.

that the surface reactions of the inorganic material (stages 1–5) can lead to biologically specific binding sites for protein molecules. The molecular-orbital calculations show differences in specific adsorption on the inorganic surface that depend on different binding sites on the protein molecules. This finding may lead to an understanding of the selective adsorption of proteins that act as growth factors or enzymes (stages 6–8). Such studies may also aid in the interpretation and optimization of new, hybrid inorganic–biological systems, such as alkaline phosphatase enzymes (48) or other optically active organic molecules (49) trapped within sol–gel silica, porous-glass matrices.

Conclusions

Results from molecular-orbital modeling calculations, combined with experimental investigations of the adsorption of biological growth factors and other biological species, should make it possible to design a new generation of gel glasses or use techniques such as ion-beam sputter coating to make surface-modified layers that enhance the rates of interfacial bonding of even aged or diseased tissues. Such surface enhancement methods may also be incorporated within the design of composites that have high toughness and a modulus of elasticity that will prevent stress shielding of bone. The potential solutions to lifetime problems of prostheses lie in the creative use of materials chemistry. Millions of people will benefit if this potential can be realized. The research direction to follow is finally apparent after many years of trial and error.

Acknowledgments

I gratefully acknowledge the financial support of the U.S. Air Force, Office of Scientific Research, Division of Chemical and Materials Sciences, and the editorial assistance of June Wilson and Alice Holt.

References

1. Hench, L. L.; Wilson, J. Mater. Res. Soc. Symp. Proc. 1986, 55, 65–75.
2. Hench, L. L. In Biomaterials and Clinical Applications; Pizzoferrato, A.; Marchetti, P. G.; Ravaglioli, A.; Lee, A. J. C., Eds.; Elsevier: Amsterdam, Netherlands, 1987; p 23.
3. Hench, L. L.; Ethridge, E. C. Biomaterials: An Interfacial Approach; Academic: New York, 1982.

4. Hulbert, S. F.; Bokros, J. C.; Hench, L. L.; Wilson, J.; Heimke, G. In *High Tech Ceramics;* Vincenzini, P., Ed.; Elsevier Science Publishers B.V.: Amsterdam, Netherlands, 1987; pp 189–213.

5. Haubold, A.; Shim, H. S.; Bokros, J. C. In *Biocompatibility of Clinical Implant Materials;* CRC Press: Boca Raton, FL, 1981, Vol. II; pp 3–42.

6. Day, D. In *Introduction to Bioceramics;* Hench, L. L.; Wilson, J., Eds.; World Scientific Publishing: New York, 1993.

7. Charnley, J. *J. Bone Jt. Surg.* **1972,** *54B,* 61.

8. Sarmiento, A.; Ebramzadeh, E.; Cogan, W. J.; McKellop, H. A. *J. Bone Jt. Surg.* **1990,** *72A,* 1470–1476.

9. Aronson, J. In *Instructional Course Lectures;* Mosby, C. V., Ed.; American Academy of Orthopedic Surgeons: St. Louis, MO, 1986, Vol. 35; pp 119–128.

10. Black, J. *Biomaterials* **1984,** *5,* 11.

11. Ham, A. W. *Histology;* J. B. Lippincott: Philadelphia, PA, 1969.

12. Christel, P; Meunier, A.; Dorlot, J. M.; Crolet, J. M.; Witvolet, J.; Sedel, L.; Boritin, P. In *Bioceramics: Material Characteristics Versus In Vivo Behavior;* Ducheyne, P.; Lemons, J., Eds.; Annals of New York Acad. Sci.: New York, 1988, Vol. 523; p 234.

13. Boutin, P. M. In *Ceramics in Clinical Applications;* Vincenzini, P., Ed.; Elsevier: New York, 1987; p 297.

14. Hulbert, S. F. In *Introduction to Bioceramics;* Hench, L. L.; Wilson, J., Eds.; World Scientific Publishing: New York, 1993.

15. Hulbert, S. F.; Matthews, J. R.; Klawitter, J. J.; Sauer, B. W.; Leonard, R. B. *Biomed. Mater. Symp.* **1974,** *5,* 85–97.

16. Holmes, R. E.; Mooney, R. W.; Bucholz, R. W.; Tencer, A. F. *Clin. Orthop. Relat. Res.* **1984,** *188,* 282–292.

17. Ducheyne, P.; Hench, L. L.; Kagan, A.; Martens, M.; Burssens, A.; Mulier, J. C. *J. Biomed. Mater. Res.* **1980,** *14,* 225.

18. *Handbook of Bioactive Ceramics, Vol. II: Calcium Phosphate and Hydroxylapatite Ceramics;* Yamamuro, T.; Hench, L. L.; Wilson, J., Eds.; CRC Press: Boca Raton, FL, 1990.

19. Hench, L. L. *J. Am. Ceram. Soc.* **1991,** *74*(7), 1487–510.

20. *Bioceramics: Material Characteristics Versus In-Vivo Behavior;* Ducheyne, P.; Lemons, J., Eds.; Annals of New York Acad. Sci.: New York, 1988, Vol. 523.

21. Hench, L. L. In *Bioceramics: Materials Characteristics Versus In-Vivo Behavior;* Ducheyne, P.; Lemons, J., Eds.; Annals of New York Acad. Sci.: New York, 1988, Vol. 523; pp 54.

22. Hench, L. L.; Splinter, R. J.; Allen, W. C.; Greenlee, T. K., Jr. *J. Biomed. Mater. Res.* **1972,** *2*(1), 117–141.

23. Gross, U.; Kinne, R.; Schmitz, H. J.; Strunz, V. *Crit. Rev. Biocompat.* **1988,** *4,* 2.

24. Hench, L. L.; Paschall, H. A.; Allen, W. C.; Piotrowski, G. *NBS Spec. Publ.* **1975,** *415,* 19–35.

25. Kitsugi, T.; Yamamuro, T.; Kokubo, T. *J. Bone Jt. Surg.* **1989,** *71A,* 264.

26. Yoshii, S.; Kakutani, Y.; Yamamuro, T.; Nakamura, T.; Kitsugi, T.; Oka, M.; Kokubo, T.; Takagi, M. *J. Biomed. Mater. Res.* **1988,** *22*(A), 327.

27. Andersson, Ö. H.; Liu, G.; Karlsson, K. H.; Niemi, L.; Miettinen, J.; Juhanoja, J. *J. Mater. Sci. Mater. Med.* **1990,** 219–227.
28. Wilson, J.; Pigott, G. H.; Schoen, F. J.; Hench, L. L. *J. Biomed. Mater. Res.* **1981,** *15,* 805.
29. Wilson, J.; Nolletti, D. In *Handbook of Bioactive Ceramics;* Yamamuro, T.; Hench, L. L.; Wilson, J., Eds.; CRC Press: Boca Raton, FL, 1990, Vol. I; pp 283–302.
30. Hench, L. L.; Paschall, H. A. *J. Biomed. Mater. Res.* **1974,** *5*(1), 49–64.
31. Clark, A. E.; Pantano, C. G.; Hench, L. L. *J. Am. Ceram. Soc.* **1976,** *59*(1–2), 37–39.
32. Ogino, M.; Ohuchi, F.; Hench, L. L. *J. Biomed. Mater. Res.* **1980,** *14,* 55–64.
33. Kim, C. Y.; Clark, A. E.; Hench, L. L. *J. Non-Cryst. Solids* **1989,** *113,* 195–202.
34. Seitz, T. L.; Noonan, K. D.; Hench, L. L.; Noonan, N. E. *J. Biomed. Mater. Res.* **1982,** *16*(3), 195–207.
35. Vrouwenvelder, W. C. A.; Groot, C. G.; de Groot, K. *J. Biomed. Mater. Res.* **1993,** *27,* 465–475.
36. *Handbook on Bioactive Ceramics, Vol I: Bioactive Glasses and Glass-Ceramics;* Yamamuro, T.; Hench, L. L.; Wilson, J., Eds.; CRC Press: Boca Raton, FL, 1990.
37. Kasuga, T.; Nakajima, K.; Uno, T.; Yoshida, M. In *Handbook of Bioactive Ceramics;* Yamamuro, T.; Hench, L. L.; Wilson. J., Eds.; CRC Press: Boca Raton, FL, 1990, Vol. I; pp 137–142.
38. Wilson, J. *J. Appl. Biomater.* **1993,** *4,* 103–105.
39. Bonfield, W. In *Bioceramics: Materials Characteristics vs. In-Vivo Behavior;* Ducheyne P.; Lemons, J. E., Eds.; Annals of New York Acad. Sci.: New York, 1988, Vol. 523; pp 173–177.
40. Hench, L. L.; Andersson, Ö. H.; LaTorre, G. P. In *Bioceramics;* Bonfield, W.; Hastings, G. W.; Tanner, K. E., Eds.; Butterworth-Heinemann: Guildford, England, 1991, Vol. 4; pp 155–162.
41. Hench, L. L.; LaTorre, G. P. In *Bioceramics 5;* Yamamuro, T., Ed.; Kobunshi Kankokai Publishers: Kyoto, Japan, 1992; pp 67–75.
42. *The Bone-Biomaterial Interface;* Davies, J. E., Ed.; University of Toronto Press: Toronto, Ontario, Canada, 1991.
43. Li, R.; Clark, A. E.; Hench, L. L. *J. Appl. Biomater.* **1991,** *2,* 231–239.
44. Brinker, C. J.; Scherer, G. W. *Sol-Gel Science;* Academic: San Diego, CA, 1990.
45. Hench, L. L.; West, J. K. *Chem. Rev.* **1990,** *90,* 33–72.
46. West, J. K.; Hench, L. L. In *Bioceramics 5;* Yamamuro, T., Ed.; Kobunshi Kankokai Publishers: Kyoto, Japan, 1992; pp 75–87.
47. West, J. K.; Hench, L. L. *J. Biomed. Mater. Res.* **1994,** *29,* 625–633.
48. Braun, S.; Rappaport, S.; Zusman, R.; Avnir, D.; Ottolenghi, M. *Mater. Lett.* **1990,** *10*(1,2), 1–5.
49. Zink, J. I.; Dunn, B.; McKiernan, J.; Preston, D.; Yamanaka, S.; Knobbe, E. *Mater. Sci. Eng.* **1989,** *61,* 2134.
50. Ohtsuki, C.; Kokubo, T.; Yamamuro, T. *J. Non-Cryst. Solids* **1992,** *143,* 84–92.
51. Andersson, Ö. H.; Karlsson, K. H.; Kangasniemi, K.; Yli-Urpo, A. *Glastech. Ber.* **1988,** *61,* 300–305.

52. Strunz, V.; Bunte, M.; Stellmach, R.; Gross, U. M.; Kühl, K.; Brömer, H.; Deutscher, K. *Dtsch. Zahnärztl. Z.* **1977**, *32*, 287.
53. Kokubo, T. In *Multiphase Biomedical Materials;* Tsuruta, T.; Nakajima, A., Eds.; VSP: Utrecht, Netherlands, 1989.
54. Berger, G.; Sauer, F.; Steinborn, G.; Wishsmann, F. G.; Thieme, V.; Kohler, St.; Dressel, H. In *Proceedings of XV International Congress on Glass;* Mazurin, O. V., Eds.; Nauka: Leningrad, Russia, 1989, Vol. 3a; pp 120–126.
55. Vogel, W.; Holland, W. *Angew Chem. Int. Ed. Engl.* **1987**, *26*, 527–544.
56. Jarcho, M.; Bolen, C. H.; Thomas, M. B.; Bobick, J.; Kay, J. F.; Doremus, R. H. *J. Mater. Sci.* **1976**, *11*, 2027–2035.
57. Akao, M.; Aoki, H.; Kato, K. *J. Mater. Sci.* **1981**, *16*, 809–812.
58. deWith, G.; Dijk, H. J. A.; Hattu, N.; Prijs, K. *J. Mater. Sci.* **1981**, *16*, 1592–1598.
59. Jarcho, M.; Salsbury, R. L.; Thomas, M. B.; Doremus, R. H. *J. Mater. Sci.* **1979**, *14*, 142–150.
60. Akao, M.; Aoki, M.; Kato, K.; Sato, A. *J. Mater. Sci.* **1982**, *17*, 343–346.
61. Nakamura, T.; Yamamuro, T.; Higashi, S.; Kokubo, T.; Ito, S. *J. Biomed. Mater. Res.* **1985**, *19*, 685–698
62. Nakamura, T.; Yamamuro, T.; Higashi, S.; Kakutani, Y.; Kitsugi, T.; Kokubo, T.; Ito, S. In *Treatise on Biomedical Materials, 1;* Yamamuro, T., Ed.; Research Center for Medical Polymers and Biomaterials at Kyoto University: Kyoto, Japan, 1985; pp 109–117.
63. Kotani, S.; Yamamuro, T.; Nakamura, T.; Kitsugi, T.; Fujita, Y.; Kawanabe, K.; Kokubo, T.; Ohtsuki, C. In *Bioceramics;* Heimke, G., Ed.; German Ceramic Society: Cologne, Germany, 1990, Vol. 2; pp 105–112.
64. Kotani, S.; Fujita, Y.; Kitsugi, T.; Nakamura, T.; Yamamuro, T.; Ohtsuki, C.; Kokubo, T. *J. Biomed. Mater. Res.* **1991**, *25*, 1303–1315.
65. Fujita, Y.; Yamamuro, T.; Nakamura, T.; Kotani, S.; Ohtsuki, C.; Kokubo, T. *J. Biomed. Mater. Res.* **1991**, *25*, 1991–1003.
66. Ducheyne, P.; Hench, L. L. *J. Mater. Sci.* **1982**, *17*, 595–606.
67. deWith, G.; Corbijn, A. T. *J. Mater. Sci.* **1989**, *24*, 3411–3415.
68. Tamari, N.; Kondo, I.; Mouri, M.; Kinoshita, M. *J. Ceram. Soc. Jpn.* **1988**, *96*, 1200–1202.
69. Li, J.; Forberg, S.; Hermansson, L. *Biomaterials* **1991**, *12*, 438–440.
70. Kasuga, T.; Nakajima, K.; Uno, T.; Yoshida, M. *J. Am. Ceram. Soc.* **1992**, *75*, 1103–1107.
71. Muller-Mai, Ch.; Schmitz, H.-J.; Strunz, V.; Fuhrmann, G.; Fritz, Th.; Gross, U. M. *J. Biomed. Mater. Res.* **1989**, *23*, 1149–1168.
72. Bonfield, W. In *Bioceramics: Material Characteristics Versus In Vivo Behavior;* Ducheyne, P.; Lemons, J. E., Eds.; Academy of Science: New York, 1988; pp 173–177.

RECEIVED for review November 9, 1992. ACCEPTED revised manuscript March 3, 1993.

INDEXES

Author Index

Armor, John N., 321
Bell, Christine M., 211
Bianconi, Patricia A., 509
Burwell, David A., 231
Cavanaugh, Marge, ix
Chaudhari, Praveen, 12
Droske, John P., 66
Eckelmeyer, Kenneth H., 76
Ellis, Arthur B., 71
Epstein, Arthur J., 161
Farrington, Gregory C., 107
Ficalora, Peter J., 62
Fister, Loreli, 425
Gates, Bruce C., 301
Giannelis, E., 259
Good, Mary L., 28
Hancock, Kenneth G., 18
Hench, Larry L., 523
Hixson, S. H., 60
Jensen, Klavs F., 397
Johnson, David C., 425
Lee, Charlotte F., 231
Lieber, Charles M., 479

Mallouk, Thomas E., 211
Marder, Seth R., 189
Mendolia, Michael S., 107
Miller, Joel S., 161
Myers, Lori K., 231
Novet, Thomas, 425
Ozin, Geoffrey A., 335
Pearson, Glen H., 55
Pinnavaia, Thomas J., 283
Reichmanis, Elsa, 85
Roy, Rustum, 44
Sears, C. T., 60
Seyferth, Dietmar, 131
Sleight, Arthur W., 471
Steigerwald, Michael L., 373
Thompson, Mark E., 231
Thompson, Larry F., 85
Valentine, Kathleen G., 231
White, Robert M., 4
Wnek, Gary E., 62
Wrighton, Mark, 37
Yang, Huey C., 211
Zhang, Zhe, 479

Affiliation Index

AT&T Bell Laboratories, 85, 373
Air Products and Chemicals, Inc., 321
Allied Signal Inc., 28
Beckman Institute, 189
California Institute of Technology, 189
Carnegie Mellon University, 4
Cornell University, 259
DuPont Science and Engineering
   Laboratories, 161
Eastman Kodak Company, 55
Harvard University, 479
Massachusetts Institute of Technology,
   37, 131, 397
Michigan State University, 283
National Science Foundation, ix, 18, 60

Ohio State University, 161
Oregon State University, 471
Pennsylvania State University, 44, 509
Princeton University, 231
Rensselaer Polytechnic Institute, 62
Sandia National Laboratories, 76
T. J. Watson Research Center, 12
University of California—Davis, 301
University of Florida, 523
University of Oregon, 425
University of Pennsylvania, 107
University of Texas at Austin, 211
University of Toronto, 335
University of Wisconsin—Madison, 71
University of Wisconsin—Stevens Point, 66

Subject Index

A

Academic preparation, industry environment, 59t
Academic scientists, advisors to various federal agency programs in materials science, 8
Acid–base reactions, use to promote intercalation, 233
Active filler-controlled pyrolysis, reduction in shrinkage of ceramic, 145
Activity, catalyst, definition, 302
Acyl chloride, intermediate in conversion from $Zr(O_3PCH_2COO^-)_2$, 248–249
Adhesion, practical application of alkylsilane monolayers, 217
Adsorbents, general properties, 328t
Adsorption capacity, CMS versus zeolite 5A, 327t
Adsorption rates, oxygen and nitrogen with CMS, 329f
Adsorption-reaction technique, See Sequential adsorption-reaction technique
Advanced Materials and Processing Program implementation, 18–27
strategic objectives for materials science, 7–8
Advanced-performance materials, development and commercialization, 28–36
Advanced tactical fighter, "test beds" for advanced-performance materials, 31–32
Advanced technology program, role in supporting materials science and other critical technologies, 8–9
Aerospace industry, area of opportunity for advanced-performance materials, 31–32
Air–liquid interface, formation of Langmuir–Blodgett monolayers, 213
Air separation, use of molecular sieves, 321–333
AlGaAs, importance as OMCVD reagents, 402

AlInGaP, importance as OMCVD reagents, 402
Alkyl selenides, research into new organoselenium precursors for OMCVD, 411–412
Alkylthiol monolayers, studies of fundamental electrochemical phenomena, 218
Allyl selenides, research into new organoselenium precursors for OMCVD, 411–412
Alternating donor–acceptor salts, structure and magnetic properties, 165–172
Alumina, use as implants and prostheses, 527–530
Aluminum, alternate ceramics, 149–150
Aluminum chlorohydrate, use as pillaring reagent, 288f–290
American Chemical Society, educational initiative in materials science, 47
Amines, effect on sensitivity of chemically amplified resists, 101
Amorphization reactions, solid-state, 432
Amorphous intermediate, direct formation of ternary product, 456–461
Amorphous poly(ethylene oxide), value as electrolyte, 115
Antiferromagnet, creation by spin alignment, 164
Antiferromagnetic coupling of alternating spin sites, model for ferromagnetic spin alignment, 172–173, 174f, 176–177
Applied field, third-order effects, 193–195
Applied field data versus magnetization, 166–167f
Aqueous ion exchange, nanochemistry technique for synthesizing zeolite-encaged semiconductor clusters, 345
Aromaticity, molecular nonlinearities, 196–199
Arrested precipitation, formation of II–VI clusters, 377
Arrested thermolysis, formation of semiconductor clusters, 381–382

As-grown crystals, atomic resolution images and Fermi level effects, 496, 497f, 498f

Aspect ratio(s)
fully exfoliated MTSs, 261
role in determining volume fraction of percolation threshold, 277

Atom-based magnetic materials, representative examples, 163t

Atomic force microscope, precise measurements of surfaces, 15

Atomic layer epitaxy, current trend in OMCVD, 416–417

Atomic precipitation, advantages over ionic precipitation, 383–384

Atomic resolution images of BiO layer, comparison of nonsuperconducting and superconducting crystals, 493, 494f

Atomic resolution map, creation by STM, 482

Axially symmetric powder pattern, principal components, 238

Azo dyes, incorporation into multilayer films, 225

B

Bardeen–Cooper–Schrieffer, theory of superconducting materials, 471–472, 499

Barrier height, tunneling, determination from current vs tip–sample separation, 490

Basic vapors, effect on sensitivity of chemically amplified resists, 101

Battery, advantages of solid polymeric electrolytes, 107

Beidellite, typical unit cell formula of typical smectite clays, 289t

$Bi_2Sr_2CaCu_2O_8$, elucidation of structural and electronic properties by STM, 480–504

Binders, role in preparation of catalytic materials, 308

BiO layer, exploitation of cleavage in STM, 487

Bioactive ceramics, use as implant and prostheses, 531–538

Bioactive composites, interfacial stability and mechanical properties, 538–541

Bioceramics, use as implants and prostheses, 527–530

Bioglass 45S5, bioactive ceramic in clinical use, properties, 532–534t

Biological fixation, occurrence with porous implants, 530

Biomaterials
attempts to mimic in synthetic materials, 46–47, 509–512
emerging areas of materials science, 25
inorganic, 523–544
scope of the MS&P program, 21

Biomimetic mineralization, application to composites, 509

Biotin-terminated alkylthiols, ability to recognize and bind streptavidin, 219–220

Bioverit, bioactive ceramic in clinical use, 533t

Bis(dithiolato)metallate salts of decamethylferrocenium, Curie–Weiss constants and effective magnetic moments, 169, 171–172

Body fluids, effect on inorganic biomaterials, 524

Bond-length alternation, hyperpolarizabilities and molecular nonlinearities, 196–203

Bonding rates, characteristic of bioactive ceramics, 534–536

Boron, alternate ceramics, 149–150

Bragg reflections, occurrence in diffraction pattern of superlattice, 439–440, 442f

Brunauer–Emmett–Teller measurements, inappropriateness for CMS materials, 327

Buckminsterfullerene, intercalation with fluorohectorite, 266–267f

C

^{13}C nuclei, NMR background, 235–238

$CaCO_3$, formation from polymer matrices with specific functional groups, 516

Cadmium selenide, growth of crystallites within inverse micelles, 379–380

Capacity
key property for air-separation sorbent, 323–324t,f
optimization of O_2 versus N_2 capacity in molecular sieves, 325–327

β-$Ca_3(PO_4)_2$, bioactive ceramic in clinical use, 533t

Carbene intermediate, role in proposed mechanism for carbon incorporation, 404f

Carbon incorporation, effect on OMCVD reagents, 403–405

Carbon kickout reaction, preparation of nitride ceramics, 142

Carbon molecular sieves
novel catalytic material, 318–319
separation of air by kinetic selectivity, 325–330

Carbonaceous deposits, effect on zeolite during petroleum cracking, 312–314

Carbosilane resins, early developments, 133–136

Carrier component, constituent of catalyst particles, 306

Catalytic materials, types and functions, 301–319

Cation-exchange resin, example of catalytic material, 309

Cations within a zeolite, effect on use as molecular sieve, 323–325

CdHgTe, use for far-IR detectors, 401

Cerabone A–W, bioactive ceramic in clinical use, properties, 532–534t

Ceramic fibers, research directions in preceramic polymers, 147–149

Ceramic–metal composite, structure, 428f

Ceramic supports in catalysts, preparation, transition aluminas, 307

Ceramics containing heteroelements other than silicon, research directions in preceramic polymers, 149–150

Ceramics, materials science as a specialization in chemistry, 46–47

Ceravital, bioactive ceramic in clinical use, properties, 532–534t

Chabazite-type zeolite, improvement as sorbent for bulk air separation, 323, 325f

Chalcogenides, formation via low-valent complexes, 393

Charge density, effect on orientation of molecular assemblies, 265–266

Charge displacement, second-order effects, 191

Charge-transfer, effect on ferromagnetic coupling and relative energy of states, 173–175

Charge-transfer resonance forms, effects on bond-length alternation, 198–199

Chemical adsorption, step in surface catalysis, 305

Chemical beam epitaxy, chemical perspective, 397–419

Chemical-shielding tensor, effect on environment on NMR resonance frequency, 235–236

Chemical-shift anisotropy, effect on NMR linewidths of dipolar nuclei in solids, 235

Chemical shift, crystal in an applied magnetic field, 236–237

Chemically amplified resists
high-resolution imaging, 102
process issues, 100–102
process sequence, 94, 95t

Chemistry courses
new curricular materials for introducing polymer topics, 66–70
shift toward materials chemistry, 62–64

Chemists, role in oxide superconductors, 476

Clusters of II–VI semiconductor materials
formation via arrested precipitation, 377–381
properties and importance to technology, 375–377

Coal, preparation of carbon molecular sieves, 327

Coating, function as a biomaterial, 526t

Cobalt telluride, initial steps in formation of solid, 387–389

Coconut shells, preparation of carbon molecular sieves, 327

Coercive fields, summary for metamagnets, 168t

Coke, effect on zeolite during petroleum cracking, 312–314

Comb-branch polymers, reduced crystallinity, 115–116

Commercial ceramic fiber, need, 148

Competitiveness, U.S., importance of materials science, 4–10

Complex materials, characterization by scanning tunneling microscopy, 479–504

Composite fabrication, comparison to biomimetic materialization, 510

Composites
bioactive, interfacial stability and mechanical properties, 538–541
ceramic–ceramic, research directions in preceramic polymers, 143
metal-matrix, technological applications, 427–428

Composition control, nucleation, 455–456

Compound semiconductors, vapor deposition, 397–419

Computational studies
correlation of β and bond-length alternation, 199–200
correlation of γ and bond-length alternation, 202–203

Conductance, insight into superconducting state of $Bi_2Sr_2CaCu_2O_8$, 501–504

Conductivity measurements, polyaniline multilayer hybrid, 269

Configuration mixing for a donor–acceptor chain, model for ferromagnetic spin alignment, 172–175

Constant-current mode, use in creating STM images, 482–483

Contaminants, chemically amplified systems, 100

Contrast, requirement in resist design, 91

Contrast value, high, requirements of materials, 96

Conventional polymer-processing techniques, requirement for preceramic polymer, 137–138

Cooperative Research and Development Agreement (CRADA), government and Industrial cooperation, 8–10

Copolymers, reduced crystallinity, 115–116

Copper–molybdenum–selenium, diffraction data and thermograms, 459f–460f

Copper oxide superconductor, elucidation of structural and electronic properties by STM, 480–504

Cracking of petroleum, use of zeolite-containing catalysts, 312–314

Critical technologies, importance of materials science, 4–10

Critical temperature
effect on oxide superconductors, 471–476
salts with donor–acceptor structures, 170t
summary for metamagnets, 168t
time evolution for increases, 185t
See also Curie temperature

Cross-linked, network-type polymers, conversion to ceramic fibers, 147–148

Cross-linked polystyrene, use as support in catalytic material, 309

Cross-linking bridges, formation in layered materials, 296–297f

Cross-linking olefinic groups, improvement of LB film stability, 221–222

Cross-linking polymers
effect on ceramic yield and pyrolysis, 139
effect on T_g, 115

Cross-polarization, use in improved NMR spectroscopy, 235

Crystal chemistry, brief history, 50

Crystal modification, molecules that poison growth on specific faces, 516

Crystal nucleation and growth, matrix mediation, 512–515

Crystalline aluminosilicates, example of catalytic material, 309

Crystalline phases, metal powder–polysilazane composites, 145t, 146t

Crystallinity
copolymers and comb-branch polymers, 115–116
PEO, disadvantage for solid electrolytes, 109
physical and mechanical properties of polymers, 272, 274–275f
plasticized systems, 118–125
polymer blends, 116
requirement for synthesis of II–VI semiconductor clusters, 376

Crystallochemical mediation, biological features in deposition of particles within polymers, 511

Cuprate superconductors, importance as oxide superconductors, 472–476

Cures of melt-spun fibers, search for alternate types, 148–149

Curie–Weiss constants, insight into a structure–function relationship, 169–172

Curie–Weiss law, use to parametrize higher temperature data, 166, 168t

Curie susceptibility, fit by Curie–Weiss expression, 169

Curie temperature
calculation, 177–180
determination of ferromagnetic order and magnetic fields, 166–172

Curing operation, step in organic polymer technology, 137

Curriculum pressure point, general chemistry, 71–75

D

Decomposition pathways, possible pathways for Te compounds, 410f

Deep-UV lithography, resist design, 93–94
Deep-UV photolithography, as alternative
  to conventional photolithography, 88–89
Deep-UV resists, etching resistance and
  optical density, 98
Defense budgets, effect on the future of
  preceramic polymer research, 152–153
Delaminated clays, formation and function,
  290–291
Density of interfaces, advantage of
  superlattice reactants, 435
Density of state, variation with
  tip–sample separation, 491–492
Density, properties of polymer-derived
  ceramics, 144
Department of Defense, support for
  materials science, 37–39
Deposition of compound semiconductors,
  chemical perspective, 397–419
Deposition of elemental layers,
  preparation of superlattice reactants,
  436–437, 438f
Deposition techniques, improved
  nanophysics fabrication methods,
  336, 337f
Diagnostics, in situ, current trends in
  OMCVD, 414
Dielectric constant
  plasticizer, 118–125
  polyethers, 111
Differential scanning calorimetry
  interfacial reactions probe, 445–450
  use in tracking melting and glass
    transition of polymers, 274, 275f
Diffraction pattern, as-deposited
  iron–silicon superlattice, 439f
Diffusion couples, use in study of
  interfacial reactions, 430–432
4-N,N-Dimethylamino-4'-nitrostilbene,
  prototypical example of NLO
  chromophores, 197
Dipolar nuclei, NMR background, 235–238
Dipole moment, dot product with β for
  thiobarbituric acid acceptors, 201f
Direct crystalization method, synthesis of
  nanoporous solids, 285
Direct formation, ternary compound from
  amorphous intermediate, 456–461
Direct polymer intercalation, advantages
  and synthetic approaches, 272, 274
Disk-shaped, metallo macrocyclic anions,
  potential use as pillaring agents,
  295–296f

Dispersion of ceramics, production of
  nanocomposites, 275–278
Donor–acceptor bond, effect on
  crystallization, 383
Donor–acceptor interactions, polar sheet
  structures, 205
Donor–acceptor molecules, computational
  studies of hyperpolarizabilities and
  bond-length alternations, 199–200f
Donor–acceptor polyenes, bond-length
  alternation, 198
Donor–acceptor salts, structure and
  magnetic properties, 165–172

E

Early transition metals, alternate
  ceramics, 149–150
Economic impact of nanoporous materials,
  U.S. and world, 283–284
Edge-to-edge aggregation, effect on
  pillared clays, 290f
Educational degrees in materials science,
  content, 48, 49t, 50f
Educational issues, importance for
  materials science, 43–82
Effective exchange integral, calculation
  of Curie temperature, 177–179
Effective exchange interaction,
  room-temperature magnet system,
  183–184
Effective moment, potential
  molecular-polymeric materials,
  function of temperature, 181f
Electric-field gradient, NMR spectra of
  quadrupolar nuclei, 239
Electric field of light, relationship to
  polarization, 190–191
Electric field, second-order effects, 191
Electrolytes, polymeric, 107–126
Electron-beam lithography, alternative to
  conventional photolithography, 90–91
Electron spin, creation of magnetism,
  162–164
Electron-spin resonance, spectra of
  oriented films of molecular
  assemblies, 263–265
Electron-transfer donor–acceptor salts,
  structure and magnetic properties,
  165–172
Electron-transfer kinetics, study with
  alkylthiols on gold, 219

Electronic absorption spectra, of
    polyaniline multilayer hybrid, 269
Electronic character of BiO layer,
    characterization by STM, 490–492
Electronic compound semiconductor devices,
    OMCVD applications, 400–401
Electronic materials, scope of the MS&P
    program, 21
Electronics-communications, opportunity
    for advanced-performance materials,
    31–32
Electronics, emerging areas of materials
    science, 25
Elemental carbon and silicon, reactions to
    give SiC, 142
Elemental layer thicknesses, effect on
    diffraction maxima, 440, 441f
Endosemiconductor, new type of
    nanomaterial, 335–367
Endotin(IV) sulfide, quantum size effects
    and electronic coupling strengths,
    360–363
Energy difference between transitions for
    quadrupolar nuclei, relation to
    orientation of sample, 239–241
Energy gap, superconducting state of
    $Bi_2Sr_2CaCu_2O_8$, 499–504
Engineering degrees, desire for a more
    macroscopic focus to freshman
    chemistry courses, 63–64
Epitaxy, variations to MOCVD, 346
Etching resistance
    deep-UV resists, 98
    requirement in resist design, 86f, 91–92
Ethylene oxidation, use of selective
    catalyst, 314–315f
Exosemiconductor, new type of
    nanomaterial, 335–367
Exotin(IV) sulfide, quantum size effects
    and electronic coupling strengths,
    360–363
Extended inorganic solids, molecule-based
    syntheses, 373–394
Extended X-ray absorption fine structure,
    structure determination of metal
    chalcogenide nanoclusters, 347–348f

                            F

Fabrication
    synthetic composite alloys, comparison
    to biomimetic materialization, 510

Fabrication—Continued
    tetrahedral quantum dot structures by
    OMCVD, 418f
Face-to-face aggregation, effect on
    pillared clays, 290f
Faujasitic zeolites, relation to
    nanoporous materials, 284
$[FeCp*_2]^{\cdot+}[TCNQ]^{\cdot-}$, structure–function
    relationship, derivatives, and
    properties as a metamagnet, 165–184
Federal Government, role in supporting
    materials science, 6–10
Fermi level, apparent gap in DOS, 491
Ferri- and ferromagnets, creation by spin
    alignment, 164
Ferrimagnetic coupling of differing spins,
    model for ferromagnetic spin
    alignment, 172–173, 176–177
Ferrocenylalkylamino intercalation
    compounds, structure, 250–254
Ferromagnetic coupling
    bis(dithiolato)metallate salts of
    decamethylferrocenium, 169
    schematic illustration, 174f
Fiber pullout, failure mode of
    metal–ceramic composites, 427–428f
Film morphologies, unique properties of Te
    compounds, 408–409f
First hyperpolarizability
    correlation with bond-length
    alternation, 199–200f
    definition, 190, 197
    experimental efforts in optimization,
    200–202
    molecular nonlinearities, 196–199
Fluorohectorite, orientation to porphyrin
    guest molecules, 263–266
Forming operation, final step in organic
    polymer technology, 137
Four-wave mixing, third-order effects,
    193–195
Fracture toughness
    characteristic of implant materials,
    527t, 529t, 532–533t, 540t
    polyaniline multilayer hybrid, 270–271f
Free carbon, effect on ceramic product
    after pyrolysis, 140
Free energy, effect on nucleation,
    433, 434f
Fullerenes, notable advances in materials
    science, 15
Funding opportunities, materials science
    education, 60–61

G

$Ga_{1-x}Al_xAs$, classic case of bulk
  semiconductor alloy, 363
GaInPAs, importance as OMCVD
  reagents, 402
Gallery height
  comparison to van der Waals thickness in
    pillared lamellar solids, 291
  definition of pores in a pillared
    lamellar solid, 287
Gallium arsenide
  proposed mechanism for OMCVD,
    415f
  transistor, 426f
Gallium species, research into new
  precursors for OMCVD, 412–413
Gas-phase reactions, formation of thin
  films, 398
Gas-phase technique, advantages over other
  adsorption techniques, 225–226
Gas separation, use of molecular sieves
  for air separation, 321–333
Gas stream, importance to pyrolysis of
  preceramic polymer, 141–142
Gaseous products, problems caused during
  pyrolysis, 140–142
General chemistry
  curriculum pressure point, 71–75
  See also Chemistry courses
General powder pattern, principal
  components, 238
Glass–ceramics, demonstration of bioactive
  material flexibility, 537–538
Glass-transition temperature
  plasticized systems, 118–125
  requirements for polymeric
    solvent, 110
Glass fibers for optical communication,
  future impact, 14
Glass transition, effect on polymer
  stiffness, 274
Goethite, formation from polymer matrices
  with specific functional groups, 516
Gold–sulfur bond, formation of
  self-assembled monolayers, 219
Gold, surface for alkylthiol and organic
  disulfide monolayers, 218
Government, role in supporting materials
  science, 6–10
Grades K–12, science education, 76–82
Graphite, evolution from carbonaceous
  material, 326f

Ground transportation, area of opportunity
  for advanced-performance materials,
  31–32
Group II–VI compounds, molecule-based
  syntheses, 374–384
Group II–VI semiconductor clusters,
  synthesis by inverse micelles, 514
Group II and Group III compounds,
  research into new precursors for
  OMCVD, 412–413
Group III–V materials, nanoscale
  clusters, 383
Group IV metal phosphates, use in
  pillaring reactions, 296–297f
Growth and habit modification, biological
  features in deposition of particles
  within polymers, 511
Growth morphology, effect on thin-film
  properties in OMCVD process, 399

H

^2H nuclei, NMR background, 238–241
Head-group charge, modification of
  nucleated crystals, 517
Heat of adsorption, key property for air
  separation sorbent, 323–324t,f
Hectorite
  layers, orientation to porphyrin guest
    molecules, 263–266
  typical unit cell formula of typical
    smectite clays, 289t
Heider–London spin-exchange, model for
  ferromagnetic spin alignment, 172–175
Heterojunction lasers, OMCVD
  application, 402
Heteronuclear decoupling, use in improved
  NMR spectroscopy, 235
High-angle X-ray diffraction, probe of
  solid-state reactions, 445–447
High-spin molecules, model for
  ferromagnetic behavior, 175–176
High-temperature superconductors, notable
  advances in materials science, 15
High-vacuum environment, preparation of
  superlattice reactants, 436–437, 438f
Homoepitaxy, synthesis of high purity
  material, 400–401
Homopolymers, simple, solubility of
  inorganic salts, 113
Host–guest inclusion chemistry,
  semiconductor nanochemistry, 340–344

Human body, repair with inorganic biomaterials, 523–544

Hund's rule, predictions for ground and excited states, 174–175

Hydrides, toxic, gaseous, effect on OMCVD reagents, 405–407f

Hydrogen bonding
polar sheet structures, 205
stability of self-assembled materials, 263
use in self-assembling approach, 260

Hydrogen selenide, research into new organoselenium precursors for OMCVD, 411–412

Hydrothermal chemistry, synthesis of endo- and exosemiconductors, 338–349

Hydroxyapatite
bioactive ceramic in clinical use, 533t
biomaterial composite with polyethylene, 538–541

Hydroxycarbonate apatite, use as bioactive ceramic, 531–536

Hysteresis loops
molecular-polymeric material at room temperature, 182f
plot of magnetization versus applied field data of a metamagnet, 166–167f

I

Ilmaplant L1, bioactive ceramic in clinical use, 533t

Implants, inorganic biomaterials for repair of human body, 524, 525f

In-registry, of $[FeCp*_2]^{\cdot+}[TCNQ]^{\cdot-}$, 178f

In situ diagnostics, current trends in OMCVD, 414

In situ synthesis, matrix mediation of crystal nucleation and growth, 512–515

Inclusion chemistry, synthesis of endo- and exosemiconductors, 338–349

Indium species, research into new precursors for OMCVD, 412–413

Industrial catalytic materials, types, 302–303t

Industrial perspective, materials science education, 55–59

Industrial scientists, advisors to various federal agency programs in materials science, 8

Industry environment, academic preparation, 59t

Infinite-layer structure, formation with oxide superconductors, 472–473

Innovation, importance to U.S. industry, 55–56

Inorganic–organometallic preceramic polymer chemistry, route to nonoxide ceramics, 131

Inorganic biomaterials, use in repair of the human body, 523–544

Inorganic chemistry, role in history of materials science, 44–46

Inorganic salts
effects on polymer electrolytes and glass-transition temperature, 110
solubility in simple homopolymers, 113
solvation by PEO, 108

Inorganic solids, extended, molecular-based syntheses, 373–394

Insulators, synthesis from sodalite supralattices, 351–357

Integrated circuits, formed by lithographic process, 85

Inter-head-group spacing, modification of nucleated crystals, 517

Intercalation
control of properties of assembly, 266–267f
description and application, 231–232
formation of a multilayer, 260

Intercalative polymerization of single polymer chains, approach, 267–275

Interdiffusion, effect on Bragg reflections and subsidiary maxima, 440, 442f

Interfaces and interfacial reactions, importance in technology, 425–466

Interfacial instability, cause of failure in prostheses, 524–526

Interfacial reactions, formation of bond between tissue and bioactive ceramics, 535

Interfacial widths, control of solid-state reactions, 450–461

Internal crystallinity, requirement for synthesis of II–VI semiconductor clusters, 376, 380

Intrazeolite MOCVD, nanochemistry technique for synthesizing zeolite-encaged semiconductor clusters, 345

Ion-beam etching, improved nanophysics fabrication methods, 336, 337f

Ion-channel sensor, use of LB monolayer-lipid bilayer membrane similarity, 212, 214f

Ion-exchange method, use to promote intercalation, 233
Ion-flux regulation, biological features in deposition of particles within polymers, 511
Ion pairing of salt−polyether systems, techniques for determination, 111
Ionic conductivity
  PEO, effect of crystallinity, 109
  plasticized systems, 118−125
  polyelectrolytes, 116
  polymer blends, 116
  polymer electrolytes, related to temperature, 125
Ionic lamellar solids, properties, 286−287
Ionic transport, plasticized systems, 118−125
Iron−silicon
  composite, low-angle diffraction pattern, 437, 439f
  diffusion couple, use in probing phase diagrams, 431f
  multilayer, representative calorimetry data, 446
  superlattices, calculated diffraction patterns, 441f, 442f, 443f, 444f, 445t

J

Job skills, requirements of materials science, 57−58

K

Keggin ions, use as reagents for pillaring, 293−296
Kinetic approach to solid-state synthesis, role in interfacial reactions, 429
Kinetics, current trends in OMCVD, 413−414

L

Lamellar solids, formation, 286−287
Langmuir−Blodgett monolayers, formation, uses, and disadvantages, 212−214
Langmuir mono- or bilayers of surfactants, effect on air−water interfaces of crystallizing solutions, 516−517

Laser, example of compound semiconductor device, 399f
Laser pyrolysis, technique to form ceramic coatings, 146−147
Lateral free separation, definition of pores in a pillared lamellar solid, 287
Layer-by-layer preparation
  advantage of superlattice reactants, 435
  synthesis with oxide superconductors, 473
Layer thicknesses, effect on solid-state reactions, 450−455
Layered double hydroxides, layer lattice structures, 293−296
Layered materials
  determination of structure by NMR spectroscopy, 231−254
  nanoporous, types and functions, 283−298
Lecture "snapshots", format, 68−69
Life-cycle dynamics, commercialization dilemma of advanced-performance materials, 30f−31
Light
  emission, nanoscale silicon-based structures, 343
  polarization and second- and third-order effects, 191−194
Linear chains of donor−acceptor salts, effect on metamagnetic properties, 169−172
Linear polarization
  definition, 190, 197
  modulation and generation, 189−191
Lithographic process
  formation of circuit patterns, 86f
  optimization, 98−100
Lithographic resist materials, design, 85−104
Lithographic techniques, improved nanophysics fabrication methods, 336−343
Lithographically patterned substrate, 464−465
Local density of sample electronic states, qualitative measure by STM, 485
Low-angle X-ray diffraction, study of interfacial structure, 437−450
Low-energy excitation, superconducting state of $Bi_2Sr_2CaCu_2O_8$, 499−504
Low growth temperature, effect on organometallic reagents for OMCVD, 407−412
Lubrication, practical application of alkylsilane monolayers, 217

M

Macropores, definition by pore diameter, 304

Macroscopic, second-order optical nonlinearities, 189

Magic angle spinning, use in improved NMR spectroscopy, 235

Magnetic behavior, synopsis, 162–164

Magnetic materials
attributes, 162
representative examples, 163*t*
scope of MS&P program, 22

Magnetic susceptibility, fit by Curie–Weiss expression, 169

Magnetization, versus applied field data, 166–167*f*

Magnets
application of oxide superconductors, 475–476
uses in society, 162

Manufacturing
requirement in resist design, 92
successful progression from innovation, 56

"Market pull-through" philosophy, effects on material development, 30

Mass-transport selectivity, description of zeolite applications in catalysis, 312,313*f*

Materials
development and commercialization, 28–36
optimal properties for resists, 92, 96

Materials chemistry
comparison to materials science, 62
courses, Rensselaer Polytechnic Institute, 62–65
creation of a resource book for teachers, 71–73
emergence of molecular magnets, 161–184
organic monolayer and multilayer thin films, 211–227
See also Materials science

Materials education
print and video teaching aids, 52
recommendations for improvements, 52–53
relationship to chemists, 44–53

Materials science
aspects to any materials-related activity, 13

Materials science—*Continued*
comparison to materials chemistry, 62
critical technology for U.S. competitiveness, 4–10
major practical and technical advances, 14–15
role of U.S. Government, 6–20
teaching and research chemistry in the twentieth century, 45*t*
vitality to industry and defense, 13–15
See also Materials chemistry

Materials science and engineering
federal funding, 13
relationship to industry, 12–13
study conducted by the National Research Council, 5–6, 12–17

Materials Science and Engineering for the 1990s, report for the National Research Council, 5–6, 12–17, 18–19

Materials science education
funding opportunities, 60–61
industrial perspective, 55–59

Materials Synthesis and Processing (MS&P) Initiative, NSF, support for materials science, 20–23

Matrix mediation of crystal nucleation and growth, in situ synthesis, 512–515

Mean-field model, calculation of Curie temperature, 177–179

Mean pore diameter, general properties of adsorbents, 328*t*

Mechanical properties in polymers, optimization, 112–113

Mechanical stability
plasticized systems, 118–125
polymer blends, 116
polymer electrolytes, 112

Melting, effect on polymer stiffness, 274

Mercury, initial discovery of superconducting properties, 471

Mesopores
definition by pore diameter, 304
relation to nanoporous materials, 284

Mesoporous molecular sieves, example of novel catalytic material, 318–319

Mesoscopic materials, design, synthesis, and characterization, 259–279

Metal and metal oxide clusters on supports, example of novel catalytic material, 318

Metal catalysts, types and functions, 314–316

Metal–ceramic composite, structure, 428*f*

Metal chalcogenide nanoclusters, synthesis inside the supercages of zeolite, 346–349

Metal ion exchange, nanochemistry technique for synthesizing zeolite-encaged semiconductor clusters, 345

Metal-matrix structural composite materials, technological applications, 427–428

Metal–nonmetal materials, nanoscale clusters 382

Metal-organic chemical vapor deposition, synthesis of endo- and exosemiconductors, 338–349

Metal-organic molecular beam epitaxy, chemical perspective, 397–419

Metal–polymer composites, research directions in preceramic polymers, 143–146t

Metal sulfides and selenides, early preparations of semiconductor crystallites, 377

Metallic implants, orthopedic applications, 524–526f

Metallocene intercalation compounds, study by orientation-dependent NMR spectroscopy, 233

Metals, synthesis from sodalite supralattices, 351–357

Metamagnet, definition, 165

Metastability, creation in high-temperature superconductors, 474–475

Metastable silica clusters, tailoring surface reactions of inorganic biomaterials, 542–544

Mica-type layered silicates, 261–262

Micelle, use of reverse micelles in arrested precipitation of II–VI crystallites, 378–380

Microcrystalline materials, preparation for NMR studies, 241–242f

Microelectronics
application of organic monolayer and multilayer films, 226
feature size, 87f
importance of interfaces, 426–427

Micropore distribution, for carbon molecular sieves, 330, 331f

Micropores
definition by pore diameter, 304
relation to nanoporous materials, 284

Microscopic–macroscopic merger, approach to alternate chemistry courses, 64–65

Microscopic, second-order optical nonlinearities, 189

Mineralization zone, confinement and physical shaping in biomineralization, 512

Models for molecule-based magnetic materials, descriptions, 172–177

Modulus of elasticity
comparison of metallic implants with bone, 526, 527t, 529t
stress shielding of glass-ceramic bioactive composites, 538

Molecular-orbital calculations, tailoring surface reactions of inorganic biomaterials, 542–544

Molecular-sieve hosts, role in assembly of endosemiconductors, 344–351

Molecular-sieving carbons, example of novel catalytic material, 318–319

Molecular assemblies, highly organized, 263–267

Molecular clusters, stepping stones to solid-state compounds, 383–384

Molecular dimensions, of various gaseous molecules, 326t

Molecular dispersion of ceramics, production of nanocomposites, 275–278

Molecular electronics, application of organic monolayer and multilayer films, 226

Molecular magnets, emerging area of materials chemistry, 161–184

Molecular sieves, use for air separation, 321–333

Molecular tailoring, surface chemistry, 541–544

Molecule-based magnetic materials, representative examples, 163t

Molecule-based syntheses, extended inorganic solids, 373–394

Molecules-to-solids reactions, preparation of solid-state products, 373–394

Molybdenum–selenium superlattice
differential scanning calorimetry data, 451f, 453f, 454f
diffraction patterns, 452f

Monolayer thin films, materials chemistry, 211–227

Monomers, intercalation, 267–275

Montmorillonite, pillaring lamellar solid, 288–289t

MoSe$_2$, proposed "designer compound", 466f

Motional averaging, effects on ^{2}H NMR spectroscopy, 240–241

Multilayer, comparison with natural crystals, 439

Multilayer thin films, materials chemistry, 211–227

Multiphase ceramics, research directions in preceramic polymers, 143

N

Nanochemistry, nanomaterials and nanophysics, 336–344

Nanocomputer, possible future technology, 366

Nanomaterials, endosemiconductors and exosemiconductors, 335–367

Nanophysics, nanomaterials and nanochemistry, 336–344

Nanoporous layered materials, types and functions, 283–298

Nanoporous semiconductor, nanophysics fabrication compared to nanochemistry synthesis, 342f

Nanoscale, two-dimensional organic–inorganic materials, design, synthesis, and characterization, 259–279

Naphtha reforming, use of metal catalyst and active support, 315–316

National Research Council

impact of materials science on competitiveness, 5

study on materials science and engineering, 5–6, 12–17

National Science Foundation's program in materials science, new frontiers, new initiatives, new programs, and new prospects, 18–27

Natural crystal, comparison with multilayers, 439

NbSe$_2$, proposed "designer compound", 466f

Necking, route to single-crystal films, 464–465

Negative resist, effect on polymer solubility, 93

Networks

plasticized systems, 124

cross-linked polyethers as media for ionic conduction, 113–115

Neutral resonance forms, effects on bond-length alternation, 198–199

Newman projection, reflection of the microcrystalline nature of zirconium phosphonates, 247

Nicalon fibers

commercially available, polymer-derived ceramic fibers, 148

properties, 135–136

Nickel telluride, large molecular clusters as small solids, 389–393

p-Nitroaniline, prototypical example of NLO chromophores, 197

Nitrobenzyl ester acid generators, possible chemically amplified resist system, 96

Nitrogen gas, production and transport in industry, 321–322

NMR spectroscopy

structural tool for layered materials, 231–254

use in determining polymers, 133–134

Nonlinear optical materials

fabrication by biomimetic mineralization, 513

optimization, 189–208

origin, 194, 196

Nonlinear optics, molecular layers, possibilities, 217

Nonlinear polarization, modulation and generation, 189–191

Normal metal–insulator–superconductor junctions, superconducting state of Bi$_2$Sr$_2$CaCu$_2$O$_8$, 499–503

Nucleation

barrier, effect on crystalline products, 433

crystal, matrix mediation, 512–515

transformation of intermediate into stable crystalline product, 433

O

Octadecyltrichlorosilane, chemisorption and lateral polymerization, 216

Olefinic groups, improvement of LB film stability, 221–222

Oligomer aerosols, technique to form ceramic coatings, 146–147

One-layer structure, formation with oxide superconductors, 473

Optical density

deep-UV resists, 98

requirement in resist design, 91

Optical fiber, successful
   commercialization, 31
Optical materials, fabrication by
   biomimetic mineralization, 513
Optical nonlinearities, optimization,
   189–208
Optical rectification, occurrence, 192
Optical semiconductor devices, OMCVD
   applications, 400–401
Optics, molecular layers, possibilities, 217
Organic chemistry, role in history of
   materials science, 44–46
Organic–inorganic interface, molecular
   recognition, 515–518
Organic–inorganic materials, nanoscale,
   two-dimensional, design, synthesis,
   and characterization, 259–279
Organic monolayer and multilayer thin
   films, materials chemistry, 211–227
Organic salts, optimization of macroscopic
   second-order optical nonlinearities,
   203–207
Organometallic chemical vapor deposition
   of compound semiconductors, chemical
   perspective, 397–419
Organometallic reagents, alternative to
   OMCVD hydride reagents, 406
Organometallic vapor-phase epitaxy
   thin-film preparation of II–VI
   compounds, 374, 397–419
Orientation-dependent NMR spectroscopy,
   structural tool for layered materials,
   231–254
Oscillating dipole, second-order
   effects, 191
Out-of-registry interactions,
   $[FeCp^*_2]^{\cdot+}[TCNQ]^{\cdot-}$, 178f
Oxidation state of copper, effect on
   superconductivity of sheets, 472–473
Oxide superconductors, metastability,
   theory, and applications, 471–476
Oxygen-annealed crystals, atomic
   resolution images and Fermi level
   effects, 496–498f
Oxygen doping, electronic and structural
   effects, 492–499
Oxygen gas, production and transport in
   industry, 321–322

P

π-electron bridge, effect on
   polarization, 202

^{31}P nuclei, NMR background, 235–238
π–π stacking, between the donor and
   acceptor rings in organic salts, 205
Packed density, general properties of
   adsorbents, 328t
Palladium membrane, example of novel
   catalytic material, 317–318f
Palladium telluride, formation of
   clusters, 384–387
Paramagnet, creation by spin
   alignment, 164
Particle density, general properties of
   adsorbents, 328t
Passivation, requirement for synthesis of
   II–VI semiconductor clusters, 377
Patterning methods, chemical approaches to
   nanoscale device production, 338
Pedagogical aids, use for teaching of
   materials science, 51–52
PEO–salt systems, improvement, alternate
   polymeric media, 112
Percolation threshold, interconnected
   network formed by reinforcing
   phase, 277
Phase conjugate, third-order effects,
   193–195
Phase equilibria, information contained in
   phase diagram, 50–51
Ph.D. programs, contents of solid-state
   sciences, 49t
Photoactive compound, use in positive
   resists, 93
Photogenerated acid, reactions in
   chemically amplified resists, 96, 100
Photolithography, alternatives and trends,
   87–88
Photonic materials, scope of the MS&P
   program, 21
Photonics, emerging areas of materials
   science, 25
Phototopotaxy, nanochemistry technique for
   synthesizing zeolite-encaged
   semiconductor clusters, 345
Physical adsorption, use in surface-area
   determination, 304–305
Pillared lamellar solids, definition,
   types, and function, 287–296
Plasticized systems, use as polymeric
   electrolytes, 118–125
Platinum, role in naphtha reforming,
   315–316
Pockets effect, importance as second-order
   nonlinear optical effects, 190–193

Polar sheets, organic salt, arrangements, 205–206

Polarization
first-, second-, and third-order effects, 191–194
modulation and generation, 189–191

Polyaniline, properties and synthesis, 268–272

Polycarbosilanes, early developments, 134

Polyelectrolytes, attempts to increase cationic transport, 116

Polyethylene, biomaterial composite with hydroxyapatite, 538–541

Poly(ethylene oxide)
strong interaction with inorganic salts, 108, 109f
synthesis by direct polymer intercalation, 272

Polyimide–silicate composite, synthesis by molecular dispersion, 275–278

Polymer-based battery, typical, 108f

Polymer blends, use as polymeric electrolytes, 116–117

Polymer membranes, means of producing N_2, 322

Polymer–metal composites, research directions in preceramic polymers, 143–146t

Polymer nanocomposites, types and synthetic approaches, 267–278

Polymer pyrolysis, creation of ceramics and ceramic fibers, 138–151

Polymer science, emerging areas of materials science, 25

Polymer-supported catalysts, example of catalytic material, 308–309

Polymeric electrolytes, solid, high-conductivity, 107–126

Polymeric materials topics, actions to increase attention, 66–68

Polymers
chemical formulas, 114t
in resists, physical and chemical properties, 97–98

Polymers with inert fillers, ionic conductivity and mechanical stability, 117

Poly(methyl methacrylate), use as self-curing "bone cement", 524

Polyoxometalate anions, use as reagents for the pillaring of LDHs, 293–295

Polyphosphazenes, use as polymeric electrolytes, 116

Poly(propylene glycol), effect of $CoBr_2$ on T_g and conductivity, 121–122f

Polysiloxanes, use as polymeric electrolytes, 116

Pore shape, general properties of adsorbents, 328t

Pore volume, importance to catalytic properties, 305

Porosity
interrelationship between endosemiconductors and exosemiconductors, 358
properties of polymer-derived ceramics, 144

Porous implants, use as implant and prostheses, 530

Porous materials, types of pores, 285

Porphyrin guest molecules
orientation to host layers of fluorohectorite, 263–266
orientation to host layers of hectorite, 263–266

Positive photoresist, reactions, 88

Positive resist, effect on polymer solubility, 93

Powder, function as a biomaterial, 526t

Preceramic polymers, past, present, and future, 131–153

Presidential initiative, strategic objectives for materials science, 7–8

Pressure, effect on PSA working capacity, 331–332f

Pressure–swing adsorption, production of N_2 or O_2, 322–332

Principal axis system, orientation of chemical shift, 236

Project conflict, causes, 58t

Promoters, constituent of catalyst particles, 306

Prostheses, inorganic biomaterials for repair of human body, 524, 525f

Protein cavities, use as host synthetic media for inorganic crystals, 514

Pulse compression technique, study of chemical reactions and electrical phenomena, 15

Purity, requirement in resist design, 92

Pyrolysis, polymer, creation of ceramics and ceramic fibers, 138–151

Q

Q-state materials, effect on electronic spectra, 376–377

Quadrupolar nuclei, NMR background, 238–241

Quantum confinement structures, synthesis by OMCVD application, 417–418f

Quantum dot array
effect on electronic spectra, 376–377
improved nanophysics fabrication methods, 336, 337f
nanophysics fabrication compared to nanochemistry synthesis, 341f

Quantum mechanical spin, creation of magnetism, 162–164

Quantum size effects, control and exploitation in nanoscale devices, 338

Quantum wire structures, formed on grooved substrates by OMCVD, 418f

R

Radicals, role in magnetic behavior, 164

Rapid condensation, technique to form ceramic coatings, 146–147

Rate-limiting step, formation of Fe–Si crystalline phase, 456

Reactivity, effect on thin-film properties in OMCVD process, 398

Reconstitution, bulk semiconductor into exo- and endosemiconductors, 339f

Redox-interconvertible tungsten oxides, synthesis inside supercages of zeolite Y, 350

Redox method, use to promote intercalation, 233

Reflectance-difference spectroscopy, current trend in OMCVD, 416–417

Refractive index, relationship to polarization, 190–191

Refractive index grating, third-order effects, 193–195

Regeneration
catalyst, definition, 302
importance to catalytic properties, 305

Relative energy, effect of charge transfer on relative energy of states, 173, 175f

Relative strength, characteristic of bioactive ceramics, 534

Rensselaer Polytechnic Institute, materials chemistry courses, 62–65

Repeat distance of elemental layers, advantage of superlattice reactants, 435

Research costs, comparison with commercial launch and development expenses, 28

Research directions, preceramic polymers, 139–143

Resist design
deep-UV lithography, 93–94
requirements, 91–93
technology trends, 87–88, 102–103
See also Chemically amplified resists

Resolution, requirement in resist design, 91

Resonance frequency, relationship with PAS, 237–238

Resorbable bioceramics, use as implant and prostheses, 530–531

Resource aids, types, for teachers of materials chemistry, 51–52, 71–75, 79–82

Restricted transition-state selectivity, description of zeolite applications in catalysis, 312

Reverse micelle, arrested precipitation of II–VI crystallites, 378–380

Rhenium–platinum, role in naphtha reforming, 315–316

Rheological properties, requirements for preceramic polymers, 139

Rietveld refinement, structure determination of metal chalcogenide nanoclusters, 347–348f

Room-temperature polymeric magnet, requirements, 180–184

S

S53P4 Glass 9, bioactive ceramic in clinical use, properties, 532–534t

Salt–polyether systems, ion pairing, techniques for determination, 111

Saponite, typical unit cell formula of typical smectite clays, 289t

Scanning electron-beam lithography, alternative to conventional photolithography, 90–91

Scanning probe microscopes, use of sharp tips for atomic and nanoscale objects, 338

Scanning tunneling microscopy
characterization of complex materials, 479–504
precise measurements of surfaces, 15

Scholar program, attempt to develop lectures, 68–70

School partnership program, cooperation between educators and industry, 76–79

Science education in grades K–12, enhancement by scientists and engineers, 76–82

Second-order optical nonlinear materials fabrication by biomimetic mineralization, 513 optimization, 189–208

Second harmonic generation efficiency, largest reported to date, 204*f*

importance as second-order nonlinear optical effects, 190–193

use of multilayer thin films, 220–221

Second hyperpolarizability correlation with bond-length alternation, 202–203 definition, 190, 197

Seeding, nucleation control via addition of impurities, 464

Selectivity catalyst, definition, 302 key property for air separation sorbent, 323–324*t*,f

Self-assembled organic–inorganic nanostructures, design, synthesis, and characterization, 259–279

Self-assembly process, strategy for formation of thin layers, 214

Semiconductor clusters formation via arrested thermolysis, 381–382 nanometer-sized, 344

Semiconductor nanoclusters, effect on electronic spectra, 376–377

Semiconductors, synthesis from sodalite supralattices, 351–357

Sensitivity chemically amplified resists, 100 requirement in resist design, 91

Sequential adsorption-reaction gas-phase technique, 225–226 production of polar mono- or multilayer films, 222–225

Shape-selective catalysis, description of zeolite applications in catalysis, 312

Shape-selective catalysts, example of catalytic material, 309

Shape control, requirement for synthesis of II–VI semiconductor clusters, 377

Shrinkage in polymer pyrolysis, effect on preceramics, 143–145

Signal damping, effect of DSC data on superlattices, 446

Silicon-based, nanoscale structures, ability to emit visible light, 343

Silicon carbide composites, 143 initial interest as preceramic polymers, 132–136 target single-phase ceramics, 139–142

Silicon nitride initial interest as preceramic polymers, 132–136 target single-phase ceramics, 139–142

Silver halides, formation of sodalite supralattices insulator, 356

Silver sodalites, nanoscale device ideas, 356, 357*f*

Single-phase ceramics, research directions in preceramic polymers, 139–142

Single-source compounds, advantages over other II–VI compounds, 375

Single polymer chains, intercalation, 267–275

Size-selective gates, formation in CMS materials, 329–330*f*

Size control, requirement for synthesis of II–VI semiconductor clusters, 377

Size of crystallites, effect on physical and mechanical properties of polymers, 272, 274–275*f*

Slurry spinning method, approach to spinning of SiC fibers, 149

Smectite clays, layer lattice structures, 288

Social sciences, role in materials science, 41

Sodalite supralattices, uses for nanomaterials, 351–357

Sodawrite, zeotype-based devices, 356, 357*f*

Sodium chlorosodalite, structure of archetypal sodalite, 353, 354*f*

Sol–gel-derived glasses, tailoring surface reactions of inorganic biomaterials, 542–544

Sol–gel-derived silica, comparison with biomimetic materials, 510

Sol particles, use in preparing pillared clays, 291–292

Solid polymeric electrolytes, high-conductivity, 107–126

Solid-state amorphization reactions, nucleation of crystalline compound at interface, 432–433

Solid-state compounds, formation from molecular clusters, 383–384

Solid-state reaction, occurrence at interfaces, 428–429

Solvation, salt by PEO chain, 109f

Spatially organized reaction environments, analogous approach to biomimetic mineralization, 514

Spectroscopy, current trends in OMCVD, 413–414

Spin of an electron, creation of magnetism, 162–164

Stability
catalyst, definition, 302
preceramic polymers, requirements, 138

Starting materials, requirements for preceramic polymers, 136–137

State diagram, effect of charge transfer on relative energy of states, 173, 175f

Storage modulus, polyaniline multilayer hybrid, 272

Strength-to-density ratio of materials, increase over time, 14

Stress shielding, problem of bone resorption, 526

Structural composite materials, technological applications, 427–428

Structural materials, scope of the MS&P program, 21

Structure, effect on thin-film properties in OMCVD process, 398

Structure–function relationship, study of metamagnets, 165–172

Structure–property predictions, correlation of γ and bond-length alternation, 202–203

Sum frequency generation, occurrence of polarization, 192

Superconducting materials, scope of MS&P program, 22

Superconducting quantum interference devices, application of oxide superconductors, 475–476

Superconductor–insulator–superconductor junctions, superconducting state of $Bi_2Sr_2CaCu_2O_8$, 499–503

Superconductors, time evolution of increasing critical temperatures, 185f

Superlattice of superlattices, proposed "designer compound", 466f

Superlattice reactants, importance in technology, 425–466

Superlattices, creation by a new multilayer film-growth technique, 226–227

Support component, constituent of catalyst particles, 306

Surface-controlled growth, current trend in OMCVD, 416–417

Surface analytical chemistry, effect of improvements on thin-film research, 212

Surface area
general properties of adsorbents, 328t
importance to catalysts, 303–305

Surface chemistry, molecular tailoring, 541–544

Susceptibility, fit by Curie–Weiss expression, 169

Synchrotron powder X-ray diffraction, structure determination of metal chalcogenide nanoclusters, 347–348f

Synthesis, basic subject matter needed to work in materials science, 48–51

T

Talc layered silicates, structure and intercalation properties, 261–262

Technology, critical, importance of materials science, 4–10

"Technology-push" philosophy, effects on material development, 30

Tellurides, formation of clusters, 384–393

Tellurium reagents, formation of Group II–VI compound semiconductors, 407–412

Temperature–conductivity models, polyether network systems, 109

Temperature
effect on conductivity of polymer electrolytes, 125
effect on O_2 capacity and selectivity of CMS, 329t
effect on PSA working capacity, 331–332f
effect on semiconductor cluster formation, 381–382f
importance to pyrolysis of preceramic polymer 141–142
glass-transition, plasticized systems, 118–125

Templating methods, chemical approaches to nanoscale device production, 338

Ternary products, direct formation from amorphous intermediate, 456–461

Tetraethyl orthosilicate, hydrolysis and polymerization as layered system, 298

Textural pores, nanoporous materials, 284

Thermogravimetric analysis, thermal and oxidative stability of intercalated compounds, 266

Thickness of amorphous phase, amorphous phase crystallization, 435

Thin-film systems, necking procedure for single-crystal films, 464–456

Thin films of semiconductor materials, properties and importance to technology, 375–377

Thiobarbituric acid acceptors, improvements in charge-transfer forms, 200

Third-harmonic generation, occurrence, 193–194

Three-layer structure, formation with oxide superconductors, 473

Time scale of interfacial reactions, advantage of superlattice reactants, 435

Tin(IV) sulfide, semiconducting properties, 358–365

Tip–sample separation, function in scanning tunneling microscopy, 484–485

Tissue attachments, different biomaterials, 528t

Titanium, alternate ceramics, 149–150

Titanium carbide
composites, 143
shrinkage, 145

Topological incoherence, effect on Bragg reflections and subsidiary maxima, 440, 443f

Topotaxy, variations to MOCVD, 346

Training materials for precollege teachers, assistance from scientists, 79–82

Transition aluminas, preparation of catalytic materials, 307–308

Transition metal chalcogenides, formation via low-valent complexes, 393

Transport processes, effect on thin-film properties in OMCVD process, 398

Tribology, practical application of alkylsilane monolayers, 217

Tubular-silicate-layered silicate nanocomposites, new type of pillared clay, 292–293

Tungsten oxides, synthesis inside supercages of zeolite Y, 350

Tunneling barrier height, 490

Tunneling microscope, schematic illustration, 480–482

U

Ultimate tensile strength, characteristic of implant materials, 527t, 529t, 532–533t, 540t

Ultrathin-film composites, reaction mechanism, 455

U.S. competitiveness, weaknesses and strengths, 14

U.S. Government, role in supporting materials science, 6–10

U.S. patents, number by company in 1991, 56t

V

van der Waals bonding
use in self-assembling approach, 260
stability of self-assembled materials, 263

van der Waals thickness, comparison to gallery height in pillared lamellar solids, 291

Vapor-phase impregnation, nanochemistry technique for synthesizing zeolite-encaged semiconductor clusters, 345

Vapor deposition of compound semiconductors, chemical perspective, 397–419

Vegard law behavior, nanoporous semiconductors, 363, 365f

Very high spin multiplicity radicals, model for ferromagnetic spin alignment, 172–176

Viscosity, plasticizer, 118–125

Vogel–Tamman–Fulcher equation, modeling of temperature dependence of conductivity, 109

Void filling, mechanism of particle size control, 378–380

W

Wettability, properties of organic thin films, 217–218

Williams–Landel–Ferry equation, modeling of temperature dependence of conductivity, 109

Wire
function as a biomaterial, 526t
preparation from oxide superconductors, 476

Work force, importance to U.S. industry,
    56–59
World Biomaterials Congresses, number of
    research centers and countries presenting
    studies on bioactive ceramics, 539f
Wurtzite, limiting structure of II–VI
    compounds, 374

                    X

X-ray diffraction
    disadvantages compared with NMR
        spectroscopy for microcrystalline
        materials, 241
    patterns, polyaniline multilayer hybrid,
        268–269f
    study of interfacial structure, 437–450
    thermal and oxidative stability of
        intercalated compounds, 266
X-ray lithography, alternative to
    conventional photolithography, 90
X-ray powder diffraction, use in
    structural characterization, 51

                    Y

Y-type LB multilayer film, formation,
    220–221
Yield, ceramic residue, requirements,
    138–139
Young's modulus
    characteristic of implant materials,
        527t, 529t, 532–533t, 540t
    See also Modulus of elasticity

                    Z

Z-type LB multilayer film, formation,
    220–221
Zeeman effect, effect on ^2H NMR spectra,
    238–239
Zeolaser, zeotype-based devices, 356, 357f
Zeoled, zeotype-based device idea, 367f
Zeolites
    example of catalytic material, 309
    mechanism of size control for II–VI
        clusters, 378
    role in assembly of endosemiconductors,
        344–351
    use in separation of air by PSA
        processes, 322–325
Zeotrans, zeoled, 367f
Zig-zag chains, donor–acceptor salts,
    effect on metamagnetic properties,
    169–172
Zig-zag pores, examples in zeolites,
    310–311
Zincblende, limiting structure of II–VI
    compounds, 374
Zirconia, use as implant and
    prostheses, 530
Zirconium alkanebisphosphonates,
    preparation of thin films, 222–225
Zirconium phosphonate, formation and use
    in intercalation reactions, 233–235
ZnSe films, research into new
    organoselenium precursors for OMCVD,
    411–412
$Zr[O_3P(CH_2)_nCOOH]^2$, structure and
    reactivity, 247–250f
ZSM–5, use as shape-selective catalyst, 312

Copy editing and indexing: Janet S. Dodd
*   and Scott Hofmann-Reardon*
Production: Donna Lucas
Acquisition: Rhonda Bitterli
Cover design: Alan Kahan

Printed by United Book Press, Inc., Baltimore, MD
Bound by American Trade Bindery, Baltimore, MD

Bestsellers from ACS Books

The ACS Style Guide: A Manual for Authors and Editors
Edited by Janet S. Dodd
264 pp; clothbound ISBN 0–8412–0917–0; paperback ISBN 0–8412–0943–X

The Basics of Technical Communicating
By B. Edward Cain
ACS Professional Reference Book; 198 pp;
clothbound ISBN 0–8412–1451–4; paperback ISBN 0–8412–1452–2

Chemical Activities (student and teacher editions)
By Christie L. Borgford and Lee R. Summerlin
330 pp; spiralbound ISBN 0–8412–1417–4; teacher ed. ISBN 0–8412–1416–6

Chemical Demonstrations: A Sourcebook for Teachers,
Volumes 1 and 2, Second Edition
Volume 1 by Lee R. Summerlin and James L. Ealy, Jr.;
Vol. 1, 198 pp; spiralbound ISBN 0–8412–1481–6;
Volume 2 by Lee R. Summerlin, Christie L. Borgford, and Julie B. Ealy
Vol. 2, 234 pp; spiralbound ISBN 0–8412–1535–9

Chemistry and Crime: From Sherlock Holmes to Today's Courtroom
Edited by Samuel M. Gerber
135 pp; clothbound ISBN 0–8412–0784–4; paperback ISBN 0–8412–0785–2

Writing the Laboratory Notebook
By Howard M. Kanare
145 pp; clothbound ISBN 0–8412–0906–5; paperback ISBN 0–8412–0933–2

Developing a Chemical Hygiene Plan
By Jay A. Young, Warren K. Kingsley, and George H. Wahl, Jr.
paperback ISBN 0–8412–1876–5

Introduction to Microwave Sample Preparation: Theory and Practice
Edited by H. M. Kingston and Lois B. Jassie
263 pp; clothbound ISBN 0–8412–1450–6

Principles of Environmental Sampling
Edited by Lawrence H. Keith
ACS Professional Reference Book; 458 pp;
clothbound ISBN 0–8412–1173–6; paperback ISBN 0–8412–1437–9

Biotechnology and Materials Science: Chemistry for the Future
Edited by Mary L. Good (Jacqueline K. Barton, Associate Editor)
135 pp; clothbound ISBN 0–8412–1472–7; paperback ISBN 0–8412–1473–5

For further information and a free catalog of ACS books, contact:
American Chemical Society
Distribution Office, Department 225
1155 16th Street, NW, Washington, DC 20036
Telephone 800–227–5558

Highlights from ACS Books